Quantum Theory Of Gravity

Essays in honor of the 60th birthday of Bryce S DeWitt

Bryce S DeWitt

Quantum Theory of Gravity

Essays in honor of the 60th birthday of
Bryce S DeWitt

EDITED BY STEVEN M CHRISTENSEN

Institute of Field Physics
Department of Physics and Astronomy
University of North Carolina

Adam Hilger Ltd, Bristol

© Adam Hilger Ltd, 1984

British Library Cataloguing in Publication Data

Christensen, Steven M
 Quantum theory of gravity
 1. Quantum gravity—Addresses, essays, lectures
 I. Title
 531′.5 QC178

 ISBN 0-85274-755-1

Published by Adam Hilger Ltd, Techno House, Redcliffe Way, Bristol BS1 6NX.

The Adam Hilger book-publishing imprint is owned by The Institute of Physics.

Printed in Great Britain by J W Arrowsmith Ltd, Bristol.

Contents

List of Contributors

S ADLER

The Institute for Advanced Study
Princeton, New Jersey 08540
USA

and

Institut des Hautes Etudes Scientifiques
F-91440 Bures-sur-Yvette
France

J BEKENSTEIN

Physics Department
Ben Gurion University of the Negev
Beer Sheva 84105
Israel

D BOULWARE

Department of Physics, FM-15
University of Washington
Seattle, Washington 98195
USA

M BROWN

Department of Astrophysics
Oxford University
South Parks Road
Oxford OX1 3RQ
United Kingdom

P CANDELAS

Center for Theoretical Physics
University of Texas
Austin, Texas 78712
USA

and

Department of Astrophysics
Oxford University
South Parks Road
Oxford OX1 3RQ
United Kingdom

and

International School for Advanced Studies
Trieste, Italy

S CHRISTENSEN   Institute of Field Physics
                                  Department of Physics and Astronomy
                                  University of North Carolina
                                  Chapel Hill, North Carolina 27514
                                  USA

P DAVIES   Department of Theoretical Physics
                                  The University
                                  Newcastle upon Tyne NE1 7RU
                                  United Kingdom

S DESER   Physics Depatment
                                  Brandeis University
                                  Waltham, Massachusetts 02254
                                  USA

D DEUTSCH   Department of Astrophysics
                                  Oxford University
                                  South Parks Road
                                  Oxford OX1 3RQ
                                  United Kingdom

                                  and

                                  Center for Theoretical Physics
                                  University of Texas
                                  Austin, Texas 78712
                                  USA

C DeWITT-MORETTE   Department of Astronomy and Center for Relativity
                                  Theory
                                  University of Texas
                                  Austin, Texas 78712
                                  USA

D DOERING   Department of Physics
                                  University of Texas
                                  Austin, Texas 78712
                                  USA

S DOWKER   Department of Theoretical Physics
                                  The University
                                  Manchester M13 9PL
                                  United Kingdom

L FORD Department of Physics
Tufts University
Medford, Massachusetts 02155
USA

S FULLING Department of Mathematics
Texas A&M University
College Station, Texas 77843
USA

L HALPERN Department of Physics
Florida State University
Tallahassee, Florida 32306
USA

J HARTLE Enrico Fermi Institute
University of Chicago
Chicago, Illinois 60637
USA

C ISHAM Theoretical Physics
The Blackett Laboratory
Imperial College of Science and Technology
Prince Consort Road
London SW7 2BZ
United Kingdom

R JACKIW Center for Theoretical Physics
Laboratory for Nuclear Science and Department of
Physics
Massachusetts Institute of Technology
Cambridge, Massachusetts 02139
USA

K KUCHAŘ Department of Physics
University of Utah
Salt Lake City, Utah 84112
USA

L PARKER Department of Physics
University of Wisconsin-Milwaukee
Milwaukee, Wisconsin 53201
USA

R PENROSE Mathematical Institute
Oxford University
Oxford OX1 3LB
United Kingdom

D SCIAMA   Department of Astrophysics
Oxford University
South Parks Road
Oxford OX1 3RQ
United Kingdom

and

International School for Advanced Studies
Trieste, Italy

and

International Center for Theoretical Physics
Trieste, Italy

L SMARR   Astronomy Department
University of Illinois
Urbana, Illinois 61801
USA

L SMOLIN   The Institute for Advanced Study
Princeton, New Jersey 08540
USA

K STELLE   Theoretical Physics
The Blackett Laboratory
Imperial College of Science and Technology
Prince Consort Road
London SW7 2BZ
United Kingdom

A STROMINGER   The Institute for Advanced Study
Princeton, New Jersey 08540
USA

J TAYLOR   Department of Mathematics
King's College
Strand, London WC2R 2LS
United Kingdom

C TEITELBOIM   Center for Theoretical Physics
University of Texas
Austin, Texas 78712
USA

and

Istituto Nazionale di Fisica Nucleare
Sezione di Torino
10125 Torino
Italy

E TOMBOULIS Joseph Henry Laboratories
Princeton University
Princeton, New Jersey 08544
USA

W UNRUH Physics Department
University of British Columbia
Vancouver, British Columbia V6T 2A6
Canada

G VILKOVISKY State Committee of Standards
Moscow
USSR

R WALD Enrico Fermi Institute
University of Chicago
Chicago, Illinois 60637
USA

P WEST Department of Mathematics
King's College
Strand, London WC2R 2LS
United Kingdom

J WHEELER Center for Theoretical Physics
University of Texas
Austin, Texas 78712
USA

J YORK Institute of Field Physics
Department of Physics and Astronomy
University of North Carolina
Chapel Hill, North Carolina 27514
USA

Preface

Consider the explorer. He decides that one direction in some uncharted territory is a better way to travel than some other. He blazes new trails for others to follow and discovers new peaks and valleys no one knew existed. Years may go by before a few more brave souls find the explorer's route. They may not even realize that someone else was there before them. Eventually, a flood of people head down the path, settling here and there, claiming pieces of it for their own. The original adventurer is often forgotten or ignored.

Physics can be just like this. Farsighted physicists—ahead of their time—quietly create new ideas, write papers and then move on to new work, leaving the rest of us to discover or 'rediscover' their ideas. Ultimately, these ideas become part of physics lore, usually with someone's name attached to them. Occasionally these names are *not* those of the scientists who had the ideas in the first place.

Bryce DeWitt is a true adventurer. When he is not hacking through some jungle, riding a camel in a desert or floating down a river, he is pushing his way through the enormous complexities of quantum gravity theory. His classic works are much like the journals of an explorer. They may sit on the shelves of libraries along with the questionable or confused 'scholarship' of others. Those willing to dig through the wasted paper in their search for something new, find and study DeWitt's work and are treated to a phenomenally exciting world. Lately, DeWitt's papers are being rediscovered and recognized more and more; especially now on the occasion of his 60th birthday.

This volume contains articles by many of the most active workers in quantum gravity theory and related topics. Some of these papers are historical reviews of various areas of gravitational research, while others give technical discussions of the latest research findings. The reader will find hard physical reasoning and enlightened speculation, difficult and deep mathematics, as well as personal feelings about DeWitt and his work. In every case, Bryce's energy and inspirational presence are evident. A book many times this size would not cover all of the subjects in theoretical physics that Bryce has touched in some valuable and fundamental way.

The compilation of the articles in this book and the other birthday celebrations came about with the help and encouragement of many people. Special thanks go to Joyce Patton, Philip Candelas, Winnie Schild and the members of the Center for

Relativity Theory and Center for Theoretical Physics at the University of Texas at Austin. Jimmy York, the relativity graduate students and other members of the Physics and Astronomy Department at the University of North Carolina at Chapel Hill were also very helpful. Jim Revill and Margaret Martin of Adam Hilger Ltd, were very patient and understanding with a new 'editor'. Part of this work was supported with funds from a National Science Foundation Grant and the Bahnson Trust Fund at the University of North Carolina.

I want to thank all the physicists who contributed to this book and to acknowledge especially the contributions of Chris Isham of Imperial College in London and Larry Smarr of the University of Illinois at Urbana given in their brilliant summaries of Bryce's work at the December 1982 Austin celebration and now in this volume.

Two colleagues, R Kallosh and E Fradkin, who very much wanted to write something for this book were unable to due to illnesses. Their essays will be missed by those of us who know and respect their work.

Now, I hope the reader will forgive me for making a few personal comments. This book took many hours of work putting together references, writing pestering letters, and so forth. My wife, Sunny, and son, Jonathan, lovingly gave me much encouragement and support through it all. My heartfelt thanks to them. Also, I must express my gratitude to my parents, family, teachers and friends who got me to the point where I could meet and then learn from Bryce.

Bryce DeWitt was born on January 8, 1923, four days after my father was born. It sure was a great week for me.

Finally, thanks to Bryce. May his adventures never end.

S CHRISTENSEN
Chapel Hill, North Carolina
January 1983

The Contributions of Bryce DeWitt to Classical General Relativity†

LARRY L SMARR

It is with great pleasure that I review the work of Bryce DeWitt. I had the honor of working with him during my graduate student days at the University of Texas. Indeed his support in early years made my career possible. I was one of the few colleagues Bryce has had that chose *not* to do quantum gravity. I suppose it is therefore appropriate that I attempt to summarize his work in classical gravity.

I think it becomes very clear as soon as you read any of Bryce's papers that his primary concern throughout his carrier has been with the *dynamics* of gravity. Certainly general relativity is the dynamical theory *par excellence*. Bryce's work in classical relativity can be divided into (1) dynamical theory of groups and fields, (2) gravitational radiation, (3) superspace, (4) numerical relativity, and (5) experimental relativity. I will survey each of these fields and then conclude with a few remarks about Bryce as a very dynamic leader.

1. DYNAMICAL THEORY OF GROUPS AND FIELDS

The Dynamical Theory of Groups and Fields is of course the title of a book; it was also a very long chapter in the 1963 Les Houches lectures. I think anyone who is aware of Bryce's work is familiar with this great opus. However, there was an earlier article in the Reviews of Modern Physics back in 1957 on 'Dynamical Theory in Curved Spaces'. That article really foreshadowed a great number of Bryce's later works.

There is a very nice quote in that paper, in which Bryce says:

> The existence of any fundamental theoretical structure which is far from having been pushed to its logical mathematical conclusions is a situation which may have great potentialities.

DeWitt (1957a)

† Based on a talk given at an informal meeting held in Austin, Texas, on December 11 1982 to honor Bryce DeWitt's 60th birthday.

This statement is the secret to much of Bryce's success. Bryce always has been in love with formalism; he used to tell us: 'A formalism has a life of its own; it leads you, it gives you part of the answer'. I have never known anyone who has the ability to push a formalism to the limit the way that Bryce can. There is a lot to be said for this approach. The whole point to mathematical physics is that it is a self-contained kind of structure which can make its own internal predictions. As Wigner has emphasized, the most amazing thing about mathematical physics is that it describes the world at all—that's a miracle, something we don't understand. Bryce has always had great faith in the ability of mathematical formalism to lead us to greater truths about the world.

In these series of works Bryce laid out some very powerful tools, not only for him to use later on, but also for very many other scientists to use. In particular, anyone who has worked with Bryce at all understands what 'condensed notation' means. Bryce tackles problems which are too hard to deal with using the mathematical notation adopted by most people. Bryce got rid of all the things we usually worry about (subscripts, space–time dependencies, etc) in favor of one 'supersymbol'. One important result of this approach is that he was able to deal with many theories at once. Although at first it seems 'just like a formalism,' often this technique has led to new insights.

I might mention in passing that as I was preparing this lecture I went to the Physics Library at the University of Illinois to find *The Dynamical Theory of Groups and Fields*. It was not on the shelf, so I went to the librarian and asked, 'Where is this? Has some sort of fanatical relativist checked out the book?' She looked it up on the computer and, no, it was a graduate student in nuclear physics that currently has this book checked out and is working through it. The modern approach of gauge fields, of course, uses much of the formalism which Bryce developed in 1963. As a result, colleagues often remark: 'Bryce was many years ahead of his time when he wrote about gauge field theory.' This is not only true at the quantum level, but also at the classical level.

I believe that it was because Bryce got his PhD under Julian Schwinger, that he developed such a love of formalism and, in particular, of the action formalism. Much of Bryce's later research centers on the action and its uses. In *The Dynamical Theory of Groups and Fields*, Bryce develops very powerful variational tools and the Hamilton–Jacobi theory for deriving physics from an action. In addition, he laid out the theory of continuous groups, Lie groups, bi-tensors, and the theory of observation. This latter theory takes a situation which can be quite complicated and actually analyzes it mathematically into

(1) what is the instrument of observation

(2) what is the observed

(3) what is the coupling between them. This is a theory which is central in the development of not only quantum physics but also classical physics.

These early papers produced a theoretical framework. Now I want to describe how Bryce used that framework to solve a number of major problems in classical gravity. I will first discuss gravitational radiation.

2. GRAVITATIONAL RADIATION

Here there are two papers, one with Brehme in 1960 and one with Cécile in 1964 (DeWitt C M and DeWitt B S 1964). In addition, there are unpublished 1971 Stanford lecture notes. I was fortunate enough to be at Stanford when Bryce was a visiting professor there in 1971. He was lecturing on general relativity and produced a set of handwritten notes. Those notes are priceless, and I hope they can someday be published. Anyone who has ever seen Bryce's handwritten notes realizes that it is not necessary to type them; they are perfect—very meticulous, like calligraphy.

In the two papers, a large amount of mathematical formalism was developed which has turned out to be extremely useful. The major physics question posed was: if an electrical charge moves through curved space, does it radiate? After all, it's in some sense gravitationally accelerated, and accelerated charges should radiate. On the other hand, we are supposed to be able to locally transform away the gravitational field, so maybe it isn't really accelerated. There is a real paradox here. To solve this problem, Bryce used a curved space Green's function. Bryce developed the exact theory with Brehme and then made some very useful approximations to it with Cécile, for the case in which there is a weak gravitational field and a charged particle providing an electromagnetic field. One asks how the gravity makes linear perturbations of the electromagnetic field. Since it is linear, one can use Green's function techniques.

Now the reason I think that Bryce got so much out of this calculation and why so many people later on used the results of that calculation was that it was a *serious* calculation. The 1960 paper says:

> The calculation is patterned directly on Dirac's famous paper on the classical radiating electron. Just as Dirac's calculation was kept Lorentz invariant throughout, so the present calculation is maintained generally covariant throughout . . .
>
> DeWitt and Brehme (1960)

Bryce's reliance on the giants of mathematical physics, the people who made very clear calculations in mathematical physics, is another of the secrets to his success.

A follow-up occurs a sentence or two later. The relevant thing that Bryce discovered here is the importance of the 'tails' of the Green's function; that is, that some of the radiation does not lie on the light cone as one might expect in flat space. Again, Bryce says:

> As has been pointed out by Hadamard [in 1923], a plane or spherical sharp pulse of light, when propagating in a curved 4-dimensional hyperbolic Riemannian manifold, does not, in general, remain a sharp pulse, but gradually develops a 'tail'. It is this phenomenon which is responsible for the electrogravitic bremsstrahlung.
>
> DeWitt and Brehme (1960)

What Bryce realized here was that he could represent a well-posed physical problem, an electron falling in a gravitational field, by the classical mathematical results of a Green's function in the presence of curvature. So, he was able to go back to Hadamard and use some of his elegant mathematics to get the answer to the physics.

I will reproduce some of the mathematics from these two papers because it is very pretty. Consider the covariant D'Alembertian on a scalar field:

$$g^{\mu\nu}\phi_{.\mu\nu} = 0. \tag{2.1}$$

The 'dot' is DeWitt notation for the covariant derivative. If you want to know what the solution for the scalar field is, then you can instead write down the equation for the Green's function. Depending on boundary conditions, there are several Green's functions, retarded and advanced, etc. DeWitt and Brehme make extensive use of a symmetric one which satisfies

$$g^{\mu\nu}\bar{G}_{.\mu\nu} = -\bar{g}^{-1/2}\delta^{(4)}. \tag{2.2}$$

Once you have the solution for that Green's function, then from that by integration you can produce the field ϕ.

There are two parts to the symmetric Green's function: a direct part and a so-called tail part:

$$\bar{G} = \frac{1}{8\pi}[\underbrace{\Delta^{1/2}\delta(\sigma)}_{\text{direct}} - \underbrace{v\theta(-\sigma)}_{\text{tail}}]. \tag{2.3}$$

I reproduce figure 1 from DeWitt and Brehme (1960) to define the symbols used here. The fundamental variable σ is the square of the bi-scalar of geodetic interval $s(x, z)$. If you imagine going off from some event z to a point x which is space-like with respect to z, then you can calculate the arc length s and by squaring that get another bi-scalar σ. That bi-scalar, $\sigma = \pm\frac{1}{2}s^2$, you notice, depends on two events in space–time (z, x), and this is typical of bi-tensors (Synge's world function is the same thing). That σ is the argument in equation (2.3). You can see that the direct part of the Green's function says that there's a delta function of σ, that is, there is support only where $\sigma = 0$ on the light cone. This part of the Green's function makes sure that propagation occurs along the null cone, just as you would expect in flat space for electromagnetic radiation. But there is another term in equation (2.3) in which a Heaviside (step function) symbol occurs. Inside the light cone this term is non-zero, times an amplitude for the tail function v. The v part is the new part of the calculation. The coefficient Δ in the direct part is the 'scalarized van Vleck determinant':

$$\Delta = 1 - \tfrac{1}{6}R^{\alpha\beta}\sigma_{.\alpha}\sigma_{.\beta} + O(s^3) \tag{2.5}$$

and you can see that the Ricci curvature of the space–time comes in to that amplitude.

To do the electromagnetic case you simply write down the vector potential equation in curved space and you formally get a symmetric Green's function that looks just like the scalar with a direct part and a tail part:

$$g^{\nu\sigma}A_{\mu.\nu\sigma} + R^{\nu}{}_{\mu}A_{\nu} = 0 \tag{2.6}$$

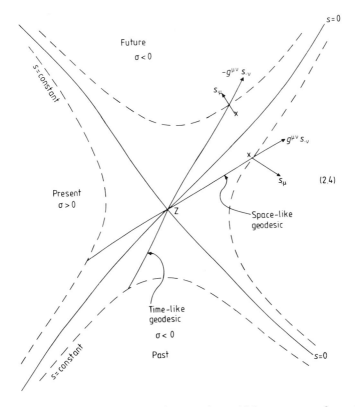

Figure 1 The geodetic structure of space–time which appears as figure 1 in DeWitt and Brehme (1960). Geodesics are drawn from a general event z to nearby events x.

on light cone   inside light
cone

$$\bar{G}_{\mu\alpha'} = \frac{1}{8\pi} \left[\Delta^{1/2} \bar{g}_{\mu\alpha'} \delta(\sigma) - v_{\mu\alpha'} \theta(-\sigma) \right].$$

bi-vector of   tail bi-vector
parallel
displacement

Here now you notice there are two indices on the Green's function and that these are evaluated in general at separate events, so again these are bi-tensors. The bi-vector of parallel displacement $\bar{g}_{\mu\alpha'}$ is very useful and comes up all the time. It transforms as the metric at each space–time point individually and can be used when you're looking at two tensors that are separated by a geodesic interval. The amplitude $v_{\mu\alpha'}$ in the symmetric Green's function physically tells you about the tails on the electromagnetic wave.

The final physical answer, about whether the charged particle follows the geodesic or not, is stated very clearly by DeWitt and Brehme (1960):

The charged particle tries its best to satisfy the equivalence principle, and on a local basis, in fact, does so. [That is, the deviation caused by the electric field scattering off the curvature is non-local.] In the absence of an externally applied electromagnetic field, the motion of the particle deviates from geodetic motion only because of the unavoidable tail in the propagation function for the electromagnetic field ... Physically, the tail may be pictured as arising from a sort of scatter process, with the 'bumps' in space–time playing the role of scatterers.

DeWitt and Brehme (1960)

There is a very elegant way of seeing that precisely. DeWitt and Brehme (1960) derived the Ponderomotive equation, the classical equation for a charged particle, in the presence of curvature:

$$
m\ddot{z}^\alpha = \frac{e}{c} F^\alpha{}_\beta \dot{z}^\beta + \frac{2}{3}\frac{e^2}{c^3}\left(\dddot{z}^\alpha - \frac{1}{c^2} z^\alpha \ddot{z}^2\right) + \frac{e^2}{c} \dot{z}^\beta \int_{-\infty}^{\tau} g^{\alpha\delta}(v_{\delta\gamma'\cdot\beta} - v_{\beta\gamma'\cdot\delta})\dot{z}^{\gamma'}\,d\tau'.
$$
$$(2.7)$$

Lorentz Force	classical radiation damping	tail induced radiation damping
local	local	non-local

The amazing thing is that it comes out just as Dirac discovered: the acceleration of the charged particle is just equal to the Lorentz force, if there is an externally applied electromagnetic field, and the classical radiation damping term. The third derivative term is the one that gives problems with runaway solutions in flat space. But in addition to these, there is the term in which you notice the curl of the function v, i.e. the tail bi-tensor $v_{\mu\alpha'}$ which was in the Green's function in equation (2.6). Its curl acts as the source of the deviation of the electron's worldline from being a geodesic. And this is integrated over the entire past history of the charged particle's trajectory.

Another way of seeing what goes on here is to assume that there's no external field $(F_{\alpha\beta} = 0)$ and we're in flat space, so that the tail goes away, then what you'd get is that the particle falls along a geodesic, as you'd expect in flat space.

No EM field + flat space	$\rightarrow \ddot{z}^\alpha = 0$	geodesic motion	
No EM field + curved space	$\rightarrow \ddot{z}^\alpha \neq 0$	non-geodesic motion	(2.8)
		electrogravitic brems.	

However, if you go to curved space but leave the electromagnetic field off, then what you see is that the tail term now forces the particle away from a geodesic and causes radiation which finally feeds back as damping. Again, this is an extremely elegant solution to this problem.

Now of course a person doesn't have an impact on a field just by doing a calculation. He really only has an impact if he provides something in that calculation that allows later scientists to do new calculations. I want to emphasize this point repeatedly in discussing Bryce's work. The calculation described above was a thought experiment, not in gravitational radiation, but in electromagnetic radiation. But what this formalism allowed was for Kip Thorne and his co-workers, in the late 70s, to calculate the full *gravitational* scattering problem of two massive particles. They included the bodies' finite gravitational self-energies and obtained an answer to the question of what kind of gravitational bremsstrahlung occurs for arbitrary mass ratios and arbitrary velocities. I include here a couple of quotes from the classic paper of Kip and his colleagues, which is one of the most beautiful calculations in relativity theory we've had in quite a while.

The first paper of the series is by Thorne and Kovács (1977). They say:

> The 'guts' of this paper, in terms of complex calculations, reside in the Green's function manipulations of Section IVc. Our particular way of handling the Green's functions is motivated in Appendix A, and has been influenced by the following papers: DeWitt and Brehme (1960), . . . Robaschik (1963), . . . John (1973a, b), Bird (1974), and especially Peters (1966).
>
> Thorne and Kovács (1977)

In the second paper in that series by Crowley and Thorne (1977), they say:

> Central to the development of their formalism is the determination of an appropriate approximation to the well-known Green's function for the scalar wave equation in curved space–time developed by DeWitt and Brehme (1960). The purpose of this paper is threefold: first, to show that an approximate form of the exact Green's function developed previously by DeWitt and DeWitt (1964) has exactly the same mathematical content (within the constraints of the approximations used) as that constructed by Thorne and Kovács.
>
> Crowley and Thorne (1977)

It is clear that Kip and his co-workers were only able to carry out this classic problem in general relativity by basing it on the tool which had been developed some 17 years earlier by Bryce in his calculation.

I will discuss briefly several other topics in gravitational radiation. If you pick up Misner *et al.* (1973) (MTW) and turn to the chapter on gravitational radiation, you'll find a number of references to Bryce, in particular, the famous Stanford lecture series (DeWitt 1971). In those lectures, I remember Bryce developing the theory of gravitational radiation, one reason being William Fairbank's experiments with the bars at Stanford. Bryce obviously wanted to describe the complete theory of how one detects gravitational radiation.

One of the important things one wants is to evaluate the energy content of a gravitational wave as it passes by some apparatus. Since that energy is not well defined locally, you have to average it over several wavelengths. Isaacson, using what was called Brill–Hartle averaging, developed the Isaacson formalism for averaging over gravitational waves and then allowing those waves to act back as the source of the gravitational field. If you look in MTW (equation (35.73′)) for the formula for how you do that averaging,

$$\langle E_{\alpha\beta}(x)\rangle \equiv \int g_{\alpha}{}^{\mu'}(x,x')g_{\beta}{}^{\nu'}(x,x')E_{\mu'\nu'}(x')f(x,x')\sqrt{-g(x')}\,d^4x' \tag{2.9}$$

you'll find that it requires the bi-vectors of parallel displacement $g_{\alpha}{}^{\mu'}$; metrics which are defined with their two indices at different events. MTW refer to Bryce's work in the paper with Brehme for the formalism of the bi-vectors.

The important use of Isaacson formalism is to show that the background gravitational field can have as its source ripples of the gravitational field averaged over large numbers of wavelengths.

$$G_{\mu\nu}^{(B)} = 8\pi T_{\mu\nu}^{(GW)} \tag{2.10}$$

Wheeler's geons are an extreme example, where the gravitational field provides its own mass energy which then traps the high-frequency gravitational waves into going around in a circle; it's a bootstrap approach.

You can derive equation (2.10) starting with the Lagrangian for the linearized gravitational field and by performing appropriate variations. Bryce did this derivation in his Stanford notes (1971). Now MTW suggest doing this as exercise 35.20, however they add: 'WARNING: the amount of algebra in this exercise is enormous'. And any of us who have sat through Bryce's derivations realize that. What MTW mean is, if you're not Bryce DeWitt, go on to the next exercise.

Let me turn finally to one last application of the Isaacson formalism. When Bryce was at Stanford, MTW were in the process of finalizing their book and they wanted to have the formula for the amount of angular momentum that is carried off by a gravitational wave. Now that's a very delicate and very, very long calculation. Peters in 1964 had gone through the calculation, but there were questions about some factors involved and whether some terms were correct. I remember Kip asking Bryce whether he'd ever worked that out and Bryce said something like, 'Well, it can't be *that* difficult—it's *only* linearized theory'. And so, some weeks of lecture later we got to the end of the calculation. One can take the reduced quadrupole moment t_{km} and by taking second and third time derivatives in appropriate combinations get exactly the density of angular momentum radiated. Then, by integrating over a two sphere find out the amount of angular momentum carried away from an isolated system by gravitational radiation (MTW equations (36.24) and (36.25)):

$$dJ_i/dt = -\int J_i r^2 \, d\Omega$$

$$J_i = \frac{1}{8\pi r^2}\,\varepsilon_i{}^{jk}\langle -6n_j\ddot{t}_{km}\dddot{t}_{mp}n_p + 9n_j\ddot{t}_{km}n_m n_p\dddot{t}_{pq}n_q\rangle \tag{2.11}$$

where the brackets $\langle\ \rangle$ are as defined in equation (2.9).

Now you notice that the above equations (2.9)–(2.11) are fundamental equations. Any process involving gravitational radiation from an isolated system will use these a thousand years from now. That's a trademark of Bryce's—he goes right to the heart of the matter and when he's finished with a problem, it's finished.

3. SUPERSPACE

I want to move on to an area which I think is not finished yet and is quite difficult, and that's the area of superspace. Again, you'll begin to notice a pattern here. In each of these major fields of classical gravity, there are usually no more than two papers by Bryce. And yet out of those a good fraction of that entire subfield usually comes. Bryce is not prolific in the number of his publications, and I think some of us might remember that these days. But the two papers he did write were very influential in this field. The first paper was in the first volume of the trilogy, *The Quantum Theory of Gravity* (DeWitt 1967), which dealt with the canonical theory. Chris Isham (in this volume) will describe the physics in the other two volumes and perhaps the quantum part of this one as well.

As John Wheeler, who really started the idea of superspace, has emphasized so many times over the last twenty years, the dynamics of the gravitational field is carried in the 3-geometry. We'll come back to this when we deal with numerical relativity below. The geometry of 3-space, the sequence of those geometries changing in time, is what makes up a space–time. And if you want to look for the state of the gravitational field, you look to the 3-geometry. Well, that realization was a long time in coming, and there were many, many people who worked on it. I'll give you a brief list of some of the names which came up while I was preparing this paper: Dirac, Pirani, Schild, Bergmann, Arnowitt, Deser, Misner, Higgs, Wheeler, DeWitt . . . .

Things were quite confused until, I would say, 1955 to 1960. There were questions being asked both from the quantum side, such as what is the appropriate state functional for quantum gravity, and also from the classical side, such as how to sort out the dynamics of gravitation. Bryce contributed to this sorting out process in a Stevens Institute of Technology talk in 1958 (unpublished). Bryce did some very important calculations which showed that the state functional had to be independent of the time–time and time–space components of the 4-metric. We would now say that the lapse and shift are not relevant, they are coordinate setting devices, and that the dynamics lies in the 3-metric. In particular, as Bryce and Wheeler kept emphasizing, it is not in the 3-metric either, but it is in the 3-geometry represented by that 3-metric.

Now as I said, Wheeler, in these wonderful articles of his for so many years, kept telling us to concentrate on the 3-geometry which had to have a place to live, superspace—the space of all possible 3-geometries. The 1963 Les Houches lectures was where superspace first started, but I believe that it first really caught fire in the Battelle Rencontres volume in 1968. In that volume, as John Wheeler (1968) begins to introduce the concept of superspace, he says, 'One climbs up to the concept of geometry only to find a new height beyond: superspace.' The only references Wheeler gives for superspace are Wheeler's (1963) own lectures at Les Houches and DeWitt's (1967) quantum theory of gravity article. So it's clear that by 1968, when the concept of superspace really came into its own, that John Wheeler certainly regarded Bryce as a sort of co-architect of that space. I remind you that one elegant formal way of defining superspace is simply (DeWitt 1970, equation (3b)):

$$S(M) \equiv \frac{\text{Riem}(M)}{\text{Diff}(M)}. \tag{3.1}$$

If you take a compact, Hausdorff 3-manifold M, such as a closed cosmology, and you imagine all possible 3-dimensional positive-definite metrics on it, you obtain the space Riem (M). Then you mod out the diffeomorphisms, Diff(M), the possible co-ordinate transformations or relabeling of points in the same 3-geometry, which leaves you with the points in superspace $S(M)$. Arthur Fischer (1970) gave a beautiful mathematical analysis of the structure of superspace and showed that it really wasn't a manifold, it was a stratified union of manifolds. This problem came up largely because of the existence of isometries in space–times which provided boundaries, from where trajectories in superspace could reflect. And so Bryce, in his article, suggested extending superspace, as defined here, to 'extended superspace', which would smooth out the kinks and curves in superspace so that one could talk about those more fruitfully.

Superspace, again, is the set of 3-geometries. Imagine you start with some 3-geometry, then a little bit later it's slightly bent and that's another point in superspace, and the next time it's somewhere else, and so you make a curve through superspace and that represents a space–time. One needs an equation for the dynamics in superspace and this is the Hamilton–Jacobi equation. The reason that, again, we arrive back at the Hamilton–Jacobi equation, something that Bryce has always stressed from 1957 on, is that the Hamilton–Jacobi equation is the proper formalism in which one is able to simultaneously do classical and quantum physics. This is because one can go back and forth very easily using the Hamilton–Jacobi functional. This allows WKB approximations to be done very nicely here. The equation one wants is for the Hamilton–Jacobi functional W in superspace (DeWitt 1970, equation (15)):

$$G_{ijkl}\left(\frac{\delta W}{\delta\gamma_{ij}}\right)\left(\frac{\delta W}{\delta\gamma_{kl}}\right) - \gamma^{1/2}\,{}^{(3)}R = 0. \qquad (3.2)$$

There is a set of coefficients G_{ijkl} coupling two first-order variations of the functional W with respect to the 3-metric γ_{ij}, minus the scalar Ricci 3-curvature. Peres, in 1962, was the one who first put forth this formula. There is a companion equation (DeWitt 1970, equation (17)):

$$\left(\frac{\delta W}{\delta\gamma_{ij}}\right)_{.j} = 0 \qquad (3.3)$$

where the 'dot' denotes covariant derivation with respect to the 3-metric. One can easily see that equations (3.2) and (3.3) are just formally the Hamiltonian and the supermomentum constraints in general relativity. One of the key ideas behind superspace is that one can get the dynamics of general relativity just out of the constraints. A pretty derivation in Bryce's quantum theory of gravity article starts with the formally-defined equations (3.2) and (3.3) and in a couple of lines recovers the familiar Einstein equation in the $3 + 1$ language, such as are used in numerical relativity; that is, you end up with equations that say that the second time derivative of the 3-metric is equal to a bunch of forcing terms.

The quantum version of equation (3.2) has been referred to by Kuchar as the 'Wheeler–DeWitt equation.' So Bryce's name is intimately hooked together with this equation both in the classical regime and in the quantum regime. The set of coefficients G_{ikjl} was considered to be just a set of coefficients until Bryce's (1967) quantum theory of gravity article. In that article he pointed out that G_{ikjl} actually is the supermetric on superspace

$$G_{ikjl} = \frac{1}{2\gamma^{\frac{1}{2}}}(\gamma_{ik}\gamma_{jl} + \gamma_{il}\gamma_{jk} - \gamma_{ij}\gamma_{kl}). \tag{3.4}$$

Because of its construction, it is an indefinite metric with one minus and five pluses. Thus the 3-metrics enter into the definition of the supermetric on superspace. Now, pointing out that G_{ijkl} is a supermetric may seem like a minor detail, until one reads from the beautiful article by Charles Misner (1972) in the Festschrift for John Wheeler. Misner wrote a review article on minisuperspace, in particular the Hamiltonian cosmologies that Misner and many others worked on in the late 60s and early 70s. While giving examples of the various 'toy models' using anisotropic homogeneous cosmologies that they had worked on, he remarks that they kept finding that the metric on superspace continually arose. Then Misner notes:

> Of course we had known of DeWitt's work pointing out the existence of a metric in superspace, but paid insufficient attention to the essential role this metric plays until we repeatedly found that it was the clue to solving specific simple examples in quantum cosmology.
>
> Misner (1972)

Another example of Bryce's deep insight into superspace is the beautiful result which Bryce found: one can consider space–times as geodesics in superspace, particularly if the extended superspace is used. That is, the geodesic equation on superspace gives a curve in superspace which represents a space–time. If it's a geodesic in superspace with the right parameterization, it turns out that the resulting space–time solves Einstein's equations, which is an amazing result. Then Bryce asked the reverse question: how is a given space–time represented in superspace? If you imagine taking a space–time and time-slicing it one way, you get a set of 3-geometries; if you time-slice it another way, using a different lapse function, you get a different sequence of 3-geometries. And yet glued together they are the same space–time. But they are different sequences of 3-geometries and therefore they are different curves in superspace. Therefore Bryce pointed out that one must consider, because of the many fingered time, a sheaf of geodesics in superspace as representing one space–time.

Again, Misner (1972) states the importance of

> ... the beautiful theorem [due to DeWitt] that solutions of Einstein's equations are geodesics in superspace which conveys succinctly the main thrust of all nonquantum study of superspace.
>
> Misner (1972)

Misner goes on to sum up, at the end of his article, the history of superspace. He said that:

The next big step in seeing Einstein's theory as a superspace structure was due to Bergmann, who pointed out that Peres was using approaches to the initial value problem of general relativity which allowed one to write the Hamilton–Jacobi equation for general relativity [equation (3.2)]. DeWitt, in private communications, made Wheeler's group aware of the importance of the Einstein–Hamilton–Jacobi equation and pointed out that it provided a basis for defining a metric in superspace. Published results [including Bryce's] appeared in due course as these ideas developed. The word 'superspace' seems to appear first in Wheeler's 1963 Les Houches lectures, but its formal baptism is not evident until the 1967 *Battelle Rencontres.*

Misner (1972)

Misner then finally ends this whole paper on minisuperspace by saying

The first minisuperspace calculations were DeWitt's for the Robinson–Walker universes.

Misner (1972)

I think it's clear that all the way through the study of superspace, which is one of the major areas of classical gravity, Bryce has shown the way over and over again. But more to the point, I think that it speaks extremely well of his work that people as bright as Charles Misner and as productive as John Wheeler felt compelled to continually point out how Bryce had made such important contributions to their understanding; they were some of the main pioneers of the subject itself.

4. NUMERICAL RELATIVITY

In the last section we saw how the dynamics of gravity can be represented as a formal structure in superspace; it's a completely formal structure. I think Bryce was never happy with just a formal structure unless it was also good for calculating particular examples. This is what led him to invent numerical relativity.

Numerical Relativity is a method for calculating, in principle, an arbitrary solution of Einstein's equations. Again there are only two published works by DeWitt on the subject, one on the maximal slicing of a black hole by Bryce and co-workers (Estabrook *et al* 1973) and one on the collision of two black holes (Smarr *et al* 1976). There is also an unpublished transcript, one which I have always found a tremendous source of inspiration, which contains Bryce's remarks at the Chapel Hill Conference on Gravitation in 1957 (DeWitt 1957c). To understand these remarks one must remember that numerical relativity requires solving for both the evolution of the matter and of the gravitational field. I will start with the former topic and show how it influenced the latter.

At Jim Wilson's Festschrift meeting held at the University of Illinois in October 1982, Bryce discussed his early 1950s days at Livermore. Most people don't realize that Bryce was one of the inventors of multidimensional Lagrangian hydrodynamics. In Lagrangian hydrodynamics, one fixes the coordinate grid on the particles of the fluid, and as the fluid moves around, it carries the grid with it, warping it as it goes.

Then one finite-differences the partial differential equations of hydrodynamics on that moving grid. Of course, the squares get bent into very peculiar diamonds after a while, and this causes a lot of trouble. In one dimension it's actually a pretty good method because, for example, in a spherical star the Lagrangian markers are just the mass shells. As the star becomes compressed, the markers get closer to each other, or if the star blows off a supernova envelope, the markers can get far apart, but they all stay in order; just the distance between the grid points changes.

By 1953, one-dimensional Lagrangian hydrodynamics had been developed quite well, but how to go successfully to higher dimensions was a mystery. Because Bryce knew about Jacobians and general coordinate transformations, which were quite foreign to most people at the lab, he was able immediately to write down the generalization for two dimensions (or *n* dimensions). Edward Teller, who was running the lab at that time, got so enthusiastic he immediately called a lab-wide meeting and Bryce became the instant lab expert on Lagrangian hydrodynamics.

The reason I relate this story is that until I prepared this lecture I didn't appreciate how much the mind-set of doing Lagrangian hydrodynamics had affected numerical relativity. If you go back to the summary of remarks made at the Chapel Hill conference in 1957, you find the following:

> Bryce DeWitt pointed out some difficulties encountered in high-speed computational techniques. 'Singularities are, of course, difficult to handle. Secondly, any non-linear hydrodynamic calculations are always done in so-called Lagrangian coordinates, so that the mesh points move with the material instead of being fixed in space. Similar problems would arise in applying computers to gravitational radiation, since you don't want the radiation to move quickly out of the range of your computer'.
>
> DeWitt (1957c)

An example of Bryce's concern is that if you are trying to solve a star collapse which bounces, in which you must follow the star inward for several orders of magnitude in radius, all your grid points are down in the center of the star when it bounces. Since gravitational radiation goes out at the speed of light, there are no grid points left outside of the star for the radiation to propagate on. Now Bryce saw that clearly in 1957—I find that amazing, because the conceptual problems in simultaneously worrying about

(1) the computer algorithm
(2) the structure of space–time
(3) the coordinate system

were things really sorted out only in the last ten years. And yet the problem was crystal clear thirty years ago to Bryce.

Although Bryce had these worries about Lagrangian hydrodynamics, people were even more afraid of Eulerian hydrodynamics—where you leave the coordinates fixed in space and then let the fluid move through them. Therefore, almost all of the early attempts to do numerical relativity were Lagrangian (for a review see Smarr *et al* 1980): all of the great work of Taub, in which he developed the formulae of general relativity in a Lagrangian coordinate system; the early codes of Misner and Sharp and of May and White, which were spherical star collapse codes; and Wilson's early work on supernovae. Until Wilson came along in the early 70s and started doing Eulerian hydrodynamics, there was this enormous mental roadblock of how one could calculate non-spherical general relativity, which is what you needed to get gravitational radiation, without having to use Lagrangian techniques for the hydrodynamics, in which case you were going to get all snarled up in your coordinates.

Bryce DeWitt and Charles Misner very clearly understood what solving Einstein's equations was going to be like in 1957, some twenty years before the first successful two-dimensional calculation. Again, from the Chapel Hill conference:

> Bryce DeWitt asked if the Cauchy problem is now understood sufficiently to be put on an electronic computer for actual calculation. 'Do we know enough about constraints and initial conditions to do this at least for certain symmetrical cases?'

DeWitt (1972)

Charles Misner, who always seems to be where Bryce is, answered that he happened to have some initial conditions for two throats and

> '. . . these can be interpreted as two particles which are non-singular, or they could be thought of as the kind of $1/r$ type singularities which one ordinarily thinks of in gravitational theory. These partial differential equations, although very difficult, can then in principle be put on a computer'. Misner thinks that one can now give initial conditions so that one would expect to get gravitational radiation, and computers could be used for this.

DeWitt (1957c)

Those two Einstein–Rosen bridges were, in fact, put on the computer in the early 60s, when the general relativistic two-body problem was first attempted numerically on a computer by Hahn and Lindquist (1964). So it was really quite rapidly from these remarks in 1957 that we moved to the beginnings of two-dimensional numerical relativity. Unfortunately what was lacking at that time was the concept of black holes. We hadn't yet understood that these weren't just throats, but that they could be considered as physical objects—black holes. The work of Wheeler, Hawking, and of other people who elucidated the concept of black holes, allowed Bryce to start the project back up again in 1969 as 'the two black hole collision' problem. Bryce tried to get students to solve it to actually see what would happen when the two holes collided and coalesced.

Bryce started out with Frank Zerilli and then with Andrej Cadez on this problem in 1969. After the Einstein equations were put on the computer, DeWitt and Cadez ran immediately into the problem of the lapse—that the time-slicing was absolutely crucial. Now Bryce understood this, as you have seen, early on; this is again the idea that the many fingered time is central to general relativity. What Bryce decided was that we had to come up with a better way of slicing space–time. Figure 2 shows the Kruskal diagram from the 1972 article on maximally slicing a single black hole (Estabrook *et al* 1973). If you would have used what people first tried, namely

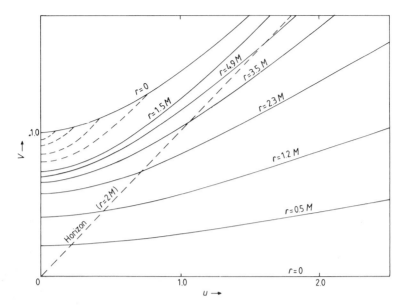

Figure 2 The Kruskal diagram for Schwarzschild space–time with a maximal time-slicing. This appeared as figure 1 in Estabrook *et al* (1973). The maximal time slices are labeled by their asymptotic Schwarzschild time coordinate values. Notice that at $t \to \infty$, the slices wrap up around $r = 3M/2$, thus avoiding the singularity at $r = 0$.

geodesic slicing, you would start off from the $t = 0$ slice and you would move forward equal increments of proper time until your slices hit the singularity in a time of the order of 3 M. If you were trying to let a star collapse, and let gravity waves be generated, and propagate out to the weak field zone, which might take 50 or 100 M, of course you'd be out of luck.

So Bryce invented the idea of using maximal time-slicing on the computer. Now Lichnerowicz already, during the War in 1944, had pointed out that one could use maximal slicing as a coordinate condition and that this would be like having an incompressible fluid for coordinates. But Bryce was the first one to really decide this was what we needed practically to build space–times. He could forsee what has now come to be known as singularity avoidance, the most important property of maximal slicing. What happens when you use maximal slicing is that as the slicing begins to approach the singularity at $r = 0$, the lapse function falls exponentially fast to zero in the strong field region. This allows the slices to wrap up around the singularity while always staying a finite distance away from it. Bryce used simple reasoning to tell us why this singularity avoidance occurs. He would tell his students: 'It's because the slice is maximal; maximal means maximal volume, and if you try to get near a singularity, the volume's going to zero, so the slice avoids that region'. Eardley and I, as well as Choquet-Bruhat, Marsden, Tipler and York, were able to prove theorems later on, which demonstrated in detail how this happens.

Steve Christensen, Elaine Tsiang, and I were young and impressionable graduate students at the time and although it sounded crazy to us, we needed a PhD so we started working on it. I think a lot of people believe that Bryce had Steve, Elaine and me slaving away on writing the codes while he would just stand up and say that maximal slicing is the way you do it. I'd like to show you that that is not the case, by publishing an artifact (figure 3). This is from the maximal hypersurface code for slicing the Schwarzschild black hole, done at the University of Texas in 1972. This is the original FORTRAN coding sheet; you can recognize the handwriting as being Bryce's because it's so neat. I remember Bryce was always infuriated with how stupid FORTRAN was as a language, in fact many people still feel that way today. So you'll notice that there's some very important comments: 'adjustable dimension subversion program'. This was to get around the fact that you might want to change your array sizes and didn't want to have to rewrite the whole program every time you did it. And further down the page: 'main program disguised as a subroutine'. This is another example of DeWitt extended FORTRAN.

Just as Bryce, Elaine, Steve and I had finished our numerical calculation, we received a letter from Frank Estabrook and Hugo Wahlquist of JPL. Independently from us, they had solved the maximal slicing of one black hole analytically! We were all disappointed because we didn't have the discovery to ourselves, but happy because we could check that we had, in fact, correctly solved the problem numerically. Many scientists, faced with this problem, would have rushed to publish their own work. This never seemed to occur to Bryce. He graciously asked Frank and Hugo to join us in publishing jointly; furthermore, he made them first authors. This professional style was a model for many of the young people in relativity and shows the true measure of Bryce.

The maximal slicing worked for one black hole and has now become the industry-wide standard for doing numerical relativity; using it Eppley and I were able to finish the two black hole collision. One of the things I always wondered about was why Bryce chose the two black hole collision problem as the GR laboratory to work in—why not a star collapse? I believe that his worries about how you were going to get around Lagrangian hydrodynamics might have led him to think: 'Well, fortunately in general relativity you don't need matter as a source for the field, so let's do a vacuum space–time and we'll worry about the matter later'. In fact, I think that was a brilliant insight, because there was plenty of work just to understand how to solve Einstein equations. In the meantime, Jim Wilson came along and developed his beautiful techniques for doing Eulerian hydrodynamics. Finally, these two methods were fused together so that today we have codes capable of solving the full matter plus Einstein's equations. I must say that program was very much influenced by Bryce, who always kept a very strong Texas–Livermore connection.

5.   EXPERIMENTAL RELATIVITY

Of course, it doesn't do any good to study the dynamics of general relativity unless ultimately relativistic gravity really exists. Now Newton knew weak gravity existed

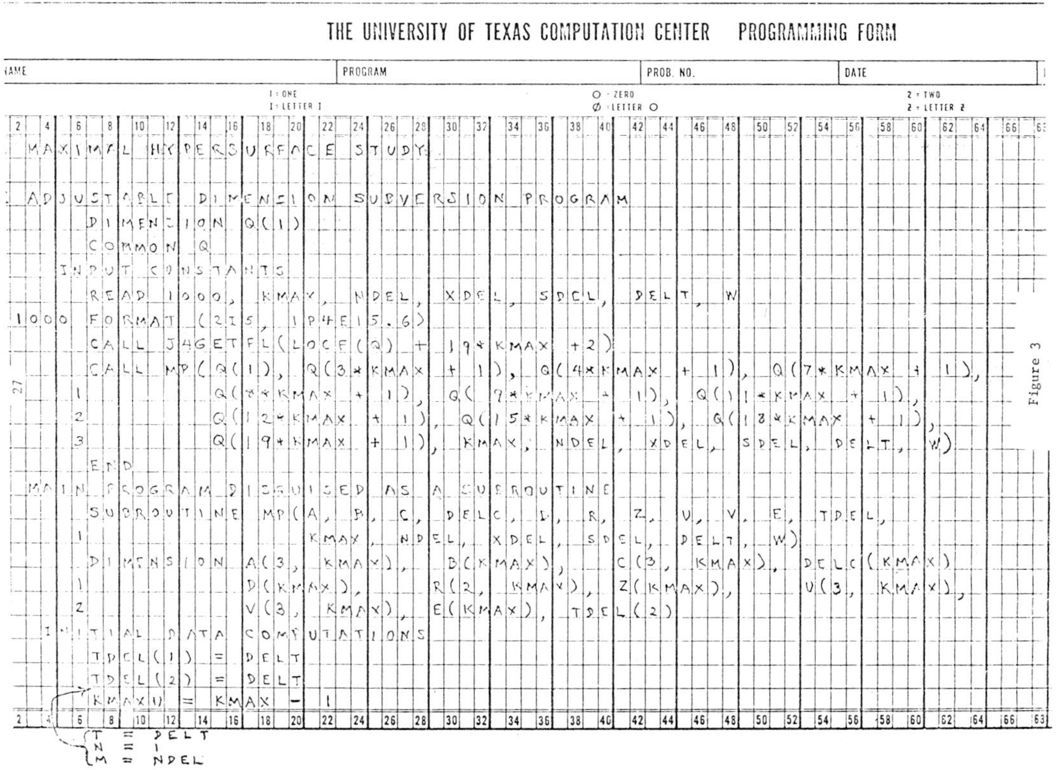
THE UNIVERSITY OF TEXAS COMPUTATION CENTER   PROGRAMMING FORM

NAME | PROGRAM | PROB. NO. | DATE

```
 MAXIMAL HYPERSURFACE STUDY

 ADJUSTABLE DIMENSION SUBVERSION PROGRAM
      DIMENSION Q(1)
      COMMON Q
C    INPUT CONSTANTS
      READ 1000, KMAX, NDEL, XDEL, SDEL, DELT, W
 1000 FORMAT (2I5, 1P4E15.6)
      CALL J4GETFL(LOCF(Q) + 19*KMAX + 2)
      CALL MP(Q(1), Q(3*KMAX + 1), Q(4*KMAX + 1), Q(7*KMAX + 1),
     1 Q(8*KMAX + 1), Q(9*KMAX + 1), Q(11*KMAX + 1),
     2 Q(12*KMAX + 1), Q(15*KMAX + 1), Q(18*KMAX + 1),
     3 Q(19*KMAX + 1), KMAX, NDEL, XDEL, SDEL, DELT, W)
      END
 MAIN PROGRAM DISGUISED AS A SUBROUTINE
      SUBROUTINE MP(A, B, C, DELC, D, R, Z, U, V, E, TDEL,
     1 KMAX, NDEL, XDEL, SDEL, DELT, W)
      DIMENSION A(3, KMAX), B(KMAX), C(3, KMAX), DELC(KMAX),
     1 D(KMAX), R(2, KMAX), Z(KMAX), U(3, KMAX),
     2 V(3, KMAX), E(KMAX), TDEL(2)
C    INITIAL DATA COMPUTATIONS
      TDEL(1) = DELT
      TDEL(2) = DELT
      KMAX1 = KMAX - 1
```

T = DELT
N = 1
M = NDEL

Figure 3 The original hand-coding sheet for the maximal slicing of one black hole. This artifact demonstrates that Bryce wrote the first FORTRAN code for the vacuum numerical relativity.

and he was able to make some very nice calculations on tides; as a result of which the British took over the world. So I've often wondered what Bryce's idea was about 'applied gravity'. I want to share with you Bryce's attitude about experimental relativity from 1957. It's in the Journal of Astronautical Sciences, from an address that Bryce gave at a meeting in 1956.

> Those of you who read newspapers will be aware that there has been a recent renewed upsurge of interest in the phenomenon of gravitation. What you may not be aware of is that the picture given by the press is by and large incorrect. The ensemble of news items to date tends to give the impression that there is a large scale program feverishly under way to 'harness the force of gravity'. This is simply not so, for the simple reason that not the slightest plausible method is known for developing practical applications of the phenomenon of gravitation . . .
>
> DeWitt (1957b)

Nonetheless, Bryce did go on to make a number of attempts to measure gravity. Bryce was always interested with how you would go about measuring something. He would always start by writing down a Lagrangian for a given apparatus. In fact, there's one famous mimeographed homework problem that he gave out to us in which you were supposed to write down the Lagrangian for a little cart that had an electric motor on it that powered a belt that went over and drove a paddle wheel through a vat of viscous fluid (apparently this problem originated with Wheeler). Well, Bryce expected us to solve this as a homework problem.

The Lense–Thirring dragging-of-inertial-frames effect was a longtime love of Bryce's. In the 1957 Journal of Astronomical Sciences article, Bryce wrote:

> There are other effects which possibly could be measured in a satellite . . . For example, Einstein's theory predicts certain induction effects. Just like electrodynamics predicts induction effects . . . Suppose the satellite is in 'free fall', and suppose a very large, heavy disc or flywheel is turning around at a high speed inside. If you take a little test particle and let it fall towards the disc, the rate of fall will be very slow, because these are relatively small objects, and the attraction of gravity is not very large, so you have a lot of time over which you can make your experiment. Now, after a while you will find that instead of falling directly toward the center of the disc, the test particle will be dragged along a little bit in the direction of the rotation of the disc. This is just an inductive effect. If the disc were motionless and not rotating, the particle would fall straight toward the center. These are things that would be very interesting to measure some day.
>
> DeWitt (1957b)

Ten years later in 1966, Bryce wrote a Physical Review Letter on *Superconductors and Gravitational Drag.* (DeWitt 1966) In it he proposed a new effect in superconductors of the Lense–Thirring term. Finally, in 1972 at Stanford (DeWitt 1972), he lectured on the detailed analysis of the Schiff gyroscope experiment which was to be orbited around the earth. Hopefully, that space experiment will be flown before the 30th year anniversary of Bryce's remarks above.

The other major experiment Bryce was involved in was the Texas Mauritania Eclipse Expedition (DeWitt *et al* 1974, DeWitt 1975, 1979, Texas Mauritanian Eclipse Team 1976) of which he was the leader. This was a grand effort to update the original 1919 observations of Sir Arthur Eddington. It was a team effort with an enormous amount of work and endurance required. Because of a dust storm during the eclipse, the accuracy was not as great as was hoped for. Nonetheless, it was a great adventure and a remarkable effort by a man known for his work in quantum gravity!

6. DYNAMIC LEADER

I come finally to the last dynamical area, namely that of Bryce himself. I think anyone who worked with Bryce is aware of the tremendous leadership quality that he has. This is of course shown up in the fact that he has been Director of the Center of Relativity at Austin since 1972; he was the Director of three of the Les Houches programs that Cécile has run for many years—the 1963, 1972, and 1983 Les Houches summer schools; he was group leader twice at Santa Barbara in 1980 and 1981; and leader of the eclipse expedition in 1973. In addition, he has always been a leader in the field of quantum gravity.

But those are just the official recognitions of what is actually the character of Bryce. Bryce fears nothing, and that attitude pervades his everyday outlook; he's always very cheerful because there's nothing he's going to come up against he can't deal with. That spirit pervades anyone who works with Bryce. I remember when we were trying to invent numerical relativity, I would come up against the situation where we didn't have any idea what was going on with the mathematical equation we were trying to solve, much less have any idea of what would be the numerical approach for it, and whether it would be stable or unstable. If you'd go to Bryce and say, 'Well, Bryce I don't know—it might be unstable, or it might not be accurate', Bryce would blow you out of the office and say, 'Don't bother me with these details. Look, these are equations, right? You can difference equations. Go difference them. Just do it!' I think that the idea of 'just do it!' comes through in everything that Bryce has ever attempted. This is the reason why many of us feel that he has had such an enormous impact on shaping not only our professional careers but our personal lives as well.

ACKNOWLEDGMENT

This work was supported in part by National Science Foundation grant No. PHY83-08826 and the Alfred P Sloan Foundation.

REFERENCES

Crowley R J and Thorne K S 1977 *Astrophys. J.* **215** 624–35
DeWitt B S 1957a *Rev. Mod. Phys.* **29** 377–97
—— 1957b *J. Astronaut.* **4** 23–8
—— 1957c *Conf. on the Role of Gravitation in Physics* Wright Air Development Center Technical Report 57–216 ASTIA Document No. AD 118180
—— 1958 Unpublished talk at Stevens Institute of Technology
—— 1963 in *Relativity, Groups, and Topology: 1963 Les Houches Lectures* ed C M DeWitt and B S DeWitt (New York: Gordon and Breach)
—— 1965 *The Dynamical Theory of Groups and Fields* (New York: Gordon and Breach)
—— 1966 *Phys. Rev. Lett.* **16** 1092–3
—— 1967 *Phys. Rev.* **160** 1113–48
—— 1970 in *Relativity: Proc. of the Relativity Conference in the Midwest* ed M Carmeli, S I Fickler and L Witten (New York: Plenum)
—— 1971 *Stanford University Lectures on General Relativity* unpublished notes
—— 1975 in *Gravitation and Relativity: Proc. of the 7th Int. Conf. on Gen. Rel.* (Jerusalem: Keter Publishing House)
—— 1979 in *Albert Einstein's Theory of General Relativity* ed G Tauber (New York: Crown Publishers)
DeWitt B S and Brehme R W 1960 *Ann. Phys., NY* **9** 220–59
DeWitt C M and DeWitt B S 1964 *Physics* **1** 3–20
DeWitt B S, Matzner R A and Mikesell A H 1974 *Sky and Telescope* **47** 301–6
Estabrook F, Wahlquist H, Christensen S, Smarr L and Tsiang E 1973 *Phys. Rev. D* **7** 2814–7
Fischer A 1970 in *Relativity: Proc. of the Relativity Conference in the Midwest* ed M Carmeli, S I Fickler, and L Witten (New York: Plenum)
Hahn S G and Lindquist R W 1964 *Ann. Phys., NY* **29** 304
Misner C W 1972 in *Magic Without Magic: John Archibald Wheeler* ed J R Klauder (San Francisco: W H Freeman)
Misner C W, Thorne K S and Wheeler J A 1973 *Gravitation* (San Francisco: W H Freeman)
Peres A 1962 *Nuovo Cimento* **26** 53
Smarr L, Cadez A, DeWitt B and Eppley K 1976 *Phys. Rev. D* **14** 2443–52
Smarr L, Taubes C and Wilson J R 1980 in *Essays in General Relativity: A Festschrift for Abraham Taub* ed F J Tipler (New York: Academic)
Texas Mauritanian Eclipse Team 1976 *Astron. J.* **81** 452
Thorne K S and Kovács S J 1975 *Astrophys. J.* **200** 245–62
Wheeler J A 1963 in *Relativity, Groups, and Topology: 1963 Les Houches Lectures* ed C M DeWitt and B S DeWitt (New York: Gordon and Breach)
—— 1968 in *Battelle Rencontres 1967: Lectures in Mathematics and Physics* ed C DeWitt and J A Wheeler (New York: Benjamin)

The Contributions of Bryce DeWitt to Quantum Gravity†

C J ISHAM

1. INTRODUCTION

I felt honored when asked to speak on Bryce DeWitt's contributions to quantum gravity and now approach with a mixture of pleasure and apprehension the task of compiling my thoughts: pleasure at the prospect of placing on record my appreciation of someone for whom I have the greatest respect; apprehension less I fail to do justice to a body of work which includes some of the most significant developments in this fascinating but very difficult, branch of theoretical physics.

As can be seen from the list of references, Bryce has written around twenty papers concerned with the quantization of the gravitational field: a number that by current standards may seem rather small until it is noticed that they occupy over eight hundred journal pages! In analyzing and assessing these contributions it is essential to deal with them as a whole; indeed, in respect of the overall cohesive plan and sense of driving purpose, there is more than a touch of Wagner in Bryce DeWitt's work. The occurrence of key themes that echo from paper to paper is also Wagnerian and these themes characterize the main thrust of his ideas; in particular he persistently emphasizes the need to:

(1) acknowledge the intrinsically non-linear features of general relativity
(2) preserve in the quantum theory those invariances possessed by the classical theory
(3) include fermions on the same footing as bosons.

To reflect this holism adequately I must attempt to cover all major areas of Bryce's research and I will base my account mainly on his papers in the order in which he wrote them. This brings the advantage that, so obvious is the significance of his original work, further comment from me is hardly necessary. However, such an approach can not do full justice to the historical evolution of the subject so I

† Based on a talk given at an informal meeting held in Austin, Texas, on December 11 1982 to honor Bryce DeWitt's 60th birthday.

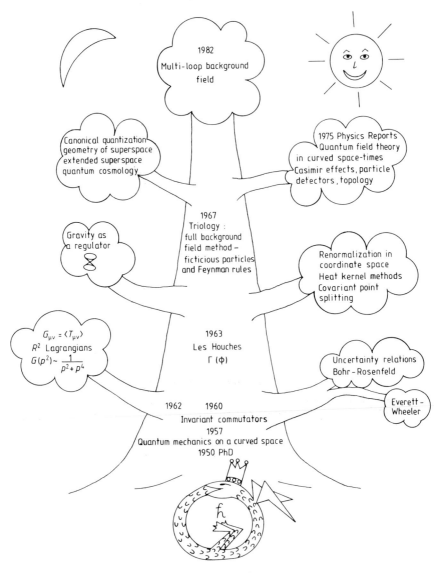

Figure 1

apologize in advance to those whose personal contributions are not sufficiently emphasized and hope that they will appreciate the spirit in which this article is written.

It seems appropriate to start my record of Bryce DeWitt's magnum opus with a tree graph (figure 1) illustrating the most important area of his research. For our purposes, the account opens in DeWitt (1957) with a discussion of dynamical theory in curved spaces. However the framework which he employed—path integrals and the WKB approximation—dates back to the original work of Morette (1951) and indeed this entire subject is one which Cécile has developed in great depth and which, especially with the realization of the significance of the fundamental group of configuration space (Laidlaw and Morette-DeWitt 1957), is becoming of ever increasing importance†.

These techniques were applied mainly to systems with a finite number of degrees of freedom and it was not long before Bryce concentrated on the complementary field theory problem. He initiated a lengthy research program that culminated in the much acclaimed background field method and whose historical development I will briefly sketch.

2. COVARIANT COMMUTATORS

There seems no better way of starting this story than to record a quotation from DeWitt (1962a, p 1073) which epitomizes his entire *Weltanschauung* for gauge theories:

First, in order to achieve the greatest possible generality, we continue our total boycott of the canonical formalism.

This antagonism, founded on the difficulty of finding and using canonical variables and on the intrinsic noncovariance of such a scheme, permeates much of his work. Thus in his early papers (1960, 1961) he concentrates on computing Poisson brackets between physical observables only; it is worth recalling this work as it contains the formal theory of the covariant Green's functions that played such a key role in later developments.

Let ψ^i be a field subject to the infinitesimal gauge transformation‡

$$\delta\psi^i = R^i_\alpha \delta\xi^\alpha \tag{2.1}$$

where R^i_α is a function of ψ and its derivatives. A physical observable is functional A of ψ satisfying

$$A_{,i} R^i_\alpha = 0 \tag{2.2}$$

where $A_{,i}$ denotes the functional derivative of A with respect to ψ^i. For example if ψ is the metric tensor $g_{\mu\nu}(x)$ and the gauge group is the group of general coordinate transformations, then a physical observable obeys

$$(\delta A/\delta g_{\mu\nu})_{|\nu} = 0.$$

---

† For a comprehensive review see DeWitt-Morette *et al* (1979, 1980). Note that the entire theory of θ vacua in quantum field theories can be understood in this way.

‡ Recall that in DeWitt's (now standard) notation indices like i or α denote both discrete and continuous (space–time) labels and, if dummies, are correspondingly both summed and integrated.

Now let A and B be two localized observables (figure 2) and consider the effect on A of subjecting the action S to the infinitesimal transformation $S \rightarrow S + \varepsilon B$. Motivated by the discussion in Peierls (1952) of theories without gauge groups, DeWitt showed that the Poisson bracket (A, B) of A and B is related to the advanced $(+)$ and retarded $(-)$ effects of B on A by

$$\delta_B{}^+ A - \delta_B{}^- A = \varepsilon(A, B). \tag{2.3}$$

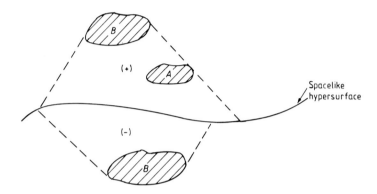

Figure 2

The advanced and retarded changes $\delta_B{}^+ A$ and $\delta_B{}^- A$ are determined by $\delta^{\pm}\psi^i$ (because of (2.2), only defined up to a gauge transformation):

$$\delta^{\pm} A = A_{,i} \delta^{\pm} \psi^i \tag{2.4}$$

whilst the field equations $S_{,i} = 0$ change by

$$\delta^{\pm}(S_{,i}) = S_{,ij} \delta^{\pm} \psi^j = -\varepsilon B_{,i}. \tag{2.5}$$

Now, if the differential operator $S_{,ij}$ were invertible, equation (2.5) could be solved for $\delta^{\pm}\psi^j$ which, via (2.4), would relate $\delta^{\pm} A$ to $B_{,i}$. However the gauge invariance of S leads to the Bianchi identities.

$$S_{,ij} R^j{}_{\alpha} = 0 \tag{2.6}$$

which shows precisely that $S_{,ij}$ is *not* invertible. To surmount this obstacle DeWitt defined the auxiliary differential operator

$$F_{ij} \equiv S_{,ij} + \gamma_{ik} \gamma_{jl} R^k{}_{\alpha} R^l{}_{\beta} \gamma^{\alpha\beta} \tag{2.7}$$

where the 'metrics' γ_{ij} and $\gamma^{\alpha\beta}$ are chosen so that F_{ij} is local and invertible. He then imposed the supplementary conditions

$$\gamma_{ik} R^k{}_{\alpha} \delta^{\pm} \psi^i = 0 \tag{2.8}$$

and inverted (2.5) as

$$\delta^{\pm} \psi^i = \varepsilon G^{\pm ij} B_{,j} \tag{2.9}$$

where G^{+ij} and G^{-ij} are respectively the advanced and retarded Green's functions of the globally hyperbolic operator F_{ij}. Green's functions of this type were discussed in Hadamard (1923) and Leray (1952) and some detailed properties developed in DeWitt and Brehme (1960); of particular importance is the combination $G^{ij} \equiv G^{+ij} - G^{-ij}$ which serves to propagate information towards and away from any Cauchy surface (see also Choquet-Bruhat 1968, Lichnerowicz 1961, 1964) and in terms of which the Peierls Poisson bracket (2.3) may be succinctly expressed (using the reciprocal relation $\delta_B{}^{\pm}A = \delta_A{}^{\mp} B$ as

$$(A, B) = A_{,i} G^{ij} B_{,j}. \tag{2.10}$$

It should be noted that the non-linearity of the field equations is no bar to finding this expression which relies solely on the theory of infinitesimal disturbances propagating on the nonvanishing background ψ: a precursor to DeWitt's full background field method for gauged quantum field theory. It can be shown that (A, B) defined by equation (2.10) is independent of the choice of γ_{ij} and $\gamma^{\alpha\beta}$ (i.e. gauge invariant) and satisfies the Jacobi identities, even for anticommuting spinor fields (DeWitt 1962c).

This operational definition of the Poisson bracket underlay DeWitt's hopes of constructing gauge theories purely in terms of physical observables[†] although to the best of my knowledge only Mandelstam's path dependent methods fully realized this goal (Mandelstam 1968a, b). DeWitt also used this definition to extend to an arbitrary gauge group the measurement theory of the quantized electromagnetic field discussed in Bohr and Rosenfeld (1933). To this end he followed Peierls in defining $D_B A = \lim_{\varepsilon \to 0} \delta_B{}^- A$ and rewrote (2.3) as $D_A B - D_B A = (A, B)$ leading to a quantum uncertainty relation of the form (DeWitt 1962b)

$$\Delta A \, \Delta B \geqslant \tfrac{1}{2}\hbar \, \langle \, D_A B - D_B A \, \rangle. \tag{2.11}$$

This form is ideally suited for discussions of the uncertainty relations in terms of mutual disturbances of simultaneous measurements and is the approach adopted by DeWitt. Parenthetically, it is intriguing to note how a theoretical physicist's philosophical views on quantum theory often reflect aspects of his personality: for example, amongst Bryce's greatest pleasures we find canoeing, rapids shooting, free-fall parachuting, treking across uncharted areas of the world etc, etc, and therefore perhaps it is no surprise to discover that he is a realist (see for example DeWitt 1970c) who, in attaching importance to the idea of a single quantum event, regards the fashionable frequency interpretations as being distinctly feeble!

DeWitt's work (1962b, c) on the extension of the Bohr–Rosenfeld measurement theory is most elegant and shows clearly the way in which the uncertainty principle for a pair of conjugate apparatus variables enforces, and is compatible with, the uncertainty relations on the system variables. This line of argument reaches its climax in DeWitt (1962d): a beautiful but sadly neglected paper concerning the use of a stiff elastic medium to measure the gravitational field. Many other important ideas are introduced in this example of one of those rather rare contributions to quantum

---

[†] Note that A and B in equation (2.10) can be either exactly gauge invariant or only when evaluated on fields that satisfy the classical equations.

gravity which manages to climb down from the esoteric heights of mathematical formalism and to tackle some of the basic physics problems that are involved.

Of course, DeWitt's ontological preoccupations extend well beyond discussions of the uncertainty relations. For example his study of model particle detectors was of considerable importance in clarifying some of the difficult conceptual problems that arise when fields are quantized in a curved background (DeWitt 1979, see also Unruh 1976). He is also an ardent expositor and defender of the Everett–Wheeler (Everett 1957, Wheeler 1957a) interpretation of quantum theory and was the author (DeWitt 1973) of the infamous aphorism:

> The mathematical formalism of the quantum theory is capable of yielding its own interpretation.

This remark has enticed and seduced generations of graduate students, much to the embarassment of their supervisors who are called upon (unsuccessfully) to explain what it means.

3. R^2 LAGRANGIANS

In 1962 Utiyama and DeWitt published a paper (Utiyama and DeWitt 1962) which inaugurated an approach to quantum gravity that still generates considerable interest. They considered the quantization of a matter field in the presence of a gravitational background and studied the semiclassical approximation (see also Møller 1962) in the form

$$G_{\mu\nu}(g) = \frac{\langle \text{in} | \hat{T}_{\mu\nu}(\text{matter}, g) | \text{out} \rangle}{\langle \text{in} | \text{out} \rangle} \equiv \langle \hat{T}_{\mu\nu} \rangle \qquad (3.1)$$

where the vacuum states $| \text{in} \rangle$ and $| \text{out} \rangle$ are defined with respect to the Minkowski background $\eta_{\mu\nu}$ which features in the metric expansion $g_{\mu\nu} = \eta_{\mu\nu} + \phi_{\mu\nu}$. Equation (3.1) is rewritten as

$$\mathcal{D}\phi_{\mu\nu} = \langle \hat{T}_{\mu\nu} \rangle + \tau_{\mu\nu} \qquad (3.2)$$

where $\mathcal{D}$ is the appropriate second-order linear differential operator and $\tau_{\mu\nu}$ includes an external source as well as the non-linear parts of $G_{\mu\nu}(\eta + \phi)$. Thus the solution of (3.2) may be expressed graphically as

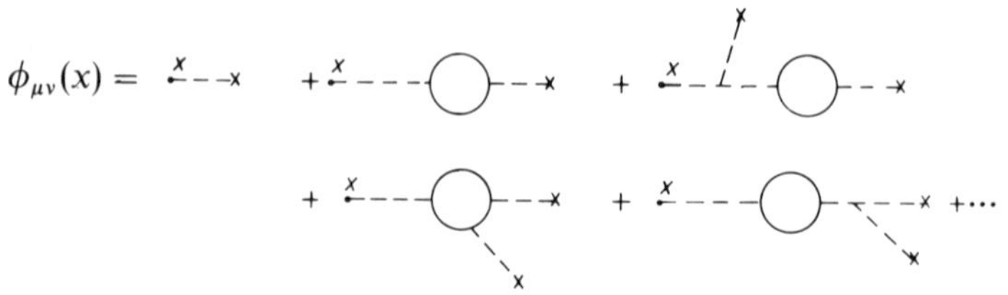

Figure 3 x denotes the external source.

In the light of recent research, one might object to such a semiclassical scheme for quantum gravity on the grounds that

(1) the neglected graviton loop ⬡ has the same size as the matter loop ◯ (Duff 1981).

(2) the stability of the entire procedure (see Kibble 1981 for a review) is open to doubt. (Horowitz and Wald 1978, Horowitz 1981).

However, of great importance is the observation by Utiyama and DeWitt that, in order to render the scheme finite, it is necessary to add counter terms to the Lagrangian including in particular terms of the form $R_{\mu\nu}R^{\mu\nu}$ and R^2. They then remark that, if such curvature-squared contributions are included from the outset, the differential operator in (3.2) will contain fourth powers of the derivatives and the corresponding propagator will behave at large momenta as p^{-4} rather than p^{-2}; this greatly improves the ultraviolet behavior of the theory, which, formally, becomes renormalizable—a result that, 15 years later, was rigorously proved using BRS techniques (Stelle 1977).

Unfortunately the ensuing theory, although renormalizable is not unitary and much recent work in this area has been aimed at surmounting this obstacle. For example attempts are made in Salam and Strathdee (1978) and Julve and Tonin (1978) to remove the ghost states by employing renormalization group techniques. Further developments have included propagating torsion (Neville 1980, Sezgin and van Nieuwenhuizen 1979) and the well developed theories of self-consistent gravity (for example, Tomboulis 1980, Adler 1982, Zee 1981).

Although this work was a mere side branch of DeWitt's main effort it generated a considerable activity which continues to the present time (see Christensen 1982); in particular one might cite Stelle's (1983) very recent result that the $N = 8$ supergravity extension of a curvature-squared Lagrangian is actually *finite*, not merely renormalizable.

4. DeWITT AND SCHWINGER

The *prima materia* of DeWitt's background field method is Schwinger's technique (Schwinger 1951, 1953) of probing a quantum field theory with a varying external source; indeed, in the early work of DeWitt (1962a, 1964b), it is difficult to unscramble the contributions of the two men. Schwinger's starting point was the vacuum–vacuum transition amplitude, $\langle \text{out} | \text{in} \rangle_J \equiv \exp iW[J]$, in the presence of an external source J. He showed that the matrix element of a quantum field $\hat{\phi}$ may be computed as:

$$\langle \text{out} | \hat{\phi}^i | \text{in} \rangle_J = \frac{1}{i} \frac{\delta}{\delta J_i} \langle \text{out} | \text{in} \rangle_J \tag{4.1}$$

with the general n-point function

$$\langle \text{out} | T \hat{\phi}^{i_1} \ldots \hat{\phi}^{i_n} | \text{in} \rangle_J = \frac{1}{i} \frac{\delta}{\delta J_{i_1}} \cdots \frac{1}{i} \frac{\delta}{\delta J_{i_n}} \langle \text{out} | \text{in} \rangle_J \tag{4.2}$$

expressed by DeWitt as the functional integral

$$\langle \text{out}| T\hat{\phi}^{i_1} \ldots \hat{\phi}^{i_n}|\text{in}\rangle_J = \int \delta[\phi]\phi^{i_1} \ldots \phi^{i_n} \exp i(\delta[\phi]+J_1\phi^i). \qquad (4.3)$$

Of particular significance in the evolution of the background field method is the definition of the classical field

$$\varphi^i[J] \equiv \frac{\delta W}{\delta J_i}[J] = \quad\bullet\!\!-\!\!-\!\!\times \;+\; -\!\!<^{\times}_{\times} \;+\; -\!\!\bigcirc\!\!-\!\!\times \;+\cdots \qquad (4.4)$$

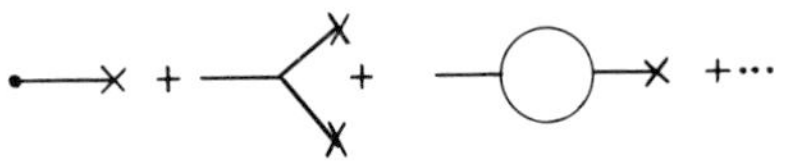

and the two-point correlation function

$$G^{ij}[J] \equiv \frac{\delta \varphi^i}{\delta J_i}[J] = \quad\bullet\!\!-\!\!-\!\!\bullet \;+\; \bullet\!\!-\!\!\top\!\!-\!\!\bullet \;+\; -\!\!\bigcirc\!\!- \qquad (4.5)$$
$$+\cdots$$

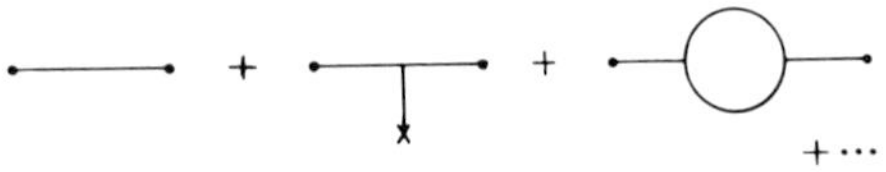

which is equal to the two-point function with disconnected parts removed:

$$-iG^{ij} = \frac{\langle \text{out}| T\hat{\phi}^i\hat{\phi}^j|\text{in}\rangle_J}{\langle \text{out}|\text{in}\rangle_J} - \varphi^i\varphi^j. \qquad (4.6)$$

The general *n*-point correlation function is expressed simply as

$$G^{i_1\cdots i_n} = \frac{\delta}{\delta J_{i_1}} \cdots \frac{\delta}{\delta J_{i_n}} W[J]$$

and the technique generalizes to yield the matrix elements of arbitrary functions of the fields as for example in figure 4.

$$\frac{\langle \text{out}|\hat{\phi}^i\hat{\phi}^i|\text{in}\rangle_J}{\langle \text{out}|\text{in}\rangle_J} = \varphi^i\varphi^i - iG^{ii}$$

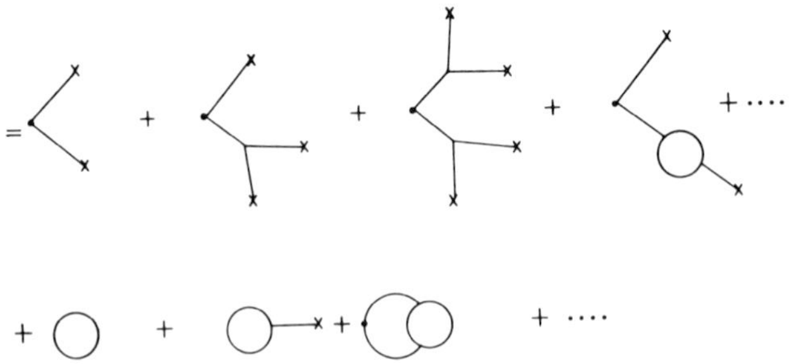

$$(4.7)$$

Figure 4

DeWitt was especially concerned with the quantum field equations in the form

$$\frac{\langle \text{out}| T(\hat{S}_{,i})|\text{in}\rangle_J}{\langle \text{out}|\text{in}\rangle_J} \equiv \langle \hat{S}_{,i}\rangle = -J_i \qquad (4.8)$$

which imply

$$\frac{\delta}{\delta \varphi^j} \langle \hat{S}_{,i} \rangle \frac{\delta \varphi^j}{\delta J_k} = -\delta_i{}^k. \tag{4.9}$$

Now, if the Green's function $\delta \varphi^j / \delta J_k = G^{jk}$ appearing in equation (4.9) is chosen to be the Feynman function $G_F{}^{jk}$, then $G_F{}^{ij} = G_F{}^{ji}$, or, more precisely, DeWitt *defines* the Feynman function in the presence of an external source to be a Green's function with this symmetry†. This property is crucial: it implies $\langle \hat{S}_{,i} \rangle_{,j} = \langle \hat{S}_{,j} \rangle_{,i}$ and hence the existence of the *C*- number functional $\Gamma[\varphi]$ of φ satisfying

$$\langle \hat{S}_{,i} \rangle = \Gamma_{,i}[\varphi]. \tag{4.10}$$

DeWitt's central remark is that this functional of φ completely determines the full quantum theory and can be regarded as the quantum analogue of the classical action. For example, the source J_i viewed as a functional of φ (i.e. the 'inverse' of equation (4.4)), may be recovered from Γ via $\Gamma_{,i}[\varphi] = J_i[\varphi]$ whilst the equation $\Gamma_{,ij} G_F{}^{jk} = -\delta_i{}^k$ illustrates precisely the sense in which $G_F{}^{jk}$ may be viewed as the Green's function of the full quantum theory. By requiring that the Feynman function $G_F{}^{ij}$ should in addition satisfy the normal 'matrix' equation $\delta G_F{}^{ij} = G_F{}^{ik} \delta(\Gamma_{,kl}) G_F{}^{lj}$, DeWitt showed how the higher correlation functions could also be related to Γ; for example

$$G^{ijk} = G^{ia} G^{jb} G^{kc} \Gamma_{,abc} \tag{4.11}$$

which demonstrates nicely the way in which an n-point function may be constructed from 'trees' with G^{ij} branches and generalized vertices $\Gamma_{,i_1 \ldots i_n}$.

The formalism above only works in the absence of a gauge group. However, at this time DeWitt was still hoping to quantize gauge theories in terms of physical observables only and hence felt that this framework might be sufficient, although, by observing that the functional integral measure is not completely determined in a gauge theory, he foreshadowed the later inclusion of fictitious fields.

Even at this early stage DeWitt emphasized the importance of expanding a quantum field around an *arbitrary* solution of the source-free quantum field equations $\Gamma_{,i} = 0$—an idea that has been of considerable value in recent developments of spontaneous symmetry breaking. He also stressed that, even if the underlying quantum field is Hermitian, the generic solution to $\Gamma_{,i} = 0$ will be complex; a property that is particularly interesting in the gravitational case and which might forge a link with Hawking's Riemannian quantum gravity program (see Hawking 1979). DeWitt suggested that, in particular, two colliding black holes would lead to a complex metric but unfortunately these intriguing ideas have as yet been little developed.

Another property of the gravitational effective action $\Gamma[g_{\mu\nu}]$, emphasized in DeWitt (1962a), is its essential nonlocality at the Planck length—a feature which, in a

---

† The problem of defining the Feynman function in the presence of an external source or background field became of great importance in the intense investigations in the mid 1970s of quantum field theory in a curved space–time.

series of interesting remarks, he relates to the possibility of a dynamical change in spatial topology:

> ... the admission of geometry as an object of quantization which will ultimately force a generalization of conventional asymptotic conditions so as to take into account the possibility of spacetime itself having unusual topology either macroscopically or microscopically. (p 1073)

> In the context of gravitation theory the admissibility of arbitrary macroscopic background fields immediately implies also the admissibility of arbitrary topology in the large. (p 1092)

> ... and hence to define them (Feynman propagators) uniquely in contexts in which the concepts of positive and negative frequencies and a unique vacuum are inappropriate, e.g. in the case of ... or spacelike cross sections of non-Euclidean and even dynamically changing topology. (p 1087)

> Must we sum over topologies as well as histories in the Feynman integral ... ? (p 1093)

Of course, these days, we have become accustomed to including topological properties in our quantum field theories and it is salutary to reflect that Wheeler (1957b, 1964) and DeWitt were seriously considering this possibility over twenty years ago.

Bryce DeWitt's article (1964b) in the Les Houches 1963 proceedings contains a comprehensive summary of his work at that time and is a masterful introduction to quantum field theory and its geometrical aspects. Even now it affords fascinating and highly instructive reading and in 1964 it was of the greatest possible significance in providing for many geometrical ideas, almost the only exposition that was usable by elementary particle physicists whose training at that time did not include differential geometry.

There are also a number of significant anticipations of later developments in various areas. Now the unraveling of the labyrinthine threads of scientific precedence often requires the magic touch of an Ariadne but in Bryce's case he was so often so far in advance of his time that the usual problems hardly arise. A good example of this is the theory of non-linear group realizations. This topic came to prominence in the middle and late 1960s and arose from the desire to construct effective Lagrangians that would transform under a full $SU(n) \times SU(n)$ chiral group and yet not involve the unwanted 0^+ meson fields. It took several years for the idea slowly to emerge that the necessary technique was the generalization of linear group representations to include groups acting on arbitrary manifolds and in particular in co-set spaces. However, even a cursory glance at DeWitt (1964) reveals most of the necessary mathematical machinery and indeed linear and 'non-linear' group actions are frequently considered together. The only reason that Bryce did not tackle and completely solve the chiral group problem seems to be simply that he had not been exposed to it as an interesting problem in physics.

Another feature of the Les Houches article is Bryce's characteristic Socratic approach to teaching in which the 'self-help' exercises form an integral part of the course and text. A delightful example may be found in Problem 77 on page 725 which requires the student to work out what is, in effect, the non-Abelian version of the old Klein–Kaluza theory: another striking anticipation of later developments.

In all aspects DeWitt (1964b) is a wonderful article in a wonderful book and the lecturers in the 1983 Les Houches Summer School (also entitled *Relativity, Groups and Topology*) face a daunting task in attempting to produce a work of anything like the same lasting value.

5. GRAVITY AS A REGULATOR

Let us take a second short break from the main story and explore another intriguing side branch of Bryce DeWitt's work. It has often been conjectured that the ultraviolet divergences that plague conventional quantum field theory will not appear if the quantized gravitational field is included from the outset (Klein 1955, 1957, Landau 1955, Pauli 1956, Deser 1957). These divergences arise from a confluence of singularities on the lightcone: a confluence which is expected to be 'smoothed away' when the lightcone itself is subject to quantum fluctuations. Of course, such a qualitative change cannot be expected to appear in any finite order of perturbation theory, indeed conventional quantum gravity plus matter is non-renormalizable, and an intrinsically non-perturbative approach must be adopted.

Such a scheme was proposed in DeWitt (1964a) (see also Khriplovich 1966) based on a summation of an infinite series of formally divergent Feynman graphs. DeWitt considered first the gravitational coupling of a scalar field and studied the Bethe–Salpeter sequence for scalar–scalar scattering with graviton exchange (figure 5). He adopted the usual leading term approximation of neglecting, in the loop integrals, the momentum q and scalar particle's bare mass. This results in a complete cancellation between the vertices and scalar propagators and the ensuing amplitudes are very similar to the type that were later handled with non-polynomial Lagrangian techniques (figure 6).

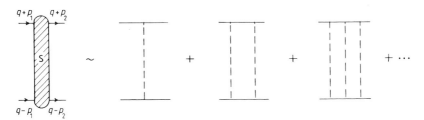

Figure 5

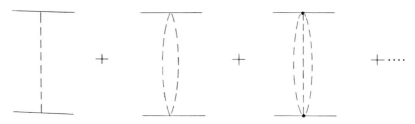

Figure 6

These approximate scattering amplitudes may be formally summed to yield a total amplitude that, at high momentum transfer $p_1 - p_2$, depends linearly on a function $Y(p_2 - p_1)$ satisfying the integral equation

$$Y(p) = \frac{1}{p^2} - \frac{i}{(2\pi)^4} \int \frac{Y(k)}{(p-k)^2} \, d^4k \tag{5.1}$$

with the solution in coordinate space

$$\tilde{Y}(x) = \frac{i}{(2\pi)^2} \frac{1}{x^2 - (4/\pi^2)L_P^2} \tag{5.2}$$

where $L_P = (G\hbar/C^3)^{1/2} \approx 10^{-33}$ cm is the usual Planck length. Thus the infinite sum is indeed finite and, as anticipated, the lightcone has been qualitatively affected by the inclusion of quantum gravity.

DeWitt applied similar methods to the self-energy graphs (figure 7) related to the non-polynomial sequence (figure 8) and to gravity modified scalar electrodynamics (figure 9), related to (figure 10).

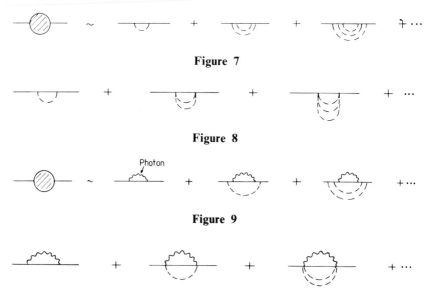

Figure 7

Figure 8

Photon

Figure 9

Figure 10

These finite scalar + photon + graviton graphs produce a finite shift in the scalar particle's self-mass of the form

$$\delta m^2 \approx \alpha/L_P^2 + \alpha \log(mL_P) \tag{5.3}$$

where the appearance of the α/L_P^2 term signals a breakdown of electromagnetic gauge invariance. In the later, more sophisticated, non-polynomial calculations (Isham *et al* 1971, 1972) this difficulty was overcome but in the crucial finiteness of the result we see another example of DeWitt's anticipation of future work by many years and with ideas that remain fruitful (for example DeWitt 1981a) even if these days, with the advent of supergravity, attention has switched to the construction of quantum field theories that are *perturbatively* finite.

6.   THE FULL BACKGROUND FIELD METHOD

We now arrive at one of the main cores of DeWitt's work: the second of the famous trilogy of papers (DeWitt 1967a, b, c) in which is developed the full background field method for quantizing a general, non-Abelian, gauge theory.

DeWitt started with an operator–Hilbert space formalism for a quantum field theory without a gauge group. There is no external source J_i but instead an external 'background' field φ^i in terms of which the quantum field $\hat{\varphi}^i$ is expanded as $\hat{\varphi}^i = \varphi^i + \hat{\phi}^i$. These fields are subjected to small variations obeying $\delta\varphi^i = -\delta\hat{\phi}^i$; these affect the vacuum–vacuum transition amplitude $\langle\text{out}|\text{in}\rangle_\varphi \equiv \exp iW[\varphi]$ in such a way that

$$\frac{1}{i}\frac{\delta}{\delta\varphi^i}\langle\text{out}|\text{in}\rangle_\varphi = S_{,ij}[\varphi]\langle\text{out}|\hat{\phi}^j|\text{in}\rangle_\varphi \tag{6.1}$$

which may be inverted to give the basic equation

$$\langle\text{out}|\hat{\phi}^i|\text{in}\rangle_\varphi = G^{ij}[\varphi]\frac{1}{i}\frac{\delta}{\delta\varphi^j}\langle\text{out}|\text{in}\rangle_\varphi \tag{6.2}$$

in which $G^{ij}[\varphi]$ is the appropriate Green's function for the differential operator $S_{,ij}[\varphi]$. Note that $G^{ij}[\varphi] \equiv \bullet\!\!-\!\!-\!\!\bullet$ may be written graphically as a sum of trees built from the propagator $G^{ij}[0] \equiv \Longrightarrow$ and the vertices $\prec\!\!\!\!<$, $\longrightarrow$ etc which describe the coupling of the quantum field $\hat{\phi}^i$ to the background $\varphi^i \equiv \wwwx$ (see figure 11).

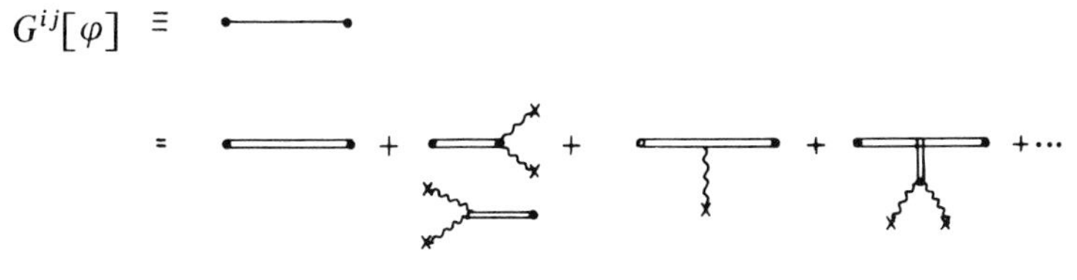

$$G^{ij}[\varphi] \equiv \quad \bullet\!\!-\!\!-\!\!-\!\!\bullet$$

Figure 11

The two-point function may be expressed as

$$\langle \text{out}|T\hat{\phi}^i\hat{\phi}^j|\text{in}\rangle_\varphi = \left(-iG^{ij}[\varphi] + G^{ik}[\varphi]\frac{1}{i}\frac{\delta}{\delta\varphi^k}\,G^{jl}\frac{1}{i}\frac{\delta}{\delta\varphi^l}\right)\langle\text{out}|\text{in}\rangle_\varphi$$

(6.3)

where ⊘ denotes the transition amplitude $\langle\text{out}|\text{in}\rangle_\varphi$; more generally the generating functional for arbitrary time ordered products is

$$\langle\text{out}|T\exp(i\hat{\phi}\cdot\lambda)|\text{in}\rangle_\varphi = \exp i\left(\sum_{n=2}^\infty \lambda_{i_1}\dots\lambda_{i_n}G^{i_1\dots i_n}[\varphi]\right)\exp(\lambda_i G^{ij}[\varphi]\delta/\delta\varphi^j)$$

$$\times\langle\text{out}|\text{in}\rangle_\varphi$$

(6.4)

where

$$G^{i_1\dots i_n}[\varphi] \equiv G^{i_1 j_1}\frac{\delta}{\delta\varphi^{j_1}}\dots G^{i_{n-2}j_{n-2}}\frac{\delta}{\delta\varphi^{j_{n-2}}}G^{i_{n-1}i_n}.$$

The virtue of equation (6.4) is the separation of the 'tree amplitudes' in the first exponential from the radiative corrections generated from $\langle\text{out}|\text{in}\rangle_\varphi$ by the second exponential—an essential ingredient in the program that is to be followed when a gauge group is present. Following Feynman (1963), DeWitt showed that the tree functions $t_{i_1\dots i_n} \equiv S_{,i_1 j_1}\dots S_{,i_n j_n}G^{j_1\dots j_n}$ provide amplitudes $t_{i_1\dots i_n}\phi_0^{i_1}\dots\phi_0^{i_n}$ (where ϕ_0^i is an asymptotic wavefunction) that are invariant under transformations in the 'metrics' γ_{ij}, $\gamma^{\alpha\beta}$ (cf §2, these functions fix the gauge of the quantum fields). Gauge invariance of any complete S-matrix element is then determined, via (6.4) and LSZ reduction, by the transformation properties of $\langle\text{out}|\text{in}\rangle_\varphi$. It is this feature that underlies DeWitt's basic philosophy for gauge theories: discard the operator–Hilbert space approach and concentrate entirely on C-numbers such as $G^{i_1\dots i_n}$ and $\langle\text{out}|\text{in}\rangle_\varphi$.

Feynman had already shown that the naive one-loop contribution $W^{(1)}[\varphi]$ to the vacuum–vacuum amplitude was wrong; it violated both gauge invariance and unitarity and needed to be corrected by the addition of a loop of fictitious particles. For example, the one-loop amplitude for the vacuum production of a single particle was found to be

$$\phi_0^i W_{,i}^{(1)}[\varphi] =$$

Fictitious
particle

(6.5)

whose gauge invariance can be proven (via the tree theorems) by showing its identity with $\phi_0{}^i S_{,\,ijk}[\varphi]G^{jk}$ where G^{jk} is a Green's function that propagates physical modes only.

Sustained only by a faith that nature must ultimately yield elegant results, DeWitt undertook the formidable task of computing the two-loop contribution to $W[\varphi]$. He aimed to reduce all amplitudes to sums of tree graphs with appropriate legs linked by the Green's functions G^{jk}. The gauge invariance of these special graphs could then be employed to demonstrate the invariance of the entire amplitude. More precisely, he could attempt to guess the corrections to the naive form of $W^{(2)}[\varphi]$ that would ensure the existence of such a reduction. The calculations involved are formidable but the end result is very encouraging: the naive two-loop contribution to $W[\varphi]$

$$W^{(2)}[\varphi] = \ominus \quad + \quad \bigcirc\!\!-\!\!\bigcirc \quad + \quad \bigcirc\!\!\bigcirc$$

can be corrected by the addition of essentially the *same* type of fictitious loop as Feynman had already found for $W^{(1)}$. This result is attractive, but naturally one wonders what will happen at higher loop levels and even Bryce baulked at the prospect of directly computing $W^{(3)}$!

DeWitt resolved this problem with the touch of a true master: he noticed that, within a functional integral framework, both the one-loop and the two-loop results could be rederived by including, in loops only, fictitious *Fermi* (but vector) *fields* ψ^α which were coupled to the quantum field ϕ^i by a simple vertex of the form $V_{\alpha i \beta}[\varphi]\psi^{*\alpha}\phi^i\psi^\beta$. DeWitt then showed that that this modified action yields gauge invariant amplitudes to *all* loop orders and the job was done.

Because of the complexity of this work there was a considerable delay between the discovery of the results and their publication and meanwhile the famous Physics Letter of Faddeev and Popov (1967) appeared in which DeWitt's functional integral was obtained by quite different, group theoretic oriented, methods. The popularity of functional integration has resulted in a tendency to regard the Faddeev–Popov paper as the foundation of quantum gauge theories but it would be a serious mistake to underestimate the significance of DeWitt's contribution. Functional methods are fine in theory and fun to use but there are several known instances where an overly naive application produces incorrect results. Indeed even some of the most ardent supporters of these methods sometimes feel the need to justify their procedures by demonstrating a somewhat abstract equivalence with the canonical formalism and so DeWitt's validation by solid calculation of a genuine physical process is of fundamental importance.

Like so many of DeWitt's ideas, the background field method is still of great interest and has been widely applied and extended. For example, in the original formalism the background field satisfied the classical field equations—a restriction that can be usefully lifted in the application to multiloop processes ('t Hooft 1975, Boulware 1981, DeWitt 1981, Abbott 1981). Development continues apace, but the widely recognized fundamental importance of gauge theories will guarantee for DeWitt's original 1967b paper a permanent place in the history of modern physics.

7.   THE CANONICAL QUANTIZATION OF GRAVITY

It is a shade ironical that, in spite of his not infrequently expressed dislike of canonical methods, DeWitt wrote one of the most interesting papers on the subject (part 1 of the trilogy, DeWitt 1967a). The canonical formalism of general relativity was born after much collective†, and sometimes painful, labor but by the mid 1960s the action had been written in the clean form

$$S = \int dt \int_\Sigma d^3x (\pi^{ij}\dot{g}_{ij} - N\mathcal{H} - N_i\mathcal{H}^i) \tag{7.1}$$

where $g_{ij}(x, t)$ is the three metric induced on the spacelike hypersurface Σ and Π^{ij} is essentially the extrinsic curvature of Σ embedded in four-dimensional space–time. The lapse and shift functions (N and N_i) appear as explicit Lagrange multipliers and generate the constraints

$$\mathcal{H}^i \equiv -2\pi^{ij}{}_j = 0 \tag{7.2}$$

and

$$\mathcal{H} \equiv G_{ijkl}\pi^{ij}\pi^{kl} - g^{1/2\,(3)}R = 0 \tag{7.3}$$

where

$$G_{ijkl} \equiv \tfrac{1}{2}g^{-1/2}(g_{ik}g_{jl} + g_{il}g_{jk} - g_{ij}g_{kl}). \tag{7.4}$$

When Σ is non compact the action in equation (7.1) must be used with caution; a total divergence has been discarded which plays an important role in the definition of asymptotic energy: indeed one of the first important results in DeWitt (1967a) is a clear statement concerning the structure of these boundary terms.

In Dirac's approach to quantizing a constrained system, the (first-class) constraints are imposed as operator equations on the state vectors. The constraint $\hat{\mathcal{H}}^i\Psi = 0$ merely reflects the invariance of the theory under spatial diffeomorphisms but the 'longitudinal' constraint

$$\hat{\mathcal{H}}\Psi = 0 \tag{7.5}$$

carries all the dynamical information. Like so many others before and after him, Bryce was strongly influenced and inspired by John Wheeler's deep investigations of canonical quantization and in particular of the significance of equation (7.5). Following Wheeler, he chose a formal representation of the canonical commutation relations in which $\hat{g}_{ij}$ was a diagonal operator (so that the state vector was a functional $\Psi[g_{ij}]$, and $\hat{\Pi}^{ij} = -i\hbar\delta/\delta g_{ij}$; substitution in (7.5) then yields the fundamental equation

$$\hbar^2 G_{ijkl}\frac{\delta^2\Psi}{\delta g_{ij}\delta g_{kl}}[g] + g^{1/2\,(3)}R\Psi[g] = 0. \tag{7.6}$$

With the modesty that is characteristic of both men, Wheeler and DeWitt refer

---

† For example, Pirani and Schild (1950), Bergmann (1956a, b), Wheeler (1957, 1964), Dirac (1958a, b), Higgs (1958, 1959), Arnowitt *et al* (1962), Baierlein *et al* (1962), Komar (1967a, b).

respectively to equation (7.6) as the DeWitt equation and the Wheeler equation: the rest of the world, quite reasonably, call it the Wheeler–DeWitt equation.

The domain space of the state functionals is, at least formally, the space Riem Σ of smooth Riemannian metrics or, after imposing the constraint $\hat{\mathscr{H}}^i \Psi = 0$ and hence factoring out the action of the diffeomorphism group Diff Σ, the quotient space Riem $\Sigma/$Diff Σ (Wheeler's superspace). One of DeWitt's main contributions was a study of the geometry of Riem Σ and its relation to the dynamical equation (7.6). The set of symmetric, positive signature, 3×3 matrices is in one-to-one correspondence with the coset space GL(3, R)/SO3 and Riem Σ is simply the space of smooth functions on Σ with values in this homogeneous space. Thus DeWitt began his investigations with the geometry of GL(3, R)/SO3 and the associated space SL(3, R)/SO3. He showed that G_{ijkl} (equation (7.4)) is a natural metric on GL(3, R)/SO3 (with signature $- + + + + +$), and with respect to which GL(3, R)/SO3 and SL(3, R) are respectively geodesically incomplete and complete—a property that he related ultimately to the existence of singularities in the gravitational field. The metric G_{ijkl} on GL(3, R)/SO3 leads naturally to a metric $G_{ijkl}\delta^{(3)}(x, y)$ on Riem Σ and hence to a comprehensive pseudo-Riemannian geometry on this basic function space. DeWitt discovered elegant relations between this geometry and the structure of the Wheeler–DeWitt equation and placed particular emphasis on the role played by singular geometries. This line of thought produced his well known suggestion that gravitational collapse would be averted if the state vector vanished on such configurations—an important idea whose full implications have yet to be unraveled (for a recent reference, see Hartle 1982). The study of Riem Σ and superspace was continued in DeWitt (1970b) where the role of geodesics in Riem Σ was developed further. For many purposes it would be helpful if superspace carried the structure of a smooth, infinite dimensional, manifold but this is vitiated by the existence of strata (Fisher 1970) arising from three geometries with isometry groups. DeWitt conceived the idea of an 'extended' superspace built from copies of ordinary superspace joined along the strata so as to yield a smooth manifold. If Diff$_*$ Σ is the group of diffeomorphisms of Σ leaving fixed a particular frame, then extended superspace is simply Riem $\Sigma/$Diff$_*$ Σ over which Riem Σ appears as a principal Diff$_*$ Σ-bundle. This construction leads to the existence of a diffeomorphism group (Isham 1981) analogue of the Yang–Mills θ vacuum structure and generalizes to quantum gravity Singer's (1978) discussion of the Gribov effect.

Many of the topics discussed in DeWitt (1967a) continue to be of considerable importance: the geometry of GL(3, R)/SO3†, inner products on the state vectors, the role of intrinsic time, the WKB approximation and Hamilton–Jacobi equation (see also Peres 1962, Gerlach 1969), minisuperspace techniques‡, interpretation of the formalism etc, etc, and it can be unequivocally affirmed that, for anyone proposing to work in canonical quantization, this paper remains obligatory reading.

† Featured in the studies of ultralocal quantum gravity by Pilati (1982) and Isham (1982).
‡ Coinvented by Misner (1969), this led to what was surely one of the most enjoyable bandwagons in quantum gravity!

8. CONCLUSION

The unifying appearance in DeWitt's work of cohering leitmotifs renders section-alization difficult if not even arbitrary and the scheme I have adopted is not free from defects; in particular, certain important papers have yet to be mentioned. For example, the third part of the trilogy (DeWitt 1967c) contains results of detailed calculations of the tree graph contributions to scalar graviton and graviton–graviton scattering—an essential ingredient in any perturbative approach to quantum gravity. There are also two short review papers on quantum gravity proper (DeWitt 1970a, 1972) a lengthy general review (DeWitt 1979) and the major review in DeWitt (1975) of quantum field theory in a curved space–time. The latter topic is one to which much of DeWitt's work is relevant and indeed it could easily qualify for a section on its own. In particular, in his study of covariant Green's functions, DeWitt emphasized Schwinger's representation.

$$G(x, y) = i \int_0^\infty K(x, s; y, 0) e^{-im^2 s} \, ds \tag{8.1}$$

with the kernel $K(x, s; y, 0)$ obeying the imaginary time diffusion equation

$$\Box K(x, s; y, 0) = -i \frac{\partial}{\partial s} K(x, s; y, 0). \tag{8.2}$$

He followed Hadamard in expanding $K(x, s; y, 0)$ as a series in the geodesic distance between x and y and thus gained control over the singular behavior of $G(x, y)$ across the lightcone: a vital step in his development of the 'covariant point splitting' regularization scheme for quantum field theories and the associated techniques for computing the covariant effective action and renormalization counter terms.

Interest in this aspect of DeWitt's work grew rapidly after Hawking's (1975) discovery of black hole radiation of quantum particles and his review of 1975 not only elegantly summarized the whole subject but also posed a number of physically motivated problems that significantly influenced the future development of this branch of research. Indeed his ability to motivate and inspire others—a gift possessed also by that other great 'Texan' John Wheeler—is an important part of Bryce's personality and is a subject on which I would like to finish with a personal note.

Once upon a time Bryce DeWitt, Karel Kuchař and I proposed to write a book on quantum gravity. The initial planning took place during a short trip to Austin in which Karel and I stayed with Bryce and Cécile at their home. Now at that time the dichotomy of three and four still occasioned lively debate and, although Karel had long since sold his soul to the Mephistophelean $3 + 1$, I was still wavering. Bryce responded to this situation with a determined evangelical mission cunningly founded on his normal daily schedule: rise at midday, skip breakfast, a quick lunch, an afternoon occupied with paperwork in the office, home for dinner and the early evening spent in pleasant conversation to a background of Mozart or Gilbert and

Sullivan. By, say, eleven o'clock Karel and I would be falling asleep and indicate a desire to retire—the signal for Bryce to attack! He would launch into a torrent of questions concerning some of the most basic problems of quantum gravity and forced us to defend our canonical position in every way. We responded as best we could but Bryce was determined and persistent† and at around three or four o'clock in the morning we would stagger off, totally defeated, leaving Bryce still working. Neither of us could adapt to Bryce's timetable and I left Austin in a state of utter exhaustion‡.

It was several years before I could bring myself to touch a spacelike hypersurface again but the lessons I learnt from Bryce during those grueling sessions made a deep impression on me and greatly influenced my subsequent research. Thus I have every reason to be grateful for the way in which he so ungrudgingly gives his time and wisdom and, in conveying my personal good wishes for his sixtieth birthday, I can only add my sincerest gratitude and thanks and express the hope that he will continue to lead us for many happy years to come.

† Karel occasionally managed to stem the tide with one of his inimitable quotations from obscure Czechoslovakian fairy tales.

‡ The book was never written but one of Bryce's appendices to one of the proposed chapters now forms a substantial part of a book on supergravity and supermanifolds that he is co-authoring with Peter Van Nieuwenhuizen and Peter West.

REFERENCES

Abbott L 1981 *Nucl. Phys.* B **185** 189–203

Adler S 1982 *Rev. Mod. Phys.* **54** 729–66

Arnowitt R, Deser S and Misner C W 1962 *Gravitation: An Introduction to Current Research* ed L Witten (New York: Wiley)

Baierlein R F, Sharp D H and Wheeler J A 1962 *Phys. Rev.* **126** 1864–5

Bergmann P G 1956a *Helv. Phys. Acta. Suppl.* **4** 79

—— 1956b *Nuovo Cimento* **3** 1177–85

Bohr N and Rosenfeld L 1933 *K. Danske Vidensk. Selsk. Mat.-Fys. Meddr.* **12** 8

Boulware D G 1981 *Phys. Rev.* D **23** 389–96

Choquet-Bruhat Y 1968 *Battelle Rencontres* ed C DeWitt and J Wheeler (New York: Benjamin)

Christensen S M 1982 in *Quantum Structure of Space and Time* ed M J Duff and C J Isham (Cambridge: Cambridge University Press) pp 71–86

Deser S 1957 *Rev. Mod. Phys.* **29** 417–23

DeWitt B S 1957 *Rev. Mod. Phys.* **29** 377–97

—— 1960 *Phys. Rev. Lett* **4** 317–20

—— 1961 *J. Math. Phys.* **2** 151–62

—— 1962a *J. Math. Phys.* **3** 1073–93

—— 1962b *J. Math. Phys.* **3** 619–24

—— 1962c *J. Math. Phys.* **3** 625–36

—— 1962d *Gravitation: an introduction to current research* ed L Witten (New York: Wiley) pp 266–381

—— 1964a *Phys. Rev. Lett* **13** 114–8

—— 1964b *Relativity, Groups and Topology* ed C DeWitt and B S DeWitt (New York: Gordon and Breach) pp 587–822

—— 1967a *Phys. Rev.* **160** 1113–48

—— 1967b *Phys. Rev.* **162** 1195–1239

—— 1967c *Phys. Rev.* **162** 1239–56

—— 1970a *Gen. Rel. Grav.* **1** 181–9

—— 1970b *Relativity: Proceedings of the Relativity Conference in the Midwest* ed M Carmeli, S I Fickler and L Witten (New York: Plenum) pp 359–74

—— 1970c *Phys. Today* **23** 30

—— 1972 *Magic Without Magic* ed J Klauder (San Francisco: W H Freeman) pp 409–40

—— 1973 *The Many-worlds interpretation of Quantum Mechanics* ed B S DeWitt and N Graham (Princeton: Princeton University Press) pp 167–218

—— 1975 *Phys. Rep.* **19** 295–357

—— 1979 *General Relativity: An Einstein Centenary Survey* ed S W Hawking and W Israel (Cambridge: Cambridge University Press) pp 680–745

—— 1981a *Phys. Rev. Lett.* **47** 1647—50

—— 1981b *Quantum Gravity* 2 ed C J Isham, R Penrose and D W Sciama (Oxford: Oxford University Press) pp 449–87

DeWitt B S and Brehme R W 1960 *Ann. Phys., NY* **9** 220–59

DeWitt B S, Hart C F and Isham C J 1979 *Physica* **96a** 197–211

DeWitt-Morette C, Maheshwari A and Nelson B L 1979 *Phys. Rep.* **50** 257–372

DeWitt-Morette C, Elworthy K D, Nelson B L and Sammelman G S 1980 *Ann. Inst. Henri Poincaré* **32** 327–41

Dirac P A M 1958a *Proc. R. Soc.* A **246** 326–32

—— 1958b *Proc. R. Soc.* A **246** 333–43

—— 1959 *Phys. Rev.* **114** 924–30

Duff M J 1981 *Quantum Gravity* 2 ed C J Isham, R Penrose and D Sciama (Oxford: Oxford University Press) pp 81–105

Everett H III 1957 *Rev. Mod. Phys.* **29** 454–62

Faddeev L D and Popov V N 1967 *Phys. Lett* **25B** 29–30

Feynman R P 1963 *Acta. Phys. Pol.* **24** 697

Fisher A E 1970 *Relativity: Proceedings of the Relativity Conference in the Midwest* ed M Carmeli, S J Fickler and L Witten (New York: Plenum) pp 303–357

Gerlach U H 1969 *Phys. Rev.* **177** 1929–41

Hadamard J 1923 *Lectures on Cauchy's Problem in Linear Partial Differential Equations* (New Haven: Yale University Press)

Hartle J B 1982 *Quantum Cosmology and the Early Universe* Enrico Fermi Institute preprint

Hawking S W 1975 *Commun. Math. Phys.* **43** 199–220

—— 1979 *General Relativity: an Einstein Centenary Survey* ed S W Hawking and W Israel (Cambridge: Cambridge University Press) pp 746–89

Higgs P W 1958 *Phys. Rev. Lett.* **1** 373–4

—— 1959 *Phys. Rev. Lett.* **3** 66–7

't Hooft G 1975 *Acta Univ. Wratislav.* **No. 38** XIIth Winter school of theoretical physics in Karpacz

Horowitz G T 1981 *Quantum Gravity* 2 ed C J Isham, R Penrose and D Sciama (Oxford: Oxford University Press) pp 106–30

Horowitz G T and Wald R M 1978 *Phys. Rev.* D **17** 414–6

Isham C J 1981 *Phys. Lett.* **106B** 188–92

—— 1982 *Quantum Geometry*—in this volume

Isham C J, Strathdee J and Abdus Salam 1971 *Phys. Rev.* D **3** 1805–17

—— 1972 *Phys. Rev.* D **5** 2548–65

Julve J and Tonin M 1978 Nuovo Cimento **46B** 137–52

Khriplovich I B 1966 *Sov. J. Nucl. Phys.* **3** 415

Kibble T W B 1981 *Quantum Gravity* 2 ed C J Isham, R Penrose and D Sciama (Oxford: Oxford University Press) pp 63–80

Klein O 1955 in *Niels Bohr and the Development of Physics* ed W Pauli, L Rosenfeld and V Weisskopf (Oxford: Pergamon)

—— 1957 *Nuovo Cimento Suppl.* **6** 334

Komar A B 1967a *Phys. Rev.* **153** 1385–7

—— 1967b *Phys. Rev.* **164** 1595–9

Laidlaw M G and Morette-DeWitt C 1971 *Phys. Rev.* D **3** 1375–8

Landau L 1955 in *Niels Bohr and the Development of Physics* ed W Pauli, L Rosenfeld and V Weisskopf (Oxford: Pergamon)

Leray J (1952) *Hyperbolic partial differential equations* Princeton University Mimeographed notes

Lichnerowicz A 1961 *Publ. Math de L'Inst. des Hautes Etudes* **10** (Paris)

——1964 in *Relativity, Groups and Topology* ed C DeWitt and B DeWitt (New York: Gordon and Breach) pp 823–61

Mandelstam S 1968a *Phys. Rev* **175** 1580–1603

—— 1968b *Phys. Rev.* **175** 1604–23

Misner C W 1969 *Phys. Rev. Lett.* **22** 1071–4

Møller C 1962 *Les theories relativistes de la gravitation* (Paris: CRNS)

Morette C 1951 *Phys. Rev.* **81** 848–52

Neville D E 1980 *Phys. Rev.* D **21** 867–73

Pauli W 1956 *Helv. Phys. Acta Suppl.* **4** 69

Peierls R E 1952 *Proc. R. Soc.* A **214** 143–57

Peres A 1962 *Nuovo Cimento* **26** 53–62

Pilati M 1982 *Phys. Rev.* **26** 2645–63

Pirani F A E and Schild A 1950 *Phys. Rev.* **79** 986–91

Salam Abdus and Strathdee J 1978 *Phys. Rev.* D **18** 4480–5

Schwinger J 1951 *Phys. Rev.* **82** 664–79

—— 1953 *Phys. Rev.* **92** 1283–99

Sezgin E and van Nieuwenhuizen P 1979 *Phys. Rev.* D **21** 3269–80

Singer I 1978 *Commun. Math. Phys.* **60** 7–12

Stelle K S 1977 *Phys.Rev.* D **16** 53–969

—— 1983 To appear

Tomboulis E 1980 *Phys. Lett.* **97B** 77–80

Unruh W G 1976 *Phys. Rev.* D **14** 870–92

Utiyama R and DeWitt B S 1962 *J. Math. Phys.* **3** 608–18

Wheeler J A 1957a *Rev. Mod. Phys.* **29** 463–5

—— 1957b *Ann. Phys., NY* **2** 604–14

—— 1964 in *Relativity, Groups and Topology* ed C DeWitt and B DeWitt (New York: Gordon and Breach) pp 317–520

Zee A 1981 *Gravity as a dynamical consequence of the strong, weak, electromagnetic interactions* to appear in proceedings of the 1981 Enrice Conference

What Have We Learned from Quantum Field Theory in Curved Space–Time?

STEPHEN A FULLING

Quantum field theory in curved space–time has had a curious history. By this I do not mean that its history is unique; I suspect that many scientific research fields of similar scope have gone through the same four stages, but I am not confident enough of that to found a historical theory upon this model. I have spent my entire scientific career concentrating primarily on this subject, and during the middle of this period I watched it pass through a remarkable episode, in which Bryce DeWitt played a major role. Consequently, I am compelled to offer a rather personal account of the course of events before pronouncing upon their significance. (If the reader is a graduate student searching for a thesis topic, I urge him or her to persevere to the end of the essay.)

The study of field quantization in gravitational backgrounds goes back at least to 1939, when Schrödinger made the 'alarming' discovery that the mathematical behavior of the solutions of the Klein–Gordon equation in an expanding universe was inconsistent with a single-particle interpretation. For thirty years, papers appeared only sporadically[†], but the subject became a serious one with Parker's paper of 1969. That is, it entered a period of sustained attention, especially in the Soviet Union (Grib and Mamayev (1969), Zel'dovich (1970)) and Princeton[‡], as well as in Milwaukee, where I went to work with Parker in 1972. But it was still very

---

[†] For evidence of activity during this period, note the abstract, DeWitt (1953). I could not attend that particular meeting of the American Physical Society, being enrolled in the second grade at the time. Later, when Parker and I had attempted a systematic development of the theory (Fulling and Parker (1974)), I was chastened to learn that Bryce had worked the same ground twenty years earlier, and had run up against the same problems—divergences and non-renormalizability (i.e., fourth-order terms in the gravitational field equations).

[‡] Bibliographies of quantum field theory in curved space–time usually contain prominently the names Bekenstein, Ford, Fulling, Hu, Unruh, and Wald, all of whom were Princeton graduate students in 1971.

much out of the mainstream of theoretical physics; very few people paid any attention to it.

That situation changed drastically in 1974. The triggering event, which everyone will recall, was Hawking's (1974, 1975) discovery of particle creation by black holes. Suddenly it seemed that everybody was writing papers about quantum fields in curved space. Whatever the actual motivations of these people may have been, in hindsight one can give a rational-sounding explanation of why Hawking's work should have had such an impact. It is necessary to recall why Schrödinger had been alarmed, and why Parker *et al* were not. Once the possibility of particle creation ('instability of the vacuum') was admitted, one had to worry about whether the theory predicted an avalanche of particles that simply is not observed in the universe in which we live[†]. When detailed calculations were done, however, the answers always turned out either infinite, or ridiculously small[‡]. It seemed likely, therefore, that after the appropriate renormalizations were understood, and naive, erroneous modes of calculation abandoned, the theory would be physically acceptable, but quite possibly rather uninteresting. The importance of the black hole, as analyzed by Hawking, Unruh (1976), and others, is that it 'amplifies' unambiguous quantum effects to a macroscopic, observable level. Henceforth the interplay of field quantization and gravity had to be taken seriously, at least in principle. (Spoilsports could still question the existence of black holes in the appropriate mass range.)

Another feature of Hawking radiation which attracted much attention is its precisely thermal nature. This stimulated a great deal of speculation about a grand synthesis of relativity, thermodynamics, and quantum theory. Although these hopes for a unifying *physical* idea have not yet been fulfilled, the curious *formal* connection between finite temperature and spatial periodicity has been a recurrent theme in the literature, and has played a major part in the establishment of connections between quantum gravity and other subjects which are better developed mathematically: on the one hand, axiomatic quantum field theory and statistical mechanics, and on the other, abstract differential geometry and algebraic topology.

The days of glory ended in 1978. Since then, although it is still a field of active research, quantum field theory in curved space–time has not been a center of

---

[†] While Davies and I were writing a manuscript (Fulling and Davies (1976)) about particle creation by accelerating mirrors, I entered the London Underground without any fear of being broiled by the braking trains, but I was uneasy until we put in the units and verified that $\hbar$ appeared in the numerator of our formula, c in the denominator.

[‡] For example, as recalled by Davies (1977), even in the early stages of a Friedmann universe the effect of particle creation was found to be negligible in comparison with the classical stress-energy needed to support the Friedmann expansion in the first place. The work of Zel'dovich and others on anisotropy damping is an exception to these statements, but at that time the measures necessary to arrive at any quantitative statements about physical processes near cosmological singularities were so *ad hoc* that the conclusions of such investigations could hardly be regarded as firmly established. The work of Hawking (like the slightly earlier work of Zel'dovich and Starobinsky and Unruh on rotating black holes [see Unruh (1974)]) was not subject to this difficulty, since it dealt with unambiguously defined particles at infinity.

attention, developments have not been as rapid, and there has been a noticeable divergence in the interests and activities of its practitioners. I see the mid-70s episode as a gigantic scattering process in which a number of subfields of theoretical physics came together and enriched each other (see figure 1): The central target, labelled (A), is classical general relativity. It encounters elementary-particle theory (B), which has been ionized during a previous collision with quantum statistical mechanics (C),

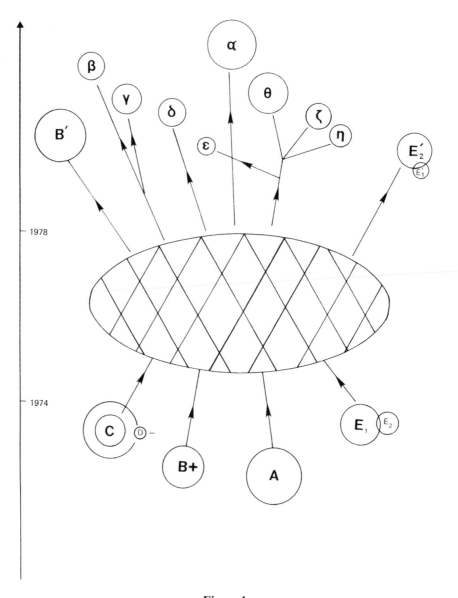

Figure 1

which stripped off the electron (D), rigorous ('axiomatic') quantum field theory. Out of mathematics comes a diatomic molecule, geometrical spectral asymptotics, uniting the geometry and topology of manifolds (E_1) with the spectral theory of differential operators (E_2). All these components came speaking their own languages, they had to learn to communicate with each other, and in the process their understanding of themselves was enriched[†].

What came out of this tremendous collision? The modern theory of elementary particles is the theory of non-Abelian gauge fields (B'), which has extensively borrowed geometrical viewpoints and tools from the theory of gravity[‡]. Spectral asymptotics is shown exiting with the relative sizes of its components reversed (E_1', E_2'); this effect is immodestly explained in Fulling (1982). The forward-scattering particle (α) represents the work still being done on the fundamentals of field theory in curved space–time (e.g. Dimock (1980) and Fulling *et al* (1981)). A major diversion of manpower has taken place (back) to mainstream quantum gravity (including supergravity), studied by background-field and loop-expansion approaches, as represented by the work of DeWitt, Christensen, Duff, and others (β). One sees (γ) branching off from this the gravitational instantons and space–time foam of Hawking and his associates. Another reaction product (so far surprisingly small, in my opinion) is the industry (δ) of applying the quantum theory of fields in curved space–time to models of genuine cosmological interest. (The names of Parker, Hartle and Hu are most prominent here.) A number of ramifications of basic theory have been pursued: boundary effects (ε) (Candelas, Kennedy); thermal effects (ζ) (Dowker, Critchley, Kennedy); twisted fields and other topological effects (η) (Ford, Isham, Toms); and the general subject of interacting fields in curved backgrounds (θ) (Panangaden, Birrell, Ford).

[†] On several occasions Bryce DeWitt has publicly exhorted the various specialists to break out of their respective jargons and talk to each other in language that can be understood. For years he has been regarded as a general relativist by particle physicists and as a particle or field theorist by relativists, and immersed in the unavoidable technicalities of quantum gravity, he has had to struggle harder than most to escape the trap of a private language and a private research problem. In the present context, his guiding influence extends beyond his widely read review articles (DeWitt (1975, 1979)). It was probably strongest during his stay in Oxford in 1975–6, when Britain was unquestionably the center of activity in curved space quantum field theory and quantum gravity. (There were research groups in Cambridge, Manchester, Cardiff, King's and Imperial Colleges in London, and both the Mathematical Institute and the Astrophysics Department in Oxford.) The weekly seminars at Oxford have for some of us the significance that Copenhagen and Göttingen had for earlier generations.

[‡] But I am supposed to be writing about *quantum field theory in curved space* specifically. Has it taught particle physicists anything? Well, maybe. Anomalies (Deser *et al* (1976), Christensen and Fulling (1977), Duff (1976)) have turned out to be related to instantons and topological conservation laws ('tHooft (1976); for a more recent development see Christ (1980) and compare Gibbons (1980)). Also, compare Bender and Hays (1976) with the tradition of study of quantum fields near boundaries by relativists (e.g. Deutsch and Candelas (1979)). As the dates suggest, these may be instances of coincidental convergent evolution, rather than direct descent.

As recently as early 1977, quantum field theory in curved space–time was a scene of confusion and controversy. This was only partly due to the variety of languages and perspectives mentioned earlier. It had more to do with the fact that none of us really understood what we were doing, while many had glimpsed enough of the truth to start shouting about it. Of course, that was what made it an exciting and rapidly progressing field. (Now, in a more mature phase, the subject attracts much less attention. Nevertheless, there is also a certain charm and satisfaction in doing the mopping-up in a field that is basically understood.) Some of the points of contention, along with what I take to be the answers now agreed upon, were: Can the particles emitted by an evaporating black hole be traced all the way back to the surface of the collapsing body? (No.) Will an accelerating observer detect particles in empty space? (Yes, in the sense that his detector will click or his thermometer will get hot.) Is the expectation value of the stress tensor in a particular quantum state well defined, independently of the motion of the observer? (Yes.) Does the renormalized stress tensor of a conformally invariant quantum field have a non-vanishing trace? (Yes.) Is this trace anomaly uniquely defined? (Yes, except for the coefficient of any term equal to the trace of a covariant, local, polynomial, conserved functional of the metric tensor involving derivatives of order less than or equal to 4—which is an arbitrary renormalization constant.) Is dimensional regularization better than point-splitting? (Yes, if you're calculating an effective Lagrangian; no, if you're calculating an expectation value of the stress tensor.) Which is more worth calculating, the effective Lagrangian or the stress tensor? (That depends on what you intend to do with it.)

Somehow the trace anomaly acquired a symbolic importance far out of proportion to its intrinsic interest, and the resolution of the controversies surrounding it seemed to mark the end of the whole period of confusion and excitement. Probably the reason was that each school of thought, in producing a completely convincing treatment of the anomaly, simultaneously achieved a sufficiently unambiguous and detailed formulation of the theory in general that its adherents felt confident in declaring victory. (Various mistakes and misunderstandings having been cleared up, the different approaches of course agreed in their essential conclusions.) For me the landmark paper which seemed to settle everything was Wald (1978); investigators of other persuasions will probably point to other papers, such as Duff (1977), Brown and Cassidy (1977), Hawking (1977), or Dowker and Critchley (1977)[†].

My original intention for this essay was to enter at this point into an exposition of the consensus arrived at by 1978, at least with respect to the identification of

[†] Perhaps even the attainment of consensus is in the eye of the beholder. In January 1978 a number of curved space field theorists were invited to a conference (supported by est) in San Francisco, along with a comparable number of particle theorists. A presentation by one of the former provoked some disagreement in the audience, whereupon one of the particle physicists, sitting behind me, said to another, 'This is a funny field. The people in it can't agree on what constitutes meaningful research.' If he thought it was bad now, what would he have thought in England in 1975, or Boston (the Texas Symposium on Relativistic Astrophysics, the impromptu session in the basement of the hotel) in December 1976?

physically reasonable states and the covariant renormalization of the stress–energy–momentum tensor. However, I now find that I shall be called upon to publish such a doctrinal statement elsewhere (Fulling (1983)). (A booklength exposition of the state of the art is now available, see Birrell and Davies (1982).) Furthermore, having announced a consensus, I would not want to abuse it by making a short pronouncement which would inevitably be biased, at least in emphasis. So I shall pass directly from thoughts of the past to thoughts of the future, in particular to speculations about the implications of quantum field theory in curved space–time for *quantum gravity* in the stricter sense.

There has always been a wide spectrum of opinion on the proper relation between quantum field theory in a classical background and quantum gravity proper. Many theoretical physicists regard the entire external-field approach to be a waste of time, irrelevant to real physics, an act of desperation committed by someone who would like to quantize gravity but doesn't know how[†]. The most well developed critique is that of Duff (private communications and 1981). Among other things, he points out that (assuming that gravitation is a field among other fields) any gravitationally induced quantum process which creates other particles will create gravitons as well; thus quantization of matter fields is doomed to be either trivial, or physically incomplete unless accompanied by quantization of gravity. At the other extreme, it has sometimes been suggested that the coupling of a quantum stress tensor to a classical metric is acceptable as an ultimate theory of nature; but most of us think this does too much violence to quantum theory to be seriously entertained. Most investigators of quantum field theory in curved space do not consider the 'semi-classical' coupling of gravity and the stress tensor,

$$R_{\mu\nu} - 1/2 R g_{\mu\nu} = \langle T_{\mu\nu} \rangle \tag{1}$$

to be anything more than a stopgap measure, with a fighting chance of being physically accurate in some situations and a certainty of being grotesquely wrong in others. We believe that the quantum theory of a field in a given gravitational background is a consistent, useful, and respectable theory, just as the Schrödinger theory of the electron is consistent, useful, and respectable although it does not include a quantum theory of the Geiger counter or photographic film which detects the electron, the ionizing apparatus that produces it, the magnetic field that bends its trajectory, and so on. We believe that such a theory, being now well defined, is well worth studying as long as there is still considerable basic uncertainty as to what a quantum theory of gravity should be like. We believe that the 'observables' of the theory are in principle observable, and in particular that the stress tensor can be observed by its effect on the gravitational field. (This does not mean taking the semi-classical equation of motion (1) literally in all situations.)

As I understand it, DeWitt's position is slightly different from this. To him, quantum gravity and quantum field theory in a curved background are in a sense *the*

[†] 'This is not worth pursuing', a certain Nobel Prize winner interjected during Wald's talk at the aforementioned San Francisco conference. 'Tell Wald not to listen to—', another Laureate said privately later. 'I think what he is doing is very interesting.'

same thing. His reply to Duff (with which I agree) is that the gravitonic contribution to (for instance) black hole radiation should be estimated by quantizing the fluctuations in the metric as just one more field[†]. But there is more than one way of interpreting this suggestion. My gut reaction is to include a graviton field (linearized?) in the semiclassical framework, without presuming to carry that theory beyond the modest goals and claims outlined above. If I knew how to do it, this would model gross graviton effects in cosmology and astrophysics, but would not bring us any closer to a fundamental quantum theory of gravity. (I don't know how to do it, because of the ambiguity in the graviton field's equation of motion and the related issues of gauge invariance, non-renormalizability, and acausal propagation.) DeWitt's proposal aims much deeper. He writes (DeWitt (1975), p 347) that *the correct way* (his italics) to quantize gravity is to split the gravitational field into a quantized fluctuation and a semiclassical background. I have another gut feeling which says that this idea ought to be correct. I would love to see it developed within the conceptual framework most congenial to me (Hilbert space, field equations, and stress tensors, as opposed to effective actions, path integrals, or *S*-matrices). Quite apart from the technical obstacles, it immediately raises some ominous questions: will the results of calculations be guaranteed to be independent of the splitting of the metric into two parts? How is the splitting of the gravitational field equation to be chosen? Can the freedom in doing that be exploited to simplify the equation for the quantized part of the metric (i.e., to keep it linear, or at least renormalizable)? How is the classical background (or the initial data for it) to be chosen, in any given physical context? Will not the linear structure of the algebra of quantum observables force the existence of many states in which the expectation value of the metric is very far from the assumed background? In such a state, does the theory break down physically, branded as an approximation which has been pushed beyond its limits of validity? Or is it our present conception of the logical structure of quantum theory which must be abandoned?

It seems to me that there are two general ways in which the insight we have gained during the past decade concerning quantum fields in curved space, together with some rigorous mathematical physics imported from constructive field theory, might help to advance, or to criticize constructively, the quantum theory of gravity itself (developed within a covariant background-field framework). The first way is internal, involving serious investigations of the consistency and validity of the semiclassical approximation itself. The second way is the development and critical appraisal of DeWitt's Correct Way.

Earlier I asserted that the semiclassical ansatz is expected to be accurate in some physical contexts and very bad in others. How does one tell which is which? Presumably it is necessary that the *dispersion* in the stress tensor, as a complex of quantum observables, be fairly small. (Perhaps there is room for some classical, statistical averaging to supplement quantum averaging.) To take an extreme example, suppose that the state of the field is one corresponding to a single particle present in the universe, with a wavefunction which has a wide spatial distribution.

[†] This is a paraphrase based on a distant memory of a conversation among the three of us.

The expectation value of the stress tensor will be spread out along with the wavefunction. But such a state does not correspond to a physical situation in which *energy* is spread out; the energy should be highly localized, with the *probability distribution* for its location being spread out. In this situation I would expect the accuracy of the semiclassical approximation to be horrible! On the other hand, I would expect (perhaps by wishful thinking) that the state of a field in the early stages of an expanding universe is a rather homogeneous, random soup, where the quantum expectation values adequately represent the aspects of objective physical reality accessible to us. Such matters deserve to be examined quantitatively. But to the best of my knowledge, nobody has yet so much as calculated an expression for the square of the renormalized stress tensor operator, which could be used to calculate dispersions in various Fock-space states[†].

A more ambitious project for testing the validity of semiclassical theories is to take a field theory whose *fully quantized* version has been rigorously constructed, work out a corresponding semiclassical theory, and compare the results with the known 'right answer'. A logical choice is the two-dimensional Yukawa theory (e.g. Dimock (1972)). Here one has a scalar field whose source is a current formed from a spinor field:

$$\Box \phi + m^2 \phi = J \equiv -\lambda \bar{\psi}\psi. \qquad (2)$$

The equation of motion of the spinor field correspondingly involves the scalar field:

$$\partial\!\!\!/\,\psi + M\psi - \lambda\phi\psi = 0. \qquad (3)$$

Let us think of ψ as a matter field, ϕ as the analogue of the metric tensor, and J as the analogue of the stress tensor, $\mathbf{T}$. Let us quantize ψ (satisfying equation (3)) in an arbitrary background ϕ, calculate a renormalized operator J, calculate the expectation value of J in some plausible state of the ψ field, substitute same into equation (2), and solve for ϕ. What does this have to do with the Yukawa field theory studied by the constructive field theorists? I have been asking constructive field theorists this question for the past ten years; by their silence they have politely suggested that I figure out the answer myself.

It is probably not necessary to go into field theory to tackle the basic issue involved here. A simple quantum-mechanical system, with finitely many degrees of freedom—but more than one—can also be treated semiclassically. (Is not this what we do when we solve the Schrödinger equation for the electron and ignore the photographic film?) Duff (1981) argues that quantization of only part of a system is 'inconsistent', because of the arbitrariness in how the system is divided into quantum and classical parts. I suspect that there is a sense in which this is true (even tautological), but that there are physical contexts in which such splits are nevertheless trustworthy and useful. It would be nice if somebody would work out some examples.

Quantum gravity has nothing to lose from critical investigations of semiclassical gravity. To the extent that the latter is vindicated, the path of the former will be made

---

[†] After writing this paragraph I became aware of a highly relevant preprint by Ford (1982).

clearer; to the extent that the latter is found wanting the urgency of the former will be more obvious.

The second way in which our mastery of curved space quantum field theory can be made relevant to quantum gravity is in the development and critical appraisal of DeWitt's Correct Way. Here too I think that there is much to be learned from simpler models. The constructive field theorists have a two-dimensional ϕ^4 model, where the scalar field is coupled nonlinearly to itself, rather than to another field:

$$\Box\phi + m^2\phi + 4\lambda\phi^3 = 0 \tag{4}$$

(Simon (1974), Glimm and Jaffe (1981)). This can be taken as an analogue of pure gravity, nonlinear but without matter fields. (Again, one might want to start with even simpler models with only finitely many degrees of freedom.) It would be prudent and instructive to try out any technique (in particular, any background-field or loop-expansion approach) on a model such as this before applying it to gravity. True, it will not teach us much about gauge invariance or non-renormalizability. But an approach which *fails* on a simpler model is unlikely to work on gravity, and one would like to know that before spending a decade plowing through tensor indices, gauge and renormalizability problems, fourth-order field equations, 3 kinds of ghosts, 76 kinds of counterterms, 200 kinds of Feynman diagrams, etc. If your favorite approach *succeeds* on $(\phi^4)_2$ (as you should be able to recognize by comparing your theory with the standard, completely rigorous approach—although much translation and working out of details will probably be needed there), then the resulting experience and encouragement will be invaluable in tackling gravity, and the clarity of a simple model will be immensely helpful in communicating with the rest of the world before, during, and after[†].

Another avenue which we should continue to pursue is the detailed theory of *interacting* quantum fields in curved space–time—not only because it is important in its own right, but because it is probably a necessary prologue to quantum gravity by the background-field approach. (In answer to a question raised earlier, I doubt that it will be possible to keep the equation of motion of the quantized field linear.) Beyond generalities, most of the quantum field theory which has been done in curved space is for free (linear) fields, although this situation is rapidly changing (see Nelson and Panangaden (1982) and references therein, also Bunch (1981) and Vilenkin and Ford (1982)). The perspective which has been gained in the work on linear field equations must not be forgotten in developing nonlinear theories.

I am far from confident that a conservative, incremental field-theoretical approach will yield an acceptable quantum theory of gravity. I think there is a strong chance that the ultimate unification of general relativity with the rest of modern physics will

---

[†] It may seem that the background-field method has already been applied to scalar field theories in the Coleman–Weinberg (1973) approach to symmetry breaking. There, however, the vacuum expectation value is taken to be a *constant* (and a solution of the field equation). It is a big jump from that to quantization of the scalar field against an *arbitrary*, space–time dependent, scalar background, which is where I deem the perspective of quantum field theory in curved space–time to be indispensable (see Hu (1982) and Vilenkin and Ford (1982)).

look very little like either contemporary general relativity or contemporary quantum field theory; it will be based on a totally new way of looking at the world. Quite possibly, in the year 2100 our present quantum field theory, with its apparatus of Fock spaces, Lagrangians, field equations, commutation relations, and S-matrices, will be seen as a misguided forced marriage of classical field theory with quantum particle mechanics, more naive than the attempts of Kelvin, young Maxwell, and their contemporaries to model the electromagnetic ether with gears and roller bearings. Let us hope that this pessimism is wrong. Genius is in short supply. Our attempt to substitute experience and hard work—building on understanding gained in the past to master a new (perhaps the last) level of nature—deserves to be rewarded.

ACKNOWLEDGMENT

Preparation of this essay has been supported by National Science Foundation Grant No. PHY79–15229 to the Texas A&M Research Foundation and by the tolerant summer hospitality of my parents.

REFERENCES

Bender C M and Hays P 1976 *Phys. Rev.* D **14** 2622
Birrell N D and Davies P C W 1982 *Quantum Fields in Curved Space* (Cambridge: Cambridge University Press)
Brown L S and Cassidy J P 1977 *Phys. Rev.* D **16** 1712
Bunch T S 1981 *Ann. Phys., NY* **131** 118
Christ N H 1980 *Phys. Rev.* D **21** 1591
Christensen S M and Fulling S A 1977 *Phys. Rev.* D **15** 2088
Coleman S and Weinberg E 1973 *Phys. Rev.* D **7** 1888
Davies P C W 1977 *Phys. Lett.* B **68** 402
Deser S, Duff M J and Isham C J 1976 *Nucl. Phys.* B **111** 45
Deutsch D and Candelas P 1979 *Phys. Rev.* D **20** 3063
DeWitt B S 1953 *Phys. Rev.* **90** 357
—— 1975 *Phys. Reports* **19** 295
—— 1979 in *General Relativity: An Einstein Centenary Survey* ed by S W Hawking and W Israel (Cambridge: Cambridge University Press) pp 680–745
Dimock J 1972 *Ann. Phys., NY* **72** 177
—— 1980 *Commun. Math. Phys.* **77** 219
Dowker J S and Critchley R 1977 *Phys. Rev.* D **16** 3390
Duff M J 1976 *Comment on Quantum Gravity and World Topology* Preprint QMC-76-29, Queen Mary College, London
—— 1977 *Nucl. Phys.* B **125** 334
—— 1981 in *Quantum Gravity 2: A Second Oxford Symposium* ed C J Isham, R Penrose and D W Sciama (Oxford: Oxford University Press) pp 81–105

Ford L H 1982 *Gravitational Radiation by Quantum Systems* Preprint TUTP-82-6, Tufts University, Medford, Massachusetts

Fulling S A 1982 *Soc. Ind. Appl. Math. J. Math. Anal.* **13** 891

—— 1983 *Gauge Theory and Gravitation: Lecture Notes in Physics No.* 176 (Berlin: Springer) pp 101–106

Fulling S A and Davies P C W 1976 *Proc. Roy. Soc.* A **348** 393

Fulling S A, Narcowich F J and Wald R M 1981 *Ann. Phys., NY* **136** 243

Fulling S A and Parker L 1974 *Ann. Phys., NY* **87** 176

Gibbons G W 1980 *Ann. Phys., NY* **125** 98

Glimm J and Jaffe A 1981 *Quantum Physics: A Functional Integral Point of View* (New York: Springer-Verlag)

Grib A A and Mamayev S G 1969 *Yad. Fiz.* **10** 1276 (*Sov. J. Nucl. Phys.* 1970 **10** 722)

Hawking S W 1974 *Nature* **248** 30

—— 1975 *Commun. Math. Phys.* **43** 199

—— 1977 *Commun. Math. Phys.* **55** 133

Hu B L 1982 *Phys. Lett.* B **108** 19

Nelson B L and Panangaden P 1982 *Phys. Rev.* D **25** 1019

Parker L 1969 *Phys. Rev.* **183** 1057

Schrödinger E 1939 *Physica* **6** 899

Simon B 1974 *The P(φ)₂ Euclidean (Quantum) Field Theory* (Princeton: Princeton University Press)

'tHooft G 1976 *Phys. Rev. Lett.* **37** 8

Unruh W G 1974 *Phys. Rev.* D **10** 3194

—— 1976 *Phys. Rev.* D **14** 870

Vilenkin A and Ford L H 1982 *Gravitational Effects upon Cosmological Phase Transitions* Preprint TUTP-82-7, Tufts University, Medford, Massachusetts

Wald R M 1978 *Phys. Rev.* D **17** 1477

Zel'dovich Ya B 1970 *Pis'ma ZhETF* **12** 443 (*JETP Lett.* 1970 **12** 307)

The World of the Schwinger–DeWitt Algorithm and the Magical a_2 Coefficient

STEVEN M CHRISTENSEN

This essay gives me an unusual opportunity. I will present a somewhat personal view of how the method known as the Schwinger–DeWitt proper-time algorithm is used and why I call the a_2 coefficient that appears in the algorithm 'magical'. I have often been asked by my students to explain how a research project in theoretical physics develops. This essay is designed with them in mind. To readers expecting a detailed discussion of the proper-time algorithm, my apologies; writing this is more fun. Most of my research papers contain long equations. This one will contain none! (For a review of the technical details, see Birrell and Davies (1982) and the references therein.) Naturally, as in all work in quantum gravity related subjects, DeWitt's influence is omnipresent in what I record here. (In addition to Bryce's support and inspiration, I want to acknowledge the very enjoyable and profitable collaborations I have had over the years, especially those with Michael Duff and Stephen Fulling.)

When I was an undergraduate at Augustana College in 1970, I bought a copy of DeWitt's *Dynamical Theory of Groups and Fields* (DeWitt 1965) and tried to read it. I understood very little of it. Becoming depressed at my failure to absorb the wealth of information and ideas in DTGF, I remember wandering into a professor's office and making some statement like, 'How could one person do all of this?' In the hope of finding something I could understand, I searched the literature for more DeWitt work and came across his three *Quantum Theory of Gravity* papers (DeWitt 1967a, b, c) or what we now call simply *The Trilogy*. Leafing through the pages of these papers I came upon enormous equations, intricate Feynman diagrams and an abstract on the second paper longer than most entire papers I had tried to write. I wondered if I would ever get anywhere with this stuff. Happily, I found myself at the right time with the right application for graduate school and became DeWitt's student in 1971. (My first contact with Bryce came one Monday night in the Spring of 1971. I was alone at home and the phone rang—'Hello, this is Bryce DeWitt. May I speak to

Steve Christensen, please?' Awestruck, like one might feel when being called by a god, I think I remembered my name and had some conversation about my becoming one of his research assistants in Europe that coming summer. It took me two years to become comfortable calling him Bryce and even then it seemed sacrilegious for a time.) I begin this story in the early 1970s.

Much of the work in quantum gravity in the past ten years was motivated by developments in Yang–Mills and black hole theory. DeWitt's Trilogy had emphasized the similarities between Yang–Mills theory and quantum gravity. When progress was made in both the physical and technical aspects of Yang–Mills theory, hope was rekindled in the rather depressed quantum gravity world. In classical gravity, the black hole was becoming better understood and it seemed that such objects might actually exist, providing a much needed 'laboratory' for studying strong gravitational fields. The success of 't Hooft and Veltman (1975) and others in studying the renormalization of Yang–Mills theory, as well as gravitation, led to renewed interest in the technical details of regularization/renormalization schemes. The work of Bekenstein (1973) and Hawking (1975) in the semiclassical treatment of particle production by the gravitational field of a black hole and the potential unification of quantum theory, general relativity and thermodynamics indicated by such calculations, led to new studies of quantum field theory in curved space–times as an approximation to a full quantum gravity theory. It was at this stage that DeWitt began to move back in the direction of quantum gravity research and away from the FORTRAN world of numerical relativity and space–time slicing, where he had been in 1971–73. I moved with him, pleased, I might add, to be free from the tyranny of the key punch and poor turnaround time.

Once we knew that a black hole could radiate particles with a thermal spectrum and that therefore the black hole should shrink and get hotter, the obvious questions were—how fast did the hole shrink and what was left when we reached the realm where the semiclassical approximation broke down? To get some idea of the answer to these questions, we decided to try to calculate the vacuum expectation value of the stress tensor for a scalar field in a static Schwarzschild background. (To avoid writing 'vacuum expectation value of' all the time, I will simply say 'the stress tensor' from now on.) DeWitt got me started on the problem by showing me how to do mode sum calculations. (Bryce was working on the manuscript of his now famous *Physics Reports* article, *Quantum Field Theory in Curved Spacetime* (DeWitt 1975) at the time.) In my naive graduate student way, I began to grind away at some big calculation, certainly not obeying Wheeler's First Moral Principle: never do a calculation before you know the answer. A few months of work made it clear that unless I either knew more about the radial mode functions or found some way of guessing what I was trying to find, I would get nowhere. Both paths gave interesting results ultimately, but the second, stimulated by Wheeler's Principle, gave by far the most information.

(I should make some comments about what it was like to be Bryce's student at the University of Texas. He was always very good to students, always calm, always patient. Nearly everyday he would go to lunch with whatever group of students was hanging around, preferring us, it seemed, to the more staid atmosphere of the faculty club. (Of course, it could just be that he liked the Mexican food at the Tower

Restaurant.) One day, two students from the particle physics group at Texas wanted to come to lunch with us. Bryce happened to choose that day to give a speculative soliloquy on why the universe appears to be four dimensional. After the lunch the two students came over to me, slightly dazed I think, thanking me for having them to lunch and hoping that the next time I thought Bryce would talk to us like that, I would make sure that they could come along. I didn't tell them that Bryce was almost always like that.

There were rare times, however, when Bryce would get mad with us. My episodes of particularly muddled thinking could try his patience and provoke his anger. It sometimes took me days to recover, but I would usually end up with a renewed sense of purpose and the strong desire not to cause Bryce ever again to wonder if he had made a mistake in choosing the 'green' kid from Illinois to be his student.)

We tried to determine what properties the stress tensor should have and came up with a number of properties that seemed reasonable. The first and most obvious was covariant conservation, that is, the covariant divergence of the stress tensor should be zero. The second requirement was spherical symmetry. There was no obvious reason why, in an exactly spherically symmetric background, the stress tensor shouldn't also be spherically symmetric. For simplicity, we chose the fields propagating on the curved background to be conformally invariant, assuming that this meant the stress tensor would have zero trace. Finally, we demanded that the stress tensor be time-independent looking only at late times after the black hole had collapsed to its 'static' state. So many restrictions made the conservation equations very easy to solve. It was possible to determine the stress tensor up to some constants of integration and one function. Unfortunately, to find these forced me to go back to the mode sum procedure which could only be done approximately. We were able to see a radiation-like piece in the solution and gain some information about the stress tensor's overall structure. It became clear that we would need to know much more about the divergence behavior of the stress tensor before we could continue.

Any object like the stress tensor constructed from a product of quantum field operators evaluated at the same space–time point is ill-defined. It will be infinite. These divergences must be separated out (regularization) and then 'discarded' (renormalization). We didn't know the exact form of the divergences, though previous experience told us that they would probably be the usual quartic, quadratic and logarithmic kinds. It was necessary to compute them exactly and then figure out how to get rid of them so that the final result satisfied our criteria.

We considered using dimensional regularization. This method had been quite successful in the Yang–Mills case. The problem with this method was that if we knew very little about the radial mode functions of the scalar wave equation in the Schwarzschild background in four dimensions, we weren't going to be able to do much with them in even more dimensions. Bryce urged me to return to the point-splitting method he had begun to develop in DTGF and complete the necessary calculations. This was in late 1974. I almost bit off more than I could chew!

Bryce is well known to be a symbolic manipulator of fearsome proportions. (A professor in Chapel Hill tells a story of how he and Bryce discussed a problem in electromagnetic theory one Friday and then decided to look at it independently over the weekend. On Monday, Bryce presented him with a long calculation ending in the solution to the problem. He held up Bryce's pages to the light and noticed not one erasure! In later years I was often humbled by the superhuman effort Bryce would devote to some grim calculation. If I am able to do any hard calculation today, it comes from observing Bryce's tricks of organization and concentration. However, I still cannot use a pocket knife to sharpen my pencils as he does.) He was convinced that the point-splitting calculations could be done in the short time I had left to finish my dissertation. It was at this stage that my first real exposure to the Schwinger–DeWitt method and the a_2 coefficient came: the beginning of a long and mostly happy relationship.

'Go read Schwinger!' was DeWitt's suggestion as a place to begin my studies. *On Gauge Invariance and Vacuum Polarization* (Schwinger 1951) became my companion for several weeks before I then moved into DTGF pages 147–159 (pages that anyone can pick out in my copy since their edges are dark from use). I became devoted to the coincidence limit and the Green's function.

The point-splitting method involves expanding the Hadamard elementary function, which is a part of the Feynman Green's function and is a function of two space–time points, in a power series in the tangent to the geodesic between the two points. (It is generally assumed, though not absolutely necessary, that there is one unique geodesic between the two points.) The tangent is the covariant derivative of the so-called bi-scalar of geodetic interval, $\sigma(x, x')$, which measures one-half the square of the geodetic distance between the end points x and x'. To determine the terms in the expansion, it is necessary to know the limit of the covariant derivatives of σ as the points approach each other along the geodesic. For the stress tensor calculation it was determined that we needed to take at least six derivatives. DeWitt had computed some of these at one time, but urged me to recompute them since I was going to need coincidence limits like $\sigma_{;\alpha\beta\gamma\delta\varepsilon\lambda}(x, x)$ which he had not done. (I seem to remember a little twinkle in his eye when he told me what I had to do. He knew what I was in for. But maybe it was just my imagination.)

Weeks of frustrating index shuffling began. Every calculation had to be done many times since sign errors and factors of two eluded me at every turn of the page. Finally, I got consistent answers that I believed and which matched with Bryce's. Now I was ready to compute something real, or should I say something 'magical', my first a_2 coefficient.

The point-splitting method is fairly straightforward when applied to the stress tensor. Consider the product of the two field operators that appear in each stress tensor term. Initially, these fields are evaluated at the same space–time point. These infinite objects can be made finite if we transport one of the fields to a nearby point in some covariant way. The vacuum expectation value of this new bi-tensor object is directly related to the Hadamard function. This function can be expanded in powers of $\sigma_{;}^{\mu}$ and so in turn the point-split stress tensor can also be expanded. It is fairly simple to isolate the divergent terms. These terms are the ones that blow up when the two points are brought back together. (There are finite terms that are also important and need to be subtracted from the stress tensor along with the divergences so that conservation laws are obeyed.)

Seeing that the stress tensor calculation was going to be long, I tried to understand just what the divergence structure of the theory looked like at the level of the action. This involved computing the one-loop counterterms to the classical action, following the outline laid out in Problems 85 and 112 of DTGF. We might expect to be able to derive the form of the divergences in the effective action; that is, the infinite counterterms. It was shown in DTGF that the effective action at the one-loop level could be written as a functional of the Hadamard function also. This is where the proper-time representation of the Green's function comes into use.

The proper-time algorithm introduces a parameter s in such a way that a Green's function can be written as an integral of a function, $K(x, x', s)$, called the kernel. It can then be shown that the kernel obeys a Schrödinger-like equation. The kernel can be approximated by an asymptotic expansion in powers of s. The expansion is plugged into the kernel's differential equation and recursion relations are produced. These relations can be used to determine the coefficients of the powers of s in the asymptotic expansion. The coefficients are labelled a_0, a_1, a_2, etc. Each is a bi-tensor whose coincidence limit can be found (after some tedious algebra) in a very straightforward way. In four dimensions, the first three a_ns determine the one-loop counterterms for the action.

It became obvious when I had computed the a_ns for several different fields propagating on an arbitrary curved background that a general formula could be found for the a_ns. The field operators were all of the general form $-\Box \delta^i{}_j + C^i{}_j$, where $\delta^i{}_j$ is a general identity matrix, $C^i{}_j$ some matrix usually constructed from the Riemann tensor and $\Box \equiv \nabla_\mu \nabla^\mu$. It turned out to be quite easy to find the general a_0, a_1 and a_2 for this operator after all the coincidence limits of σ and its derivatives had been computed. (We were not aware at the time that Gilkey (1975) was performing the same calculation using different methods and for completely different reasons.)

We immediately applied our results to the graviton one-loop counterterm problem and found that just as 't Hooft and Veltman had said, one-loop gravity was finite on-shell. That is, if we demand that the background field equations (the vacuum Einstein equations usually) be satisfied, the counterterms vanish. (Later we will discuss a modification of this result.) Seeing that the proper-time algorithm and the a_2 coefficient were giving the expected result, we were more confident that the more lengthy point-split stress tensor calculation could be pushed to completion.

Finishing the stress tensor calculation was just an exercise in tensor algebra and was carried out in the first few months of 1975. Only one hitch appeared. DeWitt began a communication with someone named Duff in England. Duff claimed that the stress tensor had a trace anomaly. A stress tensor that classically had no trace developed one after renormalization. Capper and Duff (1974) knew something about its form from some Feynman diagram computations they had done. Bryce, who is never really convinced of anything until he sees a calculation he whole-heartedly believes, asked me if I had seen anything like a trace anomaly in any of my calculations. I said yes, and he sent me away to try to figure out where it came from. Not really understanding what I had in front of me in my own calculation, I worked hard to eliminate or explain away the trace. I now wish I had just accepted what I had and studied things more carefully before making any definite statement. I told Bryce that I could get rid of the 'anomaly' and wrote so in my dissertation (Christensen

1975). It would be months before I realized I had been wrong, but my final exam was approaching and I got careless. A postdoctoral position in Chris Isham's group in London awaited and I was tired of long calculations, wanting some rest over the summer.

I arrived in London in September 1975 and was met by Chris Isham, Bryce's friend and colleague and an awesome worker in his own right. (Note Isham's *two* important contributions to this volume.) He took me to a flat in Ealing I was to share with Steve Fulling. Fulling was also interested in quantum field theory in curved space–times and had done some pioneering work with Leonard Parker on quantization in cosmological backgrounds. (Parker had been a postdoc in Chapel Hill with DeWitt in the 1960s. My DeWitt-related family was always growing, his influences always there.) The next day, Fulling and I rode the Green Line to central London and King's College. He told me about the point-splitting work that he, Paul Davies and Bill Unruh were doing and what they were planning. I was extremely pleased to find that I was no longer going to be alone in doing big calculations. I showed Fulling my dissertation and a very fruitful collaboration began. Steve urged me to write up my coincidence limit calculations so that others could use them. I spent the next few months redoing each calculation and writing the paper (Christensen 1976).

At the same time in Oxford, Bryce took up residence at All Souls College for the year and began the Monday Quantum Gravity Seminar. Fulling and I would go to these each week and we got to know the large group of talented gravity theorists residing in England at the time. It was an extremely inspiring time, one I have not seen repeated since then.

In the spring of 1976, DeWitt was to give a lecture on some work he had been doing on the black hole back-reaction problem. No one had been able to get a complete solution to the stress tensor problem and DeWitt was trying something himself. I had to go to Cambridge to give a seminar on point-splitting that Monday and Fulling promised to tape-record the talk. (At this time, my paper on point-splitting was accepted for publication. Strangely, someone at the journal decided that a point could not be split! But—it could be separated! So the name was officially changed to point-separation. Of course, no one ever called it that.) That night when we both returned home, I listened to the tapes and realized that Bryce had resurrected the postulate methods we had worked on in Texas in 1974. He was looking at what form the stress tensor should have if it satisfied the conservation equations, symmetry conditions, etc. Using what he knew about mode functions he produced a stress tensor with the above properties. He pointed out that his resultant stress tensor had *no* trace anomaly. Fulling, ever vigilant, saw that this was wrong. (This was one of the very few times I ever heard of Bryce making a statement that was not true.) The stress tensor did have a trace and that trace could not be explained away.

The trace anomaly was becoming very popular at King's at the time. I shared an office with Mike Duff, the co-discoverer of the anomaly, and Steve Fulling. Mike, Stan Deser and Chris Isham (Deser *et al* 1976) were trying to find the exact form for the anomaly and were playfully berating those 'point-splitters' who shared Mike's office. (Certain members of the King's group were very adept at 'sniffing the air' at the mention of the foul smelling split-point.) However, . . . one day Mike came over to my desk and showed me what they believed the form was for the trace anomaly. He asked if I had seen anything like it in our (sniff, sniff) point-splitting calculations. I *very* happily showed him my dissertation and the work Fulling and I were doing. There, in the a_2 coefficients, was the anomaly Mike had just shown me. Mike's face brightened and he grabbed our papers and immediately walked out of the office and down the hall to Chris Isham's office. I followed (nonchalantly). Walking into Isham's office, I remember Stan turning to me with a sly smile and saying. 'See. I told you point-splitting was good for something!'

Encouraged by our now firm belief in the anomaly, Fulling and I began to rethink the stress tensor from conservation laws technique I had failed to complete two years earlier. We now knew what the trace of the renormalized stress tensor should be and Fulling knew a lot more about mode sums and vacuum states than I did. One night, very late, after filling our white-topped kitchen table with calculations, we were convinced that we had solved the problem. We had the stress tensor needed to start the solution of the back-reaction of Hawking radiation on the black hole. We vowed to start writing for *Physical Review Letters* the next day.

Our euphoria faded into malaise. When I arrived at King's the next day, a mistake appeared in our work. Something simple, like a factor of 4π, was missing. After a respectful period of mourning, we set out to salvage what we could from the work. What we did realize was that knowledge of the trace anomaly and therefore the a_2 coefficient was crucial in the analysis. In the two-dimensional black hole model, the flux of Hawking radiation was equal to the anomaly. (In two dimensions the anomaly is proportional to the a_1 coefficient. In d-dimensions the anomaly is proportional to $a_{d/2}$.) No anomaly, no radiation. (Fulling tells the story of how the anomaly was discovered to be related to conservation in the two-dimensional black hole case. Applying the method of demanding conservation laws and zero trace, Davies *et al* (1976) had found that the conservation laws could not be satisfied. Instead, the covariant divergence of the stress tensor was equal to a constant times the covariant derivative of the Riemann scalar. They concluded that the conservation laws may not hold in the semiclassical approximation. Then one day, Fulling was taking a bath in the Ealing flat and, knowing of the work of Capper and Duff, he realized in a flash that if the two-dimensional stress tensor had an anomaly proportional to the Riemann scalar, conservation could be maintained. Unlike Archimedes, Fulling did not run into the street shouting 'Eureka!', but did dress and run up to the pay phone on the second floor to ring up Unruh and tell him the solution.)

In the four-dimensional case, while the anomaly did not fix the value of the flux, it was directly related. For the first time we saw that not only did a_2 determine the one-loop counterterms in the effective action, some of the terms in the point-splitting of the stress tensor and the trace anomaly, but also played a key role in the radiation from a black hole. This amazing quantity helped Fulling and me to find out a great deal about the properties of the stress tensor in both regions far from and near to the black hole horizon (see Christensen and Fulling (1977)).

(One unlikely resource for scientific research in London was the London

Underground at rush hours. Fulling and I lived 30 to 45 minutes from King's depending on the time of day. Much of the work we did on the Schwarzschild stress tensor problem was done on the train on our way to Ealing. Perhaps if the trains had been faster, we might still be working on the problem.)

The work on computing stress tensors and anomalies grew into an industry over the next few years. A great many people got considerable use out of what came to be known as the 'pure-thought method'. That was the name Paul Davies, Steve Fulling and I gave to the technique of finding the renormalized stress tensor by solving the conservation equations and forcing 'reasonable' symmetry properties on it. It was most easily applied to cosmological models such as the Robertson–Walker universe. (See Birrell and Davies (1982) for a review.) Wald (1977) and others were able to axiomatize the pure-thought method and prove that the various techniques of regularization/renormalization actually gave the expected results.

I will focus on the anomaly industry from here on since it seems this is where developments are still being made. The Schwarzschild back-reaction problem is still unsolved. Progress has been made recently, though from a very different direction (see York's essay in this volume).

After leaving London, I went to the University of Utah to work with Karel Kuchař for a year. (Karel is another of Bryce's close friends. In Chris Isham's review of Bryce's work in this volume, he mentions the time when he, Karel and Bryce were in Austin to plan a book on quantum gravity; a book that never got written. I was lucky enough to be able to sit in on some of the organizational meetings they held. [As I write this, I am looking at rare copies of the notes and preliminary chapters of their non-existent book. Too bad it wasn't finished; the outline still looks fascinating.] Karel is a very colorful man. His lectures have been some of the best and most entertaining I have ever heard. The only times he ever caused me any 'trouble' was during the seminars I gave at Utah. I had given a talk on point-splitting at many different places by this time and it usually took slightly over an hour to complete. However, when I gave the same talk in Salt Lake City in the Fall of 1976, it took two and one-half hours! Karel would *never* let me get by with 'physicist' mathematics or hand-waving arguments. To this day I try to think more clearly at every step of a calculation and I credit Karel with embarrassing me into being more careful, though I doubt that I'll ever come up to his standards of excellence.) During that year, in between skiing lessons, I worked to extend my earlier point-splitting results to spin-1/2 and spin-1 fields. Using the general algorithm for finding the a_ns, the stress tensor divergences were isolated *and* the anomalies were computed (Christensen 1978).

Other people were working on anomalies, using different methods and for different reasons. But in one way or another they all eventually used the proper-time algorithm and the a_2 coefficients. Mike Duff, in particular, was trying to understand why the anomaly coefficients had the values they did. He was also interested in studying the notion of a gravitational instanton, a special solution of the Euclidean space Einstein equations. The instanton idea had come from Yang–Mills theory and, as usual, was carried over to gravitation. In addition, another sort of anomaly, the axial anomaly, appeared in Yang–Mills instanton theory and then in turn became a popular topic in gravity. The equations giving the axial anomaly could be integrated.

Somewhat surprisingly, such work showed that the physics of instantons was intimately related to the topological properties of the group manifold, for example, the Pontryagin number. Duff looked at similar calculations in the gravitational case and found that the trace anomaly, the axial anomaly and gravitational instantons were all connected with the topology of the (Euclidean) space–time. A whole set of rather strange looking numbers began to appear: values for the anomalies, values for the topological characteristics of the instantons, numbers of zero and nonzero modes of various operators, and on and on. There seemed to be no pattern to any of these numbers and we felt awkward about this unsettling, yet exciting, state of affairs.

In the Fall of 1977 Duff went to Brandeis and I to Harvard. (Incidentally, Bryce received his undergraduate and PhD degrees at Harvard. The first *summa cum laude* and the second working with Schwinger. I joined another DeWitt student, Larry Smarr, in Bill Press's group. Larry stayed in the FORTRAN world and is now one of the world's experts in numerical relativity. [See his contribution on Bryce's work in classical gravity in this volume.]) Mike and I had met at the Waterloo GRG meeting in 1977 and had discussed the possibility of working together when we got to Massachusetts. In November, we were finally brought together by our interest in anomalies. (This situation was a bit peculiar in a practical day-to-day sense. Mike's appointment was at Brandeis in Waltham, Massachusetts and he lived in Cambridge, a few blocks from Harvard. I had an appointment at Harvard and lived a few blocks away from Brandeis in Waltham! Fortunately, like the long train rides on the London Underground a few years earlier, the rides in my car from Waltham to Cambridge each night when I took Mike home were very valuable work times, *except* in the dead of winter when the car doors would freeze shut and we couldn't get in.)

The connections between topology, anomalies, instantons and black holes were very interesting, but still not really understood. The other 'big' development of the 70s, supergravity (see van Nieuwenhuizen (1981) for a review), was beginning to play a role also. We knew something about spin-0, spin-1/2, spin-1 and spin-2, but almost nothing about the anomalies for spin-3/2. (Just for your amusement, go to DTGF and look at problems 79 and 80 to see what else DeWitt was thinking about in 1963!) The anomaly industry had produced some calculations of the spin-3/2 axial and conformal anomalies, but we (or at least I) didn't understand them. Mike already knew that some patterns do exist in the strange anomaly numbers and so we hoped that an overall unification, that is, a very general formula for the anomaly for arbitrary spin fields could be found. We set out to find such a formula. (I'd like to set the scene a little. Mike and I worked in an office in the Brandeis Physics building most of the time. Typically, I would sit in a dilapidated overstuffed chair next to the blackboard with a can of Dr Pepper in my hand. Mike would stand at the board or sit at the desk and smoke. Since then, Mike has courageously given up smoking. I, on the other hand, continue to have some sugary-caffeinated substance within reach whenever I work.)

Lessons from the past had told us to try to guess the answer to a problem before diving into some tedious index-shuffle, so we did some 'numerology' first. That is, we took our table of anomaly coefficients and tried to fill in the blank spaces. We already knew one general form for the a_2s. From it we found that if a field transformed according to the (A, B) representation of the Lorentz group (in Euclidean space SO(4)), then the formula for the anomaly coefficients was a polynomial of order five or less in A and B. A few more assumptions and educated guesses led to an expression with only a few undetermined coefficients. These were found by plugging in the anomalies we already knew. The value for the axial anomaly for the $(1, 1/2)$ part of the spin-3/2 field popped out. Realizing that the $(1, 1/2)$ representation gives a field with six degrees of freedom, we subtracted two spin-1/2 gauge ghost contributions (two $(1/2, 0)$ fields) giving the desired two physical degrees of freedom.

The result? The spin-3/2 axial anomaly was -21 times the spin-1/2 axial anomaly. This answer was not very different from those found previously. The discrepancy came about because previous workers ignored the so-called 'third' or 'Nielsen–Kallosh' ghost. This extra ghost, which we noticed, but did not name, is a spin-1/2 ghost. (In fact, DeWitt already knew about the need for a third ghost structure in his writing in DTGF. It comes from his $\gamma_{\alpha\beta}$ matrix discussed on page 41. I once told Bryce that *he* had really discovered the extra ghost term and he didn't believe me. I had to show him his own book. He made some comment like, 'You know my book better than I do.' I was flattered to hear him say that, but thought to myself, 'Fat chance!')

Luckily, our guesswork was giving sensible answers, but it gave us no proof. During the next few weeks, Duff and I learned more than we ever wanted to know about the representations and generators of SO(4) (Weinberg 1972). (We were highly motivated. We found ourselves in a race with other groups to find the spin-3/2 anomaly. It was my first serious competition in my work and it was a very unsettling experience at the time.) Finally, we produced a general $a_2(A, B)$. From this we were able to verify our numerology and move onto deriving the axial anomalies for the fields $\phi(A, B)$. Our results (Christensen and Duff 1978a) verified the known low spin results. A few, like the spin-3/2 anomaly, were new. A complete analysis of the structure of our arbitrary spin quantities led us to even more generalization problems. We wanted to know what happened when the curved manifold had a boundary. This difficult problem (which takes too long to explain, so I won't) led into a study of Weinberg's (1979) idea of asymptotic freedom, quantum gravity in $2 + \varepsilon$ dimensions (Christensen and Duff 1978b) and even to the problem of radiation for accelerating observers, another holdover from King's College days (Christensen and Duff 1978c).

One new thing did appear that we thought was 'cute', but we ignored it. For every axial anomaly there is a corresponding 'index theorem'. In simple terms, an index theorem relates the difference between the number of zero eigenvalue modes for the representation (A, B) and the representation (B, A) to the Pontryagin number P, which is a topological invariant. The index theorem comes from integration of the axial anomaly expression. We devised something we called a 'super theorem', sort of a square root of a regular index theorem. The regular theorems relate zero modes of representations with the same spin, while super theorems relate zero modes of representations differing by spin one-half. These super theorems were useful in computing anomalies and such in extented supergravity theories. There was one super theorem that involved the representation $(1, 0)$. This irreducible representation makes up part of an anti-symmetric tensor field. It was very rare that one could consider such a field as fundamental like the scalar $(0, 0)$, the spinor $(1/2, 0)$, the vector $(1/2, 1/2)$, etc were. It was this $(1, 0)$ theorem that we did not pay attention to, at least not then. This was a mistake, but we had to stop somewhere and write up what we had done. We had grown tired of working night and day. We began a paper we called 'New Gravitational Index Theorems and Super Theorems'. (We affectionately call this paper NGIT (Christensen and Duff 1979).) A month or two of slaving over a hot word processor (thanks to Bill Press and his computer), writing and rewriting until 3 AM each night, finished off the paper. (The manuscript was 93 pages long. We had followed Bryce's example and resisted the temptation to split the paper up into many little ones.)

In early 1978, Mike went back to London and I prepared to join Bryce again at the new Institute for Theoretical Physics in Santa Barbara, California. Fortunately, work with Duff continued. At one conference in Cambridge in the Summer of 1978 and then at another in Stony Brook in the Fall, we looked at the problem of quantizing gravity with a cosmological constant. We knew from the work of Deser and Zumino (1977) that if you add a cosmological term in the gravity action you must add a spin-3/2 'mass term' to preserve supersymmetry. Using the magical a_2, we computed the counterterms and compared our results to those obtained by others. They didn't match. We spent a week trying to figure out why. After beating our heads against the wall trying to track down differences in sign conventions, we found out that we were indeed right and began to write up the paper. (We were under the extra pressure of secrecy. Once again we were in competition with other groups and not real happy about it.)

Back at Santa Barbara, my collaboration with Duff went on via long distance telephone to London. (I called him nearly every week and talked for an hour sometimes! After a few months of this, the financial powers got justifiably annoyed with me. I agreed to cut down to thirty minutes or less.) One day, Mike and I were discussing the possible uses of the anomalies and counterterms we had computed. We were again in the situation of having lots of mysterious numbers in a big table of anomalies. It was time for more numerology. Mike and I each went away, and with the help of Martin Rocek and Gary Gibbons, computed the one-loop counterterms for all the extended supergravity models we knew. Strange things began to occur. From numbers like -233 and 848, we began to get numbers like -1, -2, -3, and amazingly, lots of zeros! We (Christensen *et al* 1980) found that extended supergravity theories with $N > 4$ were finite up to a topological term. (It was this topological term that was ignored in the renormalization theory work done in the early 1970s. Standard quantum gravity is not one-loop finite, it's one-loop renormalizable.) Later, these 'miraculous' cancellations, as we called all the zeros that appeared in the counterterms, were explained. (See Duff (1982) for a summary.) At the time however, we were nearly dumbfounded by what we had found. Excited, too. On my next call to Mike in London, one of us started the conversation with something like, 'Did you find what I found?!' The other said, 'Lots of zeros! Right?'

Like the night in Ealing when Fulling and I thought we'd found the Schwarzschild stress tensor or the day in Waltham when Mike and I confirmed our value for the spin-3/2 anomaly, this was a euphoric time. And this time, unlike the Ealing incident, we hadn't made a mistake.

Now I am at Chapel Hill, the place where Bryce did all of his phenomenal work in the 1960s. My work here centers in large measure on ideas Bryce had years ago. At this moment we are interested in fourth-order gravity theories, another topic Bryce pioneered (with Utiyama). (See Utiyama and DeWitt 1962, DTGF pages 56–60, 134–135 and 233–234.) The Schwinger–DeWitt algorithm appears in this work also. The a_2 coefficient extends its magic to many other fields, all of which Bryce has touched in some significant way. My advice to someone starting out in this field? 'Go read DeWitt!'

ACKNOWLEDGMENTS

Thanks are due to my graduate and undergraduate relativity students who inspired me to tell this story. Special thanks to N Barth, R Peterkin, B Smith and J York for their comments. This work was supported by the National Science Foundation and the Bahnson Trust Fund of the University of North Carolina at Chapel Hill.

REFERENCES

Bekenstein J D 1973 *Phys. Rev* D **7** 2333
Birrel N D and Davies P C W 1982 *Quantum Fields In Curved Space* (Cambridge: Cambridge University Press)
Capper D M and Duff M J 1974 *Nucl. Phys.* B **62** 147
Christensen S M 1975 PhD Dissertation, University of Texas
—— 1976 *Phys. Rev.* D **14** 2490
—— 1978 *Phys. Rev.* D **17** 946
Christensen S M and Duff M J 1978a *Phys. Lett.* **76** B 571
—— 1978b *Phys. Lett.* **79** B 213
—— 1978c *Nucl. Phys.* B **146** 11
—— 1979 *Nucl. Phys.* B **154** 301
Christensen S M, Duff M J, Gibbons G W and Roček M 1980 *Phys. Rev. Lett.* **45** 161
Christensen S M and Fulling S A 1977 *Phys. Rev.* D **15** 2088
Davies P C W, Fulling S A and Unruh W G 1976 *Phys. Rev.* D **13** 2720.
Deser S, Duff M J and Isham C J 1976 *Nucl. Phys.* B **111** 45
Deser S and Zumino B 1977 *Phys. Rev. Lett.* **38** 1433
DeWitt B S 1965 *Dynamical Theory of Groups and Fields* (New York: Gordon and Breach)
—— 1967a *Phys. Rev.* **160** 1113
—— 1967b *Phys. Rev.* **162** 1195
—— 1967c *Phys. Rev.* **162** 1239
—— 1975 *Phys. Rep.* **19** 295

Duff M J 1982 in *Supergravity 81* ed S Ferrara and J G Taylor (Cambridge: Cambridge University Press)

Gilkey P B 1975 *J. Diff. Geom.* **10** 601

Hawking S W 1975 *Commun. Math. Phys.* **43** 199

't Hooft G and Veltman M 1975 *Ann. Inst. Henri Poincaré* A **20** 69

van Nieuwenhuizen P 1981 *Phys. Rep.* **68** 189

Schwinger J 1951 *Phys. Rev.* **82** 664

Utiyama R and DeWitt B S 1962 *J. Math. Phys.* **3** 608

Wald R M 1977 *Commun. Math. Phys.* **45** 9

Weinberg S 1972 *Gravitation and Cosmology* (New York: Wiley)

—— 1979 in *General Relativity: An Einstein Centenary Survey* ed S W Hawking and W Israel (Cambridge: Cambridge University Press)

Particles do not Exist

P C W DAVIES

One may distinguish two approaches to the study of quantum fields propagating in a background space–time. The first is the investigation of the quantum stress tensor, $\langle T_{\mu\nu} \rangle$, with all its attendant regularization and renormalization difficulties. The second is the particle content of the field. The name of Bryce DeWitt is strongly associated with the first approach, and the DeWitt–Schwinger expansion of the effective Lagrangian remains the central feature in the development of a comprehensive covariant regularization scheme. Bryce's pioneering work in this area, a decade before the mushrooming of interest in curved space quantum field theory, is a landmark in his contribution to theoretical physics.

But Bryce has made important contributions to the particle approach too. Among these is his study of model particle detectors, an analysis of characteristic clarity and elegance. The idea of investigating the particle content of a quantum field in non-Minkowskian situations was proposed by Unruh (1976), whose famous 'box' detector proved to be the prototype for all future investigation. Bryce spotted that much of the internal structure of the Unruh box was an irrelevant complication, and a more elegant diagnostic device—a classical point-like object with internal quantum states—was invented (DeWitt 1979). By this refinement, Bryce was able to expose and clarify the essential physics that lies behind particle detection, and greatly increased our understanding of the particle concept.

Before embarking on my thesis, let me make two apologies. The first is for the low level of hard mathematics in what follows. It exists, but there would be no room to cover what I want to say if I put it all in. In any case, the details have been given elsewhere, mainly in my book with Birrell (Birrell and Davies 1982). Secondly, Bryce is a well known advocate of the Everett many-universes interpretation of quantum mechanics (DeWitt and Graham 1973). I wish to argue strongly for the conventional Copenhagen view as expressed by Bohr. No disrespect is intended, and in fact I think that as far as the forthcoming topics are concerned, the two interpretations are equally viable. What I do try to discredit is what might be called naive realism.

1. PARTICLES IN CURVED SPACE

The fundamental problem concerning the particle concept in quantum theory can be traced to the Uncertainty Principle[†]. If we wish to ascribe a definite energy (or frequency) ω, or a definite momentum k, to the particle, then the location of the particle is completely indeterminate; it can be anywhere in space. In conventional flat space quantum field theory, this problem is not too severe because Minkowski space–time is highly symmetric. Specifically, the particle states are associated with irreducible representations of the Poincaré group, the symmetry group of Minkowski space. In non-Minkowskian spaces, however, things are not so straightforward. There is, in general, no simple symmetry group that characterizes the space–time geometry, and the global and causal structure of the space–time might be complicated (e.g. non-trivial topology, horizons, singularities). If we continue to define particle states in terms of field modes, then these states probe the global structure, and the situation requires a considerably deeper analysis.

To illustrate this point, consider a massless scalar field ϕ propagating in Minkowski space. The wave equation

$$\Box \phi = 0 \tag{1.1}$$

possesses plane wave mode solutions $(16\pi^3 \omega)^{-1/2} \exp{(ik \cdot r - i\omega t)}$, so that ϕ may be expanded as follows:

$$\phi = \sum_k (16\pi^3 \omega)^{-1/2} [a_k e^{ik \cdot r - i\omega t} + a_k{}^\dagger e^{-ik \cdot r + i\omega t}].$$

The vacuum state $|0\rangle$ is then defined by the requirement

$$a_k |0\rangle = 0 \qquad \forall k \tag{1.3}$$

and from this state one may construct states containing $1, 2, 3, \ldots$ quanta in the model k by use of the creation operator

$$|n_k\rangle = (n!)^{-1/2} (a_k{}^\dagger)^n |0\rangle. \tag{1.4}$$

It follows that

$$\langle n_k | a_k{}^\dagger a_k | n_k \rangle = n_k$$

so $a_k{}^\dagger a_k$ is referred to as the number operator for the mode k. This procedure is upset as soon as complicating features are introduced into the space–time. For example, if boundary surfaces are present, or the topology is non-trivial, the plane wave modes given above (the form of which is closely linked to the properties of the Poincare group) will no longer in general be solutions of the wave equation (1.1). In the case that the space–time is curved, simple exponential modes are impossible.

Nevertheless it will usually still happen that some complete set of (more complicated) mode solutions will exist; let us call them ϕ_i where i may stand for a

[†] Units $\hbar = c = 1$

whole set of labels. We may still formally define particle states by the requirement

$$\phi = \sum_i (a_i \phi_i + a_i{}^\dagger \phi_i{}^*) \tag{1.5}$$

$$a_i |0\rangle = 0 \qquad \forall i \tag{1.6}$$

and then use (1.4) as before. The problem is, how do we know the states thereby defined have the physical properties associated with our notion of 'particle'?

Scepticism on this point arises because, in general, there will be many (usually infinitely many) *other* sets of mode solutions of (1.1) in terms of which particle states could be defined. For example, if we have another set of modes $\bar{\phi}_j$, then

$$\phi = \sum_j (\bar{a}_j \bar{\phi}_j + \bar{a}_j{}^\dagger \bar{\phi}_j{}^*) \tag{1.7}$$

$$\bar{a}_j |\bar{0}\rangle = 0 \qquad \forall\, j \tag{1.8}$$

defines another vacuum state, $|\bar{0}\rangle$, and new associated many-particle states

$$\bar{n}_j = (n!)^{-1/2} (\bar{a}_j{}^\dagger)^n |\bar{0}\rangle \tag{1.9}$$

that are in general *different* from the first set.

The relationship between the two sets of states can be found by expanding $\bar{\phi}_j$ in terms of ϕ_i:

$$\bar{\phi}_j = \sum_i (\alpha_{ji} \phi_i + \beta_{ji} \phi_i{}^*) \tag{1.10}$$

known as a Bogolubov transformation. It is readily shown that

$$\langle 0 | \bar{a}_j{}^\dagger \bar{a}_j | 0 \rangle = \sum_i |\beta_{ji}|^2$$

where $\bar{a}_j{}^\dagger \bar{a}_j$ is the number operator for particles defined in the $\bar{\phi}_j$ system (call them $\bar{\phi}$-particles for short). Thus the vacuum state $|0\rangle$ contains $\bar{\phi}$-particles. Similarly the vacuum state $|\bar{0}\rangle$ contains ϕ-particles. Which state (out of perhaps infinitely many) corresponds to the 'real, physical' vacuum?

I firmly believe that this question, which has so often been asked, is meaningless, and contrary to the Copenhagen spirit of quantum mechanics.

2. THE MYTH OF THE 'PHYSICAL' VACUUM

There seems to be a general impression that there exists a unique state—the 'physical' vacuum state—and that the multiplicity of vacua associated with the multifarious mode decompositions obscures what this 'real' vacuum state is except, it is said, in a few special cases. For example, in Minkowski space, the state defined by

(1.3) is said to represent the real vacuum. In the static Einstein universe and de Sitter space there is enough symmetry to proceed in close analogy to flat space, thus giving confidence that the 'physical' vacuum can be identified. Discussion of this topic is usually accompanied by remarks about 'natural' sets of modes ϕ_i which, seeming to be privileged in some sense, are seized upon to define 'the' vacuum state.

All this is nonsense. Bohr taught us that quantum mechanics is an algorithm for computing the results of measurements. Any discussion about what is a 'real, physical vacuum', must therefore be related to the behavior of real, physical measuring devices, in this case particle-number detectors. Armed with such heuristic devices, we may then assert the following. There are quantum states and there are particle detectors. Quantum field theory enables us to predict probabilistically how a particular detector will respond to that state. That is all. That is all there can ever be in physics, because physics is about the observations and measurements that we can make in the world. We can't talk meaningfully about whether such-and-such a state contains particles except in the context of a specified particle detector measurement. To claim (as some authors occasionally do!) that when a certain detector responds (registers particles) in somebody's cherished vacuum state that the particles concerned are 'fictitious' or 'quasi-particles', or that the detector is being 'misled' or 'distorted', is an empty statement.

Why, then, does everybody agree that the mode decomposition (1.2) is somehow the 'right' way to describe a quantum vacuum state in conventional Minkowski-space quantum field theory? The answer lies in criteria which are outside the scope of quantum theory. Quantum theory as such (even with detectors explicitly included) cannot tell us which vacuum state (or mode decomposition) to choose. When people talk of 'the' vacuum (I ignore here the complication of possible 'false' vacuum states that crop up in interacting quantum field theory—the discussion is here restricted to free fields) they have in the backs of their minds a sort of void or emptiness, featureless except for vacuum fluctuations.

Before the 1970s nobody thought very much about *for whom* the said state appears empty of stuff. It seems to have come as something of a surprise that an *accelerating* observer would disagree with an inertial observer over the absence or otherwise of 'stuff' (Davies 1975, Unruh 1976). The precise sense of this disagreement will be analyzed below. Here I wish to make the central point that, *for a very wide class of observers* there is agreement that the state defined by (1.3) is devoid of particles (I use 'observers' here interchangeably with 'detectors'). This class is, namely, the set of all *inertial* (uniformly moving) observers. That there is this consistency is, of course, a fundamental element in the definition of the Minkowski vacuum state, henceforth denoted $|0_M\rangle$, which is required to be Poincaré invariant, and hence invariant under Lorentz boosts.

Now we shall see that, in general, the set of 'emptiness-agreeing' observers is very much smaller than for the $|0_M\rangle$ case. So that although the latter class does not encompass *all* observers (they must not accelerate), nevertheless the class is still large enough that the state $|0_M\rangle$ defined in (1.3) is considered to be 'natural' or privileged (it clearly is the latter) in a way that the state (1.6) or (1.8) is not.

But there is, I think, a further point of a more subtle nature in all this. It turns out that although an accelerating observer may perceive particles in the state which his inertial colleague declares is vacuous, nevertheless the perceived particle content is in some sense small. For example, a uniformly accelerating detector registers a thermal bath of radiation in the Minkowski vacuum state $|0_M\rangle$, but the acceleration must be around 10^{20} m s^{-2} for a mere 1 K to show up. Why is this temperature so low compared to, say, room temperature?

The answer can be traced, dimensionally, to the speed of light. A typical wavelength of perceived radiation is c^2/(proper acceleration) which is very large because c is so large compared to familiar velocities and accelerations. Because of this, the class of detectors that will *very probably* remain unresponsive for a long (but finite) period of time in $|0_M\rangle$ is much wider than the inertial observers alone. Thus, in this approximate sense, $|0_M\rangle$ is the vacuum state for a very wide class of detectors—indeed almost as wide as we are likely to ever consider practically. (Though perhaps not quite—Bell and Leinaas (1982) report the possible experimental detection of 'acceleration radiation'.)

What is true of accelerations is true also of gravitational fields. Only in cases of extreme curvature will the concept of particles begin to break down at familiar wavelengths. As a rough guide, for a scalar curvature R, particles with wavelengths $\lambda \ll R^{-1/2}$ will be approximately well defined for a wide class of inertial (in this case, free-falling) observers.

There has been much discussion about 'Rindler' particles in flat space–time. It is possible to cover Minkowski space with four patches coordinatized by (ξ, η), related to Minkowski coordinates (t, x) in the region $x > |t|$ by

$$t = a^{-1}\,e^{a\xi}\sinh a\eta$$
$$x = a^{-1}\,e^{a\xi}\cosh a\eta \tag{2.1}$$

where a is a constant. Lines of constant ξ are

$$x^2 - t^2 = \alpha^{-2} \tag{2.2}$$

where α is a constant with the physical meaning that, in the region $x > |t|$ (the so-called Rindler wedge), it is the proper acceleration of a particle which moves along the trajectory $\xi = $ constant. Thus, the hyperbolae (2.2) are trajectories of constant proper acceleration. For this reason one sometimes hears people talk about the Rindler coordinate system being 'naturally adapted' to the circumstances of an accelerated observer.

In my view, this language is also badly misleading. A detector, at least as it is conceived for this sort of analysis, has a classically well defined trajectory, so it occupies a single worldline. The Rindler coordinate system, however, covers a patch (two- or four-dimensional depending on the discussion). The fact that the $\xi = $ constant lines coincide with the worldlines of a particular system of accelerating observers (and note that the acceleration must be different for each observer according to their value of ξ) is a special feature which owes its origin to the high degree of symmetry inherent in the Rindler coordinate system (the worldlines are Killing trajectories of Lorentz boosts). One could equally well choose a different coordinatization of Minkowski space which nevertheless happened to coincide with the Rindler system in the vicinity of one particular detector's worldline. In any case, in curved space–time, no such 'naturally adapted' coordinatization will generally exist so we must be careful not to base too much on it in the special case of the Rindler system.

Now it does so happen that the vacuum state defined by solving (1.1) in Rindler coordinates and using the analogous exponential modes to (1.2) for ϕ_i—usually known as the Rindler vacuum, $|0_R\rangle$—is a state for which detectors moving along $\xi = $ constant trajectories ('Rindler' detectors) remain unexcited. However, the origin of the Rindler coordinate system must be matched to the trajectory so that the null lines $x = \pm t$ are asymptotes to the hyperbola $\xi = $ constant. The Rindler vacuum $|0_R\rangle$ is not translation invariant. So, in fact, there are infinitely many Rindler vacuum states depending on these asymptotes. You have to choose the right one to get zero response from your uniformly accelerating detector.

A related special feature is that if one evaluates the Bogolubov transformation between the Minkowski and Rindler modes (this was first done by Fulling (1973)), one finds that the number operator for Rindler particles, when evaluated in $|0_M\rangle$, has a thermal spectrum with temperature $a/2\pi k_B$ (where k_B is Boltzmann's constant). It also happens that a particle detector (at least of the DeWitt variety) when following a $\xi = $ constant Rindler trajectory, responds as though it is at rest but immersed in a bath of thermal radiation of temperature $\alpha/2\pi k_B$. One notes that $\alpha = a g_{00}^{-1/2}$ and that the Tolman relation for the temperature of a thermal bath in a gravitational field is $T = g_{00}^{-1/2} T_0$. There is thus a striking concordance between the results of these two very different calculations. However, caution is necessary, for the calculations are really addressing quite different issues. One is the experiences of a detector along a given worldline, the other concerns particles that are not localized but are defined over the whole Rindler wedge. It is only because of the special symmetry properties of the Rindler system that these two results are so closely related. (For full details of Rindler quantum field theory, see the review by Sciama *et al* (1982).) This point is reinforced by the calculations of Letaw and Pfautsch (1982), who investigated more general detector motions in Minkowski space. They found that, for example, a uniformly rotating detector responds (the spectrum cannot be written in terms of elementary functions) but nevertheless the Bogolubov transformation between modes in the rotating and non-rotating systems reduces to the identity (i.e. $\beta_{ij} = 0$) and the two vacuum states are equivalent. Thus the experiences of a specific detector are in general no guide at all to the 'particle content' as defined by the Bogolubov transformation! But one must resist the temptation to divide up detector motions into 'reasonable' and 'screwy' on the basis of their compatibility with Bogolubov transformations. Let me repeat: there are quantum states, and detector measurements. What we mean by 'a particle' cannot sensibly be expressed without reference to a detector. All we can predict and discuss (as far as the *physical* world is concerned) are the experiences of detectors.

In ordinary laboratory quantum theory, we often talk about 'particles' as though they really exist, in the sense of entities with well defined properties independent of our observations. But in fact we cannot substantiate that image, for a 'particle' is merely an abstract heuristic model that provides an easy mental image of how one type of detector measurement is related to another. There is no need (and I contend it is meaningless) to regard the particle as a really existing thing skipping between measuring devices.

This strongly Copenhagen philosophy receives, in my view, powerful support from the particle analyses of curved space quantum field theory, where any attempt to hang on to the idea of particles being 'really there' is doomed to failure. Is a Rindler particle 'really there' in the state $|0_M\rangle$? Is a 'rotating' particle 'really there' when a rotating detector responds to $|0_M\rangle$? Is it the Bogolubov transformation which tells you what particles are present, or a system of detectors? If the latter, which detectors? Which trajectories? The answer is that the idea of 'particles' is just a model, which works well for some conventional situations, but is usually utterly useless as soon as one moves away from inertial observers in flat space–time.

3. THE STRESS TENSOR

Further confusion is heaped on the particle concept by the compelling mental image of a quantum as a tiny packet of energy and momentum. The modes (1.2) are indeed associated with energy ω and momentum k (in a non-localizable way) but this tidy arrangement collapses as soon as one departs from the Minkowski space case. In general, $\langle T_{\mu\nu}\rangle$ is non-zero (actually it is formally infinite and must be renormalized) even for a vacuum state. For a state with 'particles' (i.e. non-zero excitations of some particular set of modes) $\langle T_{\mu\nu}\rangle$ is usually *not* a simple sum of ωs and ks for each 'particle'.

This is dramatically illustrated by the 'moving-mirror' system in flat two-dimensional space–time which I investigated with Fulling (Fulling and Davies 1976, Davies and Fulling 1977a). A mirror which accelerates generally creates a flux of radiation. A *uniformly* accelerating mirror, however, does not radiate energy–momentum. Nevertheless, it still disturbs the field modes, and an explicit calculation shows that the Bogolubov transformation between the disturbed modes and the Minkowski modes has a non-zero β_{ij} (it is essentially a Macdonald function). Thus, we may conclude that 'particles' are created—and, indeed, a DeWitt detector at rest in the wake of these disturbed modes will certainly respond—yet the stress tensor calculation shows that $\langle T_{\mu\nu}\rangle = 0$. These 'particles' apparently carry no energy or momentum, yet they still excite a *static* detector! (Readers concerned about the law of energy conservation should note that the quantum state in this case is not an energy eigenstate, so that the expectation value of energy can jump abruptly when the detector makes a measurement.)

Similarly we know that $\langle 0_M | T_{\mu\nu} | 0_M \rangle = 0$ (by definition of Minkowski space). Thus, the 'particles' that excite a Rindler (accelerating) detector do not carry energy or momentum either.

Fulling and I showed that only in certain special cases is $\langle T_{\mu\nu}\rangle$ for a 'many-particle' state a simple sum of energy quanta ω. Specifically, for radiation to the right of a moving mirror

$$\langle T_0^{\ 0} \rangle = (2\pi)^{-1} \int_0^\infty \mathrm{d}p \int_0^\infty \mathrm{d}p' (pp')^{1/2} \left[\rho_{pp'} \mathrm{e}^{\mathrm{i}(p-p')u} - \mathrm{Re}\left(\mu_{pp'} \mathrm{e}^{\mathrm{i}(p+p')u} \right) \right] \quad (3.1)$$

where $u = t - x$ and

$$\rho_{pp'} = \int_0^\infty \beta^*_{p\omega} \beta_{p'\omega} \mathrm{d}\omega \qquad (3.2)$$

$$\mu_{pp'} = \int_0^\infty \beta_{p\omega} \alpha_{p'\omega} \mathrm{d}\omega. \qquad (3.3)$$

Now if it so happens that $\mu_{pp'} = 0$ and $\rho_{pp'}$ is diagonal, then (3.1) yields

$$\langle T_0^{\ 0} \rangle = \int_0^\infty \mathrm{d}p \int_0^\infty \mathrm{d}\omega\, p\, |\beta_{p\omega}|^2$$

$$= \int_0^\infty \langle N_p \rangle p \, \mathrm{d}p \qquad (3.4)$$

where N_p is the number operator for particles in mode p. This result has the obvious physical interpretation of a gas of particles each carrying energy p. Generally, however, such a simple relation between $\langle T_0^{\ 0} \rangle$ and $\langle N_p \rangle$ will not exist. Indeed, if the μ-term in (3.1) dominates, then it may happen that $\langle T_0^{\ 0} \rangle$ is negative.

If $\langle T_0^{\ 0} \rangle$ is integrated over all u (i.e. along $\mathscr{I}^+$) then the *total* energy is readily shown to have the form (3.4) (expressed as a total energy rather than an energy density). However, the *location* of the energy can be quite peculiar; it is not distributed evenly. For example, in the case of the uniformly accelerating mirror, whose trajectory is joined smoothly at $t = 0$ to a static trajectory, the energy flux from the mirror is zero at all times except $t = 0$, where a δ-function 'spike' is emitted. A particle detector which is adiabatically switched on at late times (and so misses the 'spike' altogether), will nevertheless respond as though a steady flux of particles is streaming from the mirror. So, in this sense, the energy is not where the particles are, and may have long since departed from the space–time region of the detector.

A good example of a controversy over what is the 'right' vacuum concerns black holes. The Schwarzschild space–time, being static, provides a 'natural' coordinate system (the Schwarzschild system) in which to solve the wave equation for modes. The resulting states were discussed initially by Boulware (1975), and the associated vacuum is sometimes called the Boulware vacuum $|0_B \rangle$. If the black hole were actually in that state there would, by definition, be no 'Boulware' particles present.

Now because the Schwarzschild coordinates tend at large distance to ordinary Minkowski coordinates, the state $|0_B\rangle$ apparently coincides there with the conventional Minkowski vacuum. Hence, there are no particles there (in the conventional Minkowski sense).

This result is in stark contrast to that of Hawking (1975) who demonstrated that black holes radiate a thermal flux of particles to infinity. Hawking's state is not $|0_B\rangle$. A thermally emitting state, sometimes called the Unruh vacuum $|0_U\rangle$, can be defined (Unruh 1976). This state mimics for the Schwarzschild space–time, the state that is obtained by starting out with Minkowski space, and the vacuum state $|0_M\rangle$, and allowing a body to collapse to a Schwarzschild black hole. By time-symmetrizing $|0_U\rangle$ a third state is obtained, variously named after Hartle and Hawking (1976), or Israel (1976), which I denote $|0_H\rangle$. This state reproduces the effect of a thermal *bath* (as opposed to flux) around the hole, such as would be obtained by enclosing the hole in thermodynamic equilibrium in a box with perfectly reflecting walls.

An examination of $\langle T_{\mu\nu}\rangle$ for the different vacuum states (Christensen and Fulling 1977, Candelas 1980) reveals that $\langle 0_B|T_{\mu\nu}|0_B\rangle$ is divergent at the Schwarzschild event horizon. The states $|0_U\rangle$ and $|0_H\rangle$ do not suffer this property. As it is a basic assumption of the semiclassical theory being discussed here that the quantum fields are test fields on a given background gravitational field, it is clearly unacceptable to have a divergent $\langle T_{\mu\nu}\rangle$. In reality, the back-reaction of an escalating $\langle T_{\mu\nu}\rangle$ would modify the gravitational field and the space–time would no longer be Schwarzschild. The Boulware vacuum is therefore unphysical in the sense that it could never be established. The space–time would simply fold up.

The conclusion is that vacuum states are not 'right' or 'wrong' but there may be physically unrealistic or impossible contenders. The state $|0_B\rangle$ is a *possible* state for a gravitational field that is Schwarzschild outside some $r > 2M$ (e.g. a static star), but it is impossible for a black hole.

A study of $\langle T_{\mu\nu}\rangle$ also opens the way to an alternative definition of 'the vacuum' as the *lowest energy* state, subject to certain constraints. Consider, for example, a scalar field in two-dimensional Minkowski space. If the space is coordinatized by (η, ξ) such that the metric is

$$ds^2 = C(\xi)(d\eta^2 - d\xi^2) \tag{3.5}$$

where $0 < C < \infty\ \forall \xi$ (i.e. a general static metric) and a vacuum state defined using the mode solutions $\exp ik\,(\eta \pm \xi)$, then it can be shown (Davies and Fulling 1977b) that for this state

$$\langle T_0{}^0\rangle = -\frac{1}{6\pi}C^{-1/2}\frac{\partial^2}{\partial\xi^2}C^{-1/2}. \tag{3.6}$$

If the spatial sections are compact, and we ask for the minimum total energy, then an integration by parts,

$$\int_{-\infty}^{\infty}\langle T_0{}^0\rangle\,d\xi = \frac{1}{6\pi}\int_{-\infty}^{\infty}\left[\frac{dC^{-1/2}}{d\xi}\right]^2 d\xi \geqslant 0, \tag{3.7}$$

reveals that $C = $ constant achieves the least value. This recovers the usual Minkowski vacuum $|0_M\rangle$.

The situation is not so straightforward, however, if the spatial sections are not compact and the total energy diverges, or if C has singularities (as in the Rindler case), or if the space–time is curved. Indeed, there is probably *no* systematic relationship between the state of minimum energy, and the state of zero particle content as perceived by some general class of detectors.

4. PARTICLE DETECTORS

Particles are what particle detectors are designed to detect. The usefulness of the particle concept (model) therefore turns on the extent to which detectors are expected to give useful results. It is at this stage that some thought has to be given to detector design.

We have seen that $|0_M\rangle$ *is* a useful concept. Simple model detectors (e.g. the Unruh box, the DeWitt detector, the DeWitt detector with more complicated couplings) all give zero response when moving inertially (Hinton 1983). If somebody's model detector failed to give a zero response in this situation, we should reject the model as 'an unreliable instrument'. There is an element of circularity here. If the state $|0_M\rangle$ is defined physically to be that which produces zero response in a 'reliable' detector, what does it mean to say a 'reliable' detector is one which does not respond to $|0_M\rangle$? This situation is common to all branches of physics, and is resolved by external criteria of a professional and philosophical nature. (How do you know when a jumpy meter is detecting something remarkable or is simply badly made? A professional judgment is necessary. A particle detector with internal noise *would* respond to $|0_M\rangle$, but such a detector we would regard as useless, on professional grounds.)

The situation is much more complicated once we allow non-inertial motions or have gravitational fields present. Generally, then, detectors give a non-zero response. But can we extract any useful information from it?

One idea is to determine the 'effective particle content' perceived by the detector. To do this one can normalize its response relative to what it would be in Minkowski space in a particular particle state. This step is crucial because differently designed detectors will respond differently to the same state (Hinton 1983). The answer may be expressed as some sort of spectrum.

But consider the following posers:

(i) What if there is no unique Minkowskian state corresponding to the particular response? This is the case if the effective particle content is anisotropic, but the detector is omnidirectional. How do we check consistency between that detector and a directionally-discriminating detector?

(ii) What if different detector-field couplings (e.g. linear, quadratic, derivative) give different responses to the same state (even when normalized and free of directional problems)?

(iii) What if differently designed detectors (e.g. Unruh box, DeWitt) give different effective responses?

(iv) Different detector trajectories usually give different responses. In curved space–time there are generally no privileged trajectories. Which trajectories are in some sense most representative of the particle content?

Calculations indicate that, generally speaking, there is *no* consistency between different detectors and/or couplings, and no obviously privileged trajectories. Only in certain very special cases is consistency obtained. Even in the Rindler case, there is inconsistency between the responses of detectors with different couplings (Hinton 1983). Moreover, directionally-discriminating detectors respond anisotropically (Davies *et al* 1983), thus rendering the simple DeWitt detector an unreliable guide even in this special case.

5. CONCLUSION

The concept of a particle is purely an idealized model of some utility in flat space quantum field theory. Away from that limited context, however, the concept becomes much less useful and has been the source of much confusion. The study of DeWitt-style particle detectors has exposed the nebulousness of the particle concept and suggests that it should be abandoned completely.

In its place one would like to study, as a probe of the field content, quantities such as $\langle T_{\mu\nu} \rangle$, $\langle \phi^2 \rangle$, etc. The major problem then arises as to how *these* quantitites are to be measured. We need a theory of model detectors for a variety of field-related physical quantities. So far as I know, there has been no attempt to tackle this interesting problem.

ACKNOWLEDGMENTS

I am grateful to Kerry Hinton for detailed discussions on the general theory of particle detectors.

REFERENCES

Bell J S and Leinaas J M 1982 *Electrons as accelerated thermometers* CERN preprint no. TH. 3363
Birrell N D and Davies P C W 1982 *Quantum Fields in Curved Space* (Cambridge: Cambridge University Press)
Boulware D G 1975 *Phys. Rev.* D **11** 1404
Candelas P 1980 *Phys. Rev.* D **21** 2185

Christensen S M and Fulling S A 1977 *Phys. Rev.* D **15** 2088

Davies P C W 1975 *J. Phys. A: Math. Gen.* **8** 365

Davies P C W and Fulling S A 1977a *Proc. R. Soc.* A **356** 237

——1977b *Proc. R. Soc.* A **354** 59

Davies P C W, Hinton K J and Pfautsch J D 1983 *Phys. Lett.* **120B** 88

DeWitt B S 1979 'Quantum gravity: the new synthesis' in *General Relativity: an Einstein centenary survey* ed S W Hawking and W Israel (Cambridge: Cambridge University Press)

DeWitt B S and Graham N 1973 *The Many-Worlds Interpretation of Quantum Mechanics* (Princeton: Princeton University Press)

Fulling S A 1973 *Phys. Rev.* D **7** 2850

Fulling S A and Davies P C W 1976 *Proc. R. Soc.* A **348** 393

Hartle J B and Hawking S W 1976 *Phys. Rev.* D **13** 2188

Hawking S W 1975 *Commun. Math. Phys.* **43** 199

Hinton K J 1983 *J. Phys. A: Math. Gen.* **16** 1937

Israel W 1976 *Phys. Lett.* **57A** 107

Letaw J R and Pfautsch J D 1982 *Phys. Rev.* D **24** 1491

Sciama D W, Candelas P and Deutsch D 1981 *Adv. Phys.* **30** 327

Unruh W G 1976 *Phys. Rev.* D **14** 870

Is There a Quantum Equivalence Principle?

P CANDELAS AND D W SCIAMA

1. INTRODUCTION

The aim of this article is to point out that the understanding that has been gained in recent years of accelerated vacuum states, motivated in large part by a desire to understand the phenomenon of black hole radiance, sheds new light on the classic issue of the radiation reaction suffered by a uniformly accelerated charge and also on the related questions of the radiation emitted by a charge that is either at rest in or freely falling through a static gravitational field. The present article draws on Bryce's work in two important respects. The first is his insistence (DeWitt 1979) that the vacuum state of quantum theory can be viewed as an aether, albeit in a new and improved form, and the second is the careful analysis that he performed together with Cécile (DeWitt and DeWitt 1964) of the problem of a classical charge falling through a gravitational field. For this problem it is shown that '. . . the particle tries its best to satisfy the naive equivalence principle . . . ,' a point that is of importance in what follows.

We begin with the case of uniform acceleration. The feature of this motion that has attracted so much attention over the years is the seemingly paradoxical relation between the radiation rate and the radiation reaction force.

In order to avoid boundary effects that would otherwise obscure the issue, we shall consider only motions of the charge such that the agency producing the acceleration has finite duration. A realization of such a motion is: the charge initially moves freely, at time $t = 0$ it enters a region where there is a uniform electric field, it remains in the field until $t = t_1$ at which time it leaves the region. Owing to the phenomenon of preacceleration the acceleration of the charge is non-zero before $t = 0$ and non-uniform before $t = t_1$. However, if the time t_1 is long compared with the characteristic time

$$\tau = \tfrac{2}{3} e^2 / mc^3$$

(the time that light takes to traverse the classical electron radius), then the motion of the charge may be considered as being effectively inertial for $t < 0$, uniformly accelerated for $0 < t < t_1$ and inertial again for $t > t_1$ though, as we shall see presently, the initial and final periods of non-uniform acceleration play a crucial role in the discussion.

2. UNIFORM ACCELERATION: THE NATURE OF THE PROBLEM

Following the detailed work of Bradbury (1962) (see also the recent article by Boulware 1980) we note that the following facts are pertinent.

(i) By direct computation from the Lienard–Wiechart potentials it is straightforward to show that at each instant of retarded time the charge radiates energy at a rate P given by the Larmor formula

$$P = \tfrac{2}{3} e^2 a^2 / c^3$$

with a the magnitude of the proper acceleration of the charge. In particular it radiates at a uniform rate whenever a is constant.

(ii) The classical radiation reaction force is given, in the instantaneous rest frame of the charge, by

$$F_{\text{rad}} = \frac{2}{3} \frac{e^2}{c^3} \frac{\mathrm{d}^2}{\mathrm{d}t^2} v$$

which depends on the second derivative of the charge's velocity and hence vanishes when a is constant.

(iii) The magnetic field of the charge is zero everywhere in the accelerated frame of the charge while it undergoes uniform acceleration. More precisely: if the charge follows the worldline $\xi = a^{-1}$ in the Rindler coordinate frame of the Appendix then the magnetic components of the electromagnetic field tensor $F_{\xi y}$, $F_{\xi z}$ and F_{yz} are zero on the Rindler manifold.

The apparent paradox is the seeming contradiction between (i) and (ii) whereby during the period of uniform acceleration there is a uniform rate of radiation of energy yet no radiation reaction force. That there is in fact no blatant violation of the principle of the conservation of energy follows from the following fact.

(iv) The radiation reaction force acts, during the initial and final periods of *non-uniform* acceleration, in just such a way as to ensure that the total work done by the agency accelerating the charge is equal to the sum of the change in the charge's kinetic energy and the total amount of energy radiated to infinity.

This last statement amounts to the assertion that the time integral of the rate at which work is done against the radiation reaction force is equal to the total amount of energy radiated and is assured, mathematically, by an integration by parts. The integrated term vanishes provided that the motion is inertial at sufficiently early and sufficiently late times.

Although there is in reality no difficulty posed by overall conservation of energy, the fact that the force of radiative reaction vanishes during the period of uniform acceleration seems counter-intuitive, especially in light of the observation first made by Callen and Welton (1951) in their celebrated paper on the Fluctuation–Dissipation theorem of the intimate relation between the force of radiation reaction and the zero point fluctuations of the electromagnetic field. Indeed, for an oscillating dipole, they show that these quantities are related by precisely such a Fluctuation–Dissipation theorem. This relation was further elucidated by Senitzky (1973) and by Milonni *et al* (1973) who showed that the decay of an excited atom can be viewed equivalently as arising from the perturbing effect of the vacuum electric field fluctuations or from the radiative reaction due to the electrons self-field or indeed any linear combination of these processes. The net effect cannot, of course, be altered, a definite answer being obtained for any observable quantity. However, the blame for a decay, say, may be apportioned at will between vacuum fluctuation and radiative reaction the relative proportion being determined by the ordering chosen for *commuting* operators (Milonni 1976). This somewhat remarkable state of affairs indicates that neither vacuum fluctuation nor radiation reaction furnishes a complete or fully consistent statement of the quantum-mechanical reality each being an oversimplification motivated by a desire to attach a physical picture to equations that arise at an intermediate stage of calculation. Thus, for example, if one thinks solely in terms of vacuum fluctuation it is difficult to understand why an atom in its ground state should never be promoted to a higher state as the result of electric field fluctuations. If, on the other hand, one thinks solely in terms of radiation reaction then one is confronted with the familiar difficulty of understanding the stability of the ground state. Despite these shortcomings the fluctuation and radiation reaction pictures furnish considerable intuitive understanding of a number of processes, a good example of this is provided by Welton's computation of the Lamb shift (Welton 1948) viewed as the shift in the energy level of the electron as the result of its interaction with the vacuum fluctuation of the electric field.

It is instructive to examine the absence of radiative reaction on a uniformly accelerated charge in the light of this duality between the fluctuation and radiative reaction pictures. We know from (ii) above that the self-force vanishes in virtue of the vanishing of the second derivative of the velocity. What we seek here is an understanding of this fact in terms of fluctuations. This is provided by the observation first made by Unruh (1976) that to a uniformly accelerated observer whose acceleration is α the Minkowski vacuum takes on the appearance of a thermal mixture of temperature $\alpha/2\pi$. We might say that the charge perceives the vacuum fluctuations as co-moving and comprising a thermal bath. Thus if the charge is constrained to move with constant acceleration *there can be no net transfer of energy or momentum* between the charge and the vacuum as seen in the accelerated frame. This explains (ii) as well as drawing a close parallel with (iii) which is the statement that the field of the charge is non-radiative as seen in the accelerated frame.

It is also worth noting that the Unruh heat bath observed in the constantly accelerating frame is subject to the Gaussian fluctuations of energy density which a conventional heat bath would possess. If the charge were acted on by a constant force, the pressure fluctuations associated with these energy fluctuations would confer on the charge an irregular motion. This motion would represent a non-constant acceleration and so would also lead to a systematic radiation damping force acting on the charge. We would expect this combination of forces to lead to a situation whose importance was often emphasized by Einstein, namely a situation in which an irregular activating force and a systematic damping one lead to a steady state in which the system's momentum distribution is that given by Maxwell for thermal equilibrium. In the present case this would mean that a charge subject to a constant external force would come to have the momentum distribution appropriate to the Unruh temperature of the ambient quantum vacuum. If, on the other hand, the charge were constrained to have exactly constant acceleration, then the external force would have to fluctuate to compensate for the pressure fluctuations in the Unruh heat bath.

A converse argument can also be made. By demanding that the spectrum of vacuum fluctuations be such that there be no net transfer of energy or momentum between the field and the charge it may be inferred that the Minkowski vacuum appears to a uniformly accelerated observer to comprise a thermal bath.

We turn now to a consideration of the radiation reaction force experienced by a freely falling charge.

3. FALLING CHARGES

For the case of a charge moving inertially at non-relativistic velocity through a static gravitational field it has been shown that (DeWitt and DeWitt 1964) 'the field in the immediate vicinity of the particle tends to fall freely with the particle, and although it suffers a local tidal distortion characteristic of an explicit occurrence of the Riemann tensor the net retarding force due to this distortion is zero when integrated over solid angle. The deviation of the particle motion from geodetic when $F_{\mu\nu}{}^{in} = 0$ is caused not by the local field of the particle but by a field which originates well outside the classical radius. . . . Physically the non-local term arises from a back-scatter process in which the Coulomb field of the particle, as it sweeps over the 'bumps' in the space–time, receives 'jolts' which are propagated back to the particle.'

This effect is analogous to the non-zero self-force which acts on a charge even when it is held at rest, for example, in the gravitational field of a stationary black hole. The distortion of the Coulomb field of the charge due to the Riemann tensor of the background gravitational field leads to the existence of such a self-force, which we could call a polarization force. Of course, for a general gravitational field which lacks a time-like Killing vector there is no invariant definition of 'at rest,' and the self-force, which will continue to depend on the charge's motion, cannot be invariantly decomposed into a radiation reaction and a polarization force.

Returning to the radiation problem, we now seek to understand from the fluctuation point of view the absence of radiation emanating from the region near a charge moving inertially at non-relativistic velocity through a static gravitational field, and for simplicity we shall confine the discussion to spherically symmetric gravitational fields. Naturally the classical analysis makes no reference to the state of motion of the electrodynamic vacuum. We know now that there are three natural vacua that are associated with the space–time geometry (for a review see Sciama *et al* 1981). These are

(i) the Boulware vacuum, this is the natural vacuum state for the space–time geometry of an extended massive body such as a neutron star, and may be thought of as a vacuum state which has come to equilibrium loaded under the action of the gravitational field in a manner not wholly dissimilar from an equilibrium atmosphere.

(ii) The Hartle–Hawking vacuum which corresponds to the natural vacuum state of a black hole enclosed by a (sufficiently small) box. This state corresponds to a black hole in equilibrium with a bath of blackbody radiation.

(iii) The Unruh vacuum which is the natural vacuum state to assign to a black hole that results from the collapse of an extended object.

The question that comes naturally to mind is whether the classical calculation applies with equal validity to a charge falling freely through each of these three vacua. To put the matter graphically we might care to think of the Hartle–Hawking vacuum as a state in which the vacuum fluctuations move inertially so that it is intuitively plausible that an inertially moving charge should find itself in harmony with the vacuum field fluctuations and hence that there should be little systematic retarding effect on the charge due to the fluctuations. However, in this picture it is not quite so evident that a charge falling freely through the Boulware vacuum, the fluctuations of which may be thought of as comprising a static aether, should not be subject to a systematic force due to its interaction with these fluctuations. Furthermore, returning to the case of the Hartle–Hawking vacuum, a fact that now requires explanation is that a charge held fixed in the gravitational field of a black hole (and hence accelerated) should not be able to extract energy from the freely-falling vacuum fluctuations and hence radiate in contravention of the classical result. That this does not occur is due to the fact that in the Hartle–Hawking vacuum the fluctuations are distributed with a thermal spectrum so that there can be no systematic exchange of energy between the charge and the field. The former case of a charge falling freely through the Boulware vacuum requires a more detailed analysis, which we present in the Appendix, the result of which is that for the case of a charge moving inertially in Minkowski space–time through an *accelerated* vacuum the spectrum of field fluctuations perceived by the charge *is the same as if the vacuum were unaccelerated*, i.e. the spectrum of field fluctuations is the same as that of the Minkowski vacuum. From this we may understand why a charge that falls freely in a gravitational field is not subject to a local reactive force due to its interaction with the field fluctuations, but only the reactive force referred to above, which arises in virtue of the field lines feeling out the 'bumps' in space–time and which originates well outside the classical radius. We may understand in the same way the fact that the radiation emitted by a freely falling charge does not originate at the charge.

In all cases therefore the classical results regarding the radiation emitted by and the radiative reaction force on an electron undergoing the various states of motion that we have discussed can be understood in terms of the spectrum of field fluctuations perceived by the charge. This is a remarkable fact since *a priori* the classical results might have been incorrect either because the quantum-mechanical equations might differ from the classical ones by terms of the order of Planck's constant or because the classical results for a given motion might apply for one vacuum state but not for another.

An understanding of the fact that neither possibility is realized can be gained under the assumption that the Heisenberg operator equations of motion take the same form as the classical equations (this point is not entirely trivial since the classical equations are not derived directly from a Hamiltonian but rather by a process of successive approximation involving, among other operations, the renormalization of the charge's mass). This is in fact known to be the case at least for a charge bound to an atom that is at rest in flat space–time (Milonni 1976). It then follows by virtue of the duality between the radiation reaction and vacuum fluctuation pictures referred to earlier that the spectrum of the field fluctuations for a given motion and vacuum must be such as to accord with the classical result. In particular since the Heisenberg operator equations of motion make no reference to the state of the field the classical results must apply with equal validity to each of the distinct vacua.

APPENDIX: CALCULATION OF THE RESPONSE OF AN UNRUH BOX THAT MOVES INERTIALLY THROUGH AN ACCELERATED VACUUM

We study, in this appendix, the extent to which an observer who moves inertially through an accelerated vacuum state is, by virtue of his interaction with the vacuum fluctuations, able to detect his motion relative to the aether. In order to concentrate on effects which might be directly attributed to the motion of the vacuum relative to the observer rather than effects which might be attributed to the effects of space–time curvature we shall perform the calculation for an observer moving inertially through the Fulling vacuum, to be denoted by $|F\rangle$, which can be thought of as the natural vacuum state above a uniformly accelerated plane mirror or equivalently as a state representing a static vacuum in the gravitational field of an infinite 'flat earth' (Sciama *et al* 1981).

A necessary complication is that an observer moving inertially can only remain in the Rindler wedge for a finite time although the observer's worldline can be chosen so as to render this time arbitrarily long (in terms of our flat earth picture a freely falling observer released from rest will strike the earth after the elapse of a finite time). We specify the initial conditions by supposing that before $t = 0$ the observer, who is equipped with an Unruh box, is subject to uniform acceleration and in fact that he follows the worldline $\xi = H$ of a Rindler coordinate system, after $t = 0$ we shall suppose that he moves inertially in such a way that his velocity suffers no discontinuity at $t = 0$. This motion is depicted in figure 1. Since the observer will leave the Rindler wedge after a time H he can only make reliable inferences about components of the vacuum fluctuations of frequency ω for frequencies such that $\omega H \gtrsim 1$. We shall therefore suppose H to be very large. At time t the acceleration of the vacuum with respect to the observer has magnitude $(H^2 - t^2)^{-1/2}$. Since we are primarily interested in effects that might be ascribed to the relative motion of the vacuum with respect to the observer and less so with transient effects that might be ascribed to the discontinuity in the observer's acceleration at $t = 0$ we shall ultimately take the limit $(H, t) \to \infty$ in such a way that the acceleration $(H^2 - t^2)^{-1/2}$ remains constant.

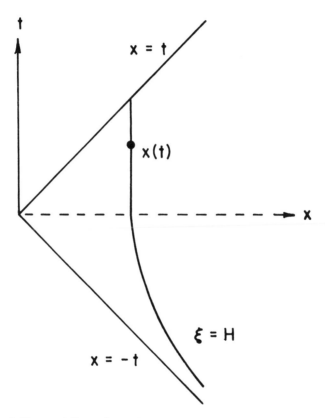

Figure 1 The worldline of an observer who is uniformly accelerated with acceleration H^{-1} before $t = 0$ and who moves inertially thereafter.

We shall assume that the Unruh box with which the observer is equipped interacts with the vacuum fluctuations of a scalar field $\hat{\phi}$ via a monopole charge $\hat{\mu}(t)$ such that the interaction Lagrangian is

$$\mathcal{L}_{\text{int}} = \hat{\mu}(t)\hat{\phi}(t)$$

evaluated at the instantaneous position of the box and that at $t = 0$ the box is in an energy eigenstate $|m\rangle$.

Standard analysis reveals that the probability of finding the box to be in an energy eigenstate $|n\rangle$ at time t is

$$\mathscr{P}_{mn}(t) = |\langle n|\hat{\mu}(0)|m\rangle|^2 \int_0^t dt_1 \int_0^t dt_2 \, \exp[-i\omega(t_1 - t_2)] \langle F|\hat{\phi}(t_1)\hat{\phi}(t_2)|F\rangle$$

where

$$\omega = \acute{E}_n - E_m$$

and $\hat{\phi}(t)$ means $\hat{\phi}(x)$ evaluated at

$$x(t) = (t, H, y, z).$$

Differentiating $\mathscr{P}_{mn}(t)$ we obtain an expression for the transition rate in the form

$$\frac{d}{dt}\mathscr{P}_{mn}(t) = |\langle n|\hat{\mu}(0)|m\rangle|^2 \Pi(\omega, t)$$

with

$$\Pi(\omega, t) = 2\mathrm{Re} \int_0^t dt' \, \exp[i\omega(t - t')] \langle F|\hat{\phi}(t)\hat{\phi}(t')|F\rangle.$$

If we now substitute an explicit expression for the Wightman function occurring on the right-hand side of this relation and take note of the fact that, in the sense of distributions,

$$\langle F|\hat{\phi}(t_1)\hat{\phi}(t_2)|F\rangle \sim -\frac{1}{4\pi^2(t_1 - t_2 - i\varepsilon)^2}$$

as $t_1 \to t_2$ where ε is an infinitesimal then we find

$$\Pi(\omega, t) = -\frac{1}{2\pi^2}\mathrm{Re}\int_0^{t-i\varepsilon} \frac{dt' \exp i\omega(t - t')}{(t^2 - t'^2)}\left\{\left[\log\left(\frac{H + t}{H + t'}\right)\right]^{-1}\right.$$

$$\left. + \left[\log\left(\frac{H - t}{H - t'}\right)\right]^{-1}\right\}$$

It remains now only to evaluate this integral. For $\omega > 0$ we deform the contour of integration into the lower half t' plane until we arrive at the contours of figure 2.

It is easy to see that the contribution to $\Pi(\omega, t)$ of the contour that extends from the origin to $-i\infty$ approaches

$$-\frac{\sin \omega t}{\omega t^2}\left\{\left[\log\left(\frac{H}{H + t}\right)\right]^{-1} + \left[\log\left(\frac{H}{H - t}\right)\right]^{-1}\right\}$$

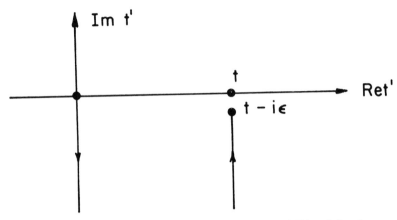

Figure 2 The contour appropriate to the evaluation of $\Pi(\omega, t)$ for the case

in the asymptotic limit. The evaluation of contribution to Π of the other contour, that extends from $t - i\infty$ to $t - i\varepsilon$, requires a certain amount of care. Setting $t' = t - iy$ we see that this contribution is the real part of

$$I = \frac{1}{2\pi^2} \int_\varepsilon^\infty \frac{dy \, e^{-\omega y}}{y(2t - iy)} \left\{ \left[\log\left(1 - \frac{iy}{H+t}\right) \right]^{-1} + \left[\log\left(1 + \frac{iy}{H-t}\right) \right]^{-1} \right\}.$$

It is convenient to separate I into the sum of three terms

$$I = J + K + L$$

with

$$J = \frac{1}{2\pi^2} \int_\varepsilon^\infty \frac{dy \, e^{-\omega y}}{y(2t - iy)} \left[\log\left(1 - \frac{iy}{H+t}\right) \right]^{-1}$$

$$K = \frac{1}{2\pi^2} \int_\varepsilon^\infty \frac{dy \, e^{-\omega y}}{y(2t - iy)} \left\{ \left[\log\left(1 + \frac{iy}{H-t}\right) \right]^{-1} - \frac{(H-t)}{iy} - \frac{1}{2} \right\}$$

$$L = \frac{1}{2\pi^2} \int_\varepsilon^\infty \frac{dy \, e^{-\omega y}}{y(2t - iy)} \left[\frac{(H-t)}{iy} + \frac{1}{2} \right].$$

We shall consider these integrals in turn. It is easy to see that in the limit $t \to \infty$

$$J = \frac{i(H+t)}{4\pi^2 t} \int_\varepsilon^\infty \frac{dy}{y^2} e^{-\omega y} - \frac{H}{8\pi^2 t^2} \int_\varepsilon^\infty \frac{dy}{y} e^{-\omega y} + O\left(\frac{1}{\omega t^2}\right)$$

the last term indicating a quantity independent of ε.

The quantity inside the braces in K has been chosen such that it is $O(y)$ as $y \to 0$. Thus it is redundant to retain the infinitesimal ε in this term and we may therefore take the lower limit of integration to be zero. Setting also $y = (H - t)u$ and taking the asymptotic limit we find

$$K \sim \frac{1}{4\pi^2 t} \int_0^\infty \frac{du}{u} \exp[-\omega(H-t)u] \left\{ [\log(1+iu)]^{-1} - \frac{1}{iu} - \frac{1}{2} \right\}$$

$$\sim -\frac{1}{8\pi^2 t} \int_{u_0}^\infty \frac{du}{u} \exp[-\omega(H-t)u]$$

$$\sim \frac{1}{8\pi^2 t} \log[\omega(H-t)].$$

The final integral is straightforward. We find

$$L = -\frac{i(H-t)}{4\pi^2 t} \int_\varepsilon^\infty \frac{dy}{y^2} e^{-\omega y} + \frac{H}{8\pi^2 t^2} \int_\varepsilon^\infty \frac{dy}{y} e^{-\omega y} + O\left(\frac{1}{\omega t^2}\right).$$

Combining the expressions for J, K and L we find that

$$\mathrm{Re}\, I \sim \frac{1}{8\pi^2 t} \log[\omega(H-t)].$$

Thus we have shown that for ω positive and $t \to \infty$, $\Pi(\omega, t)$ differs from zero only by transient terms.

The analysis for ω negative is entirely similar apart from the fact that the contour is deformed into the upper half plane (see figure 3). The portions of the contour that are parallel to the imaginary axis represent transient terms just as in the previous case

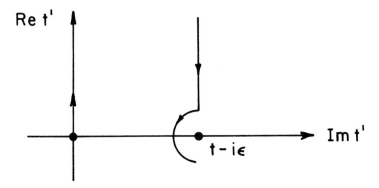

Figure 3 The contour appropriate to the evaluation of $\Pi(\omega, t)$ for the case $\omega < 0$.

while the semicircular portion of the contour yields (in the limit $\varepsilon \to 0$) the contribution $\omega/2\pi$.

Thus finally we have, in the asymptotic limit,

$$\Pi(\omega, t) \to \frac{\omega}{2\pi} \Theta(-\omega)$$

where Θ denotes the step function which is precisely the result that would obtain for an Unruh box moving inertially through the Minkowski vacuum.

ACKNOWLEDGMENTS

It is a pleasure to acknowledge fruitful conversations with D Deutsch. One of us (PC) wishes to thank K Kuchař for instructive conversations and the Department of Physics of the University of Utah, the Department of Astrophysics of Oxford University and the International School for Advanced Studies for their hospitality.

This work was supported in part by NSF grant PHY 7826592 and by a grant from the SERC

REFERENCES

Boulware D G 1980 *Ann. Phys., NY* **124** 169
Bradbury T C 1962 *Ann. Phys., NY* **19** 323
Callan H B and Welton T A 1951 *Phys. Rev.* **83** 34
DeWitt B S 1979 in *General Relativity: An Einstein Centenary Survey* ed S W Hawking and
　　W Israel (Cambridge: Cambridge University Press)
DeWitt C M and DeWitt B S 1964 *Physics* **1** 3
Milonni P W 1976 *Phys. Rep.* **25** 1
Milonni P W, Ackerhalt J R and Smith W A 1973 *Phys. Rev. Lett.* **31** 958
Sciama D W, Candelas P and Deutsch D 1981 *Adv. Phys.* **30** 327
Senitzky I R 1973 *Phys. Rev. Lett.* **31** 955
Unruh W G 1976 *Phys. Rev. D* **14** 970
Welton T A 1948 *Phys. Rev.* **74** 1157

Some Cosmological Aspects of Quantum Gravity

LEONARD PARKER

1. INTRODUCTION

In the extreme conditions of the very early universe, quantum phenomena involving gravity played an important role. Cosmology is a natural setting in which to study quantum gravity. Quantum gravity may provide answers to such deep cosmological questions as why is the expansion of the universe isotropic, can the cosmological singularity be avoided, why is the early universe hot, why is the magnitude of the cosmological constant so small?

Underlying the quest for a quantum theory of the gravitational field is the ultimate hope of unifying gravity with the other interactions. It is only from such a unified theory that one can expect to obtain a complete answer to the above questions. Although this unification may well require a major revision of the conceptual structure on which the theory is based, it is encouraging that a partial answer to some of the above questions is emerging from the presently known fundamental theories. Consideration of these cosmological questions may serve as a guide toward unification.

No one contributed more to the development of a covariant quantum theory of gravitation than Bryce S DeWitt (1967). He has recently emphasized the importance to that theory of the question of avoidance of the cosmological singularity and has produced an approximate effective action with which that question can be investigated (DeWitt 1981a,b). Such approximation is necessary at the present time because no one knows how to circumvent the infinites which appear in the full theory.

The simplest theory that one can use in order to explore the influence of quantum effects involving strong gravitational fields on the evolution of the early universe is the theory of quantized elementary particle fields propagating on a classical background curved space–time. A better approximation to the full quantum theory may be obtained by including gravitons as quantized gravitational wave perturbations on this background metric. One may call this general approach quantum field theory in curved space–time. It has yielded a number of important results which may plausibly be expected to follow from the fully unified theory and which may therefore serve as an aid in the formulation of such a theory.

Among these results is the creation of particles by the expansion of the universe (for reviews with more complete references see Parker 1977, Gibbons 1979, Birrell and Davies 1982) and by black holes (Hawking 1974, 1975). Here I will be mainly concerned with particle creation in cosmological space–times. This particle creation may explain why the early universe is hot. It also reacts back on an initially anisotropic expansion in such a way as to rapidly bring about isotropy (Zeldovich 1970, Zeldovich and Starobinsky 1971, Lukash and Starobinsky 1974, Fulling *et al* 1974, Hu and Parker 1978, Hartle and Hu 1979, 1980). These studies have been dynamical in nature, assuming a set of initial conditions involving anisotropic expansion and showing by numerical or other means that the dynamical evolution of the system rapidly isotropizes the expansion.

Here I will briefly discuss the theory of particle production by the expansion of the universe and will consider some of its implications. Rather than following the detailed dynamical evolution of the system, I will make use of a postulate concerning the equilibrium state toward which the system tends. In this way, one preserves the essence of the dynamical results without commiting oneself to a detailed theory of the underlying structure of the very early universe or to any particular initial conditions. In addition, such an approach seems to offer a possible explanation of why the magnitude of the cosmological constant is so small.

Section 2 contains a description of the use of Bogolubov transformations to analyze particle production by an expanding universe. Section 3 is devoted to the anomalous trace of the stress tensor and its relation to particle creation. Section 4 is concerned with the creation of gravitons on a Robertson–Walker background metric. Finally, Section 5 consists of a brief discussion of the reaction back of particle creation on the expansion of the universe. This is followed by a suggestion that the universe tends to a state in which the production of the dominant type (i.e., relativistic or non-relativistic) of particle present in the universe is suppressed, or in which there is a well defined physical vacuum state. This hypothesis seems to give a possible explanation of why the magnitude of the cosmological constant is observed to be so small.

2. BOGOLUBOV TRANSFORMATION APPROACH TO PARTICLE CREATION

For simplicity, consider a Robertson–Walker universe with line element

$$ds^2 = -dt^2 + a^2(t)d\sigma^2 \tag{2.1}$$

where $d\sigma^2 = \hat{g}_{ij}dx^i dx^j$ is the line element of a space of constant curvature. A free scalar field propagating in this space–time satisfies the field equation,

$$\nabla^\mu \nabla_\mu \phi - (m^2 + \xi R)\phi = 0 \tag{2.2}$$

where ∇_μ is the covariant derivative, R is the scalar curvature, and ξ is a dimensionless constant ('minimal coupling' corresponds to $\xi = 0$ and 'conformal coupling' to $\xi = \frac{1}{6}$). One may write

$$\phi = G(x)\psi(t) \tag{2.3}$$

where

$$\hat{g}^{lm}\hat{\nabla}_l\hat{\nabla}_m G = -k^2 G. \tag{2.4}$$

The eigenvalues k^2 range continuously from 0 to ∞ when the 3-space is flat, and from 1 to ∞ when the spatial curvature is negative. For positive spatial curvature, $k^2 = n^2 - 1$, with n ranging over the positive integers from 1 to ∞. The function $\psi(t)$ satisfies the equation

$$a^{-3}\frac{\mathrm{d}}{\mathrm{d}t}(a^3\frac{\mathrm{d}}{\mathrm{d}t}\psi) + (a^{-2}k^2 + m^2 + \xi R)\psi = 0. \tag{2.5}$$

The general Hermitian operator solution of equation (2.2) can be written as

$$\phi = \sum_k [A_k G_k(x)\psi_k(t) + \text{HC}] \tag{2.6}$$

where HC denotes the Hermitian conjugate of the previous term, the index k labels the modes, and Σ refers to integration over continuous values and summation over discrete values. The properties of the G_k are described in appendix A of Parker and Fulling (1974). According to equation (2.5), one may require that

$$\psi \mathrm{d}\psi^*/\mathrm{d}t - \psi^*\mathrm{d}\psi/\mathrm{d}t = ia^{-3} \tag{2.7}$$

where the factor of i on the right has been chosen so that the canonical commutation relations of the field ϕ and its conjugate momentum imply the usual commutation relations for the creation operators $A^+{}_k$ and annihilation operators A_k:

$$[A_k, A_{k'}] = 0, \qquad [A_k, A^+{}_{k'}] = \delta(k, k'). \tag{2.8}$$

When the scale factor, $a(t)$, in equation (2.1) is constant, there are natural positive and negative frequency solutions of equation (2.5). If $a(t)$ has the constant value a, then $R = 6\varepsilon/a^2$, where $\varepsilon = 1$, -1, or 0, corresponding to the sign of the spatial curvature. Then a suitable positive frequency solution is

$$\psi_k^{(+)} = (2a^3 W_k)^{-1/2} \exp(-iW_k t) \tag{2.9}$$

where $W_k = [m^2 + a^{-2}(k^2 + 6\xi\varepsilon)]^{1/2}$. The operators A_k multiplying these $\psi_k^{(+)}$ in equation (2.6) correspond to physical particles. In particular, there is a well defined physical vacuum state annihilated by the A_k.

In general, when $a(t)$ is not constant, there does not exist a unique physical vacuum state corresponding to the absence of real particles. That can be proved by considering an asymptotically static expansion of the universe in which $a(t)$ is constant at early times, changes in an arbitrary manner, and again becomes constant at late times. A positive frequency solution at early times, $\psi_k^{(+)\text{in}}$, will evolve according to equation (2.5) into a superposition of positive and negative frequency solutions at late times:

$$\psi_k^{(+)\text{in}} = \alpha_k \psi_k^{(+)\text{out}} + \beta_k \psi_k^{(-)\text{out}}. \tag{2.10}$$

Here $\psi_k^{(+)\text{in}}$ is the late time solution of equation (2.5) into which the early time positive frequency solution has evolved, the $\psi_k^{(\pm)\text{out}}$ refer to positive and negative frequency solutions at late times, and α_k and β_k are complex constants which depend on the detailed evolution of $a(t)$.

If one uses the $\psi_k^{(+)\text{in}}$ in the expansion of ϕ, equation (2.6), then the corresponding annihilation operators A_k^{in} correspond to physical particles at early times. Similarly, if one uses the $\psi_k^{(+)\text{out}}$, then the corresponding annihilation operators A_k^{out} correspond to physical particles at late times. Comparing the two expressions for ϕ and using equation (2.10), one finds that

$$A_k^{\text{out}} = \alpha_k A_k^{\text{in}} + \beta_k^* A_k^{+\text{in}}. \tag{2.11}$$

The state $|0\text{ in }\rangle$, containing no particles at early times, is annihilated by the A_k^{in}. However, if β_k in equation (2.11) is non-zero, then $|0\text{ in }\rangle$ is not annihilated by the A_k^{out}, so that the state $|0\text{ in }\rangle$ is not the same as the vacuum, $|0\text{ out }\rangle$, at late times. Thus, in a curved space–time there does not in general exist a unique physical vacuum state containing no real particles.

This lack of a unique vacuum is the essence of the phenomenon of particle creation in curved space–time. The number density of created particles in mode k is found to be proportional to $|\beta_k|^2$. The relation (given in equation (2.11)) between the creation and annihilation operators at early and late times is a Bogolubov transformation. In more complicated space–times, such as that of the mixmaster universe or that of a black hole, there is a mixing of modes, and the coefficients α and β have two indices, k and k'. The use of Bogolubov transformations to analyze particle creation in curved space–time was first carried out by Parker (1966, 1969, 1971). The technique was further developed and applied to additional space–times by Fulling (1972), Zeldovich and Starobinsky (1971), Hu *et al* (1973), Fulling *et al* (1974), Hawking (1974, 1975), DeWitt (1975), Rumpf and Urbantke (1978), Audretsch (1979), Charach and Parker (1981), and others.

Of primary interest to us here are the cases in which real particles are not created. These are the cases in which a unique physical vacuum state exists. When $m = 0$ and $\xi = \frac{1}{6}$ in equation (2.5), one can find a unique positive frequency solution for all forms of the scale factor $a(t)$. In that case, the function

$$\psi_k^{(+)} = [2a(t)^3 W_k(t)]^{-1/2} \exp\left(-i \int^t W_k(t')dt'\right) \tag{2.12}$$

with $W_k(t) = (k^2 + \varepsilon)^{1/2} a^{-1}$, is an *exact* solution of equation (2.5). (In that equation, $R = 6a^{-2}(\dot a^2 + a\ddot a + \varepsilon)$.) Furthermore, for an arbitrary asymptotically static expansion of the universe, this solution (to within a constant phase) has the positive frequency form of equation (2.9) both at early and at late times. Therefore, β_k in equation (2.11) is zero, and no real particles are created. Because the function $a(t)$ is arbitrary at intermediate times, this absence of real particle creation can not be a mere interference effect, and one can conclude that the creation rate of real particles vanishes for all scale factors $a(t)$, including those that are not asymptotically static. Hence, for the free massless field with $\xi = \frac{1}{6}$, the solution given by equation (2.12) is a positive frequency solution at all times and for all forms of $a(t)$. The unique physical vacuum state is the state annihilated by the operators A_k which multiply this positive frequency solution in the expression for ϕ.

In §3, the same result will be obtained from the renormalized energy–momentum tensor. From that discussion, it will be clear that the renormalization of infinites which appear during the expansion of the universe does not alter the above result for the free field. After renormalization, one finds that virtual pairs will be present when $a(t)$ is changing. However, these virtual pairs do not accumulate; no real particles are created (Parker 1979). These virtual particles may excite a particle detector when $a(t)$ is changing, but are not present when $a(t)$ is constant (Birrell and Davies 1982).

In the maximally symmetric de Sitter space–time, which can be represented in more than one way as a Robertson–Walker universe, geodesic worldlines possess event horizons. Gibbons and Hawking (1977) showed that a particle detector on such a worldline will detect an isotropic flux of blackbody radiation coming from the event horizon of the geodesic. Because of the existence of the event horizon it is energetically possible for the detector to be excited while a real particle appears on the other side of the event horizon. No energy need come in this case from a device carrying the particle detector (in contrast, for example, to an accelerated detector in Minkowski space (Unruh 1976)). Two detectors on geodesics in de Sitter space moving relative to one another will both detect *isotropic* blackbody radiation. Therefore, they can not both be detecting the same radiation. It seems to me that one must conclude that the presence of the detector catalyzes the production of pairs, one member of which excites the detector and the other member of which appears on the other side of the event horizon (Parker 1977). The detector is interacting with the vacuum (Candelas and Raine 1975), which has no real particles in it before the detector or some excitable body is present. Because of the presence of event horizons for geodesic worldlines, de Sitter space is evidently less stable, as far as the spontaneous production of real particles is concerned, than are space–times without event horizons.

There is an underlying symmetry behind the existence of a unique vacuum for the free massless scalar field with $\xi = \frac{1}{6}$ in a Robertson–Walker universe. Namely, the wave equation satisfied by ϕ is form invariant under a conformal transformation of the metric (Penrose 1964). If the metric $g_{\mu\nu}$ is multiplied by an arbitrary function $\Omega^{-2}(x)$ to give a new metric

$$\tilde{g}_{\mu\nu}(x) = \Omega^{-2}(x)g_{\mu\nu}(x) \tag{2.13}$$

and if the field ϕ is multiplied by Ω to give a new field $\tilde{\phi}(x) = \Omega(x)\phi(x)$, then the transformed field satisfies a wave equation of the same form as that satisfied by ϕ:

$$\tilde{\nabla}^\mu \tilde{\nabla}_\mu \tilde{\phi} - \tfrac{1}{6}\tilde{R}\tilde{\phi} = 0 \tag{2.14}$$

where $\tilde{\nabla}^\mu$ and $\tilde{R}$ are defined in terms of the transformed metric. For the Robertson–Walker metrics of equation (2.1), one can easily make a conformal transformation of the metric, equation (2.13), such that the new metric is static. Let $\Omega = a(t)$ and define a new time coordinate η such that $d\eta/dt = a^{-1}(t)$. Then the natural positive frequency solutions of equation (2.14), when transformed to the original space–time, agree with those found earlier in equation (2.12).

The two component neutrino equation, the massless Dirac equation, the Maxwell equations, and the Yang–Mills equations, when generalized to curved space–time by the minimal procedure of replacing ordinary by covariant derivatives, are all conformally invariant (see, for example, Parker 1982a, p 217 and Appendix). It can be shown that particles are not created in a Robertson–Walker universe for any of these free massless fields, with the possible exception of the Yang–Mills field. In the latter case, the self-interaction of the field would be expected to result in particle creation as has been shown to occur for the self interacting scalar field (see the next paragraph). However, the asymptotic freedom of the Yang–Mills field (Gross and Wilczek 1973, Politzer 1973, Toms 1982, Leen 1982) may imply that such particle creation is not significant in the very early universe, if the expansion is isotropic. For a cosmological metric which is anisotropically expanding, the conformal invariance of the field equations does not suppress particle creation.

If one adds a quartic self interaction, $\lambda\phi^4$, to the Lagrangian of the massless scalar field with $\xi = \frac{1}{6}$, then at the classical level the conformal invariance is preserved. However, at the quantum level, renormalization requires that counter terms be added to the Lagrangian of the interacting field which prevent one from carrying out the same argument as was used for the free field. Indeed, for the interacting field, it can be shown that real particles are created by the expansion of the universe (Birrell and Davies 1980, Birrell and Taylor 1980).

3.   STRESS TENSOR TRACE ANOMALY AND PARTICLE CREATION

The conformally invariant free scalar field satisfies the field equation,

$$\nabla_\mu \nabla^\mu \phi - \tfrac{1}{6} R \phi = 0. \tag{3.1}$$

Its energy–momentum tensor is

$$T_{\mu\nu} = \delta_\mu \phi \partial_\nu \phi - \tfrac{1}{2} g_{\mu\nu} \partial_\alpha \phi \partial^\alpha \phi + \tfrac{1}{6} \left[g_{\mu\nu} \nabla_\alpha \nabla^\alpha (\phi^2) - \nabla_\mu \nabla_\nu (\phi^2) + G_{\mu\nu} \phi^2 \right] \tag{3.2}$$

where $G_{\mu\nu} = R_{\mu\nu} - \tfrac{1}{2} g_{\mu\nu} R$. At the classical level, it follows from equation (3.1) that

$$\nabla_\nu T^{\mu\nu} = 0 \tag{3.3}$$

and

$$T_\mu{}^\mu = 0. \tag{3.4}$$

At the quantum level, expectation values of expressions like $T_{\mu\nu}$, which involve products of fields at the same point, are infinite and must be renormalized. When that is carried out, one finds that the covariant conservation law of equation (3.3) remains valid, but that the expectation value of the trace of the renormalized stress tensor is not zero. As this is contrary to the classical result, the non-vanishing trace of the renormalized stress tensor is known as the trace anomaly. It is independent of the state of the field. Any real particles which are present behave classically, contributing nothing to the trace of the stress tensor.

It can be shown that the anomalous trace is proportional to the coefficient of the second-order term of the Schwinger–DeWitt proper time expansion of the propagator (DeWitt 1964). One of the earliest works on renormalization in curved space–time is Utiyama and DeWitt (1962). The principal techniques which have been used to calculate the anomalous trace of the stress tensor are dimensional regularization, point-splitting regularization, zeta function regularization, adiabatic regularization, and Pauli–Villars regularization. A complete list of references can be found in Birrell and Davies (1982). Here I will only summarize the results obtained in the Robertson–Walker universes of equation (2.1).

For the above scalar field one finds that

$$\langle T_\mu{}^\mu \rangle = (4\pi)^{-2}(-360^{-1}G + 180^{-1}\nabla_\mu\nabla^\mu R) \tag{3.5}$$

where

$$G = R_{\alpha\beta\gamma\delta}R^{\alpha\beta\gamma\delta} - 4R_{\alpha\beta}R^{\alpha\beta} + R^2 \tag{3.6}$$

is a topological invariant in four dimensions. In the coordinates of equation (2.1), with $\varepsilon = \pm 1$, or 0, one has

$$G = 24(a^{-3}\dot{a}^2\ddot{a} + \varepsilon a^{-3}\ddot{a}) \tag{3.7}$$

and

$$\nabla_\mu\nabla^\mu R = 6(-a^{-1}\dddot{\ddot{a}} - 3a^{-2}\dot{a}\dddot{a} + 5a^{-3}\dot{a}^2\ddot{a} - a^{-2}\ddot{a}^2 + 2\varepsilon a^{-3}\ddot{a}). \tag{3.8}$$

Similarly, the expressions for the anomalous trace of the two component neutrino and photon energy–momentum tensors in the Robertson–Walker universes have the form

$$\langle T_\mu{}^\mu \rangle = (4\pi)^{-2}(BG + C\nabla_\mu\nabla^\mu R) \tag{3.9}$$

where B and C are constants (the specific values are not required below).

The anomalous trace of the stress tensor can be used to show that for the above free conformally invariant fields there is no creation of real particles by an isotropically expanding universe (Parker 1979). In the metric $g_{\mu\nu}$ of equation (2.1), the vector $\xi_\mu = -a(t)\delta_\mu{}^0$ satisfies the equation

$$\nabla_\mu\xi_\nu + \nabla_\nu\xi_\mu = a(t)g_{\mu\nu}. \tag{3.10}$$

(It is known as a conformal Killing vector field.) From the symmetry of $\langle T^{\mu\nu} \rangle$ one finds that

$$\nabla_\mu[\langle T^{\mu\nu} \rangle\xi_\nu] = \dot{a}(t)\langle T_\mu{}^\mu \rangle. \tag{3.11}$$

Integrating over the four volume bounded by constant time hypersurfaces at t_1 and t_2 and using the spatial homogeneity, one obtains the following relation between the proper energy density, $\rho = \langle T_{00} \rangle$, and the anomalous trace of the stress tensor

$$\rho(t_2)a^4(t_2) - \rho(t_1)a^4(t_1) = -\int_{t_1}^{t_2} dt\, a^3\dot{a}\langle T_\mu{}^\mu \rangle. \tag{3.12}$$

The general expression for $\langle T_\mu{}^\mu \rangle$ is given by equations (3.7)–(3.9). One finds that the integration can be explicitly carried out for arbitrary $a(t)$, with the result

$$\int_{t_1}^{t_2} dt\, a^3 \dot{a} \langle T_\mu{}^\mu \rangle = g(t_2) - g(t_1) \tag{3.13}$$

where

$$g(t) = 6(4\pi)^{-2}\left[B(\dot{a}^4 + 2\varepsilon \dot{a}^2) + C(-a^2\dot{a}\,\dddot{a} - a\dot{a}^2\ddot{a} + \tfrac{1}{2}a^2\ddot{a}^2 + \tfrac{3}{4}\dot{a}^4 + \varepsilon \dot{a}^2)\right]. \tag{3.14}$$

From equations (3.12) and (3.13) one obtains the explicit expression for the energy density

$$\rho(t) = -a^{-4}(t)g(t) + a^{-4}(t)E \tag{3.15}$$

where E is a constant that depends on the state of the field and on the global properties of the space–time. The pressure, $p = \langle T_i{}^i \rangle$ (no sum), is given by

$$p = \tfrac{1}{3}(\rho + \langle T_\mu{}^\mu \rangle). \tag{3.16}$$

The terms involving E in the energy density and pressure have the same form as for a classical gas of massless particles, and include the contribution of any real particles which may be present. The terms coming from $g(t)$ are independent of the state, and depend on the instantaneous values of $a(t)$ and its derivatives. Furthermore, those terms vanish during any period when $a(t)$ is constant. They can not correspond to the accumulation of real particles created by the gravitational field. If no real particles are present initially, then they are absent at all times. Thus, free conformal scalar particles, neutrinos, and photons are not created by an isotropically expanding universe. This is consistent with the results of the previous section.

If one adds a $\lambda\phi^4$ interaction to the Lagrangian of the scalar field, then a term proportional to R^2 appears in the expression for $\langle T_\mu{}^\mu \rangle$ corresponding to equation (3.9) (Birrell and Davies 1980). This term prevents one from carrying out an explicit integration as in equation (3.13), so that the result involves an integral which depends on the history of $a(t)$ and does not vanish during periods when $a(t)$ is constant. One concludes that renormalization of this interacting field theory in the curved space–time introduces counter terms which cause particles to be created and destroy the uniqueness of the vacuum state.

4. GRAVITON PRODUCTION IN ROBERTSON–WALKER UNIVERSES

Consider the classical Einstein equations

$$R_{\mu\nu} - \tfrac{1}{2}g_{\mu\nu}R + \Lambda g_{\mu\nu} = 8\pi G T_{\mu\nu} \tag{4.1}$$

where $T_{\mu\nu}$ is now the classical stress tensor of a perfect fluid having four velocity u^μ:

$$T_{\mu\nu} = (\rho + p)u_\mu u_\nu + p g_{\mu\nu}. \tag{4.2}$$

Here ρ and p refer to the classical proper energy density and pressure of the fluid, and no longer refer to the expectation values of the previous section. The constant Λ is the cosmological constant. To obtain the gravitational wave perturbations, one perturbs the Einstein equations without disturbing the matter, that is, one lets $\delta\rho = \delta p = \delta u^\mu = 0$ (Lifshiftz 1946). Let the metric perturbations $\delta g_{\mu\nu}$ be denoted by $h_{\mu\nu}$. In the Robertson–Walker metrics, one can impose the gauge conditions $u^\mu h_{\mu\nu} = 0$ and $h^{\mu\nu}{}_{;\nu} = 0$. It can be shown that $h_\mu{}^\mu$ also vanishes. The two components of $h_{\mu\nu}$ which are independent correspond to the two polarizations of a gravitational wave. In the coordinates of equation (2.1), only the spatial components $h_i{}^j$ do not vanish. They satisfy the perturbation equation (Lifshitz 1946, Grishchuk 1974, Ford and Parker 1977)

$$a^{-3}\partial_t(a^3\partial_t h_i{}^j) - a^{-2}\hat{g}^{lm}\hat{\nabla}_l\hat{\nabla}_m h_i{}^j + 2\varepsilon a^{-2}h_i{}^j = 0. \tag{4.3}$$

It will be recalled that $\hat{g}_{ij}$ is defined after equation (2.1) and that ε is ± 1 or 0, depending on the sign of the spatial curvature. Note that the cosmological constant does not appear in the perturbation equation.

Let

$$h_i{}^j = G_i{}^j(x)\psi(t) \tag{4.4}$$

where the $G_i{}^j$ satisfy the equations

$$\hat{g}^{lm}\hat{\nabla}_l\hat{\nabla}_m G_i{}^j = -(k^2 - 2\varepsilon)G_i{}^j \qquad \hat{\nabla}_j G_i{}^j = 0 \qquad G_i{}^i = 0. \tag{4.5}$$

For positive or negative spatial curvature ($\varepsilon = \pm 1$) these functions are tensor spherical harmonics (Lifshitz 1946). In the case of zero curvature ($\varepsilon = 0$), they may be taken to be plane waves. The eigenvalues k^2 range continuously from 0 to ∞ when $\varepsilon = 0$ and from 1 to ∞ when $\varepsilon = -1$. For $\varepsilon = 1$, one has $k^2 = n^2 - 1$ with n ranging over the positive integers from 3 to ∞. It follows that the time dependent function ψ in equation (4.4) satisfies the equation

$$a^{-3}d(a^3 d\psi/dt)/dt + a^{-2}k^2\psi = 0. \tag{4.6}$$

This is the same as equation (2.5) with $m = 0$ and $\xi = 0$. The time-dependent part of the graviton field in a Robertson–Walker universe obeys the same equation as does the time-dependent part of the minimally coupled massless scalar field. The quantized graviton field behaves like a pair of minimally coupled quantized scalar fields, one field corresponding to each polarization of the gravitational wave (Grishchuk 1974, Ford and Parker 1977). Thus, the production of minimally coupled scalar particles and of gravitons in a Robertson–Walker universe can be dealt with by the same methods.

In general, equation (4.6) causes a mixing of positive and negative frequencies, as explained in §2, so that particles are created and the vacuum state is not unique. Unlike the conformally invariant free fields considered earlier, the creation rate of real gravitons will vanish in the Robertson–Walker universes only for particular scale factors $a(t)$. For the discussion below, I am interested in the cases in which there is no production of minimally coupled scalar particles and gravitons. The free scalar field satisfying the conformally invariant equation (3.1) has well defined positive frequency solutions, given in equation (2.12), for arbitrary $a(t)$. For this conformal scalar field there is no creation of real particles and there exists a unique vacuum state. In the cases when $a(t)$ is such that $R = 0$, then equation (4.6) is the same equation as is satisfied by the time-dependent part of the conformal scalar fields. Hence, there is a solution of equation (4.6) having at all times the positive frequency form given by equation (2.12). I interpret that as implying that real minimally coupled massless scalar particles and gravitons are not created in the Robertson–Walker universes with $R = 0$. Because the time-development of the minimally and conformally coupled scalar fields is identical in this case, one would expect a particle detector on a geodesic to have the same response to each field, corresponding to zero creation rate of real particles. Similarly, each polarization of the graviton field develops in time as does a minimally coupled scalar field, so that one also expects the creation rate of real gravitons to vanish when $R = 0$. (A detector in the above cases may still respond to vacuum fluctuations which do not accumulate as do real particles.)

Thus, it appears that in the Robertson–Walker universes with $R = 0$ there exists a unique physical vacuum state for free massless scalar particles and gravitons. It is interesting that for the scalar field with quartic self-interaction, the term shown by Birrell and Davies (1980) to cause particle creation is proportional to R^2 and also vanishes in this case.

5.   EFFECTS OF BACK-REACTION

A number of studies of the back-reaction of particle production in cosmological space–times have been carried out. These studies have generally been based on the Einstein equations with a source consisting of the created particles as well as any background matter which may be present. One approach is to take the particle production into account by including in the source term the expectation value of the energy–momentum tensor of the quantum field (Zeldovich and Starobinsky 1971, Lukash and Starobinsky 1974, Parker and Fulling 1973, Fulling *et al* 1974, Hu and Parker 1977, 1978, Berger 1982). Another approach is to use an effective action which includes the Einstein–Hilbert action as well as terms coming from the matter quantum fields (Hartle 1977, 1981a,b, Hartle and Hu 1979, 1980, Hu 1980, DeWitt 1981b).

These calculations have been mainly concerned with the influence of back-reaction on an initially anisotropic expansion of the universe and on the cosmological singularity. Although it has been demonstrated that it is possible for quantum matter effects to avoid the cosmological singularity (Parker and Fulling 1973), it is not yet clear how general such effects are (DeWitt 1981a,b, Berger 1982). Concerning the effect on an initially anisotropic expansion of the universe, both methods of calculating the back-reaction agree that the expansion is rapidly isotropized by particle creation occurring at sufficiently early times. For example, for a Bianchi type I metric of the form

$$ds^2 = -dt^2 + a_1{}^2(t)dx^2 + a_2{}^2(t)dy^2 + a_3{}^2(t)dz^2 \tag{5.1}$$

if one starts the system at a time t_0 near the Planck time, then the expansion is isotropized very rapidly. This isotropization is an important result, which gives a viable dynamical explanation of the observed isotropy of the expansion.

This damping of anisotropy can be understood as a consequence of an hypothesis formulated some time ago (Parker 1969). Namely, 'that the reaction of the particle creation back on the gravitational field will modify the expansion in such a way as to reduce the creation rate'. This is a kind of quantum gravitational Lenz's law, according to which an effect acts in such a way as to oppose its cause. In the very early universe, particle masses which are small with respect to the Hubble parameter can be neglected. In §2 and §3, it was proved that spin-1/2 and spin-1 massless particles, as well as scalar particles obeying the conformally invariant free-field equation, are not created by *isotropic* expansion of the universe. Therefore, my hypothesis implies that if the expansion of the universe is initially anisotropic, then the back-reaction resulting from the creation of such particles will bring about isotropic expansion, thus reducing the creation rate of those particles to zero. Near the Planck time, when the creation rate would be very large, the back-reaction would be expected to act rapidly in bringing about isotropy. Furthermore, it has been shown (Hartle 1981b), using the effective action approach in a radiation dominated Friedmann universe, that the back-reaction of the created particles sufficiently softens the behavior near the cosmological singularity as to remove an infinity which would otherwise be present in the particle creation.

There seems to be a deep connection between the Einstein equations and the conditions for the particle creation rate to vanish, or equivalently, for a well defined physical vacuum state to exist. It has been shown (Parker 1969) that in a Friedmann universe with flat 3-space containing only massless particles in equilibrium, there will be precisely no creation of massless spin-0 particles. Similarly, for flat 3-space, it was shown that the creation rate of massive minimally coupled scalar particles (with mass large with respect to the Hubble parameter) is most strongly suppressed in a dust-filled Friedmann universe. Here, Friedmann universe refers to solutions of the Einstein equations with zero cosmological constant. Thus, it appears that the assumption (Parker 1969) that the universe evolves towards a state in which the particle creation rate vanishes, or in which there is a well defined physical vacuum state, implies that the cosmological constant is zero. Adler (1982) has noted that, within the context of the theory of induced gravitation, my hypothesis can serve as a reason why the cosmological constant vanishes.

Without making a commitment to any particular underlying theory or 'picture' of the very early universe prior to the time when the classical Einstein equations become well defined, I will assume that the absence of particle creation, in particular, of the creation of minimally coupled scalar particles and gravitons, is a kind of equilibrium condition towards which the evolution of the very early universe tends, and that the classical Einstein equations must be consistent with that condition. Another way of stating this is that the underlying fully unified theory which determines the form of the classical Einstein equations, including the value of the cosmological constant, is such that *at the time in the evolution of the universe when the classical Einstein equations become well defined as governing the macroscopic evolution of the universe, those equations must permit an expansion law in which there is a well defined physical vacuum state.* Such an hypothesis has the advantage of avoiding questions about initial conditions or the possible existence of a cosmological singularity. We have already seen how the absence of particle creation implies that the expansion is isotropic. Let me now show how the above hypothesis also implies that the cosmological constant Λ is zero (Parker 1982b,c).

For the creation rate of gravitons and minimally coupled scalar particles to vanish, the expansion must be isotropic. Therefore, consider the Robertson–Walker metrics of equation (2.1), and the classical Einstein equations with cosmological constant, as given in equations (4.1) and (4.2). It was shown in §4 that the condition for the creation rate of the above particles to vanish is that $R = 0$. To see what that implies about the value of Λ, contract the Einstein equation (4.1) and set $R = 0$. One finds that

$$\Lambda = 2\pi G (3p - \rho) \tag{5.2}$$

where p and ρ are the pressure and energy density of the classical matter. Therefore, if the matter in the early universe consists of relativistic particles and radiation, with equation of state,

$$p = \rho/3 \tag{5.3}$$

then the cosmological constant must vanish,

$$\Lambda = 0. \tag{5.4}$$

If one assumes that the equation of state of the classical matter is such that p vanishes when ρ vanishes, then both equations (5.3) and (5.4) follow from equation (5.2). Thus, the existence of a well defined vacuum state for gravitons and minimally coupled scalar particles in the early universe implies that the cosmological constant is zero.

The possibility, discussed in §2, that de Sitter space has a vacuum state which is more symmetric but is unstable to the production of particles, raises the possibility of connecting the above ideas with an inflationary universe model (Guth 1981, Brout *et al* 1980). The universe may emerge into a de Sitter-like inflationary stage which then undergoes a phase transition to the more stable radiation dominated expansion with zero cosmological constant. Such a model would also explain why the universe is nearly spatially flat.

It is hoped that considerations such as those described here may ultimately be of value in helping to guide us toward a unified theory which includes gravitation.

ACKNOWLEDGMENT

This work was supported in part by the National Science Foundation.

REFERENCES

Adler S L 1982 *Rev. Mod. Phys.* **54** 729
Audretsch J 1979 *J. Phys. A: Math. Gen.* **12** 1189
Berger B K 1982 *Phys. Lett.* **108B** 394
Birrell N D and Davies P C W 1980 *Phys. Rev. D* **22** 322
—— 1982 *Quantum Fields in Curved Space* (Cambridge: Cambridge University Press)
Birrell N D and Taylor J G 1980 *J. Math. Phys.* **21** 1740
Brout R, Englert F, Frere J-M, Gunzig E, Nardone P and Truffin C 1980 *Nucl. Phys. B* **170** 228
Candelas P and Raine D J 1975 *Phys. Rev. D* **12** 965
Charach Ch and Parker L 1981 *Phys. Rev. D* **24** 3023
DeWitt B S 1964 *Relativity, Groups, and Topology* ed B S DeWitt and C DeWitt (New York: Gordon and Breach)
—— 1967 *Phys. Rev.* **162** 1195 and 1239
—— 1975 *Phys. Rep.* **19C** 297
—— 1981a *Quantum Gravity* 2 ed C J Isham, R Penrose and D W Sciama (Oxford: Oxford University Press) p 486
—— 1981b *Phys. Rev. Lett* **47** 1647
Ford L H and Parker L 1977 *Phys. Rev. D* **16** 1601
Fulling S A 1972 *PhD Thesis* Princeton University
Fulling S A, Parker L and Hu B L 1974 *Phys. Rev. D* **10** 3905
Gibbons G W 1979 *General Relativity: An Einstein Centenary Survey* ed S W Hawking and W Israel (Cambridge: Cambridge University Press) pp 639–79
Gibbons G W and Hawking S W 1977 *Phys. Rev. D* **15** 2738
Grishchuk L P 1974 *Zh. Eksp. Teor. Fiz.* **67** 824 (*Sov. Phys.–JETP* **40** 409 (1975))
Gross D and Wilczek F 1973 *Phys. Rev. Lett.* **26** 1343
Guth A 1981 *Phys. Rev. D* **23** 347
Hartle J B 1977 *Phys. Rev. Lett.* **39** 1373
—— 1981a *Quantum Gravity* 2 ed C J Isham, R Penrose and D W Sciama (Oxford: Oxford University Press) pp 313–28
—— 1981b *Phys. Rev. D* **23** 2121
Hartle J B and Hu B L 1979 *Phys. Rev. D* **20** 1772
—— 1980 *Phys. Rev. D* **21** 2756
Hawking S W 1974 *Nature* **248** 30
—— 1975 *Commun. Math. Phys.* **43** 199
Hu B L 1980 *Recent Developments in General Relativity* ed R Ruffini (Amsterdam: North Holland)
Hu B L, Fulling S A and Parker L 1973 *Phys. Rev. D* **8** 2377
Hu B L and Parker L 1977 *Phys. Lett.* **63A** 217
—— 1978 *Phys. Rev. D* **17** 933
Leen T K 1982 *Ann. Phys., NY* (in press)
Lifshitz E M 1946 *Zh. Eksp. Teor. Fiz.* **16** 587 (also *J. Phys. USSR* **10** 116)
Lukash V N and Starobinsky A A 1974 *Zh. Eksp. Teor. Fiz.* **66** 1515 (*Sov. Phys.–JETP* **39** 742)
Parker L 1966 *PhD Thesis* Harvard University
—— 1969 *Phys. Rev.* **183** 1057
—— 1971 *Phys. Rev. D* **3** 346
—— 1977 *Asymptotic Structure of Space–Time* ed F P Esposito and L Witten (New York: Plenum) pp 107–226

—— 1979 *Recent Developments in Gravitation: Cargese* 1978 ed S Deser and M Levy (New York: Plenum) pp 219–73

—— 1982a *Fundam. of Cosmic Phys.* **7** 201–39

—— 1982b *Symposium on Gauge Theories and Gravitation* (in press)

—— 1982c *Third Marcel Grossmann Meetings* (in press)

Parker L and Fulling S A 1973 *Phys. Rev.* D **7** 2357

——1974 *Phys. Rev.* D **9** 341

Penrose R 1964 *Relativity, Groups and Topology* ed B S DeWitt and C DeWitt (New York: Gordon and Breach)

Politzer H D 1973 *Phys. Rev. Lett.* **26** 1346

Rumpf H and Urbantke H K 1978 *Ann. Phys., NY* **114** 332

Toms D J 1982 *Phys. Rev.* D **26** 2713

Unruh W G 1976 *Phys. Rev.* D **14** 870

Utiyama R and DeWitt B S 1962 *J. Math. Phys.* **3** 608

Zeldovich Ya B 1970 *Zh. Eksp. Teor. Fiz. Pis. Red.* **12** 443 (*JETP Lett.* **12** 307)

Zeldovich Ya B and Starobinsky A A 1971 *Zh. Eksp. Teor. Fiz.* **61** 2161 (*Sov. Phys.–JETP* **34** 1159)

Vacuum Energy in a Squashed Einstein Universe

J S DOWKER

This article is dedicated to Bryce DeWitt on the occasion of his 60th Birthday.

My first contact with Bryce's work was around 1963 when I came upon an incomplete set of his 1953 Les Houches lecture notes on quantum mechanics. They were very full and dealt with a number of non-standard topics that interested me such as quantization on curved spaces and the ordering problem. I began to study his 1957 *Reviews of Modern Physics* paper on functional integrals more closely. Then in 1964 his wide ranging 1963 Les Houches lectures appeared and have remained, for me and others, a standard reference for ideas and techniques.

Unlike Bryce I am not able to address myself significantly to the grander problems of physics and so I have taken refuge in a specific calculation. Since Bryce's work has always been very detailed I hope that this contribution will be seen as an appropriate one.

1.   INTRODUCTION

The picture of the vacuum as a gas of virtual particles behaving as a polarizable medium emerged with Dirac's hole theory and the formulation of a relativistic theory of interacting quantum fields (quantum electrodynamics) in the early nineteen thirties (Dirac 1934).

One important consequence is that even in 'freespace' Maxwell's equations have to be replaced by something more complicated. An external, unquantized electromagnetic field will polarize the vacuum and the field due to the induced vacuum current must be added to the impressed one self-consistently, producing an effective nonlinearity. This leads, for example, to the famous phenomenon of the scattering of light by light.

The neatest, formal description of this situation uses the concept of the effective action. This occurs already in the earliest works (Heisenberg and Euler 1936, Weisskopf 1936) for special situations and is described in detail and generality in many places (DeWitt 1964, Schwinger 1951a). Only a brief, schematic outline is needed here.

Let A represent the unquantized field and let $\hat{\psi}$ stand for a generic matter quantum field operator. Assume that the state we are interested in is the matter vacuum, $|0\rangle$, then the statement that the polarization of the vacuum acts as a source for A is written as

$$\frac{\delta S[A]}{\delta A} = -\left\langle 0 \left| \frac{\delta S_M[A,\hat{\psi}]}{\delta A} \right| 0 \right\rangle \tag{1.1}$$

where $S[A]$ is the action for the external field alone and S_M is the action for the matter field in the presence of A.

If a functional $W[A]$ can be found such that $\delta W[A] = \langle 0|(\delta S + \delta S_M)|0\rangle$ then it can be called the effective action because variation of it produces the A equation of motion (1.1).

The right-hand side of (1.1), which is the vacuum average of the source of the A field, is a functional of A possibly because the functional derivative is still explicitly A dependent (we are not particularly interested in this) but certainly because the matter vacuum $|0\rangle$ depends on A. The mode equations of the ψ field involve A, as an external field, and the mode decomposition determines the vacuum. (Note also that the modes depend on any boundary conditions so that the vacuum state $|0\rangle$ and the vacuum averages are truly *global* quantities. Many difficulties stem from trying to reconcile this with a local concept of particles based on detectors.)

Anyone can see where this development breaks down. Firstly, if the external field depends on time, the notion of a single, natural vacuum becomes impossible and one has to resort to the 'in' and 'out' formalism (see the previous references and the review works mentioned later). Since the forthcoming calculation involves a static background we ignore this complication.

Secondly, an unquantized external field does not exist in nature, at least not in the sense we have in mind here. One should really start from the complete theory where A is quantized and then appeal to standard discussions of the external field approximation (e.g. Jauch and Rohrlich 1955, chapter 15, Bogoliubov and Shirkov 1959). For all aspects of the external field problem we refer to the book by Grib *et al* (1980) and to the reviews by Rafelski *et al* (1978) and Soffel *et al* (1982).

However, another point of view is possible. A can be considered as a mathematical object introduced to obtain the vacuum's response to an external stimulus. Such is the role of A in the background field method (DeWitt 1964) which is closely related to Feynman's approach to quantum electrodynamics (Feynman 1950 §8) and to Schwinger's external source theory (Schwinger 1951b).

For example, if the Green's function for ψ is known as a functional of the external A, the exact one with all radiative corrections follows upon a functional integration over A (see e.g. Bogoliubov and Shirkov 1959).

In Feynman diagram language it is clear that only diagrams with no external ψ lines, and of course no internal A lines, contribute to the right-hand side of (1.1).

The only possibility is a single ψ loop with an arbitrary number of external A lines. This corresponds to integrating out the ψ field in the relevant functional integral with ψ vacuum boundary conditions.

Another justification for the study of quantum fields propagating in a classical background can be found in nonlinear field theories such as Yang–Mills, gravitation and the σ model. After Euclideanization it often appears that the functional integrals are dominated by solutions of the classical equations of motion (instantons). One then corrects by considering perturbations about these solutions. The linear perturbation level is equivalent to a free-field propagating in the classical background and the one-loop approximation. Also of course other fields may be involved and there is a big industry working out the theory of propagation in various instanton backgrounds (e.g. Brown and Creamer 1978, Berg and Lüscher 1979, Corrigan *et al* 1978, Schwarz 1979).

When the Euclidean theory has a direct physical relevance the instanton solutions can possess a certain physical reality. For example, in a statistical physics context they can correspond to disclinations.

Consider now the case when the external field is a gravitational one. The response of the system to variations in the metric gives the energy momentum tensor $T_{\mu\nu}$ of the ψ field and the quantity of interest will be its vacuum average. Since the gravitational field is not quantized this will lead only to a truncated and approximate theory which will not possess, nor reflect, all the properties of a fully quantized version[†] (whatever this might be) (Duff 1980). Nevertheless the one loop results are of interest, not least because they are accessible.

The formal vacuum average in (1.1) is infinite when evaluated. Various techniques have been evolved for handling this, the earliest being point-splitting (Dirac 1934, Heisenberg 1934 and Heisenberg and Euler 1936). The divergences were rendered finite and made subject to normal algebraic manipulation in the intermediate calculation. It was noticed that certain infinities could be removed by charge symmetrization (in quantum electrodynamics), others by straight subtraction of the non-interacting result and the remainder by wavefunction and electric charge renormalization, as we would now say. (Remember all this was before a systematic theory of renormalization.)

The point-splitting technique was taken up later in the gravitational context and combined with the proper time method (Fock 1937, Nambu 1950, Schwinger 1951a) by a number of workers to yield a comprehensive procedure for analyzing the divergences (DeWitt 1964, Christensen 1976, Davies *et al* 1977). However, the renormalization is now more contentious. Quadratic curvature terms occur which are not present in the original Einstein action (e.g. Utiyama and DeWitt 1962). Moreover, and relatedly, quantized general relativity is not renormalizable and this throws the whole procedure into doubt.

All this, and more, is well known. For an expansion of the points raised above and much else we refer to the review works by DeWitt (1979, 1975), Parker (1979),

[†] This is of course true for any field theory in which one field is treated as unquantized and certainly there are peculiarities, if not inconsistencies, in the back-reaction approach.

Gibbons (1979), Boulware (1979), and to the book by Birrell and Davies (1982).

Despite the many question marks it is intended in the present article to evaluate the renormalized vacuum energy for the massless spin-1/2 (neutrino) field in a frozen Mixmaster universe. Some remarks on the photon case are also made.

The value of this exercise is partly that it illustrates a number of interesting technical points concerning the zeta function method and demonstrates that, despite a rather complicated eigenvalue formula, useful and precise results can be obtained. The calculation can be carried through exactly, up to a numerical integration, to yield the vacuum energy as a function of the deformation from the spherical case (the Einstein universe). In particular the vacuum energy can be negative even for small deformations.

One can think of the renormalized vacuum energy as a Casimir energy since the presence of the external field and the boundary conditions imply a modification of the flat space, Minkowski modes in straight analogy with the original Casimir effect (cf. Dewitt 1975 § 2, 1979 § 14.2).

2. SPIN-1/2 VACUUM ENERGY. GENERAL FORMULAE

We derive an expression for the vacuum energy in a static space–time in terms of a zeta function on the spatial section.

The metric is taken in the manifestly static form

$$ds^2 = g_{00}(x)\,dt^2 + g_{ij}(x)\,dx^i dx^j$$

and the covariant spin-1/2 equation reads

$$i\gamma^\mu \nabla_\mu \psi = 0$$

with ∇_μ the covariant derivative $\partial_\mu + \Gamma_\mu$. To begin with we have chosen a four component representation.

The classical spin-1/2 action is

$$S_M[\psi, g] = \int \bar{\psi} \gamma^\mu \nabla_\mu \psi \, (-g)^{1/2}\, d^4x$$

whence, after standard manipulations (cf DeWitt 1964, Problem 112)

$$\delta W^{(1)} = \langle 0 | \delta S_M[\psi, g] | 0 \rangle$$
$$= -i\mathrm{Tr}_4(G\delta G^{-1})$$

where G is the Feynman Green's function of the squared Dirac equation,

$$\gamma^\mu \gamma^\nu \nabla_\mu \nabla_\nu G(x, x') = \delta(x, x').$$

Tr_4 stands for a spinor trace and a covariant space–time integration in the convenient notational abbreviation extensively employed by DeWitt (e.g. 1964).

Because of the static nature of the metric it is possible, and convenient, to remove a time integration and obtain an expression for the effective Lagrangian $L^{(1)}$

$$\delta L^{(1)} = -i\text{Tr}_3(G\delta G^{-1}) \tag{2.1}$$

with

$$W^{(1)} \equiv \int L^{(1)}\,dt.$$

The calculation proceeds as for the scalar case given by Dowker and Kennedy (1978). The right-hand side of (2.1) diverges and is rendered finite by replacing G by the matrix power G^s for s sufficiently positive. Then

$$\delta L^{(1)} = \lim_{s \to 1} \frac{i}{s-1} \text{Tr}_3(\delta G^{s-1})L^{-2(s-1)}$$

where we have introduced the necessary scaling length L.

The limit $s \to 1$ is achieved by continuing G^{s-1} into the complex s plane and down to $s = 1$. By definition the continuation yields the zeta function $\zeta_4(s-1)$ of the space–time operator G^{-1} and so, up to metric independent (topological) terms

$$\begin{aligned} L^{(1)} &= \lim_{s \to 1} \frac{i}{s-1} L^{-2(s-1)}\text{Tr}_3\zeta_4(s-1) \\ &= i\left[\lim_{s \to 1} \frac{1}{s-1} \text{Tr}_3\zeta_4(0) + \text{Tr}_3\zeta_4(0)\ln L^{-2} + \text{Tr}_3\zeta_4'(0) \right] \end{aligned} \tag{2.2}$$

which exhibits the divergence as a pole at $s = 1$. At some stage this must be removed by a renormalization.

A simple scaling argument, which exactly parallels that for the scalar case (Dowker and Kennedy 1978), shows that the total vacuum energy E is just the negative of $L^{(1)}$

$$E = -L^{(1)}. \tag{2.3}$$

E is defined as the spatial integral of the vacuum energy density

$$\begin{aligned} E &= \int \langle 0| \hat{T}_0{}^0 |0 \rangle \, (-g)^{1/2}\,d^3x \\ &= 2\int \frac{\delta L^{(1)}}{\delta g^{00}} g^{00}\,d^3x. \end{aligned}$$

Incidentally note that the simple result (2.3) depends on the fact that the space–time is static so that $\nabla_0 = \partial_0 - \frac{1}{8}[\gamma^i, \gamma^0]\partial_i g_{00}$. If a scalar potential V is present then there is an extra term because the potential does not scale. For example we find

$$E = \int \langle 0| V |0 \rangle \, (-g)^{-1/2}\,d^3x - L^{(1)} \tag{2.4}$$

which is Weisskopf's (1936) equation (8).

Problem 1

Show that the result (2.3) extends to finite temperatures with E replaced by F, the free energy. Prove (2.4) and extend it to finite temperatures.

The particular space–time we will be considering later is ultrastatic, i.e. $g_{00} = 1$, so attention is now restricted to this simplified situation.

The squared Dirac equation then reads

$$\left(\frac{\partial^2}{\partial t^2} + \gamma^i \gamma^j \nabla_i \nabla_j\right) G = 1$$

which enables us to relate the space–time zeta function, ζ_4, to that, ζ_3, on the spatial section:

$$(t, x', m' | \zeta_4(s) | t, x, m) = \frac{i}{(4\pi)^{1/2}} \frac{\Gamma(s - \frac{1}{2})}{\Gamma(s)} (x', m' | \zeta_3(s - \tfrac{1}{2}) | x, m) \qquad (2.5)$$

where m, m' are spinor indices. $\zeta_3(s)$ is the zeta function for the operator $\gamma^i \gamma^j \nabla_i \nabla_j$. Thus from (2.2), (2.3) and (2.5) we obtain our basic formula for computing E,

$$E = -\lim_{s \to 1} L^{-2(s-1)} \mathrm{Tr}_3 \zeta_3(s - \tfrac{3}{2}). \qquad (2.6)$$

Problem 2

Derive equation (2.6) by working at finite temperature and using $E = (\partial/\partial\beta)(\beta F)$ (cf Gibbons 1978).

$\mathrm{Tr}_3 \zeta_3(s)$ is given in terms of the eigenvalues, ω_n^2, of $\gamma^i \gamma^j \nabla_i \nabla_j$ by the continuation of the series

$$\sum_n \omega_n^{-2s}$$

from the region (here $\mathrm{Re}\, s > \frac{3}{2}$) where it converges.

One can see that (2.6) is reasonable because it is a regularized realization of the formal equality

$$E = -\sum_n |\omega_n|$$

and we recall the standard expression for the fermion vacuum energy

$$E = -(\Sigma \varepsilon_p - \Sigma \varepsilon_n) \qquad (2.7)$$

where $\varepsilon_p (\varepsilon_n)$ are the positive (negative) energies. (Energy signifies the eigenvalue of $i(\partial/\partial t) = -i\gamma^0 \gamma^i \nabla_i$.)

Equation (2.7) already includes the usual spin-$\frac{1}{2}$ degeneracy factor of 2.

It is convenient to escape from the four component formalism at this point by using the Weyl chiral split $\psi = (\varphi\chi)^T$ and

$$\gamma^i = i\begin{pmatrix} 0 & \sigma^i \\ -\sigma^i & 0 \end{pmatrix} \qquad \text{with} \quad \sigma^i(\sigma^j) = -g^{ij}$$

so that in flat space the σ^i are the usual Pauli matrices. The Dirac equation then separates into

$$\sigma^i\nabla_i\varphi = -\frac{\partial\varphi}{\partial t}, \qquad \sigma^i\nabla_i\chi = \frac{\partial\chi}{\partial t}.$$

Because φ and χ transform similarly under dreibein rotations their covariant derivatives are identical. Thus the energy eigenvalues for φ are opposite to those of χ and φ and χ have the same positive and negative energies but with the signs switched. The total vacuum energy is the sum of the vacuum energies of φ and χ, assuming both fields exist. These energies are clearly each equal to $E/2$ and the doubling due to addition is the above mentioned degeneracy factor of 2, in this representation.

We shall therefore concentrate on the operator $-i\sigma^i\nabla_i \equiv \Pi$ from the eigenvalues of which any necessary traced zeta function can be determined.

Since the object of the exercise is to obtain some explicit formulae the general discussion must be abandoned in favor of a description of the particular spatial geometry that has taken our fancy.

3.   SPECTRUM ON SQUASHED EINSTEIN UNIVERSE

Generally it is easiest to find the eigenvalues of a system possessing symmetry. An obvious symmetrical candidate for the spatial section is the three-sphere S^3 which has the maximal symmetry group $SU(2)\otimes SU(2)$. A fair amount of work has been done on field theories defined $T\otimes S^3$, the Einstein Universe.

The symmetry and the fact that S^3 is isometric to $SU(2)$ mean that the mode problem reduces to group theory, in particular entirely to angular momentum theory. For scalar fields the theory is more or less mathematically identical to the quantum mechanics of the spherical top which dates back to Reiche and Rademacher (1926) (see the review by Van Winter 1954), and is to be found in any decent quantum mechanics textbook (e.g. Landau and Lifshitz 1965).

In these books one can also find the theory of the *asymmetric top*. This corresponds to breaking the symmetry group of the configuration space to $SU(2)\otimes D_2'$ where the remaining body fixed symmetry is the discrete double-four, or quaternion, group D_2', (Dowker and Pettengill 1974). (It should be explained here that the top in question is an *ideal* top and not a physical one for which the symmetry group is the Z_2 quotient, $SO(3)\otimes D_2$.)

It is not possible to find the eigenvalues in this general case without the pain of diagonalizing larger and larger secular determinants. The theory is well developed for integral angular momentum modes particularly by molecular spectroscopists (e.g. King *et al* 1943) and has been used in the construction of the scalar field theory in a Mixmaster Universe (Hu 1973, Hu *et al* 1973). For half-integral angular momentum modes a discussion of the diagonalization can be found in Dowker and Pettengill (1974).

However, if the symmetry is broken only to $SU(2) \otimes U(1)$ corresponding to the 'symmetric' top, the eigenvalues follow without extra diagonalization. It is this case that will be investigated in the rest of this article.

The space with symmetry $SU(2) \otimes U(1)$ can be pictured as a particular squashed three-sphere. Gibbons (1980) gives an interesting discussion of the metric deformations and their mode decompositions. It is shown that the space can be thought of as filled with a homogeneous, circularly polarized gravitational standing wave of right-handed polarization and with the longest possible wavelength (Grischuk 1976). There is a built-in handedness and we will see a consequence of this when the energy eigenvalues of the neutrino are evaluated.

The left invariant metric is written as (e.g. Brill and Cohen 1966)

$$\mathrm{d}r^2 = \sum_{A=1}^{3} l_A^2 (\omega^A)^2 \tag{3.1}$$

where the ω^A are left invariant one-forms. The numbers l_A determine the deformation from the spherical situation and our normalization is such that $l_1 = l_2 = l_3 = 1$ corresponds to a three-sphere of radius $a = 2$.

Using the Cartan moving frame method the Weyl operator Π is written $-i\sigma^A \nabla_A$ where now the σ^A are the usual Pauli matrices and the covariant derivative is given by

$$\nabla_A = l_A^{-1} Y_A + \tfrac{1}{2} i t_A \sigma^A$$

with Y_A the right generators of $SU(2) \otimes SU(2)$ and the t_A given by

$$t_A = \frac{l_B^2 + l_C^2 - l_A^2}{2 l_A l_B l_C} \qquad (A, B, C \text{ cyclic}).$$

The Y_A satisfy the commutation rules

$$[Y_A, Y_B] = -\varepsilon_{ABC} Y_C.$$

Hitchin (1974) has determined the eigenvalues of $-i\sigma^A \nabla_A$ using a two-spinor approach. We will use angular momentum theory (Dowker 1972, Dowker and Pettengill 1974) as being more natural.

The operator $-iY_A \equiv L_A$ has the commutation rules of angular momentum so that in the spherical case ($l_A = 1 = 2t_A$, $\forall A$) we can write $-i\nabla = L + \tfrac{1}{2} j$ with $j = \sigma/2$ and

$$\Pi_S = \sigma \cdot (L + \tfrac{1}{4} \sigma).$$

To obtain the eigenvalues of Π_S treat this as a problem in spin–orbit coupling and construct the total angular momentum $J = L + j$. Then a complete set of commuting observables will be J^2, J_3, L^2, j^2 together with just $iX_3 = \tilde{L}_3$ from the left set of generators, X_a (remember $X^2 = Y^2$). Hence the eigenvalues of Π_s are $(J-L)(J+L+1)$ and the eigenvectors, $|JMLj(N)\rangle$. Since $J = L \pm \frac{1}{2}$ we have the spherical eigenvalues

$$\omega_S = \tfrac{1}{2}[\tfrac{1}{2} \pm (2L+1)].$$

The degeneracy is $(2J+1)$, for the different M values, times $(2L+1)$ for the different N values. N is a left label, the eigenvalue of L_3, and is a spectator in all manipulations (unless left operators occur, which they don't here). The eigenvectors can be treated exactly as ordinary angular momentum addition states so that all standard results are available to us.

The spin–orbit coupling is effected by Clebsch–Gordan coefficients.

$$|JMLj(N)\rangle = |LN'(N)\rangle (2J+1)^{-1/2} \begin{pmatrix} L & M & j \\ N' & J & m \end{pmatrix} |jm\rangle. \qquad (3.2)$$

In coordinate representation this gives the 'spinor hyperspherical harmonics',

$$\langle m, g | JMLj(N)\rangle = \left[\frac{(2L+1)(2J+1)}{16\pi^2} \right]^{1/2} \mathscr{D}^L{}_N{}^{N'}(g) \begin{pmatrix} L & M & j \\ N' & J & m \end{pmatrix}$$

in terms of the representation matrices $\mathscr{D}^L$ of SU(2) $\ni g$.

Turning now to the symmetric case, $l_1 = l_2 = 1$, $l_3 \neq 1$, algebra easily gives

$$\Pi = \Pi_s + \tfrac{1}{4}(l_3 - 1) - (l_3^{-1} - 1)\sigma_3 J_3.$$

The last term removes most of the degeneracy on the M values and one can rapidly check that it is diagonal in L, M and (N) using the unperturbed states (3.2), $|JML\frac{1}{2}(N)\rangle$, as a basis. The secular determinant is thus a simple 2×2 one labelled by $J = L \pm \frac{1}{2}$. The eigenvalue ω of Π for fixed L, M and (N) satisfies the characteristic equation

$$\omega^2 - \operatorname{Tr} V\omega + \det V = 0$$

where $\operatorname{Tr} V = \lambda(+) + \lambda(-)$.
$\det V = \lambda(+)\lambda(-) - 2(2L+1)^{-1}M^2(l_3^{-1} - 1)[\lambda(+) - \lambda(-)] - M^2(l_3^{-1} - 1)^2$
and $\lambda(+)$, $\lambda(-)$ are the eigenvalues of $\Pi_s + (l_3 - 1)/4$ namely

$$\lambda(\pm) = \tfrac{1}{2}[\tfrac{1}{2}l_3 \pm (2L+1)].$$

(The characteristic equation also follows if Π^2 is constructed and σ_3 terms eliminated in favor of Π. This is basically Hitchin's (1974) method.)
One easily finds

$$\omega_{\pm} = \tfrac{1}{2}\{\tfrac{1}{2}l_3 \pm [(2L+1)^2 + 4M^2(l_3^{-2} - 1)]^{1/2}\} \qquad (3.3)$$

with $-(L \pm \frac{1}{2}) \leqslant M \leqslant L \pm \frac{1}{2}$ and $L \geqslant 0$ for ω_+, $L \geqslant \frac{1}{2}$ for ω_-. ω_+ is positive and ω_- negative if $l_3 < 4$. As l_3 becomes larger than 4 the highest negative energy level, which is still doubly degenerate (Kramer) $(L = \frac{1}{2}, M = \pm\frac{1}{2})$, passes through zero and becomes positive. As l_3 increases further there are more and more level crossings.

Problem 3
Obtain the eigenstates of Π as linear combinations of $|L+\frac{1}{2}, M, L, \frac{1}{2}, (N)\rangle$ and $|L-\frac{1}{2}, M, L, \frac{1}{2}, (N)\rangle$.

4. ZETA FUNCTIONS

In order that comparison with the calculation of Hitchin (1974) should be easier, the eigenvalues (3.3) are rewritten, after setting $n \equiv 2L + 1$ and $q \equiv n/2 - M$, as

$$\omega_\pm = (2l_3)^{-1}\{\tfrac{1}{2}l_3^2 \pm [n^2 + 4(l_3^2 - 1)q(n-q)]^{1/2}\}.$$

The traced zeta functions for the positive and negative eigenvalues of $-i\sigma^A\nabla_A$ are constructed separately,

$$\zeta_+(s) = (2l_3)^s \sum_{n=1}^{\infty} \sum_{q=0}^{n} n\{[n^2 + 4\beta^2 q(n-q)]^{1/2} + \tfrac{1}{2}l_3^2\}^{-s}$$

$$\zeta_-(s) = (2l_3)^s \sum_{n=1}^{\infty} \sum_{q=1}^{n-1} n\{[n^2 + 4\beta^2 q(n-q)]^{1/2} - \tfrac{1}{2}l_3^2\}^{-s}$$

where $\beta^2 \equiv l_3^2 - 1$.

The zeta function for the squared operator $(-i\sigma^A\nabla_A)^2$ is

$$\mathrm{Tr}_3\zeta_3(s) = \zeta_+(2s) + \zeta_-(2s)$$

and will yield E using (2.6) after multiplication by the chiral degeneracy of 2.

Although E is our main 'physical' objective we can peripherally touch on a number of interesting mathematical questions. One of these concerns the spectral asymmetry function of Atiyah *et al* (1973)

$$\eta(s) = \zeta_+(s) - \zeta_-(s)$$

in particular its value at $s = 0$ which gives a renormalized value for the difference in the number of positive and negative eigenvalues. Hitchin (1974) has evaluated $\eta(0)$. We shall simplify his calculation and extend it to spin one.

It might seem that the complicated dependence of the eigenvalues on n and q would prevent the extraction of any exact information about $\zeta_\pm(s)$. We will now see that this is not true.

The general procedure in these cases is to employ the Euler–Plana summation formula exactly as when determining the analytical structure of the Riemann zeta function (e.g. Lindelöf 1905). Typically certain terms exhibit the poles and others are analytic.

The algebraic structure of E does however present certain difficulties. To remove the awkward square root an expansion in l_3^2 is made following Hitchin (1974). After some rearrangement one finds

$$\zeta_+(s) = (2l_3)^s \left\{ 2[\zeta(s-1,w) - w\zeta(s,w)] + f\left(\frac{s}{2}\right) - wsf\left(\frac{s+1}{2}\right) \right.$$
$$\left. + w^2 \frac{s(s+1)}{8} f\left(\frac{s+2}{2}\right) - \cdots \right\}$$

$$\zeta_-(s) = (2l_3)^s \left\{ f\left(\frac{s}{2}\right) + wsf\left(\frac{s+1}{2}\right) + w^2 \frac{s(s+1)}{8} f\left(\frac{s+2}{2}\right) \right.$$
$$\left. + w^3 \frac{s(s+1)(s+2)}{48} f\left(\frac{s+3}{2}\right) + \cdots \right\}$$

where $w = l_3^2/2$ and $f(s)$ is defined by

$$f(s) = \sum_{n=2}^{\infty} \sum_{q=1}^{n-1} n[n^2 + 4\beta^2 q(n-q)]^{-s}.$$

This expansion limits what we can do but it will enable us to determine the analytical structure of the zeta function from that of $f(s)$ and also to find the values $\zeta_\pm(-n)$, $n = 0, 1, 2, \ldots$, without too much trouble.

The analytical structure of $f(s)$ follows a double application of the Plana summation formula. A few details will be given. The summation formula applied to the q sum yields almost immediately,

$$f(s) = -\zeta_R(2s-1) + \zeta_R(2s-2) \int_0^1 \frac{dy}{[1+4\beta^2 y(1-y)]^s} - 2i \int_0^\infty \frac{dt}{\exp(2\pi t) - 1} \times$$
$$\left\{ \sum_{n=1}^{\infty} n[n^2 + 4\beta^2(t^2 - it\,n)]^{-s} - (t \to -t) \right\}$$

explicitly showing poles at $s = \frac{3}{2}$ and $s = 1$ coming from the Riemann zeta functions, ζ_R. The integral contains a further series of poles starting at $s = \frac{1}{2}$ which are revealed after another application of the summation formula. One finds that the poles at $s = \frac{3}{2} - m$, $m = 1, 2, \ldots$ have residues

$$r_m = \frac{2^{2m-2}\Gamma(m-\frac{1}{2})}{m!\,\Gamma(\frac{1}{2})} (l_3^2 - 1)^m l_3^{2m-2} B_{2m}. \tag{4.1}$$

B_{2m} is a Bernoulli number.

From these facts the analytic structures of $\zeta_\pm$ can be found. In particular one can check that the pole residues in $\mathrm{Tr}_3\zeta_3(s)$ agree with the results known from general zeta function theory for operators on Riemannian spaces (Minakshisundaram and Pleijel 1949).

The relevant result here is that the traced zeta function $\mathrm{Tr}_d\zeta_d(s)$ for an elliptic, positive second-order operator on a d-dimensional space without boundary has poles at $s = d/2 - m$, $m = 0, 1, 2, \ldots$ with residues

$$[(4\pi)^{d/2} \Gamma(d/2 - m)]^{-1} C_m. \tag{4.2}$$

The C_m are the coefficients in the time expansion of the heat kernel associated with the operator. In the present case they are integrals over the manifold of local expressions involving the curvature (e.g. Gilkey 1974).

Pioneer calculations of explicit forms for some of these coefficients were performed by DeWitt (1964). For example, for the squared Dirac operator, the expression in Problem 85 of that reference gives for the residue at $s = -\frac{1}{2}$

$$[(4\pi)^{3/2} \Gamma(-\tfrac{1}{2})]^{-1} C_2 = \frac{1}{360\pi^2} \int (R_{ij} R^{ij} - \tfrac{1}{3} R^2) \, \mathrm{d}^3 x. \tag{4.3}$$

From (2.6) we see the importance of the coefficient C_2 in determining the divergence of the effective Lagrangian. This came out of the point-splitting method as first described by DeWitt (1964) and extensively elaborated by Christensen (1978). The same quantity also occurs in the work of 't Hooft and Veltman (1974) in dimensional regularization but its mathematical significance was not realized. The zeta function derivation of the pole term can be found in Dowker and Critchley (1976b). In Hawking's (1977) approach the pole is defined away.

The recognition that the quantity determining the divergences is something well known to mathematicians enables us to use many of their results and also to enlist their help. It is only relatively recently though that mathematicians have been interested in explicit forms. The first evaluations were actually by a physicist (DeWitt 1964). In d dimensions the relevant coefficient is $C_{d/2}$ and in six dimensions the explicit forms for C_3 derived by Sakai (1971) and Gilkey (1975a) have been used to determine the one-loop divergences of various fields (Dowker 1977). (See corrections and extensions by Critchley 1978.)

Insertion of the curvature for the special metric (3.1) into (4.3) produces

$$\tfrac{1}{30}(l_3{}^2 - 1)^2 l_3 \tag{4.4}$$

for the residue.

To check this it is helpful to write out $\mathrm{Tr}_3 \zeta_3(s) = \zeta_+(2s) + \zeta_-(2s)$,
$$\mathrm{Tr}_3 \zeta_3(s) = 2(2l_3)^{2s}[\zeta(2s - 1, w) - w\zeta(2s, w) + f(s) + w^2 s(2s + 1) f(s + 1)$$
$$+ \tfrac{1}{12} w^4 s(2s + 1)(s + 1)(2s + 3) f(s + 2) + \ldots]. \tag{4.5}$$

The use of (4.1) yields the residues of the poles in $\mathrm{Tr}_3 \zeta_3(s)$ at $s = \frac{3}{2} - m$, $m = 2, 3, \ldots$ as

$$\Gamma(2m - 2) \sum_{p=2}^{m} [\Gamma(2m - 2p + 1)\Gamma(p + 1)\Gamma(p - 1)]^{-1} \left(\frac{l_3}{2}\right)^{2m - 2p + 1} (l_3{}^2 - 1)^p B_{2p} \tag{4.6}$$

which, for $m = 2$, agrees with (4.4). Such agreements provide very useful practical checks of the calculation.

Note that the residues (4.6) all vanish in the spherical limit $l_3 = 1$ leaving just the residues at $s = \frac{3}{2}$ and $\frac{1}{2}$ coming from the Hurwitz–Lerch zeta functions $\zeta(\cdot, w)$ in (4.5) (see Lindelöf 1905). These residues are, respectively, $8l_3$ and $l_3(l_3^2 - 4)/6$. We leave the reader to check these against the general result (4.2). (Remember the spatial volume is $16\pi^2 l_3$.)

For the conformally coupled scalar field all residues except that at $s = \frac{3}{2}$ vanish. This is related to the exactness of the WKB formula on group manifolds (e.g. Dowker 1970, 1971).

In general $\zeta_+(s)$ does not equal $\zeta_-(s)$. This is a reflection of the handedness of the manifold which induces an asymmetry of the energy spectrum of Weyl's equation, the positive energies corresponding to positive helicity particles and negative energies to negative helicity antiparticles. The APS asymmetry function is given as the series

$$\eta(s) = 2(2l_3)^s \left[\zeta(s-1, w) - w\zeta(s, w) - wsf\left(\frac{s+1}{2}\right) \right. $$
$$\left. - \frac{1}{6}w^3 s(s+1)(s+2)f\left(\frac{s+3}{2}\right) - \cdots \right]. \tag{4.7}$$

Setting $s = 0$ and using the known singularities of $f(s)$ at $s = \frac{3}{2}$ and $\frac{1}{2}$ as well as the values of $\zeta(-n, w)$ (e.g. Lindelöf 1905, Whittaker and Watson 1963) one easily finds Hitchin's result

$$\eta(0) = -\tfrac{1}{6}(l_3^2 - 1)^2.$$

The values $\eta(-n)$, $n = 1, 2, \ldots$ can also be found. For example, explicit calculation shows that $\eta(-1) = 0$, in agreement with the general result that $\eta(s)$ vanishes at all negative odd integers.

If the analogue of the integrated heat kernel is constructed

$$G(t) = \sum_i \omega_i \exp(-t\omega_i^2) \qquad t > 0$$

then

$$\eta(s) = \left[\Gamma\left(\frac{s+1}{2}\right) \right]^{-1/2} \int_0^\infty t^{(s-1)/2} G(t)\, dt.$$

The standard argument now is firstly to note that the singularities of $\eta(s)$ come from the lower limit behavior of the integrand so that we can write

$$\eta(s) = \left[\Gamma\left(\frac{s+1}{2}\right) \right]^{-1/2} \int_0^1 t^{(s-1)/2} G(t)\, dt + \text{entire remainder.}$$

The asymptotic series for $G(t)$ in d dimensions,

$$G(t) \sim (4\pi t)^{-d/2} \sum_{n=0}^\infty \alpha_n t^n \qquad t \downarrow 0,$$

can now be substituted and yields

$$\eta(s) = \left[(4\pi)^{d/2} \Gamma\left(\frac{s+1}{2}\right) \right]^{-1/2} \sum_{n=0}^{\infty} 2(s-d+2n+1)^{-1} \alpha_n + \text{remainder}$$

showing the possible analytic structure of $\eta(s)$. In particular one sees $\eta(-2n-1) = 0$ and also poles at $s = d - 2n - 1$, $n = 0, 1, \ldots$. Actually $s = 0$ is not a pole for the integrated η function (e.g. Gilkey 1979, Atiyah *et al* 1973) although it is for the local (i.e. untraced) one and there are complications if the manifold has a boundary.

From (4.1) and (4.7) we can work out the coefficients for our special geometry. The expressions are given for completeness:

$$\alpha_0 = \alpha_1 = 0$$

$$\alpha_n = 8\pi \frac{\Gamma(\frac{3}{2}-n)}{\Gamma(2n-2)} \sum_{p=2}^{n} \frac{\Gamma(2p-1)\Gamma(p-\frac{1}{2})}{\Gamma(2n-2p+2)\Gamma(p+1)} \left(\frac{l_3}{2}\right)^{2n-2p+2} (l_3^2 - 1)^p B_{2p}.$$

General methods exist for calculating the α_n (Gilkey 1979).

Because $\eta(-1)$ vanishes $\zeta_+(2s)$ tends to $\zeta_-(2s)$ as $s \to -\frac{1}{2}$. This means that, despite the spectral asymmetry, the positive and negative energy levels contribute equally to the vacuum energy according to equation (2.6).

It is indeed fortunate that $\eta(-1)$ is zero. If it were not, Weisskopf's (1936) method of summing just the negative energies (the occupied sea) would lead to different answers in general.

$\eta(-1)$ vanishes whatever the value of l_3 in contrast to the behavior of $\eta(0)$ which changes by an integer every time an eigenvalue passes through zero.

5. VACUUM ENERGY. NUMERICAL RESULTS

The calculation substitutes (4.5) into (2.6). The first problem is the infinity due to the $-\frac{1}{2}$ pole. Our operational procedure is simply to drop this pole. In principle one should assume a divergent term proportional to C_2 in the bare gravitation action which is then renormalized by the pole term to a finite value to be determined by experiment. This renormalization would also absorb any finite term proportional to C_2 remaining in the one-loop expression. Since C_2 is also the coefficient of the ln L^{-2} term in the effective Lagrangian it is possible to think of L as a constant to be determined by experiment (e.g. DeWitt 1979, Gibbons 1979, Birrell and Davies 1982).

In presenting the numerical results a value for L has to be chosen more or less at random. How much of the finite term proportional to C_2 is removed is therefore arbitrary. Of course this is not satisfactory but it is the best that can be done within the limitations of the one-loop theory. We record here simply that the specific quantity computed was

$$E = - \lim_{s \to 1} L^{-2s+2} \mathrm{Tr}_3 \zeta_3 (s - \tfrac{3}{2}) + \frac{1}{30} l_3 (l_3^2 - 1)^2 \left(\frac{1}{s-1} + \frac{25}{32}\right). \qquad (5.1)$$

There are two technical points concerning the display of the results. Firstly (5.1) has been derived for $a = 2$. A general scale is easily restored by noting that the overall factor of $\frac{1}{2}$ in (3.3) is really a^{-1}. Following this through yields the scaling relation

$$aE(a) = a'E(a') - \tfrac{2}{15}l_3(l_3{}^2 - 1)^2 \ln(a/a')$$

which enables $E(a)$ to be found from $E(2)$, expression (5.1).

Secondly, varying l_3 with fixed a means that the volume of the spatial section changes. To work at constant volume say V, which may be preferable, it is necessary to fix $2\pi^2 a^3 l_3 = V$.

Problem 4

Work out the trace anomaly $\int \langle 0| T_\mu{}^\mu |0 \rangle$ from the various scaling properties and show it agrees with the general result $C_2/16\pi^2$ (DeWitt 1979, Birrell and Davies 1982).

The actual computation is not particularly interesting. There are no approximations apart from a necessary numerical integration.

Figure 1 shows $E(a)$ plotted against the deformation l_3.

For the spherical case $E = 17/480\, a$, a positive value (Ford 1975, Dowker and Critchley 1976a) but this soon turns negative for deformations in the prolate direction (increasing l_3). There is a local minimum at a position $l_3 > 1$ that varies for different a (and L). If $L = 1$ this minimum is least deep for $a \sim 1$. It deepens, broadens and moves further out in l_3 for larger a. However the more interesting behavior is for smaller a. The minimum deepens quite sharply and moves down towards the spherical limit $l_3 = 1$.

It is tempting to view $E(a, l_3)$ as a potential well in which the system configuration point rolls. Tentatively adopting this picture the vacuum Casimir energy would drive the universe to the spherical limit assuming it had been created in a not too asymmetrical configuration.

The limit $l_3 \to 1$ is achieved for zero radius, a, when the well is infinitely deep and narrow. Back-reaction would prevent this catastrophe but the calculation is not presented here. We just note that, if $l_3 = 1$ from the start, self-consistency produces a radius of typically quantum geometric dimensions (Dowker and Altaie 1978).

6. FINITE TEMPERATURE

If the field is at a finite temperature, $\beta_0{}^{-1}$, the average of $\hat{H}$ in a Gibbs state yields the internal energy E,

$$E = \sum_{\omega_i} d_i |\omega_i| [\exp(\beta_0|\omega_i|) + 1]^{-1} + E(\beta_0 = \infty)$$

where d_i is the degeneracy of the ω_i level ($\omega_i = \{\varepsilon_p, \varepsilon_n\}$). $E(\beta_0 = \infty)$, the previously

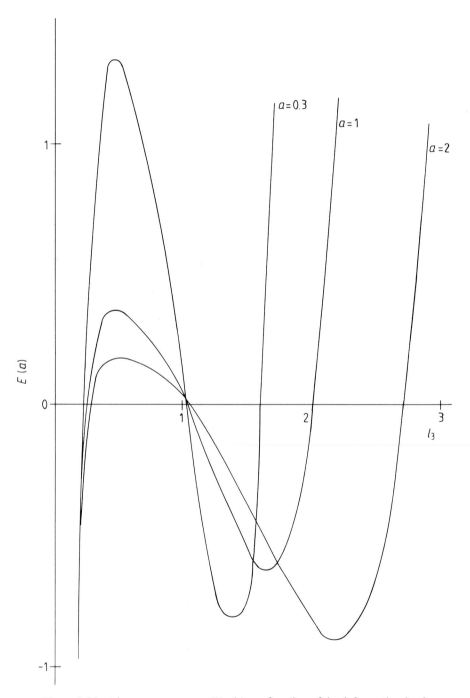

Figure 1 Neutrino vacuum energy $E(a, l_3)$ as a function of the deformation l_3 of a frozen symmetrical mixmaster universe ($l_1 = l_2 = 1$) for different scale lengths a and renormalisation length $L = 1$. Units are $\hbar = c = G = 1$ and the fixed volume result is similar.

evaluated zero temperature vacuum energy, can be negative for certain ranges of the deformation and, although it can always be cancelled by the thermal contribution for sufficiently high temperatures to yield a positive E, l_3 and β_0 can be chosen to make the entropy to energy ratio S/E as big as one likes.

A finite upper limit to this ratio has been suggested by Bekenstein (1981). It is known that his inequality is violated for fields confined to cuboid cavities (Dowker 1976 unpublished, Mamaev and Trunov 1979, Grib *et al* 1980, Unwin 1982) but the present example has the advantage that there are no walls, i.e. no boundary conditions, to cloud the issue (Bekenstein 1982).

E is related to the free energy F by $E = \partial/\partial\beta_0 (\beta_0 F)$ and F equals the negative of the effective Lagrangian, $L_{\beta_0}^{(1)}$, constructed from the finite temperature Green's function G_{β_0} in the same way that $L^{(1)}$ is constructed from G. For free fields G_{β_0} can be written as an image sum of zero temperature spinor Green's functions G (Dowker and Critchley 1977)

$$G_{\beta_0}(x, x'; t - t') = \sum_{m=-\infty}^{\infty} (-1)^m G(x, x'; t - t' + im\beta_0). \tag{6.1}$$

The chemical potential is assumed to be zero for massless particles.

The relevant expressions are given in Dowker and Kennedy (1978) for a scalar field but the alternating sign in (6.1) can be accounted for easily by combining the scalar results as follows

$$F_{\text{spinor}}(\beta_0) = 2F_{\text{scalar}}(2\beta_0) - F_{\text{scalar}}(\beta_0).$$

(This relation applies only to the general forms of the quantities and not to specific numerical values.) In particular the high temperature series for the internal energy is (cf G Kennedy 1979 *PhD Thesis* Manchester University)

$$E \sim \frac{7\pi^2}{240} \frac{C_0}{\beta_0^4} + \frac{3}{8}\pi^{-3/2} \zeta_R(3) \frac{C_{1/2}}{\beta_0^3} + \frac{1}{48} \frac{C_1}{\beta_0^2} + \frac{C_2}{16\pi^2} \left[\ln(\beta_0/\pi L) + \gamma + \tfrac{7}{32}\right]$$

$$+ \tfrac{1}{8}\pi^{3/2} \sum_{l=5/2} C_l \Gamma(l - \tfrac{1}{2}) \zeta_R(2l - 3)\pi^{-2l}(2^{l-1} - 1)\left(\frac{\beta_0}{4}\right)^{l-2}.$$

For our geometry there are no half odd-integral coefficients such as $C_{1/2}$ and the remaining coefficients have all been effectively determined from the residues of the poles in the zeta function. The high temperature behavior is thus explicitly known and the first few terms are $(a = 2)$

$$\frac{E}{V} \sim \tfrac{7}{60}\pi^2\beta_0^{-4} + \tfrac{1}{288}(l_3^2 - 4)\beta_0^{-2}$$

$$- \tfrac{1}{240}\pi^{-2}(l_3^2 - 1)\left[\ln(\beta_0/\pi L) + \gamma + \tfrac{7}{32}\right] + O(\beta_0)$$

for four component massless spinors, showing a correction to the standard T^4 Planck term. We notice the absence of a term proportional to β_0^{-1}.

The effect of a finite temperature on the approach to sphericity has not been investigated.

7. SPIN ONE ZETA FUNCTIONS

It is always worthwhile extending any analysis as far as possible even if there is no immediate motivation. One possibility that I have always been fond of is the extension to arbitrary spin. Unfortunately the problem of deciding on the appropriate field equations and the algebraically complicated analysis prevent a discussion here. We content ourselves with spin-one, i.e. Maxwell theory.

Even for photons there is a choice of descriptions; tensors, forms, spinors, etc. An amusing and convenient way of writing Maxwell's equations in Dirac-like form was given by Rumer (1930) and has since been rediscovered many times. (Some of the references are given by Dowker and Dowker 1966.) Rumer's equation, generalized to static curved space is

$$\left(\frac{\partial}{\partial t} + \boldsymbol{a} \cdot \boldsymbol{\nabla}\right)\Phi = 0$$

where Φ is a four-row column vector behaving under local Lorentz transformations as the direct sum $(1, 0) \oplus (0, 0)$, i.e. as a spin-one plus a spin-zero field. The latter can be chosen to vanish without loss of generality and

$$\Phi = \begin{pmatrix} H - iE \\ 0 \end{pmatrix}.$$

The $\boldsymbol{a}$ are a set of 4×4 matrices satisfying the same algebra as the Pauli $\boldsymbol{\sigma}$, which is very convenient for squaring the equation. If the $\boldsymbol{a}$ are split up corresponding to the division of Φ into $(1,0) \oplus (0,0)$ the blocks can be written as $3j$ symbols (Dowker and Goldstone 1968). In flat space the operator $\boldsymbol{a} \cdot \partial$ may be more recognizable as

$$\begin{pmatrix} \text{curl} & \text{grad} \\ \text{div} & 0 \end{pmatrix}.$$

Using the same notation as before, the eigenstates of the spherical Hamiltonian $H_S = H(l_3 = 1)$ are the spin-one hyperspherical harmonics, $|JML1(N)\rangle$, with eigenvalues

$$\omega_s = \tfrac{1}{2}(J^2 - L^2) = \tfrac{1}{2}(J - L)(J + L + 1)$$

where $J = L + j$ and with J restricted by $J = L \pm 1$. The modes with $J = L$ have been eliminated by the condition that the fourth component of Φ be zero, which turns out to be essentially a gauge condition. The degeneracy is $(2J + 1)(2L + 1)$.

If l_3 is not equal to one it is easiest to note that $H = -i\boldsymbol{\alpha} \cdot \boldsymbol{\nabla}$ satisfies the characteristic equation

$$H^2 - l_3 H - L^2 - (l_3^{-1} - 1)J_3^2 = 0$$

when acting on a Φ of the special form $\begin{pmatrix} \varphi \\ 0 \end{pmatrix}$. The eigenvalues are

$$\omega_{\pm} = \tfrac{1}{2}\{l_3 \pm [n^2 + 4(l_3^{-2} - 1)M^2 + l_3^2 - 1]^{1/2}\} \tag{7.1}$$

as is easily shown using the unperturbed spherical states as a basis. The degeneracy is $n = 2L+1$.

Comparison with the eigenvalues ω_s shows that M runs from $L+1$ to $-(L+1)$ for the upper sign and from $L-1$ to $-(L-1)$, $L \geqslant 1$ for the lower. ω_+ is always positive and ω_- always negative.

Equation (7.2) agrees with the result of Gibbons (1980) obtained by a different method.

Problem 5
Derive the eigenvectors and discuss the secular equation method.

The zeta functions

$$\zeta_\pm(s) = \sum(\omega_\pm)^{-s}$$

can be constructed and their singularity structure investigated. No details are needed here but a few results will be mentioned. One can show that the residues of $\mathrm{Tr}_3\zeta_3(s) \equiv \zeta_+(2s) + \zeta_-(2s)$ at $s = \frac{3}{2}, \frac{1}{2}$ and $-\frac{1}{2}$ are, respectively $8l_3, \frac{2}{3}l_3(l_3^2-4)$ and $\frac{1}{5}l_3(l_3^2-1)^2$. These values check against the general results. Note that the vanishing fourth component of Φ has to be accounted for. It turns out that the relevant coefficients are those for the combination $\mathrm{Re}[(1,0)\oplus(0,0)] = \frac{1}{2}[(1,0)\oplus(0,0)\oplus(0,0)\oplus(0,1)]$ where the spin-zero fields are minimally coupled and come in with a minus sign. Using the table in Dowker (1978), which was culled from other sources, one finds, in particular, the C_2 coefficient

$$C_2 = \frac{16}{45}\chi + \frac{1}{90\pi^2}\int[(R^{\mu\nu}R_{\mu\nu} - \frac{1}{3}R^2) + \frac{1}{2}\Box R].$$

The Euler characteristic term χ takes the same value (namely zero) as in the undistorted case.

Problem 6
This last statement is something of a cheat. Why?

It is amusing to see if equivalent results would have occurred if the vector potential description for the photon had been used. The well known formula (e.g. Gilkey 1975b)

$$C_2(1,0) + C_2(0,1) + 2C_2(0,0) - 2C_2(\tfrac{1}{2},\tfrac{1}{2}) = \chi$$

can be rearranged,

$$\tfrac{1}{2}[C_2(1,0) - C_2(0,0) + C_2(0,1) - C_2(0,0)] = [C_2(\tfrac{1}{2},\tfrac{1}{2}) - 2C_2(0,0)] + \tfrac{1}{2}\chi.$$

On the left is the C_2 coefficient for the description used in this section while on the right occurs the correct combination of vector potential and ghost terms. The two forms differ by a topological quantity, a behavior typical of distinct descriptions of the 'same' field theory, (e.g. Duff 1980).

Problem 7

(i) show that the value of the spectral asymmetry function $\eta(s) \equiv \zeta_+(s) - \zeta_-(s)$ at $s = 0$ is $\eta(0) = \frac{2}{3}(l_3{}^2 - 1)^2$ for spin one.

(ii) Find the residues of the poles in $\mathrm{Tr}_3 \zeta_3(s)$ at $s = \frac{3}{2} - m$, $m = 2, 3, \ldots$ and show that they vanish in the spherical limit. Pursue the significance of the vanishing of the coefficients C_n ($n > 1$) for spins $\frac{1}{2}$ and 1 in relation to WKB approximations and Macdonald identities.

Problem 8

Work out the finite temperature theory and derive the high temperature series for the vacuum energy

$$\frac{E}{V} \sim \tfrac{1}{15}\pi^2 \beta_0{}^{-4} + \tfrac{1}{72}\beta_0{}^{-2}(l_3{}^2 - 4) + \ldots$$

(Note that $\mathrm{Tr}_3 \zeta_3(0) \neq 0$ because of a zero mode for the minimal scalar. This complicates matters slightly.) Investigate the thermodynamics of the photon gas on the squashed hypersphere. What is the effect of the conformal anomaly and does '$PV = \frac{1}{3}E$'?

Problem 9

Calculate some other quantities of interest on the squashed hypersphere. The useful review by Baltes and Hilf (1976) might suggest some likely topics.

Problem 10

Work out the vacuum energy for a scalar field defined on the (higher dimensional) space–time $T \otimes G$ where G is a compact Lie group. Try $G = \mathrm{SU}(n)$ as an example and let $n \to \infty$.

Replace G by a squashed group manifold.

REFERENCES

Atiyah M F, Patodi V K and Singer I M 1973 *Bull. Lond. Math. Soc.* **5** 229
Baltes H P and Hilf E R 1976 *Spectra of Finite Systems* (Mannheim: Bibligraphisches Institut)
Bekenstein J D 1981 *Phys. Rev.* D **23** 287
—— 1982 *Phys. Rev.* D **26** 950
Berg B, Lüscher M 1979 *Commun. Math. Phys.* **69** 57
Birrell N and Davies P C W 1982 *Quantum Fields In Curved Space* (Cambridge: Cambridge University Press)
Bogoliubov N N and Shirkov D V 1959 *Introduction to Quantum Field Theory* (New York: Interscience)

Boulware D G 1979 in *Recent Developments in Gravitation. Cargese* 1978 (New York: Plenum Press)

Brown L S and Creamer D B 1978 *Phys. Rev.* D **18** 3695

Brill D R and Cohen J M 1966 *J. Math. Phys.* **7** 238

Christensen S M 1976 *Phys. Rev.* D **14** 2490

—— 1978 *Phy Rev.* D **17** 946

Corrigan E, Goddard P, Fairlie D B and Templeton S 1978 *Nucl. Phys.* B **140** 31

Critchley R 1978 *Phys. Rev.* D **18** 1849

Davies P C W, Fulling S A, Christensen S M and Bunch T S 1977 *Ann. Phys., NY* **109** 108

DeWitt B S 1953 *Les Houches Lectures* (unpublished)

—— 1957 *Rev. Mod. Phys.* **29** 377

—— 1964 *Relativity, Groups and Topology* ed B S DeWitt and C DeWitt (New York: Gordon and Breach)

—— 1975 *Phys. Rep.* **19** 297

—— 1979 *General Relativity. An Einstein Centenary Survey* ed S W Hawking and W Israel (Cambridge: Cambridge University Press)

Dirac P A M 1934 *Report of 7th Solvay Conference* p 203

Dowker J S 1970 *J. Phys. A: Gen. Phys.* **3** 451

—— 1971 *Ann. Phys., NY* **62** 361

—— 1972 *Ann. Phys., NY* **71** 577

—— 1977 *J. Phys. A: Math. Gen.* **10** L63

—— 1978 *J. Phys. A: Math. Gen.* **11** 347

Dowker J S and Altaie B 1978 *Phys. Rev.* D **18** 3557

Dowker J S and Critchley R 1976a *J. Phys. A: Math. Gen.* **9** 535

—— 1976b *Phys. Rev.* D **13** 3224

—— 1977 *Phys. Rev.* D **15** 1484

Dowker J S and Dowker Y P 1966 *Proc. R. Soc.* A **294** 175

Dowker J S and Goldstone M 1968 *Proc. R. Soc.* A **303** 381

Dowker J S and Kennedy G 1978 *J. Phys. A: Math. Gen.* **11** 895

Dowker J S and Pettengill D 1974 *J. Phys. A: Math., Nucl. Gen.* **7** 1527

Duff M 1981 in *Quantum Gravity 2* (Oxford: Oxford University Press)

Feynman R P 1950 *Phys. Rev.* **80** 440

Fock V A 1937 *Phys. Z. Sow.* **12** 404

Ford L H 1975 *Phys. Rev.* D **11** 3370

Gibbons G 1978 *Lect. Notes in Math.* **676** 513

—— 1979 in *General Relativity* ed S W Hawking and W Israel (Cambridge: Cambridge University Press)

—— 1980 *Ann. Phys., NY* **125** 98

Gilkey P B 1974 *The Index Theorem and the Heat Equation* (Boston: Perish)

—— 1975a *J. Differ. Geom.* **10** 601

—— 1975b *Adv. Math.* **15** 334

—— 1979 *Math. Ann.* **240** 183

Grib A A, Mameev S G and Mostepanenko V M 1980 *Quantum Effects in Strong External Fields* (Moscow: Atomizdat) in Russian

Grischuk L P 1976 *Sov. Phys.–JETP* **42** 943

Hawking S W 1977 *Commun. Math. Phys.* **55** 133

Heisenberg W 1934 *Z. Phys.* **90** 209

Heisenberg W and Euler H 1936 *Z. Phys.* **38** 714

Hitchin N 1974 *Adv. Math.* **14** 1

't Hooft G and Veltman M 1974 *Ann. Inst. H. Poincaré* **20** 69

Hu B-L 1973 *Phys. Rev.* D **8** 1048

Hu B-L, Fulling S A and Parker L 1973 *Phys. Rev.* D **8** 2377

Jauch J M and Rohrlich F 1955 *The Theory of Photons and Electrons* (Cambridge, MA: Addison–Wesley)

Kennedy G 1979 *PhD Thesis* Manchester University

King G W, Hainer R M and Cross P C 1943 *J. Chem. Phys.* **11** 27

Landau L D and Lifshitz E M 1965 *Quantum Mechanics* (London: Pergamon Press)

Lindelöf E 1905 *Le calcul des Residus* (Paris: Gauthier-Villars)

Mamaev S G and Trunov N N 1979 *Izv. Vys. Uch. Zav. Fiz.* No. 7 p 88

Minakshisundaram S and Pleijel A 1949 *Can. J. Math.* **1** 242

Nambu Y 1950 *Prog. Theor. Phys.* **5** 82

Parker L 1979 in *Recent Developments in Gravitation Cargèse* 1978 (New York: Plenum Press)

Rafelski J, Fulcher L P and Klein A 1978 *Phys. Rep.* **38** No. 5

Reiche F and Rademacher H 1926 *Z. Phys.* **39** 444

Rumer G 1930 *Z. Phys.* **65** 244

Sakai T 1971 *Tohoku Math. J.* **23** 589

Schwinger J 1951a *Phys. Rev.* **82** 664

—— 1951b *Proc. Nat. Acad. Sci. USA* **37** 452

Schwarz A 1979 *Commun. Math. Phys.* **64** 233

Soffel M, Muller B and Greiner W 1982 *Phys. Rep.* **85** No. 2

Unwin S 1982 *Phys. Rev.* D **26** 944

Utiyama R and DeWitt B S 1962 *J. Math. Phys.* **3** 608

Van Winter C 1954 *Physica* **20** 274

Weisskopf V 1936 *K. Dansk. Vid. Sel. Mat.-fys. Medd. XIV* (6)

Whittaker G T and Watson G N 1963 *A Course of Modern Analysis* (Cambridge: Cambridge University Press)

Aspects of Interacting Quantum Field Theory in Curved Space–Time: Renormalization and Symmetry Breaking

L H FORD

1. INTRODUCTION

Interacting quantum field theory in curved space–time is the study of particle interactions in the presence of an unquantized background gravitational field. As such, it is a less ambitious undertaking than the as yet unattained full quantum theory of gravity. The viewpoint which will be adopted in this essay is that quantum field theory in curved space–time represents a limit of a more complete theory and is of importance both because of its fundamental significance and because of its possible applications to cosmology.

Its fundamental significance lies partly in the fact that it allows one to break the stranglehold which Lorentz invariance has so long held upon field theory. This introduces a number of interesting consequences which must be confronted in theories which involve both quantum theory and gravitation, among which are the non-uniqueness of the vacuum state, the need to formulate the theory in terms of generally covariant Green's functions, the possible effects of non-trivial space–time topology, and more complicated discussions of renormalizability. Indeed, this was noted by Bryce in his famous 1967 paper (DeWitt 1967) on covariant quantization of gravity. Although he called it the background field method, he was in fact one of the first authors to discuss interacting quantum field theory in curved space–time.

Field theory in curved space–time is also of interest—as the theory which describes the physical effects which arise when matter fields interact in the presence of a strong gravitational field. In the vicinity of black holes and in the early universe one expects to find gravitational fields sufficiently strong so that particle interactions can no longer be described by quantum field theory in flat space–time. Among the resulting effects are particle creation (Parker 1969, Hawking 1975, DeWitt 1975, Birrell and Ford 1979) and spontaneous symmetry breaking or restoration. A sampling of the many papers concerned with gravitation and symmetry breaking is given by the following list: Abbott (1982), Allen (1982), Coleman and DeLuccia (1981), Denardo and Spallucci (1980a, b, 1981) Ford (1980a, b), Ford and Toms (1982), Gibbons (1978), Grib and Mostepanenko (1977), Hut and Klinkhamer (1981), Kennedy (1981), Shore (1980), Toms (1980a, b, c) Vilenkin (1982) and Vilenkin and Ford (1982).

In this essay I will not attempt to give a complete review of the subject of interacting quantum field theory in curved space–time, but will concentrate upon two topics, the renormalization of such theories and the role of gravitation in phase transitions in cosmology. (For recent reviews of the subject, see Birrell 1981, and Birrell and Davies 1982.)

2. RENORMALIZABILITY IN CURVED SPACE–TIME

Because the ultraviolet divergences of quantum field theory are associated with short distance behavior, and one can always transform the space–time metric to the Minkowski metric at a given point, one might expect the ultraviolet behavior of a theory to be unaffected by space–time curvature. That the situation is not quite simple may be seen by examining the regularized propagator in a curved space–time. Consider a free scalar field which satisfies the wave equation

$$\Box\phi + m^2\phi + \xi R\phi = 0 \tag{2.1}$$

where R is the scalar curvature and ξ is a constant. In dimensional regularization, the coincidence limit $(x = x')$ of the Feynman Green's function $G_F(x, x')$ is given by (Birrell 1980, Bunch and Parker 1979):

$$G_F(x, x') = -\frac{i}{8\pi^2} \frac{m^2 + (\xi - 1/6)R}{n - 4} + \text{finite part}, \tag{2.2}$$

where n is the number of space–time dimensions. The term proportional to $m^2/(n-4)$ is the usual pole term found in Minkowski space–time; however, there is now an additional pole term proportional to R which shows that ultraviolet divergences can be influenced by space–time curvature. In $\lambda\phi^4$ theory, equation (2.2) enters the first-order self-energy diagram shown in figure 1(a). The pole term requires an additional renormalization beyond those in Minkowski space–time in which the parameter ξ is renormalized.

The appearance of new divergences does not in itself threaten the renormalizability of a theory as long as we can find corresponding counterterms. (Curvature dependent divergences in the stress tensor were first treated by Utiyama and DeWitt 1962.) A more serious complication arises when we consider diagrams with two or more loops, such as the two-loop self-energy graph in $\lambda\phi^4$ theory shown in figure 1(b). The finite part of G_F depends both upon the choice of quantum state and upon

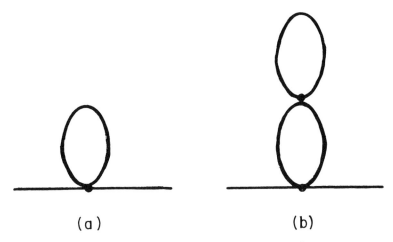

Figure 1 (a) The one-loop self-energy graph in $\lambda\phi^4$ theory.
(b) One of the two-loop self-energy graphs in $\lambda\phi^4$
theory.

the global structure of space–time (i.e. not just the curvature in a neighborhood of x).
As a consequence, the graph of figure 1(b) will have the form

$$\text{figure } 1(b) = \frac{\text{local}}{(n-4)^2} + \frac{\text{nonlocal}}{(n-4)} + \text{finite}. \tag{2.3}$$

The leading pole term has a local coefficient, that is, one which is a local function of
m^2 and R. However, the next pole term has a nonlocal coefficient which arises from
the combination of the pole term from one loop and the finite part of the other loop
of the diagram. This term cannot be removed by a local counterterm in the
Lagrangian. For example, if one calculates this graph in the locally flat but
topologically non-trivial space–time $S^1 \times R^3$ obtained by periodically identifying
one spatial coordinate with period L (Birrell and Ford 1980, Toms 1980a), one finds
that the local pole term is the same as in Minkowski space–time but that the nonlocal
pole term depends upon L.

However, when one calculates all of the second-order self-energy graphs in $\lambda\phi^4$
theory and adds them together, the nonlocal pole terms cancel out and one is left
only with infinities which can be absorbed by counterterms (Birrell 1980, Birrell and
Ford 1980, Toms 1980a, Bunch et al 1980, Bunch and Panangaden 1980, Bunch and
Parker 1979). Bunch (1981) has given a proof that this cancellation occurs to all
orders in perturbation theory in $\lambda\phi^4$ theory and hence that this theory remains
renormalizable in curved space–time. However, a corresponding proof has not yet
been given for other theories. Freedman and Pi (1975) have also shown that in a
perturbation expansion of the metric around flat space, no non-renormalizable
infinities arise in $\lambda\phi^4$ theory. Although this approach is not applicable to
space–times which are not representable by such an expansion (such as topologically
non-trivial space–times), it does support the view that theories which are renormalizable in Minkowski space–time remain so in curved space–times. The

renormalizability of $\lambda\phi^3$ theory has been partially analyzed by Birrell and Taylor (1980), and by Drummond (1975, 1979). The one-loop infinities of quantum electrodynamics have been studied by Drummond and Shore (1979) in de Sitter space and by Panangaden (1981) for general space–times; the one-loop infinities of Yang–Mills theories in curved space–times have been studied by Toms (1982) and by Leen (1982). In all of these cases, no non-renormalizable infinities were encountered. The closely related problem of scaling behavior and the renormalization group in curved space–time has been treated by several authors, including Birrell and Davies (1980), Brown and Collins (1980), Gass (1981), Nelson and Panangaden (1982), Ford and Toms (1982) and Leen (1982).

The problem of nonlocal divergences arises in a somewhat milder form even in Minkowski space–time. If one calculates a multiloop diagram, one finds that the leading divergence is proportional to a polynomial in the external momenta of the diagram but that there are also divergences which are proportional to non-polynomial (e.g. logarithmic) functions of the external momenta. It is only the former which can be immediately absorbed by counterterms. It is the difficult task of a proof of renormalizability to show that the latter divergences (the overlapping divergences) can also be removed by some subtraction procedure which introduces only a finite number of undetermined renormalization constants into the theory. For example, the BPH method defines a recursive subtraction procedure (the R–operation) which when applied to a given diagram renders it finite by a set of local subtractions within the subdiagrams of the given diagram. Furthermore, the sum of the R–operations for all diagrams of a given order which contribute to a given process can be implemented by local counterterms in the Lagrangian, which means that the nonlocal divergences cancel when all such diagrams are summed. (For a good recent account of BPH renormalization, see Caswell and Kennedy (1982).)

Although the complications introduced by nonlocal, state dependent divergencies are greater in curved space–time (for example, figure 1(b) has only a simple local divergence in Minkowski space–time), I think that it is unlikely that they pose any serious threat to the formulation of quantum field theories in curved space–time. Bunch (1981) has shown that they may be overcome for $\lambda\phi^4$ theory, and it is likely that with enough diligence similar proofs could be constructed for other theories.

One of the crucial assumptions in all of the treatments of renormalizability in curved space–time is that the singular part of the Green's functions is state independent and depends only locally on the curvature, as in equation (2.1), at least for states which can be regarded as physically acceptable. There is a considerable body of evidence to support this assumption, which includes

(1) explicit calculations in particular space–times, such as de Sitter space (Tagirov 1973, Candelas and Raine 1975) or the Einstein universe (Dowker 1971),

(2) the DeWitt–Schwinger expansion (DeWitt 1965, see also the article by Christensen in this volume) which is an asymptotic expansion of a Green's function obtained by heat kernel techniques,

(3) asymptotic expansions of Green's functions obtained from high frequency approximations in momentum space representations (Birrell 1980, Bunch and Parker 1979).

However, one can construct examples of Green's functions (Rumpf 1981) whose singular parts are not given by the asymptotic expansions of (2) and (3). One is tempted to regard the states associated with such Green's functions as unphysical, that is, as containing in some sense an infinite number of particles in high frequency modes. So long as a sufficiently wide class of states is available which do fulfill our notions of physical reasonableness, this is probably a safe assumption. However, because of the ambiguity inherent in defining quantum states in curved space–time, it is worth bearing in mind that it *is* an assumption.

An example of the complications introduced by the choice of quantum state is the phenomenon of infrared divergences in curved space–time (Ford and Parker 1977). There are choices of the state in which the energy density is finite, but $G_F(x, x')$ does not exist even for $x \neq x'$. Such states can be regarded as containing an infinite number of long wavelength particles and cannot arise by dynamical evolution from a more well behaved initial configuration (Fulling *et al* 1978); consequently they can be excluded as unphysical. However, this exclusion has some interesting consequences. For example, in de Sitter space–time it is not possible to define a state for the massless, minimally coupled scalar field ($m = \xi = 0$) which is de Sitter invariant (Vilenkin and Ford 1982). This gives rise to some effects which may be important in the evolution of the universe and which will be discussed in the next section.

The renormalizability problems which have been discussed in this section are less severe than those of full quantum gravity, where the complications posed by graviton self-interactions arise. However, the issues of the locality of the singular parts of Green's functions and of the cancellation of state-dependent divergences must also be dealt with in the full theory, and it is reassuring that they seem to pose no insurmountable barriers.

3. PHASE TRANSITIONS IN COSMOLOGY

One of the regimes in which the coupling of fields to space–time curvature may be important is in the early universe. However, a simple calculation suggests that the effects of such coupling will be important only near the Planck time, when the effects of quantum gravity must also be considered. Consider the quantity $\langle \phi^2 \rangle$, the finite part of figure 1(a); the contribution due to gravitational coupling is expected on dimensional grounds to be of the order of $R_{\mu\nu\alpha\beta}$, a typical component of the Riemann tensor measured in a co-moving orthonormal frame, and the thermal contribution is $\frac{1}{12} T^2$. In a radiation-dominated Robertson–Walker universe, $T^2 \sim t^{-1}$ whereas $R_{\mu\nu\alpha\beta} \sim t^{-2}$, so the thermal effects dominate for $t \gtrsim 1$, i.e. after the Planck time. This argument implies that the early universe can be accurately described by quantum field theory in flat space–time at finite temperature, where the expansion only has the effect of decreasing the temperature. However, there are two possible loopholes. The first is that the gravitational effects may not always be characterized by the Riemann tensor, but may be much larger. As will be seen below, this can indeed be the case if the coupling of the field to the curvature is non-conformal. The second situation in which gravitational effects can dominate thermal effects is in variations of the standard model in which the time dependence of the curvature and the temperature differs from that assumed above.

The most interesting example of such a model is the inflationary universe proposed by Guth (1981). In this model it is assumed that in the early universe there is a cosmological constant term in the Einstein equations; this term is generated by the vacuum energy of a Higgs field associated with a gauge theory, such as the grand unified (GUT) models. At sufficiently early times, i.e. shortly after the Planck time, the universe is radiation-dominated, but the radiation is rapidly red-shifted and the cosmological term begins to dominate the expansion. The metric is taken to be that of a spatially flat Robertson–Walker universe:

$$ds^2 = dt^2 - a^2(t)dx^2. \tag{3.1}$$

The solution of Einstein's equations for a source which is a sum of the stress tensor for radiation and a cosmological term is

$$a(t) = \sinh^{1/2} 2Ht \sim \begin{cases} t^{1/2}, t \leqslant H^{-1} \\ e^{Ht}, t \gtrsim H^{-1} \end{cases} \tag{3.2}$$

where H is a constant. Thus for $t \gtrsim H^{-1}$, the universe is approximately described by the de Sitter metric.

Guth has shown that if this inflationary phase lasts for a time of the order of $60\, H^{-1}$ or longer, a number of cosmological puzzles may be solved, in particular the horizon, flatness and monopole problems. In short, these three problems can be stated as follows: the horizon problem is to explain why the temperature of the microwave background radiation is so isotropic even though in non-inflationary models it is arriving from causally disconnected regions of the universe: the flatness problem is the problem of explaining why the matter density of the universe was extremely close to the critical density in the early universe; the monopole problem arises because grand unified models appear to predict an enormous density of monopoles which is not observed. In the original version of the model, Guth assumed that the end of the inflationary era would occur by barrier penetration and bubble formation (Coleman 1977). However, this leads to an excessively cold and inhomogeneous universe. More recently, Albrecht and Steinhardt (1982) and Linde (1982a) have proposed a variation of the model which avoids this problem; these authors assume that the end of the inflationary era occurs not by tunnelling from a local minimum of the effective potential but by a transition from a local maximum. This helps to resolve the difficulty of the original version because the phase transition occurs more nearly simultaneously throughout the universe. (A somewhat different scenario has been proposed by Hawking and Moss (1982).) However, there still seems to be a problem of fluctuations producing overly large inhomogeneities (Guth and Pi 1982, Hawking 1982, Bardeen *et al* 1982).

The main interest in the inflationary model from the viewpoint of this essay is that effects of interacting quantum field theory in curved space–time can be important in the evolution of the universe if there is an inflationary era. Thermal effects are rapidly redshifted, but the space–time curvature remains constant, so the argument given above breaks down.

Let us now discuss briefly the stability of a self-coupled scalar field in a curved space–time. Our approach will follow that of Ford (1980b), Ford and Toms (1982), and Vilenkin and Ford (1982). Take the Lagrangian to be

$$\mathscr{L} = \tfrac{1}{2}(\partial_\alpha\phi\partial^\alpha\phi - m^2\phi^2 - \xi R\phi^2) - \tfrac{1}{12}\lambda\phi^4 \tag{3.3}$$

so the associated equation of motion is

$$\Box\phi + m^2\phi + \xi R\phi + \tfrac{1}{3}\lambda\phi^3 = 0. \tag{3.4}$$

Let $\phi = \phi_c + \phi_q$ where ϕ_c is a c-number and ϕ_q is a quantum field with zero vacuum expectation value, so that $\langle\phi\rangle = \phi_c$. Equation (3.4) is equivalent to a pair of coupled equations for ϕ_q and ϕ_c. Although one can write down exact equations, one must resort to perturbation theory to obtain solutions. Here we will restrict our attention to first-order perturbation theory. In this case, the equation for ϕ_c becomes

$$\Box\phi_c + (m^2 + \xi R + \lambda\langle\phi_q^2\rangle)\phi_c + \tfrac{1}{3}\lambda\phi_c^3 = 0 \tag{3.5}$$

where ϕ_q is now a free-field solution of equation (2.1). A solution of equation (3.5) will be considered to represent a stable configuration provided that the equation

$$\Box\Phi + (m^2 + \xi R)\Phi + \lambda(\phi_c^2 + \langle\phi_q^2\rangle)\Phi = 0 \tag{3.6}$$

has no growing solutions. (The quantity Φ may be interpreted as either a small perturbation of ϕ_c or as the expectation value of ϕ_q in a coherent state.) If ϕ_c is independent of space and time, then this stability criterion is equivalent to the more familiar effective potential approach in which one requires the vacuum to be associated with a local minimum of the effective potential. However, in a time-dependent problem, such as one encounters in cosmology, the approach outlined here is more suitable.

The quantity $\langle\phi_q^2\rangle$ is of course formally divergent. In lowest order, its pole terms in dimensional regularization are given in equation (2.2), and may be absorbed into renormalization of m^2 and ξ. In accordance with the discussion in §2, we expect that this can be done consistently to all orders. The remaining finite part carries the information about the coupling of the quantized field to the space–time curvature which effects spontaneous symmetry breaking. Let us focus our attention upon the analysis of the stability of the symmetric phase ($\phi_c = 0$), so $\langle\phi_q^2\rangle = \langle\phi^2\rangle$, which we take to be the finite, renormalized value.

The quantity $\langle\phi^2\rangle$ depends, of course upon the choice of quantum state and is not a local function of the space–time curvature. However, it does contain a local, state-independent contribution which is studied in Ford and Toms (1982) for the case $m = 0$; it is

$$\langle\phi^2\rangle_{\text{local}} = (16\pi^2)^{-1}(\xi - \tfrac{1}{6})R\ln|R\mu^{-2}|. \tag{3.7}$$

This term may be thought of as a finite residual of the renormalization process which is determined solely by the pole term in equation (2.2). Here, μ is an arbitrary unit of mass. Because the theory is invariant under redefinition of μ, $\langle\phi^2\rangle$ alone ceases to be physically measurable, and all physical quantities will depend upon the combination $\xi R + \lambda \langle \phi^2 \rangle$, which is invariant under rescaling of μ. However, to facilitate the present discussion, let us suppose that μ has been fixed so $\langle \phi^2 \rangle$ becomes measurable. Equation (3.7) illustrates some important characteristics of $\langle \phi^2 \rangle$.

(1) $\langle \phi^2 \rangle$ can be negative. Thus the quantum fluctuations can act in the opposite direction from the thermal fluctuations and can have a destabilizing effect upon the symmetric phase.

(2) The quantum fluctuations are most pronounced if the field is not conformally coupled ($\xi \neq \frac{1}{6}$).

(3) It is not necessarily the case that $\langle \phi^2 \rangle \approx R$, i.e., a typical component of the Riemann tensor. Because ξ and μ are free parameters, it is in principle possible to have $\langle \phi^2 \rangle$ much longer than the curvature.

Let us now return to the specific context of the inflationary universe model. To understand the effects of coupling to curvature, we need $\langle \phi^2 \rangle$ in de Sitter space–time, including both thermal and vacuum contributions, which has been discussed by Vilenkin (1982), Vilenkin and Ford (1982), Linde (1982b), and Starobinsky (1982). The thermal part $\langle \phi^2 \rangle_T$ initially behaves as would a thermal distribution in flat space–time which is being redshifted by the expansion of the universe: $\langle \phi^2 \rangle_T \approx \frac{1}{12} (\theta/a)^2 \propto e^{-2Ht}$. If $m^2 + \xi R = 0$, $\langle \phi^2 \rangle_T$ approaches a non-zero constant proportional to θH. This behavior arises because those modes associated with wavelengths of the order of, or longer than the radius of curvature of space–time (here H^{-1}, the horizon size) do not redshift as rapidly as do shorter wavelength modes. When $m^2 + \xi R = 0$, the contribution of a given mode ceases to decrease after its wavelength becomes of the order of H^{-1}.

The vacuum part $\langle \phi^2 \rangle_0$ can exhibit even more bizarre behavior. However, we must first decide upon a definition of the vacuum state. It is natural to try to choose a de Sitter invariant state in which $\langle \phi^2 \rangle_0 = \langle \phi^2 \rangle_{inv} = $ constant (Bunch and Davies 1978). If, for example, we consider the conformal field, $m = 0$, $\xi = \frac{1}{6}$, then

$$\langle \phi^2 \rangle_{inv} = \frac{R}{576\pi^2} = \tfrac{1}{12} T_{GH}^2 \qquad (3.8)$$

where $T_{GH} = H/2\pi$. This can be interpreted as a thermal contribution at the temperature T_{GH} which Gibbons and Hawking (1977) have associated with the vacuum state in de Sitter space. However, one can also take the viewpoint that vacuum and thermal effects cannot be unambiguously distinguished. In any case, such contributions are automatically accounted for when one calculates $\langle \phi^2 \rangle$ in an appropriately chosen state.

If we now turn to the case of a massless, minimally coupled field, we find that no de Sitter invariant state exists. This can be seen by examining the asymptotic form of $\langle \phi^2 \rangle_{inv}$ as $m^2 + \xi R \to 0$:

$$\langle \phi^2 \rangle_{inv} \sim -\frac{R}{64\pi^2} \left[m^2 + (\xi - \tfrac{1}{6}) R \right] (m^2 + \xi R)^{-1} \qquad (3.9)$$

so if $\xi = 0$, $\langle \phi^2 \rangle_{inv} \sim (384\pi^2)^{-1} R^2/m^2$ as $m \to 0$. This is an infrared divergence of the sort discussed in §2. It does not prevent us from defining physically allowable states in de Sitter space–time; it simply requires that such states break the de Sitter symmetry. In fact, one can show (Vilenkin and Ford 1982, Linde 1982b, Starobinsky 1982) that if $m^2 + \xi R = 0$, then $\langle \phi^2 \rangle_0$ must grow linearly in time

$$\langle \phi^2 \rangle_0 \sim (4\pi^2)^{-1} H^3 t + \text{constant.} \tag{3.10}$$

This growth of quantum fluctuations in an expanding universe may seem rather startling, but it is, nonetheless, a rigorous consequence of quantum field theory in curved space–time. Its origin lies in the anomalously large contribution which long wavelength modes make to $\langle \phi^2 \rangle$ in this case. This is another case where $\langle \phi^2 \rangle$ can be much larger than the curvature.

Although the details of how the coupling of quantum fields to space–time curvature effect the evolution of an inflationary universe have not yet been worked out, it seems clear that they are not negligible. The main point of the discussion of this section has been to illustrate some aspects of these effects and show why they may be important. In particular, quantum effects associated with the gravitational field itself, may well have influenced the evolution of the universe long after the Planck time and possibly even have resulted in some of the observable features of the present-day universe. If such a connection can be established, it would truly illustrate the unity of physics on the smallest and largest of scales.

REFERENCES

Abbott L F 1982 *Nucl. Phys.* B **185** 233–8
Albrecht A and Steinhardt P J 1982 *Phys. Rev. Lett* **48** 1220–3
Allen B 1982 "Phase Transitions in De Sitter Space" to be published
Bardeen J, Steinhardt P J and Turner M J 1982 "Spontaneous Creation of Almost Scale-Free Density Perturbations in an Inflationary Universe" to be published
Birrell N D 1980 *J. Phys. A: Math. Gen.* **13** 569–84
—— 1981 in *Quantum Gravity 2: A Second Oxford Symposium* ed C J Isham *et al* (London: Oxford) 164–82
Birrell N D and Davies P C W 1980 *Phys. Rev.* D **22** 322–9
—— 1982 *Quantum Fields in Curved Space* (Cambridge: Cambridge University Press) Ch 9
Birrell N D and Ford L H 1979 *Ann. Phys., NY* **122** 1–25
—— 1980 *Phys. Rev.* D **22** 330–42
Birrell N D and Taylor J G 1980 *J. Math. Phys.* **21** 1740–60
Brown L S and Collins J C 1980 *Ann. Phys., NY* **130** 215–48
Bunch T S 1981 *Ann. Phys., NY* **131** 118–48
Bunch T S and Davies P C W 1978 *Proc. R. Soc.* A **360** 117–34
Bunch T S and Panangaden P 1980 *J. Phys. A: Math. Gen.* **13** 919–32
Bunch T S, Panangaden P and Parker L 1980 *J. Phys. A: Math. Gen.* **13** 901–18
Bunch T S and Parker L 1979 *Phys. Rev.* D **20** 2499–510
Candelas P and Raine D J 1975 *Phys. Rev.* D **12** 965–74
Caswell W E and Kennedy A D 1982 *Phys. Rev.* D **25** 392–408
Coleman S 1977 *Phys. Rev.* D **15** 2929–36
Coleman S and DeLuccia F 1981 *Phys. Rev.* D **21** 3305–15

Denardo G and Spallucci E 1980a *Nucl. Phys.* B **169** 514–26
—— 1980b *Nuovo Cimento* A **58** 243–53
—— 1981 *Nuovo Cimento* A**64** 15–26, 27–38
DeWitt B S 1965 *The Dynamical Theory of Group and Fields* (New York: Gordon and Breach)
—— 1967 *Phys. Rev.* **162** 1195–238
—— 1975 *Phys. Rep.* **19** 297–357
Dowker J S 1971 *Ann. Phys., NY* **62** 361–82
Drummond I T 1975 *Nucl. Phys.* B **94** 115–44
—— 1979 *Phys. Rev.* D **19** 1123–33
Drummond I T and Shore G M 1979 *Ann. Phys., NY* **117** 89–120
Ford L H 1980a *Phys. Rev.* D **21** 933–48
—— 1980b *Phys. Rev.* D **22** 3003–11
Ford L H and Parker L 1977 *Phys. Rev.* D **16** 245–50
Ford L H and Toms D J 1982 *Phys. Rev.* D **25** 1510–8
Freedman D Z and Pi S Y 1975 *Ann. Phys., NY* **91** 442–9
Fulling S A, Sweeny M and Wald R M 1978 *Commun. Math. Phys.* **63** 257–64
Gass R 1981 *Phys. Rev.* D **24** 1688–9
Gibbons G W 1978 *J. Phys. A: Math. Gen.* **11** 1341–5
Gibbons G W and Hawking S W 1977 *Phys. Rev.* D **15** 2738–51
Grib A A and Mostepanenko V M 1977 *Zh. Eksp. Teor. Fiz Pis.* **25** 302–5 (1977 *JETP Lett.* **25** 277–9)
Guth A H 1981 *Phys. Rev.* D **23** 347–56
Guth A H and Pi S Y 1982 *Phys. Rev. Lett.* **49** 1110–3
Hawking S W 1975 *Commun. Math. Phys.* **43** 199–220
—— 1982 *Phys. Lett.* **115B** 295–7
Hawking S W and Moss I G 1982 *Phys. Lett.* **110B** 35–41
Hut P and Klinkhamer F 1981 *Phys. Lett.* **104B** 439–43
Kennedy G 1981 *Phys. Rev.* D **23** 2884–900
Leen T K 1982 'Renormalization and Scaling Behavior of Non-Abelian Gauge Fields in Curved Space–time' to be published
Linde A D 1982a *Phys. Lett* **108B** 389–93
—— 1982b 'Scalar Field Fluctuations in Expanding Universes and the New Inflationary Universe Scenario' to be published
Nelson B L and Panangaden P 1982 *Phys. Rev.* D **25** 1019–27
Panangaden P 1981 *Phys. Rev.* D **23** 1735–46
Parker L 1969 *Phys. Rev.* **183** 1057–68
Rumpf H 1981 *Phys. Rev.* D **24** 275–89
Shore G M 1980 *Ann. Phys., NY* **128** 376–424
Starobinsky A A 1982 to be published
Tagirov E A 1973 *Ann. Phys., NY* **76** 561–79
Toms D J 1980a *Phys. Rev.* D **21** 928–32
—— 1980b *Phys. Rev.* D **21** 2805–17
—— 1980c *Ann. Phys., NY* **129** 334–57
—— 1982 "The Background Field Method and Renormalization of Non-Abelian Gauge Theories in Curved Space–time" to be published
Utiyama R and DeWitt B S 1962 *J. Math. Phys.* **3** 608–18
Vilenkin A 1982 *Phys. Lett.* **115B** 91–4
Vilenkin A and Ford L H 1982 *Phys. Rev.* D **26** 1231–41

What Happens to the Horizon when a Black Hole Radiates?

JAMES W YORK, JR

1. INTRODUCTION

Quantum field theory in curved space–time is a subject in which Bryce DeWitt has long been a leader. One of the most interesting phenomena that can occur in such problems is the creation of particles by the gravitational field. This subject has been widely studied, beginning with Parker (1968, 1977) in recent years, and is the subject of an important review by DeWitt (1975) and of a recent monograph (Birrell and Davies 1982).

Probably the most important example of particle creation by the gravitational field is the quantum radiance of black holes found by Hawking (1975). In this article, which is dedicated to Bryce DeWitt on his sixtieth birthday, I study in detail the changes the horizon undergoes when a black hole emits energy. I show that the basic reaction is the formation of a 'quantum ergosphere' that results from the coincidence, for spherical holes, of the apparent horizon and quasi-static limit surface and from the fact that these surfaces are outside the event horizon, in contrast to the case for static or accreting holes. The consistency of the solution is checked by studying the Raychadhuri equation for null geodesics. Perhaps the most interesting findings concern the existence of two mass-independent characteristics of the quantum ergosphere. One shows that there is a maximum local Unruh (1976) temperature for a quasi-static observer at the quantum ergosphere. The other shows that the areas of the apparent and event horizons do not become equal (as they are for Schwarzschild holes) in the zero Hawking-temperature limit. Both of these results depend quantitatively on the number of species and the spins of massless quanta that exist in nature. They also suggest an important role for low energy quantum gravity in checking the consistency of quantization of external fields on a 'fixed' classical background space–time with an event horizon (York 1983). I also deal with a conjecture by Hawking (1976) asserting the indistinguishability, because of quantum radiance, of black and white holes. The conjecture is found to be untenable.

2. HORIZONS

Properly speaking, a black hole does not have a single 'horizon' that fully characterizes its structure: it has three important horizon-like loci. The future event horizon (EH) is a null three-surface which is the locus of outgoing future-directed null geodesic rays that never manage to reach arbitrarily large distances from the hole. The apparent horizon (AH) is the outermost marginally trapped surface for the outgoing photons. I shall consider the history of the AH and regard it as a three-dimensional surface. Classically, the AH can be either null or spacelike, that is, it can 'move' causally or acausally. Allowing for quantum radiance, we will see that the AH can also be timelike. The third horizon-like locus can be called the timelike limit surface (TLS). For black holes that possess in the exterior region an exact timelike Killing vector $\partial/\partial t$, the TLS is where $g(\partial_t, \partial_t) = g_{tt} = 0$. For black holes with a small dimensionless accretion or luminosity (L), one can define the TLS (or quasi-static limit) as the locus where $g(\partial_v, \partial_v) = g_{vv} = 0$, where $\partial/\partial v$ is the timelike vector of an observer at rest a large distance from the hole and with respect to which he measures luminosity $L = -dM/dv$, $M = $ mass of hole. In general, the TLS can be null, timelike, or spacelike as a three-dimensional surface. I want to stress at the beginning that when L refers to quantum radiance (explicit form $L = L_H = B\hbar M^{-2}$), it is usually very small and corresponds to a rather small number of quanta leaving the hole in any reasonable time interval. Hence its fluctuations could be very large, at least comparable to L itself. Thus L is to be thought of as an average value.

In this article, I shall confine my attention to spherical black holes. In this case, if $R_{ab} = 0$, then from Birkhoff's theorem one has that there is an 'accidental degeneracy' in which all three horizons coincide at the usual locus $r = 2M$. However, if there is a small accretion or luminosity ($R_{ab} = O(L)$), then the degeneracy is partially lifted even if spherical symmetry remains. As we shall see, we have then AH = TLS but EH ≠ AH. (If we preserve $R_{ab} = 0$ but break spherical symmetry, preserving stationarity (as for a Kerr hole), then EH = AH but EH ≠ TLS. In general the three horizons do not coincide; they are very sensitive to small perturbations.)

I treat the situation as follows. Let $v = $ constant be an ingoing null surface ($v = $ 'advanced time') with future-directed null tangent (normal) β^a. The metric of $v = $ constant will be of course degenerate (two-dimensional): call it γ_{ab} ($\gamma_{ab}\beta^b = 0$) and let the space–time metric be g_{ab}. (Here I am following the notation of Carter 1979).) Then define the future-directed (outgoing) null geodesics by their tangent vectors l^a, where $l_a l^a = 0$, $l_a\beta^a = -1$, $l^a\gamma_{ab} = 0$. One has that

$$g_{ab} = \gamma_{ab} - l_a\beta_b - \beta_a l_b. \tag{2.1}$$

We shall be interested in the outgoing null geodesic congruence. The defining equation is

$$l^b\nabla_b l^a = \mathcal{K} l^a \tag{2.2}$$

where the rays have 'time' parameter v and the affine parameter λ is related to v by $\mathcal{K} = \ddot{\lambda}(\dot{\lambda}^{-1})$, $\cdot = d/dv$. On a horizon, $\mathcal{K}$ is called the surface gravity and can be calculated conveniently from

$$\mathcal{K} = -\beta^a l^b \nabla_b l_a. \tag{2.3}$$

The expansion Θ is given by

$$\Theta = \nabla_a l^a - \mathcal{K}. \tag{2.4}$$

The AH is the locus where $\Theta = 0$. In the case of spherical symmetry, the vorticity and shear of l^a are zero.

We shall be interested in the areas A of the various horizons. These will be defined and compared by intersecting the EH, AH, and TLS with the incoming null surface $v = $ constant.

3. EVAPORATING METRIC

The space–time of a spherical black hole that is emitting energy into 'empty' space can be described by a metric of the type discussed by Bardeen (1981),

$$ds^2 = -(\exp 2\psi)(1 - 2mr^{-1})dv^2 + 2(\exp \psi)dv\,dr + r^2(d\theta^2 + \sin^2\theta\,d\varphi^2) \tag{3.1}$$

where ψ and m are functions of the advanced time v ($v \cong t + r$ at large r) and the radial coordinate r. The area A of spheres $r = $ constant measured on ingoing null surfaces $v = $ constant is given by the standard formula $4\pi r^2$. Note that $\partial/\partial r$ is a null vector. If $\psi = 0$ and $m = M = $ constant > 0, then (3.1) is the vacuum ($R_{ab} = 0$) Schwarzschild black hole in advanced time Eddington–Finkelstein coordinates. In this case $\partial/\partial v$ is a Killing vector that is timelike for $r > 2M$, null at $r = 2M$, and spacelike for $r < 2M$. The locus $r = 2M$ is the EH $=$ AH $=$ TLS in the Schwarzschild geometry. If $\psi = 0$ and $m = m(v)$, (3.1) is a Vaidya metric.

The Einstein equations for (3.1), with $G = c = 1$, are simple and are dictated essentially by spherical symmetry:

$$\partial m/\partial v = 4\pi r^2 T^r{}_v \tag{3.2}$$

$$\partial m/\partial r = -4\pi r^2 T^v{}_v \tag{3.3}$$

$$\partial \psi/\partial r = 4\pi r T_{rr} \tag{3.4}$$

where T_{ab} is the effective stress–energy tensor that describes the radiation. I assume that the dimensionless luminosity L is small so that terms of $O(L^2)$ are negligible. I shall not assume, initially, that L has the Hawking form $L_H \propto \hbar M^{-2}$. Near the horizon the T_{ab}s are assumed to be regular and of order LA^{-1}. Physically, this is equivalent to regularity of the orthonormal frame components of the stress–energy tensor of an observer falling freely across the horizon (the 'Unruh (1976) vacuum').

The mass of the black hole in the zeroth order of L is denoted $M(v)$ and is defined as the value of $m(v, r)$ such that $g_{vv} = 0$. One can set, for convenience, $\psi = 0$ at $r = 2M$ and from (3.4) one sees that $\partial \psi/\partial r \sim Lr^{-1}$. A quasi-static observer is one at rest at $r \gg 2M$. His four-velocity is approximately $\partial/\partial v$ and one can show that he observes a luminosity $L \cong -\partial m/\partial v = -dM/dv$, where $\cong$ denotes equality through the first order in L. One sees that $g_{vv}(r = 2M) = 0$ implies that $r = 2M$ is the TLS.

The horizon structure is examined by using the null vectors

$$l^a = [l^v, l^r] = [1, \tfrac{1}{2} (\exp \psi) (1 - 2mr^{-1})] \tag{3.5}$$

$$\beta^a = [0, -(\exp -\psi)] \tag{3.6}$$

with angular components zero and normalization $\beta_a l^a = -1$. Note that the outgoing radial null geodesics are parametrized by the time v of the distant observer, rather than by an affine parameter. On the other hand, because it turns out that the effects in $\theta(L)$ of ψ and $\partial\psi/\partial r$ are negligible near $r = 2M$, the ingoing null rays are affinely parametrized by the area radial coordinate r.

I will exhibit the exact outgoing radial null geodesic equation for $r(v)$, using $d/dv = l^a \nabla_a = \partial/\partial v + \tfrac{1}{2} (\exp \psi) (1 - 2mr^{-1}) \partial/\partial r$ $(dr/dv = \tfrac{1}{2}(\exp\psi)(1 - 2mr^{-1}))$:

$$d^2 r/dv^2 = (dr/dv)[\partial\psi/\partial r - (\exp\psi)r^{-1}(\partial m/\partial r) + (\exp\psi)mr^{-2}] +$$
$$(dr/dv)^2 (\partial\psi/\partial r) - (\exp\psi)r^{-1}(\partial m/\partial v). \tag{3.7}$$

The purpose of writing out (3.7) is merely to show that near $r = 2M$ in $O(L)$, the radial null geodesics are just the same *as if* we had used a simple Vaidya metric as described previously. Hence, as long as regularity of T_{ab} and spherical symmetry are good approximations, then the horizons can be described accurately in $O(L)$ by very simple equations. Thus, the horizon loci in $O(L)$ are determined simply by T^r_v and are insensitive to the other components until the extreme situation $L \sim 1$ ensues in late stages of evaporation. This is a fortunate circumstance because it follows that energy conservation in a small (and therefore essentially steady) flow situation determines the results uniquely in order L.

For an outgoing null geodesic ray near $r = 2M$, one has $(\dot r = dr/dv)$

$$\dot r \cong \tfrac{1}{2}(1 - 2mr^{-1}). \tag{3.8}$$

Differentiation of (3.8) with respect to v produces the $O(L)$ approximation to (3.7) near $r = 2M$:

$$\ddot r \cong \dot r(4M)^{-1} + L(2M)^{-1}. \tag{3.9}$$

At the TLS, $r = 2M$, one has $\dot r = 0$ and therefore $\ddot r > 0$ for $L > 0$. Hence photons will escape from $r = 2M$ and reach arbitrarily large distances (unless someone has thrown in a shell of matter that will turn out 'later' to confine these photons; I exclude such situations in this paper). This surface is therefore not the event horizon; it is the apparent horizon, as can be seen by computing the rate-of-change of area along l^a. For spherical metrics one has $\dot A = \Theta A$, where Θ is the expansion ($\Theta = -2\rho$, $\rho = $ convergence). From $A = 4\pi r^2$ and (3.8), one finds

$$\Theta \cong r^{-1}(1 - 2mr^{-1}). \tag{3.10}$$

Hence $\Theta \cong 0$ at $r = 2M$, which is therefore the outermost 'trapped' surface or apparent horizon. The AH and the TLS thus coincide. The conclusion to be drawn from $\ddot r(\text{AH}) > 0$ is not, however, that the AH will subsequently expand; rather, the photons are only momentarily 'at rest' at this locus. They do not remain at the apparent horizon.

The event horizon is defined by the locus of outgoing photons that can never reach large r. This definition, as is well known, has the unfortunate 'teleological' property of requiring knowledge of the entire future history of the hole. In our present state of ignorance, we do not know what the final state is with any great precision. However, this difficulty is readily circumvented in the present case by recognizing that the question of the escape *versus* trapping of null rays is, physically, a matter of quantitative degree. A working definition of the event horizon is as follows.

(1) The event horizon is strictly null in the order of L in which one is working (here $O(L^1)$).

(2) Photons are imprisoned by the event horizon for times long compared to the dynamical scale $\cong 4M$ of the hole.

In practice this means one evaluates and solves (3.7) near $r = 2M$, keeping all terms in the order of L desired. These criteria enable us to distinguish the AH and the EH to the necessary accuracy without knowing the full global space–time structure. The 'long times' referred to in (2) are quantified below.

The above requirements in $O(L)$ are met by the simple physical condition that the photons at the EH are 'stuck' or 'unaccelerated' in the sense that $\ddot{r}(\text{EH}) \cong 0$. We then find readily that

$$r_{\text{EH}} \cong 2M(1 - 4L) \tag{3.11}$$

$$A_{\text{EH}} \cong 16\pi M^2(1 - 8L) \tag{3.12}$$

$$\dot{r}_{\text{EH}} \cong -2L \tag{3.13}$$

$$\Theta_{\text{EH}} \cong -2LM^{-1}. \tag{3.14}$$

4. QUANTUM ERGOSPHERE

The result (3.11) agrees with Bardeen's (1981) result, which was obtained in order to refute the claim of Tipler (1980) that the EH decays exponentially as seen by a distant observer. (Hajicek and Israel (1980) also showed that exponential decay does not occur, though in a less general way than in Bardeen (1981) or the present work.) Actually, (3.11) slightly overestimates r_{EH} by an amount of order ML^2, which is negligible here. However, this does show that the photons at r_{EH} given by (3.11) would reach $r = 2M$ in a time $\sim ML^{-1}$, which is $\gg 4M$ and is comparable to the expected lifetime of the hole. We see that in contrast to the usual (classical) case or accretion, the EH is inside the AH $=$ TLS. (This qualitative fact was noted without further study by Hajicek and Israel (1980).)

The ordering of the horizons and the invariant geometrical measure of this effect are both determined by a comparison of *areas*:

$$\delta A = A_{\text{AH}} - A_{\text{EH}} = A_{\text{TLS}} - A_{\text{EH}} = 16\pi M^2(8L). \tag{4.1}$$

In analogy to the definition of the ergosphere of a Kerr black hole, where $A_{\text{TLS}} > A_{\text{EH}}$ whenever $(aM^{-1})^2 = (JM^{-2})^2$ is not negligible in comparison to unity, we can define a *quantum ergosphere* as the region between the TLS and the EH. ('Quantum' because classically L cannot be positive for a neutral spherical hole.)

Note that one has shown by solving the relevant dynamical equations that the EH decays in order L according to the formula

$$\frac{\mathrm{d}M}{\mathrm{d}v} \cong \frac{\mathscr{K}}{8\pi} \frac{\mathrm{d}A_{\mathrm{EH}}}{\mathrm{d}v} \qquad (4.2)$$

where, as shown later, the surface gravity $\mathscr{K} = (4M)^{-1} + \mathrm{O}(LM^{-1})$. This, of course, is just the same as the so-called 'first law of black hole mechanics', obtained simply by differentiation of the static Schwarzschild relation $M = (4\pi)^{-1}\mathscr{K} A$. The coincidence of the two formulas in order L might be taken to imply the physical equivalence, for small L, of an evaporating hole to a sequence of static Schwarzschild holes. However, this attitude is misleading. For, as I have shown, the formula (4.2) is valid for a decaying black hole if and only if there is a gap—the quantum ergosphere—of order L between the AH = TLS and the EH. Hence, in order L the two situations are physically different. One cannot hope to reach an accurate understanding of black hole radiance if this difference is ignored.

Because the apparent horizon has radius $r_{\mathrm{AH}} = 2M(v)$, one has that $\dot{r}_{\mathrm{AH}} \cong -2L$. This means that the generating vector $t^a = (1, \dot{r})$ of the AH is timelike: $g_{ab}t^a t^b \cong -4L$. Thus the AH is timelike. This can also be seen by substitution of $r = 2M$ into the metric. One obtains

$$\mathrm{d}s^2(\mathrm{AH}) \cong -4L\,\mathrm{d}v^2 + (2M)^2\,(\mathrm{d}\theta^2 + \sin^2\theta\,\mathrm{d}\varphi^2). \qquad (4.3)$$

Intersecting the AH with a null slice $v = \mathrm{constant}$ shows that the area is $\cong 16\pi M^2$, as used above.

It is clear that radially outgoing massless quanta and ultra-relativistic 'free' massive particles (with relativistic γs such that $\gamma^2 \sim L^{-1}$) can escape from the quantum ergosphere. For example, an outgoing null ray at $r = 2M(1 - 2L)$ has $\dot{r} \cong -L$ and $\ddot{r} \cong +L(4M)^{-1}$. Therefore, in the characteristic time $4M$, it reaches the AH and subsequently escapes. At the AH, it has $\cong 0$ energy-at-infinity and it can gain very little more as it passes through the quasi-static region (the quantum ergosphere is not quasi-static because in it $g_{vv} > 0$). Therefore this effect could only enhance slightly (in $\mathrm{O}(L^2)$) the assumed luminosity L. Hence, the quantum ergosphere in the present treatment must be regarded *only* as the signature imprinted on the metric of black hole radiance, but not as its direct cause, the direct cause being the quantal processes that give rise to L in the first place. Nevertheless, we shall see that the quantum ergosphere is of fundamental significance.

5. PROPER ACCELERATION AND LOCAL TEMPERATURE OF THE APPARENT HORIZON

In the case of an evaporating hole, in contrast to a static or accreting hole, it is possible to construct a timelike worldline (non-geodesic) that remains at the AH and from which two-way communication to large distances is possible in principle. Because the AH coincides with the quasi-static limit (TLS), an 'observer' who follows this worldline is the quasi-static observer who is 'nearest to the black hole'. (The region $r < 2M$ is not quasi-static.) It is clearly of interest to compute his proper acceleration and the local temperature he observes. (Unfortunately, he is likely to be incinerated!)

The proper four-velocity of the observer at $r = 2M$ is given by

$$u^a = [(2L^{1/2})^{-1}, -L^{1/2}, 0, 0]. \tag{5.1}$$

It is straightforward to compute his proper acceleration α^a and find that its magnitude α is given by

$$\alpha(2M) \cong \mathcal{K}\gamma \qquad \gamma = (2L^{1/2})^{-1}. \tag{5.2}$$

Now, for the first time in this paper, I invoke the particular form of the Hawking luminosity: $L_H = B\hbar M^{-2}$, where B is a dimensionless number that depends on the masses and available angular momentum states of the emitted quanta. Consider large holes, where the Hawking temperature $T_H = \mathcal{K}\hbar(2\pi)^{-1}$ is very small. Then B depends only on the number of species and spins of the strictly massless quanta that exist in nature. It is convenient to express B as NB_ν, where N is a positive number and B_ν is the factor for one massless neutrino, i.e., one spin-$\frac{1}{2}$ helicity state. (According to the calculations of Page (1976), $B_\nu = 4.092 \times 10^{-5}$.) We see that α is *independent* of M. (Of course for hot mini-holes, N can depend implicitly on M, but this is of no importance here.) We find an acceleration

$$\alpha = (64NB_\nu\hbar)^{-1/2}. \tag{5.3}$$

If we take $\alpha_p = \hbar^{-1/2}$ as the Planck acceleration, one has that

$$\alpha \approx 20N^{-1/2}\alpha_p. \tag{5.4}$$

We may speculate that this is the largest possible physically meaningful acceleration. If we assume that the only strictly massless particles are photons and gravitons, we find from the results of Page (1976) that $N = 9.181 \times 10^{-1} \approx 1$. ($N$ will be larger, and thus α will be *smaller*, for hot mini-holes because more particles will behave as if they were massless.) This gives a representative value for the acceleration

$$\alpha_{max} \approx 21\alpha_p \approx 13 \times 10^{54} \text{ cm s}^{-2} \approx 14 \times 10^{51} g_\oplus. \tag{5.5}$$

In a similar way, one can regard $\alpha\hbar(2\pi)^{-1}$ as a local Unruh temperature T_U in the limiting quasi-static accelerated frame at the AH. If $T_p = \hbar^{1/2}$ is the Planck temperature, then one has

$$T_U = (256\pi^2 NB_\nu)^{-1/2}T_p. \tag{5.6}$$

Again, this will be largest for holes with low Hawking temperature and thus small N. Using only photons and gravitons as above, one finds

$$T_{U,max} \approx 3.25 T_p \approx 4.61 \times 10^{32} \text{ K}. \tag{5.7}$$

It is interesting that the existence of such maximum values can be inferred merely from the existence of black holes and Hawking effect. In particular it calls to attention the possibly fundamental nature of the zero Hawking-temperature limit and a possible intimate relation among gravity, quantum theory, and the number and spin of the massless particle species that exist in nature (those that can propagate to large distances on the outgoing physical lightcone). An outstanding problem is whether quantum gravity plays a role here. (I mean quantum gravity apart from the consideration of massless spin-two quanta on a fixed classical background.) Below (§9) I shall give an argument that quantum gravity must play such a role.

6.   RAYCHADHURI EQUATION, SURFACE GRAVITY, AND MASS

The Raychadhuri equation for *null* geodesics is

$$d\Theta/dv = \mathscr{K}\,\Theta - R_{ab}l^a l^b - \tfrac{1}{2}\Theta^2 - \sigma_{ab}\sigma^{ab} + \omega_{ab}\omega^{ab}. \tag{6.1}$$

(Cf. Carter (1979), but note a typographical error in the sign of $\mathscr{K}$ in his (6.57).) Here the shear σ_{ab} and vorticity ω_{ab} vanish. The second term on the right is $-8\pi T_{ab}l^a l^b \cong -8\pi T^r{}_v$ in order L. If our solution for the EH is accurate in $O(L)$, we must find $d\Theta/dv \cong 0$ at the EH. (This equation cannot be applied at the $r = 2M$ surface, because the $r = 2M$ surface is not null!)

Because $\Theta_{\mathrm{EH}}{}^2 = O(L^2)$ by (3.14), we need only calculate the surface gravity $\mathscr{K}$ according to formula (2.3). This yields

$$\mathscr{K} \cong \frac{m}{r^2} - \frac{1}{r}\frac{\partial m}{\partial r}. \tag{6.2}$$

We have not needed $\partial m/\partial r$ heretofore; all we need now is to recall from (3.3) that $\partial m/\partial r = O(L)$. Hence, $\mathscr{K} = (4M)^{-1} + O(LM^{-1})$ And we need only use $(4M)^{-1}$ in (6.1). Thus (6.1) becomes at the EH

$$d\Theta/dv \cong (4M)^{-1}(-2LM^{-1}) - 8\pi(-L)(16\pi M^2)^{-1} \tag{6.3}$$

$$\cong -L(2M^2)^{-1} + L(2M^2)^{-1} \cong 0.$$

To associate a definite mass to the black hole, I choose the 'irreducible mass'

$$M_{\mathrm{irred}} \equiv (A/16\pi)^{1/2} \tag{6.4}$$

that was shown by Christodoulou (1970) and Christodoulou and Ruffini (1971) to be the fundamental measure of mass in describing black hole energetics (also in agreement with Hawking's area theorems). We find that

$$M_{\mathrm{irred}}(\mathrm{EH}) \cong M - 4LM \tag{6.5}$$

$$M_{\mathrm{irred}}(\mathrm{AH}) \cong M_{\mathrm{irred}}(\mathrm{TLS}) \cong M. \tag{6.6}$$

Hence, with the quantum ergosphere we may associate the positive mass (positive because $A_{\mathrm{AH}} > A_{\mathrm{EH}}$)

$$\delta M_{\mathrm{QE}} = M_{\mathrm{irred}}(\mathrm{AH}) - M_{\mathrm{irred}}(\mathrm{EH}) \cong 4ML. \tag{6.7}$$

Not surprisingly, we see that in each dynamical time scale $\mathcal{K}^{-1} \cong 4M = \delta v$, the energy δM_{QE} is radiated by the hole:

$$L \cong \mathcal{K} \delta M_{\mathrm{QE}}. \tag{6.8}$$

This is not to say, of course, that the actual 'particle creation' (that implies the existence of a positive luminosity L) occurs physically in the quantum ergosphere.

7. BLACK HOLES AND WHITE HOLES

Hawking (1976, p 196) has observed that an interpretation of black hole decay in terms of reversible thermodynamics (Bekenstein 1975) implies that the formation and evaporation of black holes are statistically the time-reverses of each other. From this, together with the fact that a white hole is (by definition) the time-reverse of a black hole (of the same positive mass), he has conjectured that there must exist a quantum-mechanical process whereby a white hole can evaporate in such a way as to be in principle indistinguishable from an evaporating black hole by an outside observer.

Hawking's argument has been criticized cogently by Penrose (1979). Here, I shall show that the idea is untenable in a different way. In the first place, note that although the formula (4.2)

$$\frac{\mathrm{d}M}{\mathrm{d}v} \cong \frac{\mathcal{K}}{8\pi} \frac{\mathrm{d}A_{\mathrm{EH}}}{\mathrm{d}v} \cong \frac{\mathcal{K}\hbar}{2\pi} \left(\frac{1}{4} \frac{\mathrm{d}A_{\mathrm{EH}}}{\mathrm{d}v} \hbar^{-1} \right) \tag{7.1}$$

with $T_{\mathrm{H}} = \mathcal{K} h (2\pi)^{-1}$ being the Hawking temperature, applies both to evaporation and to accretion by a black hole ($\mathrm{d}M/\mathrm{d}v < 0 \to \mathrm{d}M/\mathrm{d}v > 0$), physically the two situations are quite different, in as much as for accretion one has the well known classical horizon structure $A_{\mathrm{EH}} > A_{\mathrm{AH}}$, while for evaporation one has the quantal horizon structure $A_{\mathrm{EH}} < A_{\mathrm{AH}}$. Only in the second case can there exist in principle a quasi-static observer at the physically well defined apparent horizon who can communicate in a two-way manner with an observer well outside the hole. In other words, the more-or-less 'locally' definable configuration of horizons is observably different in principle. Thus, it is not true that one can regard black hole formation and evaporation as statistically the time-reverses of each other as defined by external observers. In as much as statistical time reversibility would be a necessary consequence of a reversible thermodynamic interpretation of (7.1), it follows that for an external observer there does *not* exist a consistent interpretation of (7.1) in terms of *reversible* thermodynamics of black holes.

The argument can be given in another way. Assume that we have a white hole that is the time reverse of some black hole. Now for white holes, it is 'easy to escape, but hard to get in'. This idea is realized by writing the white hole metric just as in (3.1), but with 'retarded time' u replacing 'advanced time' v:

$$\mathrm{d}s^2 = -(\exp 2\psi)(1 - 2mr^{-1})\mathrm{d}u^2 - 2(\exp \psi)\mathrm{d}u\,\mathrm{d}r + r^2(\mathrm{d}\theta^2 + \sin^2\theta\,\mathrm{d}\varphi^2). \tag{7.2}$$

The infalling photons are described by

$$\tilde{l}^a = (1, \dot{r}). \tag{7.3}$$

This vector is null; therefore,

$$\dot{r} = \frac{dr}{du} \cong -\tfrac{1}{2}(1 - 2mr^{-1}) \tag{7.4}$$

where I use the same approximation as in the case of black holes. The *past* event horizon is where

$$\ddot{r} \cong -\frac{1}{4M}\dot{r} + \frac{1}{2M}\left(\frac{\partial m}{\partial u}\right) \cong 0. \tag{7.5}$$

Therefore,

$$r_{\text{EH}} \cong 2M\left(1 - 4\frac{\partial m}{\partial u}\right) \tag{7.6}$$

and $r_{\text{AH}} \cong 2M$. Hence, if $\partial m/\partial u > 0$, $r_{\text{EH}} < r_{\text{AH}}$, that is, the 'condensation' of a white hole is a quantal process and is 'locally' the time-reverse of black hole evaporation. (This is implicit in Penrose's (1979) argument, which goes on to argue that *globally* the two situations are not the time-reverses of each other. However, here I am not trying to deal with the fully global problem for reasons I have explained in giving a workable definition of the event horizon.)

What I have shown can be described as follows. Let B denote a black hole and W a white hole with equal mass. Assume that they both can have the same temperatures and absolute values of rate-of-change of mass. Let a denote accretion and e evaporation. Let $\mathcal{T}$ denote the operation of time-reversal ($\mathcal{T}\mathcal{T} =$ identity). Let $\sim$ denote physical equivalence (in particular of the horizon configurations) in order L. Then $\mathcal{T}(B_e) \sim W_a$, $\mathcal{T}(B_a) \sim W_e$ (which follows from using $\partial m/\partial u < 0$ in (7.6) and the fact that $\partial/\partial u = \partial/\partial v$ at large r), $\mathcal{T}(W_a) \sim B_e$, and $\mathcal{T}(W_e) \sim B_a$. Now the claim $B_e \sim W_e$ is necessary and sufficient for $\mathcal{T}(B_a) \sim B_e$. But $\mathcal{T}(B_a) \not\sim B_e$ so $B_e \sim W_e$ is impossible. (Note that $\mathcal{T}$ (classically allowed configuration of black hole) $\sim$ (classically allowed configuration of white hole) and that $\mathcal{T}$ (black hole with quantal evaporation) $\sim$ (putative quantally condensing white hole) as is not surprising.)

8.   HIGH TEMPERATURE EXTRAPOLATION

As a black hole evaporates, its temperature $T_{\text{H}} = \mathcal{K}h(2\pi)^{-1} = h(8\pi M)^{-1}$ and luminosity $L_{\text{H}} = B\hbar M^{-2}$ can become very large. (One expects B to increase as M decreases because at higher T_{H}, the number of radiatable species increases.) If one ignores the back-reaction, the quantities T_{H} and L_{H} become unboundedly large as $M \to 0$. However, already in order L, we can see that the back-reaction must significantly modify these conclusions both quantitatively and qualitatively. It would be desirable to use a more accurate description of the metric and stress–energy tensor in order to have complete confidence in the picture of an evaporating micro-hole. Nevertheless, let us examine the extrapolation in O (L) to see what it says; as the present results are, as I have argued, only weakly dependent on details of the stress–energy tensor.

The EH is inside the AH and both are shrinking at a 'velocity' $dr/dv \cong -2L$ as seen by a distant observer. In O (L^2), it is clear (and can easily be verified) that the EH is receding slightly faster than the AH because it is null and the AH is timelike. Thus, the AH cannot 'catch' the EH. The EH must disappear first, with $M(v) > 0$. Let us define, heuristically, 'disappearance' as when the gravitational radius of the apparent horizon equals that of the quantum ergosphere. This implies that $L \cong \frac{1}{4}$. This value is very crude and is not a serious quantitative estimate. The point, however, is that L is not large in comparison to unity. At the 'last moment' in the lifetime of a black hole that is assumed to run the entire course of evaporation, there will be an apparent horizon of small area enclosing a small amount of mass, but no event horizon. The putative 'naked singularity' that will be 'revealed' in the next instant ($\sim 4M$) will have no very dramatic, that is, infinite, features. (Of course, if what I have defined as the EH *does* completely disappear, and if one insists upon the technical teleological definition of an event horizon, then there 'never was' an event horizon in the first place. I think this illustrates very well that the text book definition of an event horizon has to be abandoned in studies of black hole radiance.)

We can further check the picture of black hole evaporation in the present approximation, and at the same time consider a speculative possibility for connecting gravitational and particle physics mass scales, in the following manner. Note that the 'red shift' factor γ in (5.2) becomes equal to unity when $L = \frac{1}{4}$, the same value as the moment when the event horizon 'disappears'. Effectively, then, in this approximation, when the EH disappears there is no red shift. When there is no effective red shift the Hawking temperature (which depends on the mass of the 'hole', meaning here the mass M of the AH) is equal to the local Unruh temperature (which can be expressed independently of M, in terms of the equivalent number N of massless spin-$\frac{1}{2}$ helicity states). By equating these two temperatures we arrive at a 'smallest mass' M_* and 'final temperature' $T_* = \hbar(8\pi M_*)^{-1}$ of the hole. We find

$$M_* = 2(NB_v)^{1/2} M_P \tag{8.1}$$

$$T_* = (256\pi^2 NB_v)^{-1/2} T_P. \tag{8.2}$$

It is amusing to put some numbers in. A great deal of further work will be required in order to have any solid confidence in the quantitative estimates; nevertheless, we proceed. Take, for example, the case of $N = 8$ supergravity. At high temperatures we take all of its $2^8 = 256$ helicity states as massless. This involves spins $0, \frac{1}{2}, 1, \frac{3}{2}$, and 2. Treat them all as external fields and convert them to equivalent black hole luminosity units N as explained before. Page (1976) gave results for spins $\frac{1}{2}, 1$, and 2. For spin 0 (real), Page has kindly supplied me with a recent (January 1983) result of T Elster (Jena). This gives $N = 1.819$ for spin 0 ($N = 1$ for spin $\frac{1}{2}$). For spin $\frac{3}{2}$, one can estimate using the other data $N \approx 1.5 \times 10^{-1}$. Adding up everything, for $N = 8$ supergravity, I obtain $N \approx 265$, from which $M_* \approx 0.2 M_P$ and $T_* \approx 0.2 T_P$. The same procedure can be followed taking the minimal GUT Georgi–Glashow theory plus Einsteinian gravity and one finds $N \approx 127$, $M_* \approx 0.15 M_P$, and $T_* \approx 0.3 T_P \approx 3 \times 10^{18}$ Gev. These temperatures are on the low side of the Planck value. I would expect more accuracte work to lower them further, but this remains to be seen.

9.   LOW TEMPERATURE LIMIT

As the mass of a black hole becomes large, T_H and L_H become arbitrarily small, so that there would appear to be essentially no quantum radiance. However, in this limit, the formula $L_H \approx B\hbar M^{-2}$ becomes strictly correct, that is, $B = B_v N$ is independent of M (N and B_v are defined as before). Let us look at what happens to the quantum ergosphere in this limit. This is best measured, as I have emphasized earlier, by comparing areas. Here

$$\delta A = A_{AH=TLS} - A_{EH} = 16\pi M^2 (8L). \tag{9.1}$$

Using $L = L_H$, we find that δA is independent of the mass:

$$\delta A = 128\pi B\hbar = 128\pi N B_v \hbar. \tag{9.2}$$

Because N cannot be zero, δA cannot be zero. In this sense, the idea that one should obtain a standard vacuum static Schwarzschild geometry in this limit for black hole in 'empty' space is not accurate. The classical limit, as always, is $\hbar \to 0$. We see, as in §5, that this mass independent limit is determined by the number of species and the spins of zero rest mass quanta. Using $N = 9.181 \times 10^{-1}$ as described in §5, we see that a representative value is

$$\delta A \approx (0.02)\hbar. \tag{9.3}$$

A mass-independent linear measure associated with the quantum ergosphere in this limit is provided by the radius of curvature of the worldline of the limiting quasi-static observer. This number (α_{max}^{-1}) is the same as the local proper time $\delta\tau$ this observer takes to travel a distance equal to the width of the quantum ergosphere:

$$\delta\tau = \alpha_{max}^{-1} = (64 B\hbar)^{1/2} = (64 N B_v \hbar)^{1/2}. \tag{9.4}$$

The representative value corresponding to (9.3) is about one twentieth of the Planck value.

The results of this section suggest the possibility that a 'zero-temperature area gap' like (9.2) could result from intrinsic zero temperature gravitational quantum noise in the metric g, i.e., $\delta g \sim \hbar^{1/2} M^{-1}$ such that the mean value $\langle \delta g \rangle$ is zero but there is a residual effect $\sim \langle (\delta g)^2 \rangle \sim \hbar M^{-2}$ that would account for a result like (9.2). Such noise should occur as zero-point energy (non-radiative) in the fundamental 'ringing' modes of the hole. (For the classical ringing modes, see, for example, Detweiler (1979).) In fact, this is the case (York 1983). It should be noted that these fluctuations are not the same as the uncertainty in the metric owing to the uncertainty of 'particle number' in the created (external) fields that was noted by Hawking (1975) and Bardeen (1981). The latter fluctuations have amplitude $\delta g \sim L \sim \hbar M^{-2}$. They are basically the fluctuations in the back-reaction to creation of propagating external fields that I have studied in this article. My conclusion is that quantum radiance cannot be understood thoroughly even at very low luminosities without a verification that the results are compatible with low energy quantum gravity, whose effects, as it were, turn out to be 'amplified' (see Fulling's article in this volume) by the event horizon.

ACKNOWLEDGMENTS

I have discussed the work reported in this article with numerous colleagues. For helpful remarks, I wish to thank D Boulware, D Eardley, P Frampton, T Kephart, R Matzner, M Perry, T Piran, S Teukolsky, W Unruh, H Van Dam, J Wheeler, and especially S Christensen. Research support was received from the National Science Foundation.

REFERENCES

Bardeen J 1981 *Phys. Rev. Lett.* **46** 382
Bekenstein J 1975 *Phys. Rev. D* **12** 3077
Birrell N D and Davies P C W 1982 *Quantum Fields in Curved Space* (Cambridge: Cambridge University Press)
Carter B 1979 in *General Relativity: An Einstein Centenary Survey* ed SW Hawking and W Israel (Cambridge: Cambridge University Press)
Christodoulou D 1970 *Phys. Rev. Lett.* **25** 1596
Christodoulou D and Ruffini R 1971 *Phys. Rev. D* **4** 3552
Detweiler S 1979 in *Sources of Gravitational Radiation* ed L Smarr (Cambridge: Cambridge University Press)
DeWitt B 1975 *Phys. Rep.* **19** 295
Hajicek P and Israel W 1980 *Phys. Lett.* **80** 9
Hawking S W 1975 *Commun. Math. Phys.* **43** 199
—— 1976 *Phys. Rev. D* **13** 191
Page D 1976 *Phys. Rev. D* **13** 198
Parker L 1968 *Phys. Rev. Lett.* **21** 562
—— 1977 in *Asymptotic Structure of Space–Time* ed F P Esposito and L Witten (New York: Plenum)
Penrose R 1979 in *General Relativity: An Einstein Centenary Survey* ed S W Hawking and W Israel (Cambridge: Cambridge University Press)
Tipler F 1980 *Phys. Rev. Lett.* **45** 949
Unruh W G 1976 *Phys. Rev. D* **14** 870
York J W *Dynamical Origin of Black Hole Radiance* to appear

Black Hole Fluctuations

JACOB D BEKENSTEIN

1. THE SIGNIFICANCE OF BLACK HOLE FLUCTUATIONS

The present resurgence of interest in quantum gravity, to which Bryce DeWitt's fundamental work has contributed so much, was triggered by the discovery of Hawking's radiance (Hawking 1975), a black hole phenomenon later found to be relevant in other contexts as well. It should not be surprising if the black hole, that geometrodynamical entity *par excellence*, continues to play a crucial role in the quest for understanding of the quantum aspects of the gravitational field. This is one reason for studying the thermal fluctuations of black holes. In this connection we may recall that it was the study of thermal fluctuations of electromagnetic energy (Einstein 1924) that clinched the evidence for a quantized electromagnetic field. One may thus hope that studies of black hole fluctuations, like the pioneering investigations of Candelas and Sciama (1977), will advance our conception of quantum gravity.

The present report is, however, not concerned with quantum fluctuations of the gravitational field, but with thermal fluctuations of black hole *parameters*. According to Hawking a hole emits quanta stochastically. Thus even if its mass M, say, is known precisely now, one will later on only be able to talk of a probability distribution for M, not of its precise value. Everybody's hope is that the distribution remains a sharp one, and that the mean mass $\langle M \rangle$ is a good estimate of the actual mass. Is this expectation supported by direct evidence?

The other black hole parameters, angular momentum L, charge Q and momentum P, must similarly be subject to a probability distribution $\Pi(M, L, Q, P; Z)$ where Z lumps all parameters governing the distribution. One approach is to assume that Z stands *only* for $\langle M \rangle$, $\langle L \rangle$ . . . and, in particular, that Π does not depend on initial conditions (Bekenstein 1981). This sounds reasonable since one is dealing with a thermal distribution which should prove forgetful of initial conditions. Further, Wheeler's principle 'black holes have no hair' is antagonistic to the appearance of initial conditions (a kind of 'hair') in the description of the present *state* of the hole. Yet this reasonable approach leads to a paradox: the hole's mass may not exceed a certain limit. The prospect is so strange that we are led to reconsider the possibility that initial conditions may, after all, show up late in the evaporation of the hole. Our main conclusion here is that Z stands for initial conditions as well as $\langle M \rangle$, $\langle L \rangle$ ....

The emphasis here is on fluctuations of an *isolated* hole. The counterpart of this problem, the fluctuations of a hole in equilibrium with blackbody radiation, has previously received some attention (Hawking 1976a, Zurek 1980). Even Page's (1977) discussion of charge fluctuations of a nearly neutral hole seems to refer to a hole in equilibrium with radiation since his balance argument is one of detailed balancing (which requires equilibrium), and not merely a CP invariance argument. By contrast the isolated hole problem has been less explored (Gerlach 1976, Bekenstein 1981). To avoid obscuring the essential points with details, we shall consider only fluctuations in M and L, assuming Q and P to be precisely zero. One can argue, in line with Page (1977, 1980), that fluctuations in Q for a neutral (in the mean) hole should be of order $\hbar^{1/2}$, and those of P for a hole at rest should be of the same order (units with $G = c = k = 1$ are used throughout).

The plan of our discussion is as follows. We first investigate the precision with which black hole mass may be measured. This sets the threshold for interesting fluctuations in M (§2). The spreading of the mass distribution in the course of time is illustrated by a concrete calculation (§3) in which we model the emitted quanta as scalars, and use the geometric approximation. Just after the dispersion becomes large enough to be directly measurable, $\Pi(M)$ becomes approximately Gaussian. Accordingly, in §4 we develop Ben-Yaacov's (1981) idea of using only second-order moments to describe the evolution of Π.

As an application we discuss in §5, in a model-free way, the coupled evolution of $\langle L \rangle$ and the covariance of L with M. The evolution of $\langle L \rangle$ is more complicated than commonly thought, but there is a tendency for both $\langle L \rangle$ and the covariance to 'evaporate' *provided* the hole does not emit too many scalar species. The variance of L is the subject of §6 where we show that it has the general form $\hbar \langle M \rangle^2$ (which might have been guessed by intuitive arguments) *provided* the variance of M is small. However, in §7 it becomes clear that the variance of M, far from being always of order $\hbar$ as one might have thought intuitively, grows as the *fourth* power of black hole temperature. It can become large when the hole is still many orders of magnitude above the Planck–Wheeler scale. As a result, the law of decrease of $\langle M \rangle$ differs from the Hawking–Page picture in the last stages of evaporation.

2. MONITORING BLACK HOLE EVAPORATION

When monitoring the evaporation of a black hole by *direct* measurements of M, one is limited by the time–energy uncertainty relation as well as by the systematic decrease of M.

If the measurement interval is τ one intuitively expects a measurement uncertainty $\Delta M = \xi \hbar / \tau$. It is not clear whether one can approach the ideal $\xi = \frac{1}{2}$ suggested by the oft-quoted relation $\Delta E \Delta t \geqslant \frac{1}{2}\hbar$. Ruffini and Ohanian (1974) have argued that if M is inferred from the small angle scattering of a mass $m \ll M$ with velocity $v \ll 1$ having impact parameter b, then ξ is at best b/m. (They identify τ with the flyby time b/v.) It can also be shown (see Appendix) that if one infers M from measurements of the energy and angular momentum of a mass $m \ll M$ *bound* to the hole, then ξ is at best $a/(\sqrt{2}\,m)$ where a is the pericenter distance. Here τ can be any arbitrary small interval. Both arguments are nonrelativistic and ignore recoil effects. With appropriate corrections one might expect the numerical coefficients to change by, perhaps, an order of magnitude. Then by going to the limits a (or b) $\to$ a few M, $m \to M$, one might hope to make ξ of order unity in the wide sense.

During the same interval τ evaporation reduces M by $\delta M = \alpha \hbar M^{-2}\tau$ (see Page (1976a) for calculation of α). Therefore, M can differ from its nominal value by perhaps $\frac{1}{2}\delta M$. Any direct determination of M thus carries a total error $(\Delta M^2 + \frac{1}{4}\delta M^2)^{1/2}$. This error is minimal when $\tau = (2\xi/\alpha)^{1/2} M$; it then equals $(\xi\alpha)^{1/2}\hbar M^{-1}$. Using Page's result $\alpha \cong 2 \times 10^{-4}$ and the generous range $\frac{1}{2} < \xi < 20$, we find that, to within a factor 3, the optimal mass resolution is $\sim 0.0025\hbar M^{-1}$ corresponding to $\tau \approx 88 \times 2M$. The question is whether fluctuations in M are large enough to be resolved.

3. SPREADING OF SCHWARZSCHILD MASS

Suppose that at some initial time $t = 0$ (defined only to about $2 \times 10^2 M$) M is known (to a precision of some $3 \times 10^{-2}\hbar M^{-1}$). What is the probability distribution $\Pi(M)dM$ after some time Δt? This is the same as asking what the probability distribution for the emitted energy $E, f(E)dE$, is. We shall here confine our attention to $\Delta t \ll M^3/\hbar$ so that M changes little by evaporation in time Δt.

The $f(E)dE$ is nothing but the sum of probabilities for all processes in which the sum of energies of the emitted quanta is between E and $E + dE$. For massless quanta the critical parameter controlling the emission in mode i of frequency ω_i is $x_i \equiv \hbar\omega_i/T_{bh} = 8\Pi M\omega_i$. For a Schwarzschild hole the absorptivity Γ_i depends only on x_i and the angular momentum quantum number l. Defining $\exp \beta_i$ by

$$\exp \beta_i - 1 = (\exp x_i - 1)/\Gamma_i \tag{3.1}$$

we can write the joint probability for emission of n_1 quanta in mode 1, n_2 in mode 2 . . . as (Bekenstein 1975, Parker 1975, Wald 1975, Hawking 1976b)

$$p(n_1, n_2 \ldots) = \Pi_i [1 - \exp(-\beta_i)] \exp(-\beta_i n_i). \tag{3.2}$$

To sum $p(n_1, n_2 \ldots)$ over all combinations of occupation members for which $\hbar\Sigma n_i\omega_i$ lies between E and $E + dE$ is hard because β_i depends intricately on ω_i. However, by sacrificing a bit of accuracy in exchange for clarity we can readily make progress.

Let us model the emitted quanta as scalar particles. Such a device was employed, for example, by Bryce DeWitt (1975) in his lucid exposition of the Hawking radiance.

He also relied on the geometrical optics approximation. The idea is to set $\Gamma = 1$ for all modes with ω large enough to surmount the centrifugal potential barrier guarding the hole, and $\Gamma = 0$ otherwise. The actual cut-off of Γ is more gradual. One can check on the effect of these approximations by comparing DeWitt's scalar model estimate $\alpha \approx 9/(10 \times 8^4 \times \pi) \approx 7 \times 10^{-5}$ multiplied by 6 (there are 6 helicities—two in photons, two in gravitons, and one each in the neutrinos) with Page's accurate value 2 $\times 10^{-4}$. The discrepancy is only a factor 2. We thus adopt the geometric approximation in what follows. For scalar waves the threshold for ω is, to a good approximation, $\omega_t = [l(l+1)/27]^{1/2} M^{-1}$. Only for $l = 0,1$ is this estimate off by more than 5%. For $l \geqslant 5$ it is good to 1%.

Because $\beta_i = x_i$ for $\Gamma_i = 1$ while $\exp(-\beta_i) = 0$ for $\Gamma_i = 0$ we have

$$f(E)dE = \sum_{\{n_i\}} \prod_i [1 - \exp(-x_i)] \exp(-x_i n_i). \tag{3.3}$$

Here the product extends over modes with $\omega > \omega_t$, and the sum over configurations with $E < \Sigma \hbar \omega_i < E + dE$. Notice that n_i enters only in the form $\Sigma n_i x_i$. Thus if $N(E)dE$ is the number of configurations with energy in $(E, E + dE)$ occupying only modes with $\omega < \omega_t$,

$$f(E) = \prod_i [1 - \exp(-x_i)] \times N(E)\exp(-E/T_{bh}). \tag{3.4}$$

To calculate the mode product in (3.4) (call it W) re-express its logarithm in the continuum approximation:

$$W = \frac{\Delta t}{2\pi} \int_0^\infty d\omega n_t(\omega) \ln[1 - \exp(-8\pi M\omega)]. \tag{3.5}$$

Here $\Delta t d\omega / 2\pi$ is the number of modes of a given angular type in the frequency interval $d\omega$ which are emitted in time Δt; $n_t(\omega)$ denotes the number of angular types at frequency ω which are above threshold. Since $2l + 1$ degenerate modes share each l,

$$n_t(\omega) = \sum_{l=0}^{l_m} (2l+1) = (l_m + 1)^2 \tag{3.6}$$

where l_m is the *largest integer* l for which $l(l+1) \leqslant 27M^2\omega^2$, or equivalently $(l + \frac{1}{2})^2 \leqslant 27M^2\omega^2 + \frac{1}{4}$. Of course, on the average l_m falls half a unit under the (non-integer) l which satisfies the equality. We thus replace $(l_m + 1)^2$ with $27M^2\omega^2 + \frac{1}{4}$ so that

$$n_t(\omega) \approx 27M^2\omega^2 \tag{3.7}$$

to good accuracy even for low ω. With this n_t (3.5) is easily integrated by parts and reduced to an integral familiar from blackbody theory:

$$W \approx \exp[-3\Delta t/(10 \times 8^3 M)]. \tag{3.8}$$

To get $N(E)dE$ we would now be invited to count all distinct configurations of quanta occupying modes above threshold with total energies in the specified interval $(E, E + dE)$, not an easy task. We note that if Δ is some typical spread in energy,

$\ln[N(E)\Delta]$ stands for the *microcanonical* entropy of a system of scalarons (the precise value of Δ is not important when one is after the entropy). Now, *canonical* entropy, which is easily computed, will closely approximate the microcanonical one for a large system. We thus propose to use canonical entropy as a stepping stone to $N(E)$.

The partition function for a system of scalarons at inverse temperature $\tilde{\beta}$ is

$$\ln Z = -\sum_i \ln[1 - \exp(-\tilde{\beta}\hbar\omega_i)] \tag{3.9}$$

where ω_i are the allowed frequencies. In our context we need include only $\omega_i > \omega_t$. But then the evaluation of (3.9) exactly parallels that of (3.5) giving

$$\ln Z = 3\pi^3 M^2 \Delta t/(10\tilde{\beta}^3 \hbar^3). \tag{3.10}$$

We stress that $\tilde{\beta}^{-1}$ has nothing to do with T_{bh} but is merely a parameter to be chosen so that the canonical mean energy, $-\partial \ln Z/\partial\tilde{\beta}$, coincides with our E. We thus discover that

$$\tilde{\beta} = (9\pi^3 M^2 \Delta t/10 E \hbar^3)^{1/4}. \tag{3.11}$$

The canonical entropy $S_c = \ln Z - \tilde{\beta}\partial \ln Z/\partial\tilde{\beta}$ is just $4\ln Z$. Substituting $\tilde{\beta}$ and equating S_c to $\ln[N(E)\Delta]$ on the assumption that our system contains many quanta, we may infer $N(E)$.

Putting all the pieces together we find

$$f(E) = \Delta^{-1} W \exp 4\left[\left(\frac{\pi^3 M^2 \Delta t E^3}{9 \times 10\hbar^3}\right)^{1/4} - \left(\frac{2\pi ME}{\hbar}\right)\right]. \tag{3.12}$$

Note that for fixed Δt f is the product of a rapidly rising function of E, and an exponentially decaying one. It thus has a sharp peak easily found to be at

$$E_{peak} = 9\hbar\Delta t/(10 \times 8^4 \pi M^2). \tag{3.13}$$

This is the *most probable* energy radiated by the hole in time Δt. It coincides with DeWitt's estimate of the *mean* energy loss to radiation (DeWitt 1975). This checks our procedure.

Near the peak $f(E)$ will be a Gaussian. Expanding the argument of the exponent (3.12) about E_{peak} to $O(E - E_{peak})^2$ we get

$$f(E) = \Delta^{-1} \exp\left[-\tfrac{1}{2}(E - E_{peak})^2/\sigma_E^2\right] \tag{3.14}$$

$$\sigma_E^2 \equiv 9\hbar^2 \Delta t/(10 \times 8^5 \pi^2 M^3). \tag{3.15}$$

Since most of the probability lies in this Gaussian core of f, we must have $\Delta \approx \sqrt{2\pi}\,\sigma_E$ in order for f to be normalized to unity. The Gaussian form will be a good approximation to (3.12) if $E_{peak} \gg \sigma_E$; otherwise the Gaussian will be truncated at low E. This inequality amounts to $\Delta t \gg 10 \times 8^3 M/9$. This same restriction happens to guarantee the validity of the more general form (3.12). After all, canonical entropy can be expected to approximate $\ln[N(E)\Delta]$ only if the system includes *many* quanta. Since a typical quantum has energy $\sim T_{bh}$, we must thus require $E_{peak} \gg T_{bh}$, or

$\Delta t \gg 10 \times 8^3 M/9$. For shorter Δt one would have to calculate $N(E)$ directly and laboriously.

When $\Delta t \gg 10 \times 8^3 M/9$, $\sigma_E \gg \hbar/8\pi M$. This σ_E is also the width of the hole's mass distribution, at least for Δt not too large. By carrying out a measurement of M over time $\tau \approx 176M \ll \Delta t$ we can, according to §2, attain an accuracy $\sim 0.0025\hbar M^{-1}$ which is much better than the width σ_E. Hence when the Gaussian approximation (3.14) is valid, the spread in M is directly measurable.

Note that

$$\sigma_E{}^2 = T_{bh} E_{\text{peak}}. \tag{3.16}$$

This may be understood by noting that the emission is stochastic and hence one may expect the *variance* in the number of emitted quanta to equal the mean number which is about E_{peak}/T_{bh}. Since the typical quantum energy is $\sim T_{bh}$ one expects $\sigma_E{}^2 \sim T_{bh} E_{\text{peak}}$. The unit coefficient in (3.16) is striking, and we conjecture that (3.16) may be valid beyond our specific approximations, and even for emission of non-scalar quanta.

4.   BEN-YAACOV'S MOMENTS FORMALISM

By conservation of energy $\Pi(M) = f(M_0 - M)$ where M_0 is the initial mass (assumed to be sharply known). The results (3.12) and (3.14) for f cannot continue to be valid for very large Δt since the implicit assumption that M in (3.12) and (3.15) is fixed begins to fail. The mass is evaporating. One can evidently take care of this by convolving f for an interval $\Delta t/n$ with itself n times, and taking care at each stage of the decrease in M. Provided $\Delta t/n \gg 8^3 M$, each f is Gaussian, and so is the full convolution which is Π. If so, the variance of M is enough to characterize Π completely. This insight motivates Ben-Yaacov's (1981) formalism describing the evolution of $\Pi(M, L, Q)$ in terms of its various second moments. In this section we develop the general formalism reserving applications for later sections.

A hole emits various conserved quantities (energy, charge . . .). Denote by $x_{\Delta t}$ and $y_{\Delta t}$ the amounts of two of them emitted in time Δt. One may discuss mean values such as $\bar{x}_{\Delta t}$ or $\overline{y_{\Delta t}^2}$ with respect to the probabilities $p(n_1, n_2 \ldots)$ of §3. Let X and Y be the hole parameters corresponding to x and y (i.e., if $x = E$, $X = M$). By the conservation laws $X(t + \Delta t) = X(t) - x_{\Delta t}$; a similar relation holds for y, Y. At any given time X and Y are distributed according to $\Pi(X, Y; t)$; one may thus consider averages with respect to Π, $\langle X \rangle$, $\langle Y \rangle$, as well as covariances like

$$\sigma(X, Y; t) = \langle X(t)Y(t) \rangle - \langle X(t) \rangle \langle Y(t) \rangle. \tag{4.1}$$

The distribution $\Pi(X, Y; t + \Delta t)$ is evidently a kind of convolution of $p(n_1, n_2 \ldots)$ with $\Pi(X, Y; t)$. Thus when we average a conservation relation, we may average the left-hand side with respect to $\Pi(X, Y; t + \Delta t)$ and the right-hand side with respect to

$p(n_1, n_2 \ldots)$ first, and then with respect to $\Pi(X, Y; t)$. In this way we sidestep the complexities of relating the two Πs explicitly. For example

$$\langle X(t + \Delta t) \rangle = \langle X(t) \rangle - \langle \bar{x}_{\Delta t} \rangle. \tag{4.2}$$

In general $\bar{x}_{\Delta t}$ has the form $f_x(X, Y) \times \Delta t$; an example is afforded by (3.13). For Δt small we may reduce (4.2) to

$$\langle \dot{X} \rangle = - \langle f_x \rangle \tag{4.3}$$

where a dot denotes time derivative. Relations like (4.3) are often used in discussing black hole evaporation, except that $\langle f_x \rangle$ is replaced by $f_x(\langle X \rangle, \langle Y \rangle)$. As we shall see, this leaves out a lot of interesting physics.

Multiplying $X(t + \Delta t) = X(t) - x_{\Delta t}$ by the analogous y relation, averaging as above, and subtracting the product of (4.2) with the analogous y relation we get, in the limit of small Δt,

$$\dot{\sigma}(X, Y) = \dot{\sigma}(x, y) - \sigma(f_x, Y) - \sigma(f_y, X). \tag{4.4}$$

By $\sigma(f_x, Y)$ we mean $\langle f_x Y \rangle - \langle f_x \rangle \langle Y \rangle$, while $\sigma(x, y)$ denotes $\langle \overline{xy} \rangle - \langle \bar{x} \rangle \langle \bar{y} \rangle$ (by x we mean $x_{\Delta t}$, of course). The example of σ_E^2 in §3, which is $\overline{E^2} - \bar{E}^2$, suggests that in general the leading term in $\sigma(x, y)$ will grow as Δt. One may thus define the 'time derivative' $\dot{\sigma}(x, y)$. Equations (4.3) and (4.4) are the basic dynamics of Π.

5. EVAPORATION OF ROTATION AND ANGULAR MOMENTUM—MASS CORRELATIONS

As an application of the general formalism consider the joint dynamics of $\langle L \rangle$ and the covariance $\sigma(M, L)$, i.e. $x = E$, $y = l$. The evaluation of the averages implicit in (4.3) and (4.4) is much facilitated by the well known result

$$\langle \bar{f}(X, Y) \rangle \approx \bar{f}(\langle X \rangle, \langle Y \rangle) + \bar{f}_{XY}\sigma(X, Y) + \tfrac{1}{2}\bar{f}_{XX}\sigma(X, X) + \tfrac{1}{2}\bar{f}_{YY}\sigma(Y, Y) + \ldots \tag{5.1}$$

where $\bar{f}$ is an arbitrary function, $\bar{f}_{XY}$ stands for $\partial^2 \bar{f}/\partial X \partial Y$ evaluated at $\langle X \rangle$, $\langle Y \rangle$, etc. The corrections in (5.1) involve higher moments. There are obvious generalizations to three or more variables. Now according to Page (1976b) $f_l = \beta \hbar L M^{-3}$ where β is positive, dimensionless and depends only on L^2/M^4. For $L = 0$ we denote β by β_0. The general form of f_l is understandable on dimensional grounds. Evaluating (4.3) with the help of (5.1) to first order both in $\langle L \rangle$ and the $\sigma(X, Y)$ we get

$$\langle \dot{L} \rangle = - B \langle L \rangle + 3B \langle M \rangle^{-1} \sigma(M, L) \tag{5.2}$$

where $B \equiv \beta_0 \hbar \langle M \rangle^{-3}$. This equation will be useful for black holes which are nearly Schwarzschild and have nearly uncorrelated M and L.

To obtain an equation for $\sigma(M, L)$ we evaluate (4.4) again to first order in $\langle L \rangle$ and to second-order moments. First we recall that $f_E = \alpha \hbar M^{-2}$; $\alpha > 0$ and for $L \neq 0$ it depends only on L^2/M^4 (Page 1976a). Because it is a sum of terms like $\hbar \omega \times \hbar m$, $\overline{El}$

must be odd in L and have a $\hbar^2$ dependence. Thus, on dimensional grounds, $\overline{El} - \overline{E}\,\overline{l} = \gamma \hbar^2 L \Delta t M^{-4}$ where γ is dimensionless and depends only on L^2/M^4. To exploit (5.1) it pays to write

$$\sigma(E, l) = \langle \overline{El} - \overline{E}\,\overline{l} \rangle + (\langle \overline{E}\,\overline{l} \rangle - \langle \overline{E} \rangle \langle \overline{l} \rangle). \tag{5.3}$$

Evidently $\langle \overline{E}\,\overline{l} \rangle$ and $\langle \overline{E} \rangle \langle \overline{l} \rangle$ are both of $O(\Delta t^2)$ so we are entitled to drop them in defining $\dot{\sigma}(E, l)$. With this precaution we find

$$\dot{\sigma}(M, L) = C \langle L \rangle - (B - 2A + 4C \langle M \rangle^{-1})\sigma(M, L) \tag{5.4}$$

where $C = \gamma_0 \hbar^2 \langle M \rangle^{-4}$ and $A = \alpha_0 \hbar \langle M \rangle^{-3}$. Here α_0 and γ_0 denote the Schwarzschild values of α and γ.

A particular solution of (5.2) and (5.4) is $\langle L \rangle \equiv 0$ and $\sigma(M, L) \equiv 0$; thus if a hole starts as being non-rotating and with M and L uncorrelated, it stays that way. If $\langle L \rangle$ and $\sigma(M, L)$ do not vanish initially, one may still draw conclusions from the system (5.2) and (5.4) for a hole much more massive than the Planck–Wheeler mass ($M^2 \gg \hbar$). Then in (5.4) $C \langle M \rangle^{-1}$ is negligible as compared to $B - 2A$. Now Page (1976b) has shown that if only photon, graviton and neutrino (or relativistic electron) contributions are considered, β_0 is an order of magnitude larger than α_0. Thus $B - 2A$ is positive and an order larger than A. We notice that the coefficient of $\langle L \rangle$ in (5.4) is smaller by a factor $\hbar \langle M \rangle^{-2}$ than that of $\sigma(M, L)\langle M \rangle^{-1}$. (By contrast the coefficients of the corresponding terms in (5.2) are of the same order.) Thus the evolution of $\sigma(M, L)$ is essentially a quasi-exponential *decay* on a timescale $(B - 2A)^{-1}$ which is an order shorter than the standard black hole lifetime A^{-1}.

We might try to cast doubt on this conclusion because when $\sigma(M, L)$ becomes very small, the term $C \langle L \rangle$ in (5.4) may become important anyway. Of course, long before that happens, the term $-B \langle L \rangle$ dominates the right-hand side of (5.2) so that $\langle L \rangle$ decays quasi-exponentially on timescale B^{-1} somewhat shorter than $(B - 2A)^{-1}$. Thus the $C \langle L \rangle$ cannot arrest the decay of $\sigma(M, L)$. Both rotation and correlation between M and L are evaporated rather quickly. This extends Page's (1976b) conclusion.

This comfortable situation will be altered if the hole can emit scalar particles (Higgs particles?). Page (1976b) has persuasively argued that in this case $\beta_0 - 2\alpha_0$ can be negative. Then according to (5.4) there is a tendency for quasi-exponential *growth* of $\sigma(M, L)$. Eventually $\sigma(M, L)$ gets big enough to affect $\langle L \rangle$ and it may interrupt the decay of $\langle L \rangle$ if the signs are right. Thus scalar fields entail a definite instability for black holes.

6. THE SPREAD IN ANGULAR MOMENTUM

We now turn to the calculation of $\sigma(L, L)$ whose square root measures the spread in L about $\langle L \rangle$. According to (4.4) we must compute $\dot{\sigma}(l, l)$. As in the example illustrated by (5.3), this only requires us to know $S = \overline{l^2} - \overline{l}^2$. Being a sum over quantities like $m^2 \hbar^2$, this variance must be even in L and have a $\hbar^2$ dependence. On dimensional grounds we write $S = \delta \hbar^2 \Delta t M^{-1}$, with δ positive, dimensionless, and depending only on L^2/M^4. Expanding out the various terms in (4.4) to the usual approximation we get

$$\dot{\sigma}(L, L) = D[1 + \langle M \rangle^{-2} \sigma(M, M)] - 2B\sigma(L, L) \tag{6.1}$$

where $D = \delta_0 \hbar^2 \langle M \rangle^{-1}$ with δ_0 being δ for $L = 0$. In writing (6.1) we have dropped a term of order $D \langle M \rangle^{-4} \sigma(L, L)$ which is a factor $\hbar \langle M \rangle^{-2}$ smaller than $B\sigma(L, L)$.

In its present form (6.1) is hard to work with. Things can be improved if we change from t to $\langle M \rangle$ as independent variable. Turning to (4.3) with $x = E$, $X = M$ we have with the usual accuracy

$$\langle \dot{M} \rangle = - A \langle M \rangle [1 + 3 \langle M \rangle^{-2} \sigma(M, M) + \mu_0 \langle M \rangle^{-4} \sigma(L, L)] \tag{6.2}$$

where μ_0 stands for the derivative of $\ln \alpha$ with respect to its argument L^2/M^4 evaluated at $L = 0$. The quotient of (6.1) and (6.2) gives

$$\sigma'(L, L) = - \frac{\delta_0 \hbar \langle M \rangle}{\alpha_0} [1 - 2 \langle M \rangle^{-2} \sigma(M, M)] + \frac{2\beta_0 \sigma(L, L)}{\alpha_0 \langle M \rangle} \tag{6.3}$$

where a prime denotes derivative with respect to $\langle M \rangle$. From (6.3) we have already dropped the term $\mu_0 \hbar \langle M \rangle^{-3} \sigma(L, L)$ in comparison with the last term because $\langle M \rangle^2 \gg \hbar$. In the regime where $\sigma(M, M) \ll \langle M \rangle^2$ (hole's mass sharply defined) (6.3) admits the solution

$$\sigma(L, L) = \frac{1}{2} \frac{\delta_0/\alpha_0}{\beta_0/\alpha_0 - 1} \hbar \langle M \rangle^2 \left[1 \pm \left(\frac{\langle M \rangle^2}{M_0^2} \right)^{\beta_0/\alpha_0 - 1} \right] \tag{6.4}$$

where M_0 and the $+/-$ choice constitute the arbitrary constant of integration.

On physical grounds only two mass scales are available out of which M_0^2 may be constructed: the natural mass scale $\hbar^{1/2}$, and the initial mass of the hole M_i. The choice $M_0^2 \sim \hbar$ can be ruled out as follows. We recall that when scalarons are *not* emitted $\beta_0/\alpha_0 > 2$ (Page 1976b). Thus for $\langle M \rangle^2 \gg \hbar$ we would find either $\sigma(L, L) \gg \langle M \rangle^4$, which is nonsense since the entire range of L^2 is only $2M^4$, or, if the negative sign is chosen, $\sigma(L, L) < 0$ which is impossible. The choice $M_0 \sim M_i$ does not raise such problems since $\langle M \rangle$ is never far above M_0. In fact, we should choose the negative sign in (6.4) and identify M_0 with the initial mass. Then $\sigma(L, L)$ starts from zero initially, grows as M evaporates, but eventually decreases as $\langle M \rangle^2$. Evidently

$$\sigma(L, L) < \tfrac{1}{2} (\delta_0/\alpha_0)\hbar \langle M \rangle^2. \tag{6.5}$$

The above discussion forces us to accept the idea that initial conditions *do* enter into the structure of Π through $\sigma(L, L)$, at least for some time until the term involving M_0 becomes negligible.

The story is hardly different if the hole emits lots of scalar species. Then β_0/α_0 is likely to be in between 0.805 and unity (Page 1976b). With the negative sign in (6.4) $\sigma(L, L)$ still starts from zero at M_0, first grows and then decreases, though somewhat slower than $\langle M \rangle^2$.

7. THE SPREAD IN MASS

The key quantity in the computation of $\sigma(M, M)$ is $\dot{\sigma}(E, E)$. In analogy with (5.3) we may start with $\overline{E^2} - \overline{E}^2$. According to (3.15) this must be of the form $\varepsilon \hbar^2 \Delta t M^{-3}$, ε being positive, dimensionless and depending only on L^2/M^4. For $L = 0$ we denote ε by ε_0; an approximation to it is given in § 3. The usual calculation starting from (4.4) gives to first order in $\langle L \rangle$ and second-order moments

$$\dot{\sigma}(M, M) = H[1 + v_0 \langle M \rangle^{-4} \sigma(L, L)] + 4A\sigma(M, M) \tag{7.1}$$

where $H \equiv \varepsilon_0 \hbar^2 \langle M \rangle^{-3}$ and v_0 is the derivative of $\ln \varepsilon$ with respect to L^2/M^4 at $L = 0$. From (7.1) we have already dropped a term of form $H \langle M \rangle^{-2} \sigma(M, M)$ in comparison with $A\sigma(M, M)$. Dividing through by (6.2), and again dropping a term a factor $\hbar \langle M \rangle^{-2}$ smaller than $\langle M \rangle^{-1} \sigma(M, M)$, we get

$$\sigma'(M, M) = -\frac{\varepsilon_0 \hbar}{\alpha_0 \langle M \rangle} \left[1 + \frac{(v_0 - \mu_0)\sigma(L, L)}{\langle M \rangle^4} \right] - \frac{4\sigma(M, M)}{\langle M \rangle} \tag{7.2}$$

If in fact $\sigma(M, M) \ll \langle M \rangle^2$, we may use (6.5) to justify neglecting the $\sigma(L, L)$ term in (7.2). The resulting equation is solved by

$$\sigma(M, M) = \frac{1}{3} \frac{\varepsilon_0}{\alpha_0} \hbar \left[\frac{M_0^4}{\langle M \rangle^4} - 1 \right] \tag{7.3}$$

where M_0^4 is the constant of integration. We have chosen its sign positive since otherwise $\sigma(M, M)$ cannot be positive. Note that $\sigma(M, M)$ is unphysical for $\langle M \rangle > M_0$. Clearly M_0 should be interpreted as the initial mass when the hole's parameters were sharply defined. It is in fact the same M_0 appearing in (6.4). Note that for $\sigma(M, M)$ the initial conditions enter in a crucial way. Even when a lot of the mass has evaporated, the very size of σ depends strongly on M_0. The result (7.2) is very different from the intuitive one. One might have argued as follows. A typical quantum has energy $\sim \hbar M^{-1}$ with variance $\sim (\hbar M^{-1})^2$. During its lifetime the hole should thus emit $\sim M_0^2 \hbar^{-1}$ quanta. The variance of the emitted energy comes out $\sim M_0^2 \hbar^{-1} \times (\hbar M^{-1})^2 = \hbar$. The hunch that this variance may be equated with $\sigma(M, M)$ is not supported by (7.2); $\sigma(M, M)$ is much larger than $\hbar$ when a lot of the mass has evaporated.

The result (7.2) leads to a novel conclusion. Suppose a population of holes of given precisely known mass is created at the same time. At some later time the spread in masses is measured and compared with (7.2). One thus infers M_0 and can work out the age of the holes by comparing M_0 with $\langle M \rangle$. One has thus evaded the spirit of 'black holes have no hair' by finding out details about the history rather than just the values of M, L and Q. However, note that the evasion is possible only for an ensemble, not for an individual.

One may now understand the paradox of the upper limit on M mentioned in § 3. Unlike previously assumed in line with 'black holes have no hair' (Bekenstein 1981), initial mass *is* a parameter in Π; M_0 is the upper limit on M, not in general, but for a particular hole.

Another striking consequence of (7.2) is that $\sigma(M, M)$ grows fast and gets to be of order $\langle M \rangle^2$ for $\langle M \rangle \sim \hbar^{1/6} M_0^{2/3}$. This critical mass need not be small. For example, for $M_0 = 10^{15}$ g the mass distribution becomes broad at $\langle M \rangle \sim 10^8$ g which is large compared to $\hbar^{1/2} \sim 10^{-5}$ g. Thus the hope that Π is a sharp distribution for a 'classical' hole is unfounded. When the assumption that $\sigma(M, M) \ll \langle M \rangle^2$ fails, our expression for $\sigma(L, L)$ (6.4) can no longer be believed. In fact the whole approach of regarding the $\sigma(X, Y)$ as small quantities is put in jeopardy. It can be seen from (6.2) that the rate of change of $\langle M \rangle$ will then deviate strongly from $-\alpha_0 \hbar \langle M \rangle^{-2}$. Hence the evolution of the hole for $\langle M \rangle < \hbar^{1/6} M_0^{2/3}$ is different from that envisaged in the Hawking–Page picture.

APPENDIX: MEASURING BLACK HOLE MASS WITH AN ORBITING PARTICLE

Consider a particle of mass m in a bound orbit of semi-major axis a_0 and eccentricity e about our black hole. If $a_0 \gg 2M$ we can temporarily ignore relativistic effects and describe the motion quantum-mechanically with the operator formalism invented by Pauli for the hydrogen atom (Pauli 1926, Schiff 1968). The eccentricity e is then replaced by the Runge–Lenz vector operator R constructed out of the angular momentum l and the momentum of the particle. These R and l commute with the (Newtonian) Hamiltonian H for the motion. In addition R^2 commutes with l^2 so H, l^2 and R^2 are all good quantum numbers related by

$$M^2 m^2 = R^2 - 2H(l^2 + \hbar^2)m^{-1}. \tag{A.1}$$

To determine M with some time resolution one must measure R^2, l^2 and H with some *phase* resolution along the orbit. Let ΔR^2, Δl^2 and ΔH be the quantum uncertainties (assumed to be independent) implicit in this measurement. It follows from (A.1) that Ms quantum uncertainty is

$$\Delta M = \left[\tfrac{1}{4} m^2 (\Delta R^2)^2 + (l^2 + \hbar^2)^2 (\Delta H)^2 + H^2 (\Delta l^2)^2 \right]^{1/2} M^{-1} m^{-3}. \tag{A.2}$$

The requirement that some phase information be acquired in the measurement forbids simultaneous infinitely accurate determination of l^2 and H: the fact that these are good quantum numbers is not very helpful here!

To see why let φ be the azimuthal coordinate of the particle about the chosen z axis. Evidently $[\varphi, l_z] = i\hbar$. Thus one has the uncertainty relation $\Delta l_z \Delta \varphi \geqslant \tfrac{1}{2} \hbar$. However, we are ultimately interested in a macroscopic problem. Thus we can choose our z axis close to l; further $\Delta l_z \ll \langle l_z \rangle$ where $\langle \; \rangle$ stands for the quantum expectation value. Because in general $\Delta l_x \neq 0$, $\Delta l_y \neq 0$ we shall have

$$\Delta l^2 > 2 \langle l_z \rangle \Delta l_z \geqslant \hbar \langle l_z \rangle \Delta \varphi^{-1}. \tag{A.3}$$

Similarly $[\varphi, H] = i\hbar \dot\varphi$ (Heisenberg picture), so $\Delta \varphi \Delta H \geqslant \tfrac{1}{2} \hbar \langle \dot\varphi \rangle$. To determine M over an interval τ, we must be able to follow the orbit over τ, i.e., $\Delta \varphi < \langle \dot\varphi \rangle \tau$; otherwise the orbital motion is 'buried' in the quantum uncertainty. Thus

$$\Delta H \geqslant \tfrac{1}{2} \hbar \tau^{-1} \tag{A.4}$$

which is the form of the time–energy uncertainty relation relevant here. We may also transform (A.3) into

$$\Delta l^2 > \hbar \langle l_z \rangle \langle \dot{\varphi} \rangle^{-1} \tau^{-1}. \tag{A.5}$$

Since $\Delta R^2 \geqslant 0$ (A.2) becomes

$$\Delta M > \frac{\hbar}{2m^3 M\tau} \left[(l^2 + \hbar^2)^2 + 4H^2 \langle l_z \rangle^2 \langle \dot{\varphi} \rangle^{-2} \right]^{1/2}. \tag{A.6}$$

In a nearly classical situation $l_z^2 \approx l^2 \approx Mm^2 a_0 (1 - e^2)$ and $H \approx -\frac{1}{2} M m a_0^{-1}$ as in the classical Kepler problem (Goldstein 1950). Also the largest $\langle \dot{\varphi} \rangle$ or $\dot{\varphi}$ occurs nearest to pericenter at radial distance $a_0 (1 - e)$. Thus $\langle \dot{\varphi} \rangle \leqslant \langle l_z \rangle m^{-1} a_0^{-2} (1 - e)^{-1}$ so that (A.6) reduces to

$$\Delta M > \frac{\hbar a_0 (1 - e)}{\sqrt{2} \, m\tau}. \tag{A.7}$$

ACKNOWLEDGMENT

I thank Uri Ben-Yaacov for many conversations.

REFERENCES

Bekenstein J 1975 *Phys. Rev.* D **12** 3077–85
—— 1981 *To Fulfill a Vision—Jerusalem Einstein Centennial Symposium on Gauge Theories and Unification of Physical Forces* ed Y Ne'eman (Reading, Mass: Addison-Wesley) pp 42–62
Ben-Yaacov U 1981 unpublished
Candelas P and Sciama D W 1977 *Phys. Rev. Lett.* **38** 1372–5
DeWitt B 1975 *Rep. Prog. Phys.* **19** 295–357
Einstein A 1924 *Berlin Ber.* **3**
Gerlach U 1976 *Phys. Rev.* D **14** 3290–3
Goldstein H 1950 *Classical Mechanics* (Reading, Mass: Addison-Wesley) pp 76–80
Hawking S 1975 *Commun. Math. Phys.* **43** 199–220
—— 1976a *Phys. Rev.* D **13** 191–7
—— 1976b *Phys. Rev.* D **13** 2460–73
Page D 1976a *Phys. Rev.* D **13** 198–206
—— 1976b *Phys. Rev.* D **14** 3260–73
—— 1977 *Phys. Rev.* D **16** 2402–11
—— 1980 *Phys. Rev. Lett.* **44** 301–4
Parker L 1975 *Phys. Rev.* D **12** 1519–25
Pauli W 1926 *Z. Phys* **36** 336
Ruffini R and Ohanian H 1974 *Phys. Rev.* D **10** 3903–5
Schiff L 1968 *Quantum Mechanics* (New York: McGraw-Hill) pp 235–9
Wald R 1975 *Commun. Math. Phys.* **45** 9–34
Zurek W 1980 *Phys. Lett.* **77A** 399–403

Black Holes, Singularities and Predictability

ROBERT M WALD

Prior to the mid 1960s, the prediction of a space–time singularity—such as occurs in homogeneous, isotropic cosmological models and in spherical collapse in general relativity—appears generally to have been viewed either as an absurdity or as a catastrophe for theoretical physics. It was generally thought that the singularities found in the cosmological case were artifacts of the high degree of symmetry of these models, while the singularities of collapse either also were artifacts of the symmetry assumptions or could be avoided by invoking mechanisms to prevent collapse. Indeed, the failure to take seriously the prediction of singularities was undoubtedly largely responsible for the fact that fully fledged efforts toward investigating gravitational collapse were not undertaken until after the discovery of quasars in the early 1960s, nearly fifty years after the formulation of general relativity and over thirty years after the physical relevance of collapse had been established by the demonstration of upper mass limits on bodies supported by degeneracy pressure.

The singularity theorems, proven between the mid 1960s and early 1970s, forced a change in attitude toward singularities. They established that under very general hypotheses, space–time singularities are an unavoidable prediction of classical general relativity in a variety of circumstances relevant to cosmology and gravitational collapse. Thus, if general relativity is not to be abandoned, the occurrence of space–time singularities must be taken seriously.

However, the view often has been expressed that the breakdown of classical general relativity due to quantum effects may still rescue one from the prediction of singularities. Now, even in the context of classical general relativity, it is not obvious how to formulate a precise definition of a space–time singularity. (The characterization of singularities used in the singularity theorems is that of timelike or null geodesic incompleteness, i.e., the existence of an inextendible timelike or null geodesic which has only a finite range of affine parameter.) Since the nature of even the basic framework of a quantum theory of gravity is presently in question, it clearly would be premature to attempt to formulate a precise notion of singularities in quantum gravity. Hence, I shall use the term 'singularity' in this context to mean, in a loose sense, a serious breakdown in 'ordinary dynamical evolution'. Some evidence in favor of the idea that singularities may be avoided in quantum gravity comes from the fact that in quantum field theory the expectation value of the stress–energy operator no longer need satisfy the classical local positive energy conditions that are required for the proof of the singularity theorems. In particular, some semiclassical calculations of cosmological models have indicated that it is possible to have a 'bouncing universe' model which avoids the presence of an initial singularity. In this article, I shall take the contrary view that we should not expect singularities to go away in quantum gravity. I shall do so by reviewing the arguments leading to the conclusion that in the process of black hole formation and evaporation an initial pure state evolves to a final density matrix, thus signaling a breakdown in ordinary quantum dynamical evolution. I also shall discuss some related issues dealing with predictability in the dynamical evolution.

The story of particle creation near black holes has been re-derived and reviewed[†] by many authors since Hawking's discovery of the effect (Hawking 1975), so I will confine myself to making only a few brief remarks here.[‡] When one calculates in the semiclassical approximation (i.e., treating gravity classically) the state of a quantum field seen by an observer at infinity at late times in a space–time where a Schwarzschild black hole has formed, one finds that this state is described precisely by a thermal density matrix (Wald 1975), corresponding to temperature $kT = \hbar\kappa/2\pi$, where κ is the surface gravity of the black hole. One obtains a density matrix in this calculation because the particles which emerge to infinity are correlated with particles which enter the black hole rather than with each other; a pure state still describes the total system consisting of particles which reach infinity together with particles which enter the black hole. However, a new feature arises when back-reaction is taken into account. Since the area of the black hole scales as M^2 and the temperature scales as M^{-1}, the energy flux from a black hole for a species of massless particles scales as $M^2(M^{-1})^4 = M^{-2}$. As a result, one predicts that the black hole will radiate away all of its mass within a finite amount of time. It appears that there are only three possible outcomes of this process:

(1) The particle creation 'shuts off' at a stage where the semiclassical approximation breaks down, and a small black hole remnant is left behind. Although the particles reaching infinity are described by a density matrix, the joint state of these particles together with the remnant is pure.

(2) The black hole 'evaporates' completely, but the correlations are restored during the evaporation process, so that the final state of the quantum field is pure.

(3) The black hole evaporates completely and the correlations are not restored, so that one has evolution from an initial pure state to a final density matrix.

With regard to the first possibility, it does not seem plausible to me that the prediction of particle creation given by the semiclassical approximation could be

---

[†] An early, comprehensive review which served as a valuable reference to many researchers was given by Bryce DeWitt in 1975.
[‡] Further details can be found in Wald (1981).

grossly wrong until Planck scales are reached. At that stage, the black hole would have mass of the order of 10^{-5} g and size of the order of 10^{-33} cm. In order to correlate with all the emitted particles, this remnant would need to have a number of 'internal states' of order $\exp M^2$ where M is the initial mass of the black hole expressed in Planck units, since the emitted particles have an entropy of the order of magnitude of the initial black hole entropy $S = \frac{1}{4} A \propto M^2$. This requirement already makes this possibility implausible, but it also appears to conflict with black hole thermodynamics, which suggests that the 'number of internal states' of a black hole should always be of order $\exp \frac{1}{4} A$ and hence should be of order 1 for a Planck mass black hole. Thus, I do not feel that possibility (1) is viable.

Possibility (2) is perhaps a more attractive means of avoiding the conclusion that a pure state evolves to a density matrix (Page 1980). In this picture the process of black hole evaporation should be much like the process of cooling an ordinary hot body down to absolute zero by radiation of photons. In that process, at early times the photons emitted by the hot body are described by a density matrix since they are correlated with the atoms which emitted them. However, these correlations are then transferred to the photons emitted at late times, so the final state of the photons is pure (assuming, of course, that the hot body began in a pure state).

I would like to emphasize here what I believe to be an essential difference between black hole evaporation and the cooling of a hot body. Figures 1 and 2 show two representations of the space–time structure expected if evaporation of a black hole occurs in a manner suggested by possibility (2). Figure 1 accurately shows that after a black hole forms from the collapse of matter, it shrinks in size as it radiates until it disappears altogether. However, unless one properly interprets the lightcones depicted in this figure, one might get the misleading impression that the horizon is a timelike surface, like the surface of an ordinary hot body. The conformal diagram of this space–time shown in figure 2 corrects this misimpression. It 'straightens out' the lightcones to 45° lines and clearly displays the fact that the horizon is expanding outward at the speed of light! (However, on account of the curvature of space–time near the black hole, the area of the horizon does not increase as it expands outward.) Thus, the transfer of correlations from inside the black hole to outside the horizon is not merely a matter of tunneling through a potential barrier; rather, it is a matter of outright causality violation. If information can pass the 'wrong way' through a horizon (i.e., from inside to outside a black hole) then it should also be possible for information to pass the wrong way (i.e., 'backward in time') through any null surface in flat space–time. Thus, if the classical picture of the space–time structure of a black hole is to retain any validity in quantum gravity and if causality is not to be violated in a blatant way, I do not see how the information needed to restore the quantum state to a pure state can pass outward through the horizon.

However, information within the black hole could escape 'upwards' in the space–time diagram of figure 2 without violating causality. In other words, the information within the black hole could be stored in a region near the classical singularity (where the classical description of space–time structure presumably breaks down) and then could escape at the final moment of black hole evaporation. However, as in possibility (1) above, we again would be faced with the requirement of

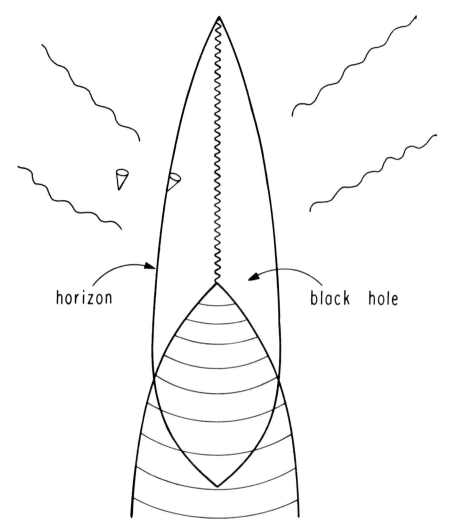

Figure 1 A space–time diagram showing the formation of a black hole by the collapse of matter and its subsequent evaporation by emission of Hawking radiation.

releasing $\sim \exp M^2$ 'bits of information' with only $\sim 10^{-5}$ g of energy. Thus, I do not feel that possibility (2) is viable.

Thus, I believe that the only viable possibility is the third one: that in the process of black hole formation and evaporation, an initial pure state evolves to a final density matrix. A conformal diagram illustrating the space–time structure corresponding to this case is given in figure 3. As illustrated by that diagram, all information inside the black hole propagates into the singularity and cannot be recovered; it is lost forever

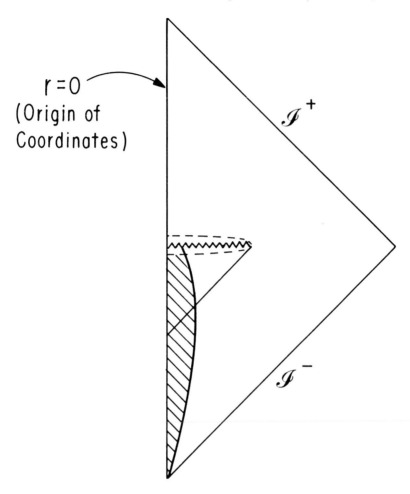

Figure 2 A conformal diagram showing the causal structure of a space–time where a black hole forms and evaporates in the manner of possibility (2) discussed in the text. Information cannot propagate outward through the black hole horizon without violating causality. The dotted lines near the classical singularity indicate a region where the classical description of space–time structure breaks down, thereby perhaps avoiding the presence of a singularity and allowing the information contained in the black hole to escape. However, this would require an enormous release of information with total energy only of order of the Planck energy. $\mathscr{I}^-$ and $\mathscr{I}^+$ are past null infinity and future null infinity.

when the black hole evaporates. Thus, I believe that this aspect of the singularities of the classical theory of black holes will persist in quantum gravity.

One of the most unpleasant features of singularities in classical general relativity is that they may cause a breakdown of predictability. This is because classical general relativity says nothing about what can emerge from a singularity. Hence, even if

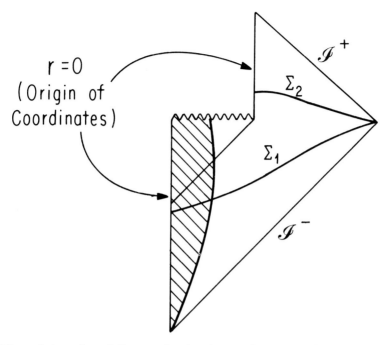

r = 0
(Origin of
Coordinates)

$\mathscr{J}^+$

Σ_2

Σ_1

$\mathscr{J}^-$

Figure 3 A conformal diagram showing the causal structure of a space–time where a black hole forms and evaporates in the manner of possibility (3).

complete information about a field is given at an initial time, the dynamical evolution equations alone may not determine what an observer sees if a singularity becomes 'visible' to the observer after that time. It should be emphasized that this sort of 'breakdown of predictability' may really only represent a breakdown in *our* ability to predict what happens knowing only the laws of classical general relativity; nature need not suffer from such a restriction. Nevertheless, it is comforting that, despite the presence of singularities, such a breakdown of predictability does not seem to occur in classical general relativity. According to the cosmic censor hypothesis, all singularities of gravitational collapse are hidden in black holes and hence do no harm to predictability outside black holes. Furthermore, according to a modified version of this hypothesis proposed by Penrose (1979), all physically reasonable space–times are globally hyperbolic and hence the dynamical evolution of fields is determined. Neither of these statements have been proven to be true—indeed, it is non-trivial even to formulate precise versions of these conjectures (see, e.g., Geroch and Horowitz 1979)—but there is some evidence that at least the first statement of cosmic censorship holds. (This evidence stems mainly from the demonstration that certain types of proposed counterexamples cannot be made to work.) If one accepts the third picture of black hole evaporation by the particle creation process as described above, one may ask whether the singularities occurring in that process should be expected to result in a breakdown of predictability, or whether a property analogous to cosmic censorship is likely to hold.

A good reason for believing in a breakdown of predictability comes from the fact that a 'naked singularity' is present in figure 3 and it is difficult to see how one could modify this figure in such a way that information is still lost into the singularity within the black hole, but the singularity does not become 'naked' at least at the final instant of black hole evaporation. In fact, in the context of a classical model of space–time structure, the impossibility of having a space–time containing a black hole which evaporates without producing a naked singularity has been established by a theorem of Kodama (1979). (This theorem is closely related to theorems of Budic *et al* (1978).) Below I shall state a modifed version of this theorem with a simplified proof using a line of argument which was suggested to me several years ago by Geroch. However I then shall argue that this result need not be interpreted as implying a breakdown of predictability.

First we remind the reader of a basic definition in the theory of causal structure of space–time. Let (M, g_{ab}) be a space–time and let Σ be a closed, achronal subset of M. The *future domain of dependence* of Σ, denoted $D^+(\Sigma)$, is defined to be the set of points $p \in M$ having the property that every past inextendible causal curve through p intersects Σ. Intuitively, for $p \in D^+(\Sigma)$ one would expect the value of physical quantities at p to be determined by appropriate data on Σ, since any information which propagates causally to p must 'register' on Σ. This expectation is born out by the study of the dynamical evolution properties of second order, diagonal hyperbolic systems of partial differential equations (see e.g., chapter 7 of Hawking and Ellis 1973). The past domain of dependence, $D^-(\Sigma)$, is defined similarly. We now state and prove our version of the theorem of Kodama (1979). Our terminology and notation will be consistent with that of Hawking and Ellis (1973) and we will use the abbreviation HE to allude to that reference in the proof.

Theorem:
Let (M, g_{ab}) be a time orientable space–time and let Σ_1 and Σ_2 be closed achronal sets such that Σ_1 is connected and Σ_2 is edgeless. Suppose there exists a point $p \in D^+(\Sigma_1)$ such that $p \notin J^-(\Sigma_2) \cup J^+(\Sigma_2)$. Suppose further that the set K is defined by $K = \Sigma_1 - D^-(\Sigma_2) \cap \Sigma_1$, has the property that $J^+(K) \cap \Sigma_2$ has compact closure. Then Σ_2 cannot be contained with $D^+(\Sigma_1)$.

Proof:
We suppose $\Sigma_2 \subset D^+(\Sigma_1)$ and obtain a contradiction. Since (M, g_{ab}) is time orientable, there exists a smooth, non-vanishing timelike vector field t^a on M (see HE, p 40). Since Σ_1 and Σ_2 are achronal, no integral curve of t^a can intersect either of these sets more than once. However, since we assume $\Sigma_2 \subset D^+(\Sigma_1)$, each integral curve which intersects Σ_2 must intersect Σ_1. Thus, by following these integral curves we obtain a one-to-one 'projection map' f from Σ_2 onto a subset, A, of Σ_1. Since t^a is smooth, $f: \Sigma_2 \to A$ will be a homeomorphism. Since Σ_2 is edgeless, by the same argument as used in the proof of proposition 6.3.1 of HE, it is an imbedded C^0 manifold (without boundary). This implies that $A = f[\Sigma_2]$ must be an open subset of Σ_1. However, we also can prove that A is closed as follows. Let $q \in \bar{A}$ and suppose $q \notin A$. Then, since $q \notin A$, we clearly have a future inextendible timelike curve from q which does not

intersect Σ_2, and hence $q \notin \overline{D^-(\Sigma_2)}$ (see proposition 6.5.1 of HE). This implies that $q \in \text{int}(K)$. On the other hand, since $q \in \bar{A}$, there exists a sequence $\{q_n\}$ in A which converges to q. Let $r_n = f^{-1}(q_n)$. Since infinitely many q_n enter int (K), infinitely many r_n lie in $J^+(K) \cap \Sigma_2$. Since this latter set has compact closure, the sequence $\{r_n\}$ must have an accumulation point r in Σ_2. By continuity of f, we must have $f(r) = q$. This contradicts the assumption that $q \notin A$ and thus proves that A is closed. Since Σ_1 is connected the fact that A is both open and closed implies that $A = \Sigma_1$. However, this is impossible since the integral curve of t^a which passes through the point p satisfying $p \in D^+(\Sigma_1)$ but $p \notin J^-(\Sigma_2) \cup J^+(\Sigma_2)$ must intersect Σ_1 at a point not lying in A. Thus $\Sigma_2 \not\subset D^+(\Sigma_1)$.

The application of this theorem to the black hole evaporation problem is straightforward. In a space–time having the type of causal structure illustrated by figure 3, the hypersurfaces Σ_1 and Σ_2 shown in that figure will satisfy the hypotheses of the theorem. (Here the point p would be taken as any event inside the black hole.) Hence, the theorem implies that the domain of dependence of Σ_1 cannot contain all of Σ_2, i.e., that there is a breakdown of predictability in going from Σ_1 to Σ_2.

However, I do not feel that this breakdown of predictability necessarily means that we must abandon deterministic evolution even in the context of a classical space–time model. To illustrate this point, consider Minkowski space–time with a point removed and consider the evolution of a massless Klein–Gordon scalar field in this space–time from smooth initial data on a hyperplane. Then, although the space–time fails to be globally hyperbolic, the evolution of the scalar field is still deterministic. This is because by Huygen's principle, the evolution equation alone uniquely determines the field from its initial data everywhere *except* on the lightcone of the removed point. However, the value of the field on the lightcone itself is determined from its values elsewhere by continuity.[†] Thus, one retains unique evolution in this case despite the lack of global hyperbolicity. (On the other hand, in Minkowski space–time with, say, a timelike line segment removed, one would not have unique evolution.) It seems plausible to me that the space–time of figure 3 may have much the same character as the example of Minkowski space–time with a point removed, so that one does not actually lose predictability from Σ_1 as a result of the evaporation of the black hole.

As already mentioned above, any breakdown of predictability which may occur due to naked singularities in classical general relativity may well represent only a breakdown of *our* ability to predict on account of our limited knowledge of the laws of nature. However, a fundamental breakdown of predictability occurs quite generally in quantum theory in that only the probabilities of outcomes of measurements can be predicted. This inability to predict definite outcomes is believed to be a true feature of the laws of nature rather than a result of the incompleteness of our knowledge. Hawking (1976) has argued that the evolution of a pure state to a

---

† Note, however, that uniqueness of evolution would fail if one admitted distributional solutions of the wave equation, since there exist distributional solutions with support on the lightcone.

density matrix in the process of black hole formation and evaporation should be interpreted as representing a significant further breakdown of predictability in quantum theory. I do not share this viewpoint, since in this case one still is able to predict the probabilities of outcomes of all measurements, and this is all that ever can be predicted in quantum theory. Thus, I do not feel that the evolution of a pure state to a density matrix should be viewed as representing a breakdown of predictability.

However, the evolution of a pure state to a density matrix *does* imply a breakdown of 'retrodictability', since it implies the existence of distinct initial states which evolve to the same final state (see Wald 1980). Thus from a knowledge of the final density matrix of the system, one cannot deduce the initial state of the system; there is information loss in this evolution. This is in accord with the picture of space–time structure shown in figure 3, since, as seen there, from a knowledge of a field at the 'final time' Σ_2 one should not be able to deduce the value of the field at the 'initial time' Σ_1. Indeed, the evolution of a pure state to a density matrix—with the information loss which occurs as a consequence—reflects in fully quantum language a key aspect of the classical notion of black holes.

Thus, considerations arising out of Hawking's discovery of particle creation by black holes lead to a picture of black holes in quantum gravity where, contrary to ordinary dynamical evolution in quantum theory, true loss of information occurs. This suggests that singularities may play a central role in quantum gravity.

ACKNOWLEDGMENTS

This research supported in part by NSF grant PHY 8026043 to the University of Chicago.

REFERENCES

Budic R, Isenberg J, Lindblom L and Yasskin P B 1978 *Commun. Math. Phys.* **61** 87
DeWitt B S 1975 *Phys. Rep.* **19** 295
Geroch R P and Horowitz G T 1979 in *General Relativity: an Einstein Centenary Survey* ed S W Hawking and W Israel (Cambridge: Cambridge University Press)
Hawking S W 1975 *Commun. Math. Phys.* **43** 199
—— 1976 *Phys. Rev.* D **14** 2460
Hawking S W and Ellis G F R 1973 *The Large Scale Structure of Space–time* (Cambridge: Cambridge University Press)
Kodama H 1979 *Prog. Theor. Phys.* **62** 1434
Page D N 1980 *Phys. Rev. Lett.* **44** 301
Penrose R 1979 in *General Relativity: an Einstein Centenary Survey* ed S W Hawking and W Israel (Cambridge: Cambridge University Press)
Wald R M 1975 *Commun. Math. Phys.* **45** 9
—— 1980 *Phys. Rev.* D **21** 2742
—— 1981 in *Quantum Gravity* 2 ed C J Isham, R Penrose and D W Sciama (Oxford: Clarendon Press)

The Gospel according to DeWitt

G VILKOVISKY

1. INTRODUCTION

Bryce DeWitt belongs to the very rare and venerable breed of scientist who, having placed before him a blank sheet of paper, then proceeds to create an entire world. He works as if the external world did not exist and his pen is not put down until the creation is complete.

But to an external observer years may pass, fashions may come and go, and suddenly there appears an enormous work hundreds of pages in length. With such a work one may do only one of two things: either become immersed in this aloof and closed world with its style and symbols, or not to read it completely. Such works are doomed to perpetual or temporary obscurity because, in the first case they have missed their time, and in the second case they have come before it. DeWitt's work falls into the latter category.

Before 1967, working alone and almost in a vacuum, he completed a huge investigation in which he laid the foundation of quantum gravity (and of quantum gauge theory in general) (DeWitt 1963, 1965, 1967a,b,c). Moreover, he did this so soundly that the next ten years of development in the field can be characterized as the gradual recognition of his work. The process occurred along the usual lines: you come up with something on your own, work it out, then open DeWitt and find that it can be done—and has already been done—better.

The strength of DeWitt's work lies in its unavoidability. That is, the development of science cannot escape it and, therefore, sooner or later it receives recognition. So it was with the Feynman–DeWitt ghosts, spiritual twins of the Faddeev–Popov ghosts. So it has been with the repeatedly-rediscovered background-field method. And yet another example: around 1973 dimensional regularization appeared and the divergences of the single graviton loop were computed. 'Hurrah!' shouted the scientific community and the attack on the one-loop counterterms was begun. All the while a book of 1963 lay on the shelf gathering dust (DeWitt 1963, 1965). Contained in this volume was a wonderful technique (the Schwinger–DeWitt technique) which was used to work out similar problems to those making all the noise in the glorious 1970s. The problems were presented as exercises for the reader.

Only two fundamental contributions were made independently of DeWitt: the discoveries of Hawking radiation and supergravity. But even here the influence of Bryce's erudition and fundamental viewpoint is brought to great fruition. (DeWitt 1975, DeWitt and van Nieuwenhuizen 1982, DeWitt draft of book).

When I meet Bryce and still from afar see his tall, immediately recognizable figure, I always think that in science he stands the same: aloof and far-removed.

A 'great poet' does not always mean a 'great man' (Rothman 1979, 1983). But everyone knows Bryce's noble and brave nature. As a personality he is also a unique specimen for our time. It is not for nothing that one colleague said, pointing out Bryce's innocence, 'Bryce is minus three years old'.

The work of DeWitt does not age and does not lose its significance; indeed there is still much in it that has not been assimilated. The collection of his works forms the 'Gospel' for true believers in quantum gravity.

The present article is a commentary and problem book for the 'Gospel according to DeWitt'. Occasionally one may find in it a small positive step between DeWitt's contribution and the edge of new frontiers. This article consists of four sections corresponding to the four distinct sections of the Gospel:

    1. The Abelianization of gauge theory and the covariant-quantization conjecture

    2. The effective action: problems in definition

    3. The effective action: problems in evaluation

    4. The effective action: applications and inherent problems

Everyone stops somewhere, even Bryce DeWitt. In his quantum theory of gauge fields (DeWitt 1967b), he made two assumptions: (i) the local generators of gauge transformations are linearly independent and, (ii) they form a Lie algebra. Both assumptions turn out to be limiting and both have been dispensed with by the present time. In addition, while DeWitt introduced the so-called covariant approach to the quantization problem, he left unclarified the connection between this approach and canonical quantization. In §2 I discuss the status of covariant methods and their generalization to the case of open gauge algebras.

The rest of the paper is devoted to the effective-action theory. Thanks to DeWitt we have a covariantly defined effective action in a gauge theory (DeWitt 1967b, 1979, 1981a). But the problem is that the effective equations for the mean field are not unique. Though being covariant, they depend on the choice of gauge conditions and field parameterizations. In §3 I suggest the solution of this problem and construct the unique effective action in field theory.

Thanks to Schwinger and DeWitt, we have a covariant technique for computing the effective action (DeWitt 1963, 1965). However, it has an essential limitation: an expansion in proper time leads to a purely local expansion for the effective action. But this expansion is not valid in the case of large or rapidly varying fields and blows up in massless theories. In §4 I show how, as a result of summing the proper-time series, the above limitation vanishes, and non-local terms arise in the effective action.

DeWitt's work of 1975 is devoted to the applications and effects of quantum gravity. In this connection, §5 is devoted to a derivation of the Hawking effect from non-local terms in the effective action. Related problems are also discussed.

In this article I do not touch upon two non-canonical texts from the Gospel (DeWitt 1964, 1981b). What indeed lies beyond the single loop is, at present, anyone's guess.

2. THE ABELIANIZATION OF GAUGE THEORY AND THE COVARIANT-QUANTIZATION CONJECTURE

The 'Gospel according to DeWitt' begins with the words: In the beginning was the action. But let me be more precise.

Let $S(\varphi)$ be the action of a field φ^i, $i = 1, \ldots, n$, in DeWitt's condensed notation. (For simplicity we assume φ^i is a bosonic field.) It is supposed that $S(\varphi)$ has a stationary point φ_0

$$\frac{\partial S(\varphi)}{\partial \varphi^i}\bigg|_{\varphi = \varphi_0} = 0 \tag{2.1}$$

and is analytic in its neighborhood. The entire examination will be carried out in a neighborhood around φ_0. This is sufficient if in quantum theory we confine ourselves to the loop expansion. All the mysteries of the WKB approximation are contained in the Hessian of $S(\varphi)$ evaluated at the stationary point:

$$\frac{\partial^2 S(\varphi)}{\partial \varphi^i \partial \varphi^j}\bigg|_{\varphi = \varphi_0}$$

For various reasons this Hessian can be degenerate. One possible cause of degeneracy is that m Noether identities hold in a neighborhood of φ_0

$$\frac{\partial S(\varphi)}{\partial \varphi^i} R^i_\alpha(\varphi) \equiv 0 \qquad \alpha = 1, \ldots, m \tag{2.2}$$

$$\operatorname{rank} R^i_\alpha(\varphi)\big|_{\varphi = \varphi_0} = m \qquad 0 \leqslant m \leqslant n \tag{2.3}$$

where the $R^i_\alpha(\varphi)$ are some analytic functions.[†] If we require that

$$\operatorname{rank} \frac{\partial^2 S}{\partial \varphi^i \partial \varphi^j}\bigg|_{\varphi = \varphi_0} + \operatorname{rank} R^i_\alpha\big|_{\varphi = \varphi_0} = h \tag{2.4}$$

then all other possible causes for the degeneracy of the Hessian will be excluded. Equations (2.1)–(2.4) define a gauge theory admitting the loop expansion (Batalin and Vilkovisky 1981).

The above definition should be supplemented by at least one more requirement: that of locality. The locality condition is

$$\varphi^i = \varphi^a(t) \qquad S(\varphi) = \int dt \, \mathscr{L}(\varphi, \dot{\varphi}, \ldots) \tag{2.5}$$

[†] I confine myself to the case when the generators R^i_α in the local basis are linearly independent. The generalization of covariant-quantization rules to theories with linearly dependent generators can be found in Batalin and Vilkovisky (1983a).

where the index a enumerates the degrees of freedom, and $\mathscr{L}$ is a function of a finite number of derivatives of φ at a single point. After the transition to phase space, equations (2.1)–(2.5) define a dynamical system with first-class constraints. Relativistic gauge fields are the most important examples of such systems.

Before the word 'locality' is uttered, the word 'quantization' is 'full of sound and fury, signifying nothing'. In a local theory, quantization means the imposition of canonical commutation relations upon the independent initial data of the field equations. Everything else, including the relativistic Feynman rules for corresponding theories, should and can be derived starting from this postulate.

There exist, however, so-called covariant methods (Faddeev and Popov 1967, Becchi *et al* 1975, Kallosh 1977, de Wit and van Holten 1979, Batalin and Vilkovisky 1981, 1983b, c) which create the illusion that the rules for quantization can be derived without any appeal to canonical formulation and the condition of locality. The shakiness of such claims will be made manifest in the following discussion. But another thing is striking: covariant methods indeed lead to the Feynman rules which are correct in all known examples and apparently are correct in general. Their validity is based on a certain conjecture which will be formulated below. The role of the canonical formalism therefore reduces to the verification of this conjecture.

The central role in the formulation of the covariant Feynman rules is played by the algebra of gauge transformations. The gauge algebra (de Wit and van Holten 1979) is described by the generating function (Batalin and Vilkovisky 1981)

$$\mathscr{S}(\varphi^i, c^\alpha; \varphi_i^*, c_\alpha^*) \tag{2.6}$$

which depends on the initial variables φ^i and auxiliary variables c^α, φ_i^* and c_α^*. If all φ^i are bosons, then c^α and φ_i^* are fermions, and c_α^* are also bosons. The generating function is defined as the bosonic solution to the equation

$$\frac{\partial_r \mathscr{S}}{\partial \varphi^i} \frac{\partial_l \mathscr{S}}{\partial \varphi_i^*} + \frac{\partial_r \mathscr{S}}{\partial c^\alpha} \frac{\partial_l \mathscr{S}}{\partial c_\alpha^*} = 0. \tag{2.7}$$

We search for a series solution in the auxiliary variables:

$$\mathscr{S}(\varphi, c; \varphi^*, c^*) = S + \varphi_i^* R^i{}_\alpha c^\alpha + \tfrac{1}{2} c_\gamma^* T^\gamma{}_{\alpha\beta} c^\alpha c^\beta$$
$$+ \tfrac{1}{4} \varphi_h^* \varphi_m^* E^{mn}_{\alpha\beta} c^\alpha c^\beta + O(c^3) \tag{2.8}$$

where each term is proportional to

$$(\varphi^*)^p (c^*)^s (c)^k$$

with the additional condition

$$p + 2s = k.$$

The coefficients of the series (2.8) are functions of the φ^i. The first two coefficients S and $R^i{}_\alpha$ satisfy postulates (2.1)–(2.4) and play the role of initial data. Under these conditions a theorem (Batalin and Vilkovisky 1981) can be proved: all coefficients of the series (2.8) exist in a neighborhood of the stationary point as analytic functions of φ^i.

Equation (2.7) is equivalent to a sequence of relations between the coefficients of the expansion (2.8). The two lowest-order relations are of the form

$$\frac{\partial S}{\partial \varphi^i} R^i{}_\alpha = 0 \tag{2.9}$$

$$\frac{\partial R^i{}_\alpha}{\partial \varphi^k} R^k{}_\beta - \frac{\partial R^i{}_\beta}{\partial \varphi^k} R^k{}_\alpha = T^\gamma{}_{\beta\alpha} R^i{}_\gamma + E^{ik}{}_{\beta\alpha} \frac{\partial S}{\partial \varphi^k}. \tag{2.10}$$

The first equation above reproduces the Noether identities (2.2). The second gives the general form of the commutator of gauge transformations. The third relation of this sequence gives the general form of the Jacobi identity, and so on.

The existence theorem for (2.8) mentioned above means, in particular, that the Lie bracket of generators satisfying postulates (2.2)–(2.4) is always of the form (2.10). Only in the case when $E^{ik}{}_{\beta\alpha} = 0$ and $T^\gamma{}_{\beta\alpha} =$ constant, do the generators $R^i{}_\alpha$ form a Lie algebra. In this case the series (2.8) terminates. In the general case the relations (2.10) are the commutation relations of what is called the open algebra.

The structure functions of the algebra (the coefficients of the expansion (2.8)) are not uniquely defined. In particular, the generators $R^i{}_\alpha$ are defined by the conditions (2.2)–(2.4) only up to a transformation

$$\mathcal{R}^i{}_\alpha = \Lambda^\beta{}_\alpha(\varphi) \left(R^i{}_\beta + K^{im}{}_\beta(\varphi) \frac{\partial S}{\partial \varphi^m} \right) \tag{2.11}$$

where $\Lambda^\beta{}_\alpha(\varphi)$ is an invertible matrix and $K^{im}{}_\beta(\varphi)$ is antisymmetric in i, m. The transformation (2.11) leads to a change of all structure functions, i.e. to a change of the basis of the algebra. A proof has been given (Batalin and Vilkovisky 1983b, c) that $\Lambda^\beta{}_\alpha$ and $K^{im}{}_\beta$ in equation (2.11) can always be chosen in such a way that the new generators $\mathcal{R}^i{}_\alpha$ will form the Abelian algebra

$$\frac{\partial \mathcal{R}^i{}_\alpha}{\partial \varphi^k} \mathcal{R}^k{}_\beta - \frac{\partial \mathcal{R}^i{}_\beta}{\partial \varphi^k} \mathcal{R}^k{}_\alpha \equiv 0 \tag{2.12}$$

and all higher-order structure functions in the new basis will vanish.

However, additional requirements such as locality and Lorentz covariance destroy the democracy of the bases. It is supposed that for a local (and Lorentz invariant) gauge action there exists a basis in which all structure functions are local (and Lorentz covariant). This is the first part of the covariant-quantization conjecture. The Abelian basis (2.12) is generally either non-local or not Lorentz covariant.

The existence of an Abelian basis means that there is a direct approach to the gauge theory: its reduction to the non-gauge theory. This approach makes all the secrets clear, and we shall exploit it for the clarification of problems inherent in the covariant-quantization methods. For simplicity, we assume the gauge algebra in a local basis is closed ($E^{ik}{}_{\beta\alpha} = 0$ in equation (2.10)). In this case equation (2.7) for the generating function admits the exact solution

$$\mathcal{S}(\varphi, c; \varphi^*, c^*) = S + \varphi_i^* R^i{}_\alpha c^\alpha + \tfrac{1}{2} c_\gamma^* T^\gamma{}_{\alpha\beta} c^\alpha c^\beta. \tag{2.13}$$

For the general case the Abelianization procedure is carried out in Batalin and Vilkovisky (1983c).

The Abelianization procedure begins with the choice of admissible gauge conditions. Functions $\chi^\alpha(\varphi)$ are admissible gauge conditions if they are analytic and

$$\det Q^\alpha{}_\beta(\varphi)\bigg|_{\varphi=\varphi_0} \neq 0 \qquad Q^\alpha{}_\beta(\varphi) = \frac{\partial\chi^\alpha}{\partial\varphi^i}R^i{}_\beta. \tag{2.14}$$

The existence of such $\chi^\alpha(\varphi)$ in a neighborhood around the stationary point is guaranteed by (2.3).

Let us introduce a new basis of generators:

$$\mathscr{R}^i{}_\alpha = R^i{}_\beta Q^{-1\beta}{}_\alpha. \tag{2.15}$$

One can verify by direct computation that if the $R^i{}_\alpha$ form a closed algebra, then the $\mathscr{R}^i{}_\alpha$ are the Abelian generators (2.12). Therefore the following Lie equations

$$\partial\varphi^i/\partial\theta^\alpha = \mathscr{R}^i{}_\alpha(\varphi) \tag{2.16}$$

with arbitrary initial data

$$\varphi^i\big|_{\theta=0} = f^i \tag{2.17}$$

are integrable and define a function

$$\varphi^i = \varphi^i(f,\theta) \tag{2.18}$$

at least for sufficiently small θ. It is guaranteed that

$$\det\frac{\partial\varphi^i(f,\theta)}{\partial f^k} \neq 0 \qquad \text{rank}\,\frac{\partial\varphi^i(f,\theta)}{\partial\theta^\alpha} = m. \tag{2.19}$$

One more property of the solution is

$$\chi^\alpha(\varphi(f,\theta)) = \theta^\alpha + \chi^\alpha(f). \tag{2.20}$$

Let us impose the gauge conditions

$$\chi^\alpha(f) = 0 \tag{2.21}$$

upon the initial data (2.17). Equation (2.21) defines some $(n-m)$-dimensional surface:

$$f^i = f^i(\xi^A) \qquad \chi^\alpha(f(\xi)) \equiv 0 \qquad \text{rank}\,\frac{\partial f^i(\xi)}{\partial\xi^A} = n-m \tag{2.22}$$

where ξ^A, $A = 1,\ldots,(n-m)$ are the coordinates on this surface. Under the conditions (2.21) the solution (2.18) of the Lie equations takes the form

$$\varphi^i = \varphi^i(f(\xi),\theta) \tag{2.23}$$

and ξ^A play the role of independent initial data.

Let us regard equation (2.23) as the equation defining a reparameterization of the field φ:

$$\varphi^i \leftrightarrow (\xi^A, \theta^\mu). \tag{2.24}$$

One can prove that this reparameterization is analytic and invertible. Therefore equation (2.23) can be solved with respect to ξ^A and θ^μ:

$$\xi^A = \xi^A(\varphi) \qquad \theta^\mu = \theta^\mu(\varphi). \tag{2.25}$$

From (2.20) and (2.21) it follows that

$$\theta^\mu(\varphi) = \chi^\mu(\varphi). \tag{2.26}$$

The properties of $\xi^A(\varphi)$ follow from the relations for the Jacobian matrix

$$\frac{\partial \varphi^i}{\partial \theta^\mu} \frac{\partial \xi^A}{\partial \varphi^i} = 0 \tag{2.27}$$

i.e.

$$R^i{}_\alpha(\varphi) \frac{\partial \xi^A(\varphi)}{\partial \varphi^i} = 0. \tag{2.28}$$

Thus $\xi^A(\varphi)$ are invariants of the gauge group. One can easily prove that these invariants are functionally independent, and that any invariant is a function of ξ^A. There are exactly $(n-m)$ invariants in any gauge theory.

Since the action of a theory is gauge invariant, it can only be a function of the $(n-m)$ invariants ξ^A:

$$S(\varphi) = \bar{S}(\xi). \tag{2.29}$$

This means that in the parameterization (ξ, θ) the action does not depend on the m field variables θ^μ and becomes a non-gauge action.

Thus a gauge theory can always be Abelianized and reduced to a non-gauge theory. An arbitrary set of gauge conditions enters the procedure of Abelization. The transition to another set is equivalent to the transition to another basis of invariants ξ, i.e. to a reparameterization of the non-gauge action $\bar{S}(\xi)$. We shall need this result in the next section. Now let us turn our attention to quantization.

Before us opens a direct path: the derivation of the Feynman rules for a gauge theory from those of the non-gauge theory. But here we fall into an abyss. In terms of the non-gauge action (2.29), the Feynman functional integral is of the form

$$Z = \int d\xi\, M(\xi) \exp\left[(i/h)\bar{S}(\xi)\right] \tag{2.30}$$

but the measure $M(\xi)$ in this integral is unknown. We know only that this measure is a gauge invariant. It would be a large error to think that this measure is unimportant. If in a local theory the number of independent variables ξ is $(n-m)$ per point, then the number of independent degrees of freedom is $(n-2m)$. The measure $M(\xi)$, compensating for this difference dynamically, should be non-local. $M(\xi)$ is generally non-local even if the basis of invariants ξ is chosen so that $\bar{S}(\xi)$ is a local action. This is a counterexample to the folk theorem that if the action is local and non-degenerate, then the measure in the Feynman integral can be neglected.

Covariant quantization is an apparent way to get something for nothing. Choose any constant invertible matrix $\beta_{\mu\nu}$ and transform expression (2.30) as follows:

$$Z = \int d\xi \, d\theta \, M(\xi) \exp \frac{i}{\hbar} \left(\bar{S}(\xi) + \tfrac{1}{2}\theta^\mu \beta_{\mu\nu} \theta^\nu \right). \tag{2.31}$$

There remains now to return from the parameterization (ξ, θ) to the initial parameterization φ. By virtue of (2.26) and (2.29) one has

$$Z = \int d\varphi \, J(\varphi) M^{\text{inv}}(\varphi) \exp \frac{i}{\hbar} \left[S(\varphi) + \tfrac{1}{2}\chi^\mu(\varphi)\beta_{\mu\nu}\chi^\nu(\varphi) \right]. \tag{2.32}$$

Here,

$$M^{\text{inv}}(\varphi) \equiv M(\xi) \tag{2.33}$$

and

$$J(\varphi) = \det \frac{\partial(\xi, \theta)}{\partial \varphi}. \tag{2.34}$$

The Lie equations and the equations defining the replacement (2.24) can be used to show that

$$J(\varphi) = N(\varphi) \det \left(\frac{\partial \chi^\alpha}{\partial \varphi^i} R^i_{\ \beta} \right) \tag{2.35}$$

where

$$-R^i_{\ \mu} \frac{\partial \ln N(\varphi)}{\partial \varphi^i} = \frac{\partial R^i_{\ \mu}}{\partial \varphi^i} + T^\gamma_{\ \mu\gamma} \tag{2.36}$$

in terms of the structure functions entering (2.10). As a result the functional integral (2.32) takes the form

$$Z = \int d\varphi \, d\bar{c} \, dc \, \mathcal{M} \exp \frac{i}{\hbar} \left(S(\varphi) + \tfrac{1}{2}\chi^\mu(\varphi)\beta_{\mu\nu}\chi^\nu(\varphi) + \bar{c}_\alpha \frac{\partial \chi^\alpha}{\partial \varphi^i} R^i_{\ \beta} c^\beta \right) \tag{2.37}$$

where the auxiliary fermionic integration varibles $\bar{c}$, c are introduced, and

$$\mathcal{M} = M^{\text{inv}}(\varphi) N(\varphi). \tag{2.38}$$

The auxiliary variables $\bar{c}$, c are the Feynman–DeWitt or Faddeev–Popov ghosts. If not for the presence of $\mathcal{M}$, expression (2.37) would be exactly the famous 1967 result of DeWitt and others. And what about $\mathcal{M}$? $\mathcal{M}$ has two multipliers. One is the initial invariant measure $M^{\text{inv}}(\varphi)$. As to the second, note that if $R^i_{\ \mu}$ and $T^\gamma_{\ \mu\nu}$ are local quantities, then the contractions on the right-hand side of (2.36) are proportional to

$$\delta^{\cdots}(0) = \frac{d}{dt} \cdots \frac{d}{dt} \delta(t - t')\big|_{t = t'}. \tag{2.39}$$

Up to such terms the second multiplier in $\mathcal{M}$ is also gauge invariant. No 'covariant-quantization method' is capable of avoiding the arbitrariness in a gauge-invariant functional, but the conjecture is that it is arbitrary up to terms proportional to $\delta'\cdots(0)$, $\mathcal{M} = 1$. Note that this can be true only if the local basis of generators is used in equation (2.37).

We have examined the case for which the algebra of local generators is closed. To formulate the conjecture for the general case, let us note that expression (2.37) can be rewritten as

$$Z = \int d\varphi \, d\bar{c} \, dc \mathcal{M} \exp \frac{i}{\hbar}\left[\mathscr{S}\left(\varphi, c; \frac{\partial \Psi}{\partial \varphi}, \frac{\partial \Psi}{\partial c} \right) + \tfrac{1}{2}\chi^{\mu}(\varphi)\beta_{\mu\nu}\chi^{\nu}(\varphi) \right] \qquad (2.40)$$

where $\mathscr{S}$ is the generating function (2.13) of the gauge algebra and

$$\Psi = \bar{c}_{\alpha}\chi^{\alpha}(\varphi). \qquad (2.41)$$

In the form (2.40) the result generalizes to the case of an arbitrary open algebra (Batalin and Vilkovisky 1981, 1983c). The covariant-quantization conjecture now reads:

For a local (and Lorentz invariant) gauge action there exists a local (and Lorentz covariant) basis for the gauge algebra. If this basis is used for the construction of the ghost Lagrangian in (2.40), and the gauge conditions $\chi^{\alpha}(\varphi)$ are arbitrary (in particular Lorentz covariant) but also local, then up to terms proportional to $\delta'\cdots(0)$

$$\mathcal{M} = 1. \qquad (2.42)$$

For many examples (Yang–Mills theory, Einstein's theory, $N = 1$ supergravity theory) this conjecture has been verified by canonical quantization (Faddeev 1969, Fradkin and Vilkovisky 1975a, 1977a, Fradkin and Vasiliev 1977).

In the canonical formalism the role of the gauge algebra is played by the algebra of first-class constraints (Fradkin and Vilkovisky 1975b, Batalin and Vilkovisky 1977, Fradkin and Fradkina 1978). This algebra defines the ghost Hamiltonian in much the same way as the gauge algebra defines the ghost action in the covariant approach. The Feynman rules in arbitrary relativistic gauges are known for any Hamiltonian system with constraints (Fradkin and Fradkina 1978). However, in the general case, '... no rigorous mathematical link has thus far been established between the canonical and covariant theories ... It is believed that the two theories are merely two versions of the same theory, expressed in different languages, but no one knows for sure' (DeWitt 1967b, p 1197).

DeWitt was one of the first investigators of the canonical theory (DeWitt 1967a) and the pioneer of the covariant approach (DeWitt 1967b). He was the first who understood that, to a certain extent, questions of gauge invariance and quantization can be studied in general terms without an appeal to the concrete form of the action. It was also DeWitt who raised the question concerning the correspondence between the canonical and covariant formalisms. Though for the case of Einstein's theory, which was his main concern, this question is now settled (Fradkin and Vilkovisky 1975a, 1977a), for the general case the words given above from the 'Gospel' remain true to this day.

3. THE EFFECTIVE ACTION: PROBLEMS IN DEFINITION

The Feynman rules enable one to construct the effective action, and the effective action, in turn, contains all the information about the quantized field. First of all, it gives the S matrix, but hopefully it should give something more, namely the effective equations for the mean field which must replace the classical equations. However, here one encounters a major difficulty: the mean field, as well as other Green's functions, are not uniquely defined. Consequently, the effective action is not unique. An example of this non-uniqueness is the dependence of the effective action on the choice of gauge conditions in a gauge theory. Here I must mention the work of DeWitt because DeWitt is an active proponent of the effective action approach (DeWitt 1975, 1979, 1981a) and because half the problem of the gauge dependence of the effective action was solved by him.

This half of the problem consists of the fact that the usual procedure of gauge fixing leads to a non-gauge-invariant effective action. This difficulty arises because by fixing the gauge of the quantized field, one automatically fixes the gauge of the mean field. The resulting gauge conditions for the mean field generally possess quantum corrections, and may be very complicated. There exists, however, a possibility of fixing the gauge of the quantized field leaving the gauge of the mean field arbitrary, in other words to obtain a gauge-invariant effective action. This possibility is realized in DeWitt's method of background (DeWitt 1967b) or mean-field (DeWitt 1981a) gauges.

The unsolved half of the problem consists of the fact that the gauge-invariant effective action thus obtained still depends (parametrically) on the choice of background gauge conditions. Different choices of these conditions lead to different covariant (!) effective actions. The solved half of the problem only stresses the acuteness of the unsolved half. It may be quite acceptable to have many non-covariant equations; they may well describe one and the same theory. But having many covariant equations surely means having many different presentations of a theory without being granted the possibility of choosing amongst them.

A more general (in fact the most general) manifestation of the non-uniqueness of the effective action is its dependence on the choice of parameterizations for the quantized fields. This problem is inherent in both non-gauge and gauge theories. Moreover, the gauge dependence of the effective action reduces to its parameterization dependence, as one may conclude from the considerations of the previous section. Indeed, as shown above, the gauge conditions can be used to find such a parameterization for the gauge field in which the action becomes a non-gauge action. The transition to another set of gauge conditions is equivalent to a reparameterization of this non-gauge action. The kernel of the whole problem is therefore the parameterization dependence of the effective action in non-gauge theories. So let us look at this problem more carefully.

Let $S(\varphi)$ be a classical non-gauge action of the bosonic field ($m = 0$ in equations (2.1)–(2.5)). The Feynman functional integral gives the following rule for obtaining the mean values of time-ordered functionals

$$\langle \cdots \rangle = \frac{\int (\cdots) \exp\left[(i/\hbar)S(\varphi)\right] \mathcal{M}(\varphi)\,d\varphi}{\int \exp\left[(i/\hbar)S(\varphi)\right] \mathcal{M}(\varphi)\,d\varphi}. \tag{3.1}$$

We assume that $S(\varphi)$ is a local action (equation (2.5)), and $\mathcal{M}(\varphi)$ in (3.1) is the measure resulting from canonical quantization. For example, if in equation (2.5)

$$\mathcal{L}(\varphi, \dot{\varphi}, \ldots) = \tfrac{1}{2} A_{ab}(\varphi)\dot{\varphi}^a \dot{\varphi}^b + B_a(\varphi)\dot{\varphi}^a + C(\varphi) \tag{3.2}$$

$\det A_{ab}(\varphi) \neq 0$, then

$$\mathcal{M}(\varphi) = \prod_t \left[\det A_{ab}(\varphi(t))\right]^{1/2}. \tag{3.3}$$

The usual (naive) definition of the mean field ϕ is

$$\phi^i = \langle \varphi^i \rangle. \tag{3.4}$$

The effective action $W(\phi)$ is a functional such that the mean field as defined in (3.4) satisfies the equation

$$\frac{\partial W(\phi)}{\partial \phi^i} = 0. \tag{3.5}$$

This functional can be constructed as

$$W(\phi) = \frac{\hbar}{i} \ln Z(J) - \phi^i J_i \tag{3.6}$$

where

$$Z(J) = \int d\phi \mathcal{M}(\varphi) \exp(i/\hbar)\{S(\varphi) + \varphi^i J_i\} \tag{3.7}$$

and the source J should be expressed as a function of ϕ with the aid of the equation

$$\phi^i = \frac{\hbar}{i} \frac{\partial \ln Z(J)}{\partial J_i}. \tag{3.8}$$

Indeed, differentiating (3.6) gives

$$\frac{\partial W(\phi)}{\partial \phi^i} = -J_i. \tag{3.9}$$

When $J = 0$, equation (3.8) reduces to the definition of the mean field (3.4), and equation (3.9) shows that this mean field satisfies (3.5).

Equations (3.4)–(3.9) can be summarized in one equation for the effective action:

$$\exp\left(\frac{i}{\hbar} W(\phi)\right) = \int d\phi \mathcal{M}(\varphi) \exp \frac{i}{\hbar}\left(S(\varphi) + (\phi^i - \varphi^i)\frac{\partial W(\phi)}{\partial \phi^i}\right). \tag{3.10}$$

This equation can be solved by expanding $W(\phi)$ in powers of $\hbar$. One goes over to the integration variables $\eta = \varphi - \phi$, and expands the measure and the argument of the exponent. The latter gives

$$S(\varphi) + (\phi^i - \varphi^i)\frac{\partial W(\phi)}{\partial \phi^i} = S(\phi) + \tfrac{1}{2}\partial^2_{ik}S(\phi)\eta^i\eta^k + \partial_i[S(\phi) - W(\phi)]\eta^i$$

$$+ \frac{1}{3!}\partial^3_{ikp}S(\phi)\eta^i\eta^k\eta^p + \cdots. \tag{3.11}$$

One also takes into account the fact that $[S(\phi) - W(\phi)]$ is already of order $\hbar$. Therefore the term

$$\partial_i[S(\phi) - W(\phi)]\eta^i$$

should be treated in the same way as all the subsequent terms in the expansion (3.11), i.e. by expanding the exponential. Then one finds that the $W(\phi)$ on the right-hand side of equation (3.10) will always be one power of $\hbar$ less than the corresponding $W(\phi)$ on the left-hand side. This makes it possible to solve for $W(\phi)$ order by order.

In the one-loop approximation the effective action is

$$W(\phi) = S(\phi) + \frac{\hbar}{i}\ln\mathcal{M}(\phi) - \frac{\hbar}{2i}\,\mathrm{Tr}\ln\left[\frac{\partial^2 S(\phi)}{\partial\phi^m\,\partial\phi^n}\right] + O(\hbar^2). \tag{3.12}$$

The problem of parameterization dependence can be expressed as follows. Nothing prevents us from redefining the mean field as

$$\phi' = f^{-1}(\langle f(\varphi)\rangle) \tag{3.13}$$

where f is an arbitrary function. At least if $\varphi \to f(\varphi)$ is a local reparameterization, the redefinition (3.13) is admissible. This redefinition corresponds to the introduction of the source J in equation (3.7) to the variables $f(\varphi)$ instead of φ. There is no way to choose among the definitions (3.13) with different local functions $f(\varphi)$. The only possibility, therefore, is that such definitions should be indistinguishable. However, denoting the effective action for the new mean field by $W'(\phi)$, we find

$$W'(\phi) = S(\phi) + \frac{\hbar}{i}\ln\mathcal{M}(\phi) - \frac{\hbar}{2i}\,\mathrm{Tr}\ln\left[\frac{\partial^2 S(\phi)}{\partial\phi^m\,\partial\phi^n} + \frac{\partial f^i(\phi)}{\partial\phi^m}\frac{\partial f^k(\phi)}{\partial\phi^n}\frac{\partial^2\phi^l}{\partial f^i\partial f^k}\frac{\partial S(\phi)}{\partial\phi^l}\right]$$

$$+ O(\hbar^2). \tag{3.14}$$

This differs from (3.12).

The opinion is widely held that when the external source is switched off, the parameterization dependence of the effective equations disappears. This opinion is mistaken. The cause of the confusion is the fact that the difference between $W'(\phi)$ and $W(\phi)$ is, in the one-loop approximation, proportional to the 'extremal'† of the classical action, $\partial S/\partial\phi$, and generally to the extremal of the full action, $\partial W/\partial\phi$. The

---

† Here and in what follows, the term 'extremal' is shorthand for the left-hand side of the field equations.

difficulty, however, comes about when the extremal in (3.14) is differentiated. One finds:

$$\phi'^k - \phi^k = \frac{\hbar}{2i} S^{-1\,mn} \frac{\partial f^i(\varphi)}{\partial \varphi^m} \frac{\partial f^j(\varphi)}{\partial \varphi^n} \frac{\partial^2 \varphi^k}{\partial f^i \partial f^j}\bigg|_{\varphi = \varphi_0} + O(\hbar^2) \qquad (3.15)$$

where

$$S_{mn} = \frac{\partial^2 S(\varphi)}{\partial \varphi^m \partial \varphi^n} \qquad \frac{\partial S(\varphi)}{\partial \varphi^i}\bigg|_{\varphi = \varphi_0} = 0.$$

The difference (3.15) survives even though the source is switched off!

The calculation of the S matrix from the effective action involves only the undifferentiated $W(\phi)$ expressed in terms of asymptotic fields. Basically, this is the reason why the difference between $W'(\phi)$ and $W(\phi)$ (which is proportional to the extremal) is irrelevant for the S matrix[†] but *is* relevant for the effective equations and the mean field.

Whether anything should exist off the mass shell is somewhat questionable. However, if the effective action 'is not merely of heuristic validity but is the correct way' (DeWitt 1975, p 347), then the problem of its definition should be taken seriously. Below I propose a solution to this problem.

Let us look at the one-loop expression (3.12). If we want to have covariance under (local) reparameterizations, then we must consider configuration space as a manifold, φ^i as coordinates on this manifold, reparameterizations as diffeomorphisms, and the classical action $S(\varphi)$ as a scalar function on this manifold. Our requirement is that if $S(\varphi)$ is a scalar then $W(\phi)$ should also be a scalar. However, this is not the case in expression (3.12) because the second derivative of a scalar is not a tensor. Clearly, we need a connection (Γ) in configuration space to convert the derivatives of $S(\varphi)$ into covariant derivatives:

$$\nabla_m \nabla_n S(\varphi) = \frac{\partial^2 S(\varphi)}{\partial \varphi^m \partial \varphi^n} - \Gamma^i{}_{mn}(\varphi) \frac{\partial S(\varphi)}{\partial \varphi^i}. \qquad (3.16)$$

If we could replace the expression (3.12) by

$$W(\phi) = S(\phi) + \frac{\hbar}{i} \ln \mathscr{M}(\phi) - \frac{\hbar}{2i} \text{Tr} \ln [\nabla_m \nabla_n S(\phi)] + O(\hbar^2) \qquad (3.17)$$

then $\text{Tr} \ln [\nabla_m \nabla_n S]$ would transform under (local) reparameterizations as the logarithm of the scalar density, and this transformation would be exactly compensated by the transformation of the measure $\mathscr{M}(\phi)$ defined in (3.3). Thus $W(\phi)$ would be a scalar. On the other hand, as seen from (3.16), the difference between the

---

[†] In fact, not every term in the effective action proportional to the extremal vanishes on the mass shell, because the coefficient, expressed as a function of asymptotic fields, may be non-local and contain a pole. For this reason only local reparameterizations are generally admissible in S-matrix theory. But even in the case of a local reparameterization one must be cautious about poles in Green's functions at coinciding points.

expression (3.17) and the naive expression (3.12) is proportional to the extremal and thus vanishes on the mass shell, provided $\Gamma^i_{mn}(\varphi)$ are local functions.

Suppose we have a connection in configuration space. Then what modification is needed in the definition of the mean field and the effective action to give the desired result (3.17)? The cause of the non-covariance of the effective action is the term $\varphi^i J_i$ in the naive definition (3.7). We introduce the source of the field φ in this term, but φ is just a coordinate; such a procedure is senseless from a geometrical point of view.

Having a connection we may construct the two-point quantity $\sigma^i(\phi, \varphi)$. It is defined as the vector, tangent to the geodesic connecting ϕ and φ. It is tangent at point ϕ, directed from φ to ϕ, and normalized by the affine parameter. The object $\sigma^i(\phi, \varphi)$ is a vector with respect to the point ϕ and a scalar with respect to the point φ. The above definition is equivalent to the equation for $\sigma^i(\phi, \varphi)$ with respect to ϕ

$$\sigma^k \nabla_k \sigma^i = \sigma^i \tag{3.18}$$

and boundary conditions

$$\sigma^i\big|_{\phi = \varphi} = 0 \qquad \det \nabla_k \sigma^i\big|_{\phi = \varphi} \neq 0. \tag{3.19}$$

Expanding the solution in powers of $(\varphi - \phi)$ with coefficients evaluated at ϕ, we find

$$-\sigma^i(\phi, \varphi) = (\varphi - \phi)^i + \tfrac{1}{2}\Gamma^i_{mn}(\phi)(\varphi - \phi)^m(\varphi - \phi)^n + \tfrac{1}{6}(\partial_k \Gamma^i_{mn}$$
$$+ \Gamma^i_{kp}\Gamma^p_{mn})\big|_\phi (\varphi - \phi)^k(\varphi - \phi)^m(\varphi - \phi)^n + O(\varphi - \phi)^4. \tag{3.20}$$

$\sigma^i(\phi, \varphi)$ is one of the basic two-point tensors from DeWitt's collection (DeWitt 1963, 1965).

The modfication needed in the definition of the effective action is the replacement of the term $\varphi^i J_i$ in (3.7) by

$$[\phi^i - \sigma^i(\phi, \varphi)]J_i = [\varphi^i + \tfrac{1}{2}\Gamma^i_{mn}(\phi)(\varphi - \phi)^m(\varphi - \phi)^n + \cdots]J_i. \tag{3.21}$$

So the correct generating functional now reads

$$Z(J) = \int d\varphi \, \mathcal{M}(\varphi) \exp\frac{i}{\hbar}\{S(\varphi) + [\phi^i - \sigma^i(\phi, \varphi)]J_i\} \tag{3.22}$$

where ϕ on the right-hand side should be considered as a function of J. With this reservation equations (3.6) and (3.8) remain valid.

The modification (3.22) corresponds to the definition of the mean field by the equation

$$\langle \sigma^i(\phi, \varphi) \rangle = 0 \tag{3.23}$$

in contrast to the naive definition (3.4). Since $\sigma^i(\phi, \varphi)$ is a scalar with respect to the point of integration φ, the definition (3.23) is completely parameterization independent. (When $\Gamma \equiv 0$, equation (3.23) coincides with the naive definition (3.4).) On the other hand, if the connection Γ^i_{mn} is local, then $\sigma^i(\phi, \varphi)$ will also be a local function of φ. Then the modification (3.22) may be regarded as the result of a very nonlinear—but nonetheless local—reparameterization of the field φ. According to the standard argument, such a modification does not change the S matrix.

The above modification leads to the following equation for the effective action

$$\exp\left(\frac{i}{\hbar}W(\phi)\right) = \int d\varphi \, \mathcal{M}(\varphi) \exp\frac{i}{\hbar}\left(S(\varphi)+\sigma^i(\phi,\varphi)\frac{\partial W(\phi)}{\partial\phi^i}\right) \tag{3.24}$$

which replaces the naive equation (3.10). The integrand of (3.24) (including the volume element $d\varphi\,\mathcal{M}(\varphi)$) is a scalar with respect to both the integration field φ and the mean field ϕ. The solution $W(\phi)$ of the new equation is now manifestly a scalar, as desired.

There remains a final step: the choice of the configuration-space connection to be used in equation (3.22)–(3.24). The connection can of course exist without a metric, but we shall see below that it is in fact the Christoffel symbol

$$\Gamma^i_{mn} = \tfrac{1}{2}G^{ik}(\partial_m G_{kn}+\partial_n G_{km}-\partial_k G_{mn}). \tag{3.25}$$

Here G_{mn} is the configuration-space metric which should now be determined.

The connection must have the following properties.

(i) The connection should be determined by the action $S(\varphi)$ itself with the aid of some universal rule.

(ii) If the field is free (non-interacting), i.e. if there exists a local reparameterization making the action $S(\varphi)$ quadratic in φ, then the effective action must coincide with the classical action up to an additive constant (there should be no quantum corrections). This means that for a free-field theory the connection should be identically zero in the parameterization in which the action is quadratic.

(iii) The connection should be 'ultralocal', that is, proportional to undifferentiated delta functions and containing only undifferentiated fields, in order not to contradict S-matrix theory.

The Lagrangian (3.2) defines three local quantities: $A_{ab}(\varphi)$, $B_a(\varphi)$ and $C(\varphi)$. The condition that the Lagrangian be a scalar defines the transformation properties of these quantities. We find that under local reparameterizations $A_{ab}(\varphi)$ is a tensor, $B_a(\varphi)$ is a vector, and $C(\varphi)$ is a scalar. The necessary and sufficient conditions for a theory of the form (3.2) to be a free-field theory are

$$R_{abcd} \equiv 0 \tag{3.26}$$

$$\mathcal{D}_a(\mathcal{D}_b B_c - \mathcal{D}_c B_b) \equiv 0 \tag{3.27}$$

$$\mathcal{D}_a \mathcal{D}_b \mathcal{D}_c C \equiv 0 \tag{3.28}$$

where R_{abcd} and $\mathcal{D}_a$ are respectively the Riemann tensor and the covariant derivative in the metric A_{ab}.

According to (3.26) and the requirement (ii), the configuration-space connection for a free-field theory of the form (3.2) must have the form (3.25) with the configuration-space metric

$$G_{mn} = A_{ab}(\varphi(t))\delta(t_a - t_b) \tag{3.29}$$

$$m = (a, t_a) \qquad n = (b, t_b).$$

If the theory (3.2) is not free-field, then the configuration-space connection may, in principle, contain tensorial additions which involve the quantities (3.26)–(3.28) and which vanish when these quantities vanish. However, if (3.2) is the particle Lagrangian, then $A_{ab}(\varphi)$ is directly the local configuration-space metric. If (3.2) is the field Lagrangian, then the quantities $B_a(\varphi)$ and $C(\varphi)$ contain spatial derivatives and cannot be used in the construction of the connection. In a field theory $A_{ab}(\varphi)$ is the only ultralocal (with respect to both space and time) tensor defined by the Lagrangian. Moreover, $A_{ab}(\varphi)$ always acts as a metric in the measure (3.3). These arguments indicate that the configuration-space metric for the theory (3.2) should always be chosen as (3.29)†.

Let us return to equation (3.24) for the effective action. To solve this equation it is convenient to make the change of integration variables

$$\varphi^i \to \sigma^i(\phi, \varphi) \tag{3.30}$$

and use the fact that any function can be expanded in the covariant Taylor series

$$S(\varphi) = \sum_{n=0}^{\infty} \frac{(-1)^n}{n!} [\nabla_{i_1} \ldots \nabla_{i_n} S(\phi)] \sigma^{i_1}(\phi, \varphi) \ldots \sigma^{i_n}(\phi, \varphi) + \text{curvature terms.} \tag{3.31}$$

Then it is evident that the solution of equation (3.24) will differ from the solution of the naive equation (3.10) in only two respects. Firstly, all partial derivatives of the action $S(\varphi)$ will be replaced by covariant derivatives (cf expansions (3.31) and (3.11)). Secondly, there will arise the Jacobian

$$I^{-1} = \left| \frac{\partial}{\partial \varphi^k} \sigma^i(\phi, \varphi) \right|. \tag{3.32}$$

Since our connection is the Christoffel symbol (equation (3.25)), we find

$$I = \frac{G^{1/2}(\phi)}{G^{1/2}(\varphi)} \Delta^{-1}(\phi, \varphi) \tag{3.33}$$

where $\Delta(\phi, \varphi)$ is the symmetric two-point scalar from DeWitt (1963, 1965, 1975):

$$\Delta^{-1} = 1 - \tfrac{1}{6} R_{ik}(\phi) \sigma^i \sigma^k + O(\sigma^3). \tag{3.34}$$

Here G is the determinant and R_{ik} is the contracted Riemann tensor of the configuration-space metric G_{ik}. With our choice of G_{ik} the measure (equation (3.33)) is

$$\mathcal{M}(\varphi) = G^{1/2}(\varphi). \tag{3.35}$$

Therefore as a result of the change of variables (3.30) we obtain

$$\mathcal{M}(\varphi) \, d\varphi = \mathcal{M}(\phi) \Delta^{-1}(\phi, \varphi) \, d\sigma. \tag{3.36}$$

---

† One possible objection is that in a field theory, $A_{ab}(\varphi)$ will always contain the factor g^{00} of the space–time metric, and thus the definition (3.29) is not relativistically covariant. In fact, for all fields except the gravitational field itself, this factor is constant (field independent), so it will drop out of the connection. As to the gravitational field, the conditions determining the connection in gauge theories are different (see below).

Thus, there is no need to expand the measure. Instead, the quantity Δ arises, for which one has the expansion (3.34).

Let us note that since the substitution (3.30) is local, the whole effect of the Jacobian (as that of the measure in general) is

$$I = 1 + \delta(0) (\ldots). \tag{3.37}$$

This is also seen from the expansion (3.34) which involves only contractions of Riemann tensors or their covariant derivatives. Since the metric is local, these contractions are all proportional to $\delta(0)$, so that

$$\Delta = 1 + \delta(0) (\ldots). \tag{3.38}$$

Thus, beginning with the two-loop order, the new diagrammatic technique for the effective action contains only elements of two types: (1) covariant derivatives of the classical action and, (2) contractions of Riemann tensors (or their covariant derivatives) which come from the expansion of Δ. If one neglects the terms $\delta(0)$, then all diagrams containing elements of the second type can be omitted. Then the new diagrammatic rules can be obtained from the naive rules simply by expanding $S(\varphi)$ according to equation (3.31). In the one-loop approximation one obtains exactly the expression (3.17).

In this way we arrive at the unique off-mass shell extension for non-gauge theories. Let us now turn to gauge theories.

In considering gauge theories I shall confine myself to the case where the generators of gauge transformations in the local basis form a closed algebra. In this case, the Feynman rules are given by the functional integral (2.37).

Introducing a connection in the space of the gauge field φ^i and constructing $\sigma^i(\phi, \varphi)$, we may formulate the modified equation for the effective action:

$$\exp\left(\frac{i}{\hbar} W(\phi)\right) = \int d\varphi \, \mathscr{M}(\varphi) \exp\frac{i}{\hbar}\left(S(\varphi) + \tfrac{1}{2}\chi^\mu(\varphi)\beta_{\mu\nu}\chi^\nu(\varphi)\right.$$
$$\left. + \frac{\hbar}{i} \operatorname{Tr} \ln Q^\alpha{}_\beta(\varphi) + \sigma^i(\phi, \varphi) \frac{\partial W(\phi)}{\partial \phi_i}\right). \tag{3.39}$$

The whole problem lies in the determination of the connection (or metric) in the gauge-field space.

In (3.39) $\mathscr{M} = 1 + \delta(0) (\ldots)$, and $Q^\alpha{}_\beta(\varphi)$ is defined in equation (2.14). Note that all quantities carrying group indices (the gauge conditions $\chi^\mu(\varphi)$, the constant matrix $\beta_{\mu\nu}$, and $Q^\alpha{}_\beta(\varphi)$) are scalars with respect to reparameterizations of the gauge field φ^i. Therefore the effective action as defined by equation (3.39) is parameterization invariant. However, now we have two more requirements. Firstly, the effective action should be gauge invariant. The necessary condition for this is the use of DeWitt's mean-field gauge conditions (DeWitt 1981a). The functions $\chi^\mu(\varphi)$ must depend parametrically on the mean field

$$\chi^\mu(\varphi) = \chi^\mu(\phi, \varphi) \qquad \chi^\mu(\phi, \phi) \equiv 0 \tag{3.40}$$

and be covariant under simultaneous gauge transformations of both arguments. Secondly, the effective action should not depend on the choice of the arbitrary elements $\chi^\mu(\phi, \varphi)$ and $\beta_{\mu\nu}$. This must be ensured by the correct choice of the configuration-space connection in (3.39).

For illustration let us write down the formulae of the one-loop approximation. The solution of equation (3.39) in the one-loop approximation is:

$$W(\phi) = S(\phi) + \frac{\hbar}{i}\ln\mathcal{M}(\phi) - \frac{\hbar}{2i}\,\mathrm{Tr}\ln D^{-1}{}_{mn}(\phi) + \frac{\hbar}{i}\,\mathrm{Tr}\ln Q^\alpha{}_\beta(\phi) + O(\hbar^2) \quad (3.41)$$

where

$$D^{-1}{}_{mn}(\phi) = \nabla_m\nabla_n S(\phi) + \chi^\mu{}_{,m}\beta_{\mu\nu}\chi^\nu{}_{,n} \quad (3.42)$$

$$\chi^\mu{}_{,m} = \left.\frac{\partial\chi^\mu(\phi, \varphi)}{\partial\varphi^m}\right|_{\varphi = \phi}$$

and where we have usd (3.40). If we compute the change in $W(\phi)$ induced by the infinitesimal change $\delta\chi^\mu$ in the gauge conditions, the result will be

$$\delta W(\phi) = -\frac{\hbar}{i}D^{mn}Q^{-1\alpha}{}_\beta\,\delta\chi^\beta{}_{,m}(\nabla_n R^p{}_\alpha)\frac{\partial S}{\partial\phi\sigma} + O(\hbar 2) \quad (3.43)$$

where the $R^p{}_\alpha$ are the local generators of gauge transformations, and where all quantities are evaluated at the point ϕ. Equations (3.41)–(3.43) are the usual expressions modified by the presence of the connection which converts all derivatives into covariant derivatives ∇. For example, the generators $R^p{}_\alpha$ transform like vectors under reparameterizations of the field φ, and

$$\nabla_n R^p{}_\alpha(\varphi) = \frac{\partial R^p{}_\alpha}{\partial\varphi^n} + \Gamma^p{}_{nk}R^k{}_\alpha. \quad (3.44)$$

Let us proceed to the formulation of conditions defining the configuration-space metric.

We have seen in the previous section that there exists a reparameterization $\varphi^i \to (\xi^A, \theta^\mu)$, such that the functions $\xi^A(\varphi)$ are gauge invariant and $\theta^\mu(\varphi) = \chi^\mu(\varphi)$ (see equations (2.15)–(2.29)). In the parameterization ξ, θ, the classical action does not depend on θ and becomes the non-gauge action $S(\varphi) = \bar{S}(\xi)$. Therefore the effective action should be determined only by some connection in the ξ space, $\bar{\Gamma}^A{}_{BC}(\xi)$. On the other hand, after the introduction of the gauge-breaking term, the action in the function integral becomes dependent on θ. But the fields ξ and θ themselves decouple, and θ enters quadratically (equation (2.31)).

We must ensure that ξ and θ decouple in the effective action as well, and that the effective action of the field θ remains the free-field action. Therefore, designating by a bar the quantities in the coordinates ξ, θ, we require the following.

(i) In ξ, θ coordinates the configuration-space connection should decompose as

$$\bar{\Gamma}^i{}_{mn} = \{\bar{\Gamma}^A{}_{BC}(\xi), 0\}. \quad (3.45)$$

The corresponding decomposition of the metric is

$$\bar{G}_{mn} = \{\bar{G}_{AB}(\xi), c_{\mu\nu}\} \qquad c_{\mu\nu} = \text{constant} \qquad (3.46)$$

$$\det \bar{G}_{AB}(\xi) \neq 0 \qquad \det c_{\mu\nu} \neq 0.$$

The constant matrix $c_{\mu\nu}$ does not enter the connection, so its value is irrelevant.

The above requirement says nothing about the form of the metric in ξ space. In this regard we must remember that in the initial parameterization φ^i we have the requirement of locality. Note, that in a gauge theory, the metric (or connection) need not be absolutely local. Indeed, when proving that the S matrix is gauge independent (see e.g. DeWitt 1967b), we make the following reparameterization:

$$\varphi \to \varphi + \delta\varphi \qquad \delta\varphi^i = R^i{}_{,\alpha} Q^{-1\alpha}{}_{\beta} \delta\chi^{\beta}. \qquad (3.47)$$

This means that we admit non-local displacements, but only in the direction of the vectors $R^i{}_{\alpha}$. Therefore we must require

(ii) The scalar product of vectors orthogonal to $R^i{}_{\alpha}$ should be defined with the aid of a local metric. In accordance with previous considerations, this local metric should be contained in the highest-derivative term of the classical action $S(\varphi)$.

Before constructing the metric satisfying the above requirements, let us derive some important consequences of requirement (i).

Let $G_{mn}(\varphi)$ and $\Gamma^i{}_{mn}(\varphi)$ be the metric and the connection in the initial parameterization φ. We have from (3.46):

$$G_{mn}(\varphi) = \frac{\partial \xi^A}{\partial \varphi^m} \bar{G}_{AB}(\xi) \frac{\partial \xi^B}{\partial \varphi^n} + \frac{\partial \theta^\mu}{\partial \varphi^m} c_{\mu\nu} \frac{\partial \theta^\nu}{\partial \varphi^n}. \qquad (3.48)$$

Multiplying this by $\partial \varphi^m / \partial \theta^\alpha$, we find that

$$\mathscr{R}^m{}_\alpha(\varphi) G_{mn}(\varphi) = c_{\alpha\beta} \frac{\partial \chi^\beta(\varphi)}{\partial \varphi^n} \qquad (3.49)$$

where the $\mathscr{R}^m{}_\alpha(\varphi)$ are the Abelian generators (2.15). On the other hand, since $\bar{G}_{mn}(\xi, \theta)$ does not depend on θ, one easily proves that the Abelian generators are the Killing vectors of the metric G_{mn}:

$$\nabla^m \mathscr{R}^n{}_\alpha + \nabla^n \mathscr{R}^m{}_\alpha = 0. \qquad (3.50)$$

The consequences of (3.49) and (3.50) are

$$\nabla_m \mathscr{R}^i{}_\alpha = 0 \qquad (3.51)$$

$$\nabla_m \nabla_n \chi^\alpha = 0. \qquad (3.52)$$

Since the Abelian generators $\mathscr{R}^i{}_\alpha$ are linear combinations of the local generators $R^i{}_\alpha$ (equation (2.15)), one has from (3.51):

$$\nabla_m R^i{}_\alpha \propto R^i{}_\beta. \qquad (3.53)$$

The property (3.53) is crucial. It makes the effective action independent of the choice of gauge conditions. In the one-loop approximation, this is seen immediately from (3.43). To see this generally, let us consider the quantity $\sigma^i(\phi, \varphi)$ constructed with the aid of the connection (3.45).

From the geodesic equation it follows that in the variables ξ, θ the quantity $\sigma^i(\varphi, \varphi')$ decomposes as

$$\bar{\sigma}^i = \{\bar{\sigma}^A(\xi, \xi'), \bar{\sigma}^\mu(\theta, \theta')\} \tag{3.54}$$

where

$$\bar{\sigma}^\mu(\theta, \theta') = \theta^\mu - \theta'^\mu.$$

Hence

$$\sigma^i(\varphi, \varphi') = \frac{\partial \varphi^i}{\partial \xi^A} \bar{\sigma}^A(\xi, \xi') + \frac{\partial \varphi^i}{\partial \theta^\mu}(\theta^\mu - \theta'^\mu). \tag{3.55}$$

Using this expression one can easily derive the transformation properties of $\sigma^i(\phi, \varphi)$ under gauge transformations of each of the arguments. One finds

$$\mathcal{R}^k_\alpha(\varphi) \frac{\partial}{\partial \varphi^k} \sigma^i(\phi, \varphi) = -\mathcal{R}^i_\alpha(\phi) \tag{3.56}$$

$$\mathcal{R}^k_\alpha(\phi) \nabla_k \sigma^i(\phi, \varphi) = \mathcal{R}^i_\alpha(\phi) \tag{3.57}$$

where ∇_k is the covariant derivative with respect to ϕ. In terms of the local generators we have

$$R^k_\alpha(\varphi) \frac{\partial}{\partial \varphi^k} \sigma^i(\phi, \varphi) \propto R^i_\beta(\phi) \tag{3.58}$$

$$R^k_\alpha(\phi) \nabla_k \sigma^i(\phi, \varphi) \propto R^i_\beta(\phi). \tag{3.59}$$

Property (3.58) realizes the impossible dream: the gauge transformation of the quantized (integration) field φ reduces to the gauge transformation of the external (mean) field ϕ. Using this property we can prove the long-soughtafter theorem: if the effective action $W(\phi)$ is gauge invariant, then it does not depend on the choice of gauge conditions. Indeed, let us make, as usual, the change (3.47) of integration variables in (3.39), which transforms the gauge conditions χ^μ into $\chi^\mu + \delta\chi^\mu$. Then the only thing that must be verified is the invariance of the source term (for all the remainder is verified in DeWitt (1967b)). The transformation of this term in (3.39) is

$$R^k_\alpha(\varphi) \frac{\partial}{\partial \varphi^k} \sigma^i(\phi, \varphi) \frac{\partial W(\phi)}{\partial \phi^i} \propto R^i_\beta(\phi) \frac{\partial W(\phi)}{\partial \phi^i} \tag{3.60}$$

by (3.58). If $W(\phi)$ is gauge invariant, then (3.60) is zero.

The proof that $W(\phi)$ is gauge-invariant uses both properties (3.58) and (3.59). One considers equation (3.39) and makes the gauge transformation of the mean field

$$\phi' = \phi'^i + R^i_\alpha(\phi')\varepsilon^\alpha$$

accompanied by the simultaneous transformation of the integration variables

$$\varphi^i = \varphi'^i + R^i{}_\alpha(\varphi')\varepsilon^\alpha$$

where the ε^α are small parameters. Then the gauge-breaking and ghost terms are invariant owing to DeWitt's mean-field gauge conditions. As to the source term, its variation reduces to the quantity

$$\frac{\partial W(\phi)}{\partial \phi^i} R^i{}_\alpha(\phi) \tag{3.61}$$

thanks to properties (3.58), (3.59) and (3.53). Thus one arrives at a homogeneous equation for the quantity (3.61). From this equation it follows that if $W(\phi)$ is gauge invariant through the first n orders in $\hbar$, then it is gauge invariant in the $(n+1)$th order as well. Since in the classical approximation $W(\phi)$ is gauge invariant (by (3.40)), the quantity (3.61) is zero.

Thus, if the configuration-space metric satisfies the above requirement (i), then the modified effective action will be parameterization independent, gauge invariant, and independent of the choice of the mean-field gauge conditions. On the other hand, if the metric satisfies requirement (ii), then the above modification will not alter the S-matrix elements between physical states.

There remains the construction of a metric which satisfies requirements (i) and (ii). Expanding the metric tensor with upper indices in the basis of linearly independent vectors

$$\partial \varphi^m / \partial \xi^A \qquad \partial \varphi^m / \partial \theta^\alpha$$

we arrive at the following explicit expression

$$G^{mn}(\varphi) = P^m{}_i(\varphi)\gamma^{ik}(\varphi)P^n{}_k(\varphi) + \mathcal{R}^m{}_\mu(\varphi)c^{\mu\nu}\mathcal{R}^n{}_\nu(\varphi) \tag{3.62}$$

where

$$P^m{}_i(\varphi) = \delta^m{}_i - \mathcal{R}^m{}_\alpha(\varphi)\frac{\partial \chi^\alpha(\varphi)}{\partial \varphi^i} \tag{3.63}$$

$c^{\mu\nu}$ is the inverse of $c_{\mu\nu}$ in (3.46), and $\gamma^{ik}(\varphi)$ is some tensor which will be defined below.

According to (2.28), the derivatives of the invariants $\partial \xi^A / \partial \varphi^i$ are covariant vectors orthogonal to $R^i{}_\alpha$ in any metric. Using that

$$P^m{}_i = \frac{\partial \varphi^m}{\partial \xi^A} \frac{\partial \xi^A}{\partial \varphi^i}$$

we find the scalar product of these vectors to be

$$\frac{\partial \xi^A}{\partial \varphi^m} G^{mn} \frac{\partial \xi^B}{\partial \varphi^n} = \frac{\partial \xi^A}{\partial \varphi^m} \gamma^{mn} \frac{\partial \xi^B}{\partial \varphi^n}. \tag{3.64}$$

Thus, if we choose $\gamma^{ik}(\varphi)$ in (3.62) as the inverse of the local metric contained in the highest-derivative term of the classical action, then requirement (ii) will be satisfied.

Let us verify whether expression (3.62) satisfies requirement (i). In the coordinates ξ, θ the metric tensor (3.62) takes the form

$$\bar{G}^{AB} = \frac{\partial \xi^A}{\partial \varphi^m} \gamma^{mn} \frac{\partial \xi^B}{\partial \varphi^n}$$

$$\bar{G}^{A\alpha} = \frac{\partial \xi^A}{\partial \varphi^m} G^{mn} \frac{\partial \theta^\alpha}{\partial \varphi^n} = 0$$

$$\bar{G}^{\alpha\beta} = \frac{\partial \theta^\alpha}{\partial \varphi^m} G^{mn} \frac{\partial \theta^\beta}{\partial \varphi^n} = c^{\alpha\beta}.$$

This will be equivalent to (3.46) if $\bar{G}^{AB}$ is invertible and if we ensure that

$$\frac{\partial}{\partial \theta^\alpha} \bar{G}^{AB} = 0. \tag{3.65}$$

One finds

$$\frac{\partial}{\partial \theta^\alpha} \bar{G}^{AB} = \frac{\partial \xi^A}{\partial \varphi^m} \frac{\partial \xi^B}{\partial \varphi^n} [\mathscr{R}^i{}_\alpha \partial_i \gamma^{mn} - (\partial_k \mathscr{R}^m{}_\alpha) \gamma^{kn} - (\partial_k \mathscr{R}^n{}_\alpha) \gamma^{km}]$$

$$= Q^{-1\beta}{}_\alpha \frac{\partial \xi^A}{\partial \varphi^m} \frac{\partial \xi^B}{\partial \varphi^n} [R^i{}_\beta \partial_i \gamma^{mn} - (\partial_k R^m{}_\beta) \gamma^{kn} - (\partial_k R^n{}_\beta) \gamma^{km}].$$

Here we have used the relation $\mathscr{R}^i{}_\alpha = Q^{-1\beta}{}_\alpha R^i{}_\beta$ (equation (2.15)) and the fact that terms with $\partial_k Q^{-1}$ do not contribute because $R^m{}_\beta \partial \xi^A / \partial \varphi^m = 0$.

To ensure condition (3.65) we must set

$$R^i{}_\beta \partial_i \gamma^{mn} - (\partial_k R^m{}_\beta) \gamma^{kn} - (\partial_k R^n{}_\beta) \gamma^{km} = 0.$$

This is Killing's equation

$$\gamma^{mk} \mathscr{D}_k R^n{}_\beta + \gamma^{nk} \mathscr{D}_k R^m{}_\beta = 0 \tag{3.66}$$

where $\mathscr{D}_k$ is the covariant derivative in the metric γ_{ik} (the inverse of γ^{ik}). Thus the local generators $R^m{}_\alpha$ should be Killing vectors of the local metric γ_{ik}. This is one further requirement that should be satisfied by γ^{ik}.

The invertibility of $\bar{G}^{AB}$ is equivalent to the invertibility of the full metric (3.62). One can prove that (3.62) is invertible if and only if

$$\det \mathscr{N}_{\mu\nu} \neq 0 \qquad \mathscr{N}_{\mu\nu} = R^i{}_\mu \gamma_{ik} R^k{}_\nu. \tag{3.67}$$

The non-degeneracy of the matrix (3.67) is the last requirement to be placed on γ^{ik}.

Thus, if γ^{ik} satisfies the above conditions, then (3.62) is the correct expression for the configuration-space metric. Deferring the determination of γ^{ik} to the last, let us now derive the expressions for the metric with lower indices and the connection.

The inversion of the matrix (3.62) convinces us that, verily, almost everything lies in the 'Gospel'! The result is

$$G_{mn}(\varphi) = (\gamma_{mn} - \gamma_{mi} R^i{}_\mu \mathscr{N}^{\mu\nu} R^k{}_\nu \gamma_{kn}) + \frac{\partial \chi^\mu}{\partial \varphi^m} c_{\mu\nu} \frac{\partial \chi^\nu}{\partial \varphi^n} \tag{3.68}$$

where $\mathscr{N}^{\mu\nu}$ is the inverse of $\mathscr{N}_{\mu\nu}$. The expression in (3.68) in parentheses exactly coincides with DeWitt's metric in the space of group orbits (DeWitt 1967a)†. However, our metric (3.68) in the full configuration space differs from that of DeWitt. (DeWitt's metric in the full configuration space is simply γ_{mn}.) The difference can be traced to our requirement (i) and is very important; from it arises the main property (3.53) of the connection.

The calculation of the connection is rather lengthy and the property (3.66) is used extensively. The result is

$$\Gamma^r_{mn} = \mathscr{T}^r_{mn} + P^r_i \left[-(\mathscr{D}_m R^i_\mu) \mathscr{N}^{\mu\nu} R^k_\nu \gamma_{kn} - (\mathscr{D}_n R^i_\mu) \mathscr{N}^{\mu\nu} R^k_\nu \gamma_{km} \right.$$
$$\left. + \tfrac{1}{2}(R^k_\beta \mathscr{D}_k R^i_\alpha + R^k_\alpha \mathscr{D}_k R^i_\beta) \times (\mathscr{N}^{\alpha\mu} R^u_\mu \gamma_{um})(\mathscr{N}^{\beta\nu} R^v_\nu \gamma_{vn}) \right]$$
$$+ \mathscr{R}^r_\alpha (\mathscr{D}_m \mathscr{D}_n \chi^\alpha) \tag{3.69}$$

where $\mathscr{T}^r_{mn}$ is the connection corresponding to the local metric γ_{ik}.

The apparent difficulty with the above expression is that the connection itself depends on the gauge conditions χ^α. However, one may notice that all terms of Γ^r_{mn} which contain gauge conditions are proportional to R^r_α. I shall prove that any term of Γ^r_{mn} proportional to R^r_α gives a vanishing contribution to the effective action. In the one-loop approximation this is seen immediately, because Γ^r_{mn} enters (3.42) in the combination $\Gamma^r_{mn} \partial S/\partial \phi^r$.

To see this generally, let us consider a new connection differing from (3.45) by an addition which, in the variables φ, has the form

$$\Delta \Gamma^i_{mn} \propto R^i_\alpha. \tag{3.70}$$

Our goal is to prove that the corresponding change in the quantity $\sigma^i(\phi, \varphi)$ will be

$$\Delta \sigma^i(\phi, \varphi) \propto R^i_\alpha(\phi). \tag{3.71}$$

For this purpose, let us find the addition to the connection in the variables ξ, θ. One finds that the addition to the components $\bar\Gamma^A_{..}$, with the superscript belonging to the ξ space, is zero. Indeed,

$$\Delta \bar\Gamma^A_{..} = \frac{\partial \xi^A}{\partial \varphi^i} \Delta \Gamma^i_{mn} \frac{\partial \varphi^m}{\partial(\xi, \theta)} \frac{\partial \varphi^n}{\partial(\xi, \theta)} = 0$$

because $R^i_\alpha \partial \xi^A/\partial \varphi^i = 0$. Then it follows from the geodesic equation that the addition to the quantity $\bar\sigma^A$, with upper index belonging to the ξ space, is also zero. Since

$$\sigma^i(\varphi, \varphi') = \frac{\partial \varphi^i}{\partial \varphi^A} \bar\sigma^A + \frac{\partial \varphi^i}{\partial \theta^\mu} \bar\sigma^\mu$$

and $\Delta \bar\sigma^A = 0$, we find

$$\Delta \sigma^i(\varphi, \varphi') = \frac{\partial \varphi^i}{\partial \theta^\mu} \Delta \bar\sigma^\mu = \mathscr{R}^i_\mu(\varphi) \Delta \bar\sigma^\mu.$$

This is precisely (3.71).

---

† DeWitt used this construction for the space of 3-geometries in the canonical formalism of gravity theory.

Since the effective action satisfying equation (3.39) with the connection $\Gamma^i{}_{mn}$ is gauge invariant, we may transform the source term identically as follows

$$\sigma^i\Big|_\Gamma \frac{\partial W(\phi)}{\partial \phi^i} = (\sigma^i\big|_\Gamma + \Delta\sigma^i)\frac{\partial W(\phi)}{\partial \phi^i} = \sigma^i\big|_{\Gamma + \Delta\Gamma}\frac{\partial W(\phi)}{\partial \phi^i} \tag{3.72}$$

where $\Delta\sigma$ and $\Delta\Gamma$ are given by (3.71) and (3.70). This proves that any addition of the form (3.70) to the connection gives a vanishing contribution to the effective action.

Omitting all terms of (3.69) which are proportional to $R^r{}_\alpha$, we obtain the final expression for the configuration-space connection in gauge theories:

$$\Gamma^i{}_{mn} = \mathcal{T}^i{}_{mn} - (\mathcal{D}_m R^i{}_\mu)\mathcal{N}^{\mu\nu} R^k{}_\nu \gamma_{kn} - (\mathcal{D}_n R^i{}_\mu)\mathcal{N}^{\mu\nu} R^k{}_\nu \gamma_{km}$$
$$+ \tfrac{1}{2}(R^k{}_\beta \mathcal{D}_k R^k{}_\alpha + R^i{}_\alpha \mathcal{D}_k R^i{}_\beta)(\mathcal{N}^{\alpha\mu} R^u{}_\mu \gamma_{um})(\mathcal{N}^{\beta\nu} R^v{}_\nu \gamma_{vn}). \tag{3.73}$$

The connection is constructed only from the local generators $R^i{}_\alpha$ and the local metric γ_{ik} and does not contain arbitrary elements like gauge conditions. One may verify by direct computation that this connection satisfies the main property (3.53).

Let us also present an expression for the second covariant derivative of any invariant, for example, the classical action $S(\varphi)$. Using (3.73) we find

$$\nabla_m \nabla_n S = \Pi^i{}_m (\mathcal{D}_i \mathcal{D}_k S)\Pi^k{}_n \tag{3.74}$$

where

$$\Pi^i{}_m = \delta^i{}_m - R^i{}_\mu \mathcal{N}^{\mu\nu} R^v{}_\nu \gamma_{vm} \tag{3.75}$$

may be called the DeWitt projection operator in contrast to the general projection operator (3.63). The essential property of the DeWitt projection operator is that any contravariant vector of the form

$$h^i_\perp = \Pi^i{}_m h^m \tag{3.76}$$

is orthogonal to $R^i{}_\alpha$ in the local metric γ_{ik}[†].

It remains to define the metric γ_{ik}. This metric must be ultralocal. It must be contained in the highest-derivative term of the classical action. It must also satisfy Killing's equation (3.66) and ensure the non-degeneracy of the matrix $\mathcal{N}_{\mu\nu}$. Since we confined our attention to bosonic theories with closed gauge algebras, I shall present the solution for γ_{ik} for only two theories: Yang–Mills theory and gravity theory. This is easy to do since, in fact, it was done for me sixteen years ago by DeWitt (1967a).

For the Yang–Mills field in the usual parameterization $A^a{}_\alpha(x)$, the local metric which satisfies Killing's equation (3.66) is:

---

† The contravariant vectors of the form $P^i{}_m h^m$, constructed with the aid of the general projection operator, are orthogonal to $R^i{}_\alpha$ in the full metric G_{ik}.

$$\gamma_{ik}\mathrm{d}\varphi^i\,\mathrm{d}\varphi^k = \int\mathrm{d}^4x\,\sqrt{g}\,g^{\mu\nu}\gamma_{ab}\mathrm{d}A^a{}_\mu(x)\,\mathrm{d}A^b{}_\nu(x) \tag{3.77}$$

$$\gamma_{ab} = C^f{}_{ag}C^g{}_{bf}$$

where the $C^f{}_{ag}$ are the structure constants of the generating Lie group. At the same time, this γ_{ik} is precisely the tensor entering the second-derivative term of the Yang–Mills Lagrangian. The non-degeneracy of $\mathcal{N}_{\mu\nu}$ can also be verified. Thus in the case of the Yang–Mills theory

$$\mathcal{T}^i{}_{mn} = 0 \qquad \mathcal{D}_m = \partial/\partial\varphi^m$$

in the parameterization $\varphi^i = A^a{}_\alpha(x)$. The connection (3.73), however, is non-trivial because the generators $R^i{}_\alpha$ are field dependent. Only in the Abelian case does $\Gamma^i{}_{mn} = 0$.

In Einstein's theory, there is only a single one-parameter family of local metrics which satisfy Killing's equation (DeWitt 1967a). In the parameterization $\varphi^i = g_{\mu\nu}(x)$, this family is given by

$$\gamma_{ik}\mathrm{d}\varphi^i\mathrm{d}\varphi^k = \int\mathrm{d}^4x\,\sqrt{g}\,(g^{\mu\alpha}g^{\nu\beta} + g^{\mu\beta}g^{\nu\alpha} + ag^{\mu\nu}g^{\alpha\beta})\,\mathrm{d}g_{\mu\nu}(x)\,\mathrm{d}g_{\alpha\beta}(x) \tag{3.78}$$

where $a \neq -\frac{1}{2}$ is the parameter of the family. To fix this parameter let us consider the second covariant derivative of the classical action.

The corresponding quadratic form is

$$h^m(\nabla_m\nabla_n S)h^n = h^m_\perp(\mathcal{D}_m\mathcal{D}_n S)h^n_\perp \tag{3.79}$$

by virtue of (3.74) and (3.76). Computing this form with the metric (3.78) for gravity, and confining our attention to second-derivative terms, we find:

$$h^m_\perp(\mathcal{D}_m\mathcal{D}_n S)h^n_\perp = \frac{1}{4}\int\mathrm{d}^4x\,\sqrt{g}\,h^\perp_{\mu\nu}[g^{\mu\alpha}g^{\nu\beta} + g^{\mu\beta}g^{\nu\alpha} + g^{\mu\nu}g^{\alpha\beta}(-a^2 - 2a - 2)]\Box h^\perp_{\alpha\beta}$$

$$+ \text{curvature terms.} \tag{3.80}$$

The tensor in square brackets must coincide with the metric (3.78). This gives the equation

$$-a^2 - 2a - 2 = a$$

which has two solutions: $a = -1$ and $a = -2$. To choose between them we use the last requirement, namely the non-degeneracy of $\mathcal{N}_{\mu\nu}$. One can easily verify that with $a = -2$ the operator $\mathcal{N}_{\mu\nu}$ is degenerate at stationary configurations of the classical action. Thus the solution for the local metric in Einstein's theory is (3.78) with

$$a = -1. \tag{3.81}$$

The configuration-space connection (and hence the modified effective action) is unique in both the Yang–Mills and Einstein theories.

4. THE EFFECTIVE ACTION: PROBLEMS IN EVALUATION

All one-loop contributions to the gravitational effective action are of one type and have the form[†]

$$\text{Tr} \ln F \tag{4.1}$$

where F is the kernel of a covariant differential operator.

The basic technique for the calculation of quantities such as (4.1) was invented by Schwinger (1951); it was then reformulated in geometric language and extended to curved space by DeWitt (1963, 1965). This technique is especially useful when F is an operator in an external gauge field. All other covariant methods ('t Hooft and Veltman 1974, Dowker and Critcheley 1977, Hawking 1977) are essentially equivalent to the method of Schwinger and DeWitt. The Schwinger–DeWitt technique admits the introduction of any version of covariant regularization (Brown and Cassidy 1977, Christensen 1976, Adler *et al* 1977, Fradkin and Vilkovisky 1976, 1977b) and may be regarded as the general framework for the evaluation of the effective action.

The Schwinger–DeWitt technique is directly applicable to second-order operators with leading derivatives of the form

$$F = C_{AB} g^{\mu\nu} \nabla_\mu \nabla_\nu + \ldots \tag{4.2}$$

where C_{AB} is a local and invertible matrix, and ∇ is a covariant derivative with any connection acting on any field. The generalization of the Schwinger–DeWitt technique to operators of any order and operators which do not have the form (4.2) can be found in Barvinsky and Vilkovisky (1983a). One can go far by considering F in very general terms, but for our purposes it is sufficient to restrict our attention to a specific example, for instance, the contribution of the scalar field:

$$W_{\text{one-loop}} = \tfrac{1}{2} \text{Tr} \ln \left(-\Box + m^2 \right) + \delta(0) \left(\ldots \right). \tag{4.3}$$

Here the ellipses (. . .) serve to remind us of additional terms proportional to $\delta(0)$ (Fradkin and Vilkovisky 1973, 1977b).

For reasons which will become clear later, we shall work in the space with Euclidean signature, so $\Box$ in (4.3) is the d-dimensional Laplacian. In this case, Schwinger–DeWitt representation is more expediently obtained via the heat equation than via the Schrödinger equation. Taking into account the fact that $\Box$ is negative-definite, one has:

$$\frac{1}{-\Box + m^2} = \int_0^\infty ds \, \exp\left[-s(-\Box + m^2) \right] \tag{4.4}$$

[†] The considerations of the previous section do not affect the contributions of matter fields to the gravitational effective action, because in these contributions the matter field is on its mass shell.

$$\mathrm{Tr}\ln(-\Box+m^2) = -\mathrm{Tr}\int_0^\infty \frac{ds}{s}\exp[-s(-\Box+m^2)]+\text{constant} \qquad (4.5)$$

where the constant does not depend on the parameters of the operator $\Box$. For the kernel of the exponentiated operator, one has the Schwinger–DeWitt representation:

$$\langle x|\exp(s\Box)|x'\rangle = \frac{1}{(4\pi)^{d/2}}\frac{D^{1/2}(x,x')}{s^{d/2}}\exp[-\sigma(x,x')/2s]\sum_{n=0}^\infty s^n a_n(x,x') \quad (4.6)$$

where d is the space–time dimension, $\sigma(x,x')$ is the world function, $D = \det(-\partial_\mu^x \partial_\nu^{x'}\sigma)$, and $a_n(x,x')$ are defined by a chain of recursion relations. All the details, including the theory of two-point tensors, can be found in DeWitt's book (DeWitt 1965)†.

Using (4.6) in (4.5) gives:

$$\mathrm{Tr}\ln(-\Box+m^2) = -\lim_{L^2\to\infty}\frac{1}{(4\pi)^{d/2}}\int_{1/L^2}^\infty \frac{ds}{s^{d/2}+1}\exp(-sm^2)\sum_{n=0}^\infty s^n\int dx\sqrt{g}\,a_n(x,x)$$

$$+\text{constant} \qquad (4.7)$$

where the ultraviolet divergence is regularized by the introduction of a positive lower limit in the proper-time integral. As a result we obtain‡ for $d = 4$:

$$W_{\text{one-loop}} = -\frac{1}{2(4\pi)^2}\int dx\sqrt{g}\left(f_0 a_0(x,x)+f_1 a_1(x,x)+f_2 a_2(x,x)\right.$$

$$\left.+\sum_{n=3}^\infty (n-3)!\frac{1}{(m^2)^{n-2}}a_n(x,x)\right) \qquad (4.8)$$

where

$$f_0 = -m^2 L^2 + \tfrac{1}{2}m^4\ln L^2 + \tfrac{1}{2}m^4(\tfrac{3}{2}-\mathbb{C}-\ln m^2) \qquad (4.9)$$

$$f_1 = L^2 - m^2\ln L^2 + m^2(\mathbb{C}-1+\ln m^2) \qquad (4.10)$$

$$f_2 = \ln L^2 - \mathbb{C}-\ln m^2. \qquad (4.11)$$

All DeWitt's coefficients $a_n(x,x)$ are local scalars constructed from the Riemann tensor and its derivatives in all possible combinations of the given dimensionality. In particular

---

† But the book itself cannot be found. When it first appeared, no one could understand anything of it. Now, twenty years later *The Dynamical Theory of Groups and Fields* has long been out of print and everybody finds they need it. Nothing in its place can be recommended except a second edition; it has no analog in the world literature.

‡ In fact, if the second-derivative term of the operator F is field dependent—as is the case in gravitation—then the Schwinger–DeWitt technique is not quite correct and requires a modification (Fradkin and Vilkovisky 1976, 1977b). The result of this modification is that divergences proportional to L^4 and terms $\delta(0)$ in (4.3) exactly cancel each other. There is no radiative cosmological term in massless theories. In the present text we neglect this modification and simply omit all volume divergences in the final result (4.8).

$$a_0(x, x) = 1 \tag{4.12}$$

$$a_1(x, x) = \alpha_1 R \tag{4.13}$$

$$a_2(x, x) = \alpha_2 \Box R + \beta_2 R_{\mu\nu\rho\delta} R^{\mu\nu\rho\delta} + \gamma_2 R_{\mu\nu} R^{\mu\nu} + \sigma_2 R^2 \tag{4.14}$$

$$a_3(x, x) = \text{constant } \Box^2 R + \text{constant } \nabla_\alpha R_{\mu\nu\rho\delta} \nabla^\alpha R^{\mu\nu\rho\delta}$$

$$+ \text{constant } R_{\mu\nu}{}^{\rho\delta} R_{\rho\delta}{}^{\gamma\sigma} R_{\gamma\sigma}{}^{\mu\nu} + \ldots \tag{4.15}$$

and so on, where

$$\alpha_1 = \tfrac{1}{6} \qquad \alpha_2 = \tfrac{1}{30} \qquad \beta_2 = \tfrac{1}{180} \qquad \gamma_2 = -\tfrac{1}{180} \qquad \sigma_2 = \tfrac{1}{72}. \tag{4.16}$$

The exact form of $a_3(x, x)$ can be found in Gilkey (1975).

The contribution of any field to the gravitational part of the effective action is of the form (4.8)–(4.15). Only the numbers $\alpha_1, \alpha_2, \beta_2 \ldots$ depend on the specific form of the operator F.

Thus we obtain the one-loop effective action (4.8) as an expansion in inverse powers of m^2. This expansion is purely local but blows up in the massless limit. More generally, the local expansion (4.8) is valid only if all dimensional background quantities (in the present case the curvatures and their derivatives) are smaller than the corresponding power of the mass parameter. With this, the corresponding chapter of the 'Gospel' concludes. Let us see if it is possible to go further.

At $m^2 = 0$ the integrals with respect to proper time in (4.7) diverge at the upper limit; moreover, each term in the series (4.7) diverges more strongly than the previous one. This problem is often claimed to be the infrared divergence, but in the present case this designation means almost nothing. In fact, the defining integral (4.5) converges at the upper limit; the whole problem lies in the fact that the series (4.7) cannot be integrated term by term. Thus a summation of the proper-time series is required.

In the summation business the trouble is always that it is difficult to find a consistent approximation scheme which corresponds to a partial summation, while the exact summation is usually impossible. Therefore, everybody sums up what he likes and defines the remainder as that part which he does not like. However, in the present case a consistent approximation scheme exists.

Indeed, let us look for an example in $a_3(x, x)$ given by equation (4.15). All terms of $a_3(x, x)$ have one and the same dimensionality, but different powers in the curvature, that is, different powers in the deviation of the metric from the flat-space metric. The terms $\nabla^2 R^2$ contribute to p-point functions beginning with $p = 2$, while the contribution of the terms R^3 begins only with the three-point function. In the same way, one distinguishes between terms $\nabla^4 R^2$, $\nabla^2 R^3$ and R^4 in $a_4(x, x)$, and so on. The suggestion is to sum all terms with a given power of the curvature and arbitrary number of derivatives in the series (4.7). Such an approximation scheme corresponds to the successive evaluation of p-point functions on a flat background, but in a completely covariant fashion.

Let us begin with terms of first order in the curvature. Clearly, these terms can only be of the form

$$a_1(x, x) = \alpha_1 R \tag{4.17}$$

$$a_2(x, x) = \alpha_2 \square R + O(R^2) \tag{4.18}$$

$$\vdots$$

$$a_n(x, x) = \alpha_n \square^{n-1} R + O(R^2). \tag{4.19}$$

All of them except the first are total derivatives. Hence we find:

$$W_{\text{one-loop}} = -\frac{1}{2(4\pi)^2} \int dx \sqrt{g} [f_0 + \tfrac{1}{6} f_1 R + O(R^2)]. \tag{4.20}$$

Let us now turn to terms of second order in the curvature. Their general form is

$$a_n(x, x) = \alpha_n \square^{n-1} R + \sum_{k=0}^{2n-4} \sum_i \text{constant}_i \underbrace{(\nabla \dots \nabla R \dots)}_{k} \underbrace{(\nabla \dots \nabla R \dots)}_{2n-k-4}$$

$$+ O(R^3) \qquad n \geqslant 2 \tag{4.21}$$

where i enumerates all possible contractions with some coefficients. Using integration by parts, the Bianchi identities, and the fact that any commutator of covariant derivatives in (4.21) gives terms of at least $O(R^3)$, one can reduce expression (4.21) to the following form:

$$\int dx \sqrt{g} a_n(x, x) = \int dx \sqrt{g} [\sigma_n R (-\square)^{n-2} R + \gamma_n R_{\mu\nu} (-\square)^{n-2} R^{\mu\nu}$$

$$+ \beta_n R_{\mu\nu\rho\delta} (-\square)^{n-2} R^{\mu\nu\rho\delta}] + O(R^3) \qquad n \geqslant 2 \tag{4.22}$$

with some coefficients σ_n, γ_n and β_n. Thus the summation of the series (4.7) to order R^2 reduces to finding the functions

$$\sigma(\xi) \equiv \sum_{n=2}^{\infty} \sigma_n \xi^{n-2} \qquad \gamma(\xi) \equiv \sum_{n=2}^{\infty} \gamma_n \xi^{n-2} \qquad \beta(\xi) \equiv \sum_{n=2}^{\infty} \beta_n \xi^{n-2}. \tag{4.23}$$

It is remarkable that the exact form of these functions is unimportant. What is important is that the integrals

$$\int^{\infty} \frac{d\xi}{\xi} \sigma(\xi) \qquad \int^{\infty} \frac{d\xi}{\xi} \gamma(\xi) \qquad \int^{\infty} \frac{d\xi}{\xi} \beta(\xi) \tag{4.24}$$

should converge at the upper limit. This is the only assumption we shall make.

Inserting (4.22) into (4.7) gives:

$$W_{\text{one-loop}} = -\frac{1}{2(4\pi)^2} \int dx \sqrt{g} \left(f_0 + \tfrac{1}{6} f_1 R + R_{\mu\nu\rho\delta} \int_{1/L^2}^{\infty} \frac{ds}{s} \exp(-sm^2) \beta(-s\square) R^{\mu\nu\rho\delta} \right.$$

$$+ R_{\mu\nu} \int_{1/L^2}^{\infty} \frac{ds}{s} \exp(-sm^2) \gamma(-s\square) R^{\mu\nu} + R \int_{1/L^2}^{\infty} \frac{ds}{s} \exp(-sm^2) \sigma(-s\square) R$$

$$\left. + O(R^3) \right) \qquad L^2 \to \infty. \tag{4.25}$$

Finally in the massless limit we obtain

$$
W_{\text{one-loop}} = -\frac{1}{2(4\pi)^2} \int dx \sqrt{g} \left(\frac{L^2}{6} R + \beta(0) \left(\ln L^3 + I_1 \right) R_{\mu\nu\rho\delta} R^{\mu\nu\rho\delta} \right.
$$
$$
+ \gamma(0) \left(\ln L^2 + I_2 \right) R_{\mu\nu} R^{\mu\nu} + \sigma(0) \left(\ln L^2 + I_3 \right) R^2
$$
$$
- \beta(0) R_{\mu\nu\rho\delta} \ln \left(-\Box \right) R^{\mu\nu\rho\delta} - \gamma(0) R_{\mu\nu} \ln \left(-\Box \right) R^{\mu\nu}
$$
$$
\left. - \sigma(0) R \ln \left(-\Box \right) R + O(R^3) \right) \tag{4.26}
$$

where

$$
\beta(0) = \tfrac{1}{180} \qquad \gamma(0) = -\tfrac{1}{180} \qquad \sigma(0) = \tfrac{1}{72} \tag{4.27}
$$

and

$$
I_1 = \int_0^1 \frac{d\xi}{\xi} \left[\beta(\xi) - \beta(0) \right] + \int_1^\infty \frac{d\xi}{\xi} \beta(\xi) \tag{4.28}
$$

$$
I_2 = \int_0^1 \frac{d\xi}{\xi} \left[\gamma(\xi) - \gamma(0) \right] + \int_1^\infty \frac{d\xi}{\xi} \gamma(\xi) \tag{4.29}
$$

$$
I_3 = \int_0^1 \frac{d\xi}{\xi} \left[\sigma(\xi) - \sigma(0) \right] + \int_1^\infty \frac{d\xi}{\xi} \sigma(\xi). \tag{4.30}
$$

We see that in the massless case the effective action certainly exists but is *non-local*. The essential parameters in (4.26) are those of DeWitt's $a_2(x, x)$ and are known, while the non-trivial integrals I_1, I_2, I_3 are absorbed by the renormalization ambiguity. As a result, modulo this ambiguity, we have completely computed $W_{\text{one-loop}}$ with an accuracy to $O(R^3)$.

The contribution of terms of order R^3 to $a_n(x, x)$ gives an expression analogous to (4.21), that is

$$
a_n(x, x) \sim \sum (\nabla \ldots \nabla R \ldots)(\nabla \ldots \nabla R \ldots)(\nabla \ldots \nabla R \ldots) \qquad n \geqslant 3. \tag{4.31}
$$

Evidently, here one can also single out a finite number of irreducible structures with $\Box^n$ insertions:

$$
\int dx \sqrt{g}\, a_n(x, x) \sim \int dx \sqrt{g}\, (\rho_n R_{\alpha\beta\gamma\sigma} R^{\gamma\sigma\mu\nu} (-\Box)^{n-3} R_{\mu\nu}{}^{\alpha\beta}
$$
$$
+ \text{a finite number of other structures)} \qquad n \geqslant 3. \tag{4.32}
$$

Here ρ_n is some numerical sequence which defines a function

$$\rho(\xi) = \sum_{n=3}^{\infty} \rho_n \xi^{n-3}. \tag{4.33}$$

We also suppose that the integral

$$I_4 = \int_0^{\infty} d\xi \rho(\xi) \tag{4.34}$$

converges at the upper limit. Then one finds that the terms of order R^3 in the effective action have the following form ($m^2 = 0$):

$$W_{\text{one-loop}} \sim -\frac{1}{2(4\pi)^2} \int dx \sqrt{g} \left[I_4 R_{\alpha\beta\gamma\sigma} R^{\gamma\sigma\mu\nu} \left(-\frac{1}{\Box} \right) R_{\mu\nu}{}^{\alpha\beta} + \text{other structures} \right] \tag{4.35}$$

where the non-local factor $1/\Box$ arises. This time the non-trivial integral I_4 enters the effective action as an essential parameter. Its calculation is of course difficult but the functional structure of W is calculable.

There the matter stands at each order in the curvature. As a result of this summation, the inverse powers of m^2 in the local expansion (4.8) become replaced by inverse powers of $\Box$ in the massless case. In the massive case one must turn to the integrals arising in expression (4.25). Introducing the dimensionless integration variable $\xi = -s\Box$ and expanding in powers of $m^2/-\Box$ one obtains non-local massive additions. On the other hand, by introducing the integration variable $\xi = sm^2$ and expanding in powers of $-\Box/m^2$, one recovers the local expression (4.8).

The above method of summation considerably improves the local expansion (4.8), but also has a limited region of validity. It is applicable only under the condition

$$\nabla^2 R_{...} \gg R^2_{...} \tag{4.36}$$

i.e. when the gradient of the curvature is larger than the absolute value of the curvature to the corresponding power. For the removal of this limitation a further summation is needed, this time of terms with different powers of the curvature. This problem is much more difficult. However, in the special case when the operator F is conformally covariant, the problem of further summation is facilitated because in this case the transformation properties of $W_{\text{one-loop}}$ under conformal transformations of the metric are known exactly and are very simple. One has either

$$g_{\mu\nu} \frac{\delta W_{\text{one-loop}}}{\delta g_{\mu\nu}} = \frac{1}{2(4\pi)^2} \sqrt{g}\, a_2(x, x) \tag{4.37}$$

which corresponds to the presence of the conformal anomaly (Deser et al 1976), or

$$g_{\mu\nu} \frac{\delta W_{\text{one-loop}}}{\delta g_{\mu\nu}} = 0 \tag{4.38}$$

which corresponds to its absence (Fradkin and Vilkovisky 1978)†.

In contrast, each term of the above partially-summed series does not possess any simple transformation properties. Therefore the requirement that conformal properties be manifest must lead to an improvement in the summation. To confirm the utility of this suggestion, I want to present one striking example: a massless field in two dimensions.

The partial summation suggested above is applicable in two dimensions as well. The main difference is that $d = 2$, the integrand in (4.7) is less negative by one power of s. In addition, the Riemann tensor in two dimensions reduce to the scalar R, where R moreover is a total derivative. In the two-dimensional case, $d = 2$, we have:

$$\int dx \sqrt{g}\, a_n(x, x) = \int dx \sqrt{g}\, [v_n R(-\Box)^{n-2} R + O(R^3)] \qquad n \geqslant 2 \quad (4.39)$$

with some coefficients v_n. The same reasoning as above leads to the result:

$$W_{\text{one-loop}} = -\frac{1}{2(4\pi)} \int dx \sqrt{g} \left(-I_0 R \frac{1}{\Box} R + \text{constant } R^2 \frac{1}{\Box^2} R \right.$$

$$\left. + \text{constant } R^3 \frac{1}{\Box^3} R + \dots \right) \tag{4.40}$$

where

$$I_0 = \int_0^\infty d\xi\, v(\xi) \qquad v(\xi) = \sum_{n=2}^\infty v_n \xi^{n-2}. \tag{4.41}$$

On the other hand, every two-dimensional space is conformally flat. Therefore, any conformally invariant functional of $g_{\mu\nu}$ in two dimensions is a constant, and any functional is uniquely determined by the trace of its variational derivative. For this reason, the contribution to the gravitational effective action of any conformally covariant field in two dimensions would be a constant if not for the presence of the trace anomaly‡ The anomaly in two dimensions is determined by the first DeWitt coefficient

$$g_{\mu\nu} \frac{\delta W_{\text{one-loop}}}{\delta g_{\mu\nu}} = \frac{1}{2(4\pi)} \sqrt{g}\, a_1(x, x) = \frac{\alpha_1}{2(4\pi)} \sqrt{g}\, R. \tag{4.42}$$

In the scalar case $\alpha_1 = \frac{1}{6}$ according to (4.16). By direct computation it can be verified that the unique W generating the trace (4.42) is of the form

† The presence or absence of the conformal anomaly depends on the choice made between two existing classes of covariant regularizations. The existence of these two classes is in turn connected with the existence of two different off-shell formulations of gravity theory (Fradkin and Vilkovisky 1978). The conformal anomaly, if present, has a universal expression (4.37) in terms of the second DeWitt coefficient. The regularization introduced in equation (4.7) belongs to the class producing the anomaly.
‡ In two dimensions anomaly-free regularizations do not exist.

$$W_{\text{one-loop}} = \frac{\alpha_1}{4(4\pi)} \int dx \sqrt{g} R \frac{1}{\Box} R. \tag{4.43}$$

Thus the one-loop effective action in two dimensions is known exactly!

Note, we have now obtained the non-local action and this action is just the first term of (4.40). But this means that all $O(R^3)$ terms of (4.40) in fact vanish! Furthermore, the non-trivial integral I_0 turns out to be exactly

$$I_0 = \tfrac{1}{2}\alpha_1 = \tfrac{1}{12}. \tag{4.44}$$

This example is of course unique. In four dimensions a manifest display of W's conformal properties can lead only to some improvement in the above partial summation.

To obtain such an improvement, one defines the following scalar Green's function:

$$(\Box + \tfrac{1}{6}R)\hat{G} = \hat{1} \tag{4.45}$$

with the zero boundary condition at Euclidean asymptotically flat infinity. Using $\hat{G}$, one defines the scalar field (Fradkin and Vilkovisky 1978):

$$\phi(x|g_{\mu\nu}) = 1 - \tfrac{1}{6}\hat{G}R \tag{4.46}$$

which depends non-locally on $g_{\mu\nu}$ and transforms as

$$\phi(x|e^{2\sigma}g_{\mu\nu}) = e^{-\sigma(x)}\phi(x|g_{\mu\nu}) \tag{4.47}$$

under conformal transformations with the parameters σ vanishing at Euclidean asymptotically flat infinity. The combination $g_{\mu\nu}\phi^2$ is conformally invariant and differs from $g_{\mu\nu}$ by terms $O(R)$. Using this fact one can construct the extension of expression (4.26), which reproduces the exact conformal properties of W (Fradkin and Vilkovisky 1978, Frolov and Vilkovisky 1982, 1983).

For example, if we consider one of the terms quadratic in the curvature

$$W_1 = -\frac{\beta}{2} \int dx \sqrt{g}\, C^{\beta\gamma\sigma}{}_\alpha \ln(-\Box) C^\alpha{}_{\beta\gamma\sigma} \tag{4.48}$$

(C is the Weyl tensor), then its conformally invariant extension is

$$W_1{}^{\text{conf}} = -\frac{\beta}{2} \int dx \sqrt{g}\, C^{\beta\gamma\sigma}{}_\alpha \ln(-H) C^\alpha{}_{\beta\gamma\sigma} \tag{4.49}$$

where

$$H = \nabla^\mu\nabla_\mu\big|_{g\phi^2} \tag{4.50}$$

is the operator in the metric $g_{\mu\nu}\phi^2$ which acts on the fourth-rank tensor $C^\alpha{}_{\beta\gamma\sigma}$. The expression (4.49) corresponds to the absence of an anomaly. The extension of the term (4.48) in the presence of the anomaly is

$$W_1{}^{\text{anom}} = W_1{}^{\text{conf}} - \frac{\beta}{2} \int dx \sqrt{g}\, C_{\alpha\beta\gamma\sigma} C^{\alpha\beta\gamma\sigma} \ln\phi^2. \tag{4.51}$$

One can verify that (4.51) generates precisely the local trace anomaly

$$g_{\mu\nu} \frac{\delta W_1^{\text{anom}}}{\delta g_{\mu\nu}} = \frac{\beta}{2} \sqrt{g} C_{\alpha\beta\gamma\sigma} C^{\alpha\beta\gamma\sigma}. \tag{4.52}$$

Both W_1^{conf} and W_1^{anom} differ from the original W_1 by terms of $O(R^3)$.

To summarize, the successive summation of terms with a given power of the curvature is quite workable. (The details are analyzed in Barvinsky and Vilkovisky (1983b), on which the discussion of the present section is based.) As to the removal of limitation (4.36), one can imagine that the partially summed series contracts into

$$W_{\text{one-loop}} = -\frac{1}{2(4\pi)^2} \int dx \sqrt{g} [\text{local terms} - \tfrac{1}{180} R_{\alpha\beta\mu\nu} \ln(-\Box + \hat{\mathscr{R}}_1) R^{\alpha\beta\mu\nu}$$
$$+ \tfrac{1}{180} R_{\mu\nu} \ln(-\Box + \hat{\mathscr{R}}_2) R^{\mu\nu} - \tfrac{1}{72} R \ln(-\Box + \mathscr{R}_3) R] \tag{4.53}$$

where the $\mathscr{R}$ are 'some curvatures'†. In fact, one can try to define the $\mathscr{R}$ in such a way as to absorb the $O(R^3)$ terms in (4.53). Then in the limit (4.36) one recovers the above non-local expansion in powers of the curvature, while in the opposite limit

$$\nabla^2 R... \ll R^2... \tag{4.54}$$

one obtains

$$W_{\text{one-loop}} = -\frac{1}{2(4\pi)^2} \int dx \sqrt{g} [\text{local terms} - \tfrac{1}{180} R_{\alpha\beta\mu\nu} (\ln \hat{\mathscr{R}}_1) R^{\alpha\beta\mu\nu}$$
$$+ \tfrac{1}{180} R_{\mu\nu} (\ln \hat{\mathscr{R}}_2) R^{\mu\nu} - \tfrac{1}{72} R (\ln \mathscr{R}_3) R + \ldots]. \tag{4.55}$$

This is the generalization of the Heisenberg–Euler Lagrangian. Such a non-analytic expression can never be obtained by perturbation theory. This is not surprising: the limit (4.54) corresponds to the calculation of the ∞-point Green's function!

The appearance of the terms (4.55) was envisaged by DeWitt. The 'some curvature' $\mathscr{R}$ is DeWitt's 'curly curvature', which appears on page 1252 of DeWitt (1967c). In general, this chapter of the 'Gospel' contains ideas which have not as yet been assimilated. Clearly, still newer revelations according to DeWitt lie in store for us.

5. THE EFFECTIVE ACTION: APPLICATIONS AND INHERENT PROBLEMS

The major gain of the summation procedure discussed in the previous section was the reconstruction of non-local terms in the effective action. These terms should be responsible for such non-local effects as particle creation in a gravitational field. The most famous example of this phenomenon is the Hawking effect.

---

†The hat indicates that $\hat{\mathscr{R}}_1$ ($\hat{\mathscr{R}}_2$) is a matrix acting in the space of fourth-rank (second-rank) tensors.

It is said in DeWitt (1975, p 303). 'There is nothing better than a concrete example to help us get a feel for whether we are doing the right things'. To get such a feel we shall therefore, in this section, consider a derivation of the Hawking effect from the non-local effective action. This discussion follows Frolov and Vilkovisky (1982, 1983).

Applications, in general, reveal new difficulties in the effective action technique. One of these problems is the following. When calculating the effective action according to the Feynman rules in a space–time with a Minkowskian signature, one finds that the effective action and the mean field are complex. The reason is that the Feynman rules define the mean field as the field operator between different vacuum states: $\langle \text{out}|g|\text{in}\rangle$, and this is not what we need. The true mean field is the vacuum expectation value $\langle \text{in}|g|\text{in}\rangle$, and a special diagrammatic technique (Buchbinder *et al* 1981a, b) is required to compute this quantity. However, this new technique has not yet led to convenient practical recipes. In the meantime, a simple procedure has been formulated for obtaining the effective equations for the true mean field (Frolov and Vilkovisky 1982, 1983). No derivation of this procedure has as yet been given. Its only justification at present is that it works and leads to correct results. This procedure can be outlined as follows.

Even before beginning any derivation, we convince ourselves that the effective action $W[g]$ for the mean field $\langle \text{in}|g|\text{in}\rangle$ must possess two properties. Firstly, it must be real and, secondly, it must satisfy the requirement of classical causality: the current $\delta W[g]/\delta g_{\mu\nu}(x)$ must not depend on $g_{\mu\nu}$ in the future of x. We shall adopt this requirement as a postulate (Wald 1977).

On the other hand, actual calculations with the effective action are usually carried out in a space with Euclidean signature. We studied precisely this in the last section. We found that non-local terms of the effective equations represent (after calculating the loops) tree diagrams of the type

$$\frac{\delta W[g]}{\delta g} \sim \sum R \frac{1}{\Box} R \frac{1}{\Box} R \ldots \tag{5.1}$$

i.e. curvatures connected by propagators. These propagators are Euclidean Green's functions.

The proposal of Frolov and Vilkovisky (1982, 1983) is: (1) compute the effective action in a space with Euclidean signature: $W_{\text{Eucl}}[g]$, (2) compute its variational derivative $\delta W_{\text{Eucl}}[g]/\delta g_{\mu\nu}$ and (3) in the current thus obtained go over to Minkowskian signature replacing all Euclidean Green's functions in expressions of the type (5.1) by retarded Green's functions:

$$\frac{1}{\Box}(\text{Euclidean}) \to \frac{1}{\Box}(\text{retarded}). \tag{5.2}$$

It should be stressed that the transition to Minkowskian signature with the replacement (5.2) must be done *not in the action W*, but in the current $\delta W/\delta g$.

Whether this procedure can be confirmed by regular methods (Buchbinder *et al* 1981a, b) remains to be seen. But we shall adopt it and examine the results.

In order to deal with an exactly solvable case we shall consider two-dimensional space–time. As shown in the previous section, the one-loop contribution to the effective action from any massless field in two dimensions is known exactly. In the Euclidean formulation it has the form (4.43):

$$W_{\text{one-loop}} = \frac{b}{2} \int d^2 x \sqrt{g} \, R \hat{G} R \tag{5.3}$$

where

$$b = \text{constant} \qquad \Box \hat{G} = \hat{1} \tag{5.4}$$

and $\hat{G}$ is the Euclidean Green's function.

The variational derivative of W_{loop} is the effective energy–momentum tensor of the corresponding quantum field:

$$\langle T^{\mu\nu} \rangle \equiv -2 \frac{1}{\sqrt{g}} \frac{\delta W[g]}{\delta g_{\mu\nu}}. \tag{5.5}$$

Calculating the variational derivative of (5.3) and making the replacement (5.2), we find:

$$\langle T^{\mu\nu} \rangle = b \{ 2\nabla^\mu \nabla^\nu \hat{G}_{\text{ret}} R + \nabla^\mu (\hat{G}_{\text{ret}} R) \cdot \nabla^\nu (\hat{G}_{\text{ret}} R) - g^{\mu\nu} [2R$$
$$+ \tfrac{1}{2} \nabla_\alpha (\hat{G}_{\text{ret}} R) \cdot \nabla^\alpha (\hat{G}_{\text{ret}} R)] \}. \tag{5.6}$$

We shall be interested in the expression (5.6) in an asymptotically flat space–time. Therefore we shall suppose that there exist two null lines, the past null infinity $\mathscr{I}^-$ and the future null infinity $\mathscr{I}^+$, near which the metric becomes flat (Penrose 1968). Since the effective energy–momentum tensor (5.6) is retarded, it gives a zero energy flux at past infinity: $\langle T^{\mu\nu} \rangle_{\mathscr{I}^-} = 0$, that is, it guarantees the absence of any incoming radiation. If there is found any outgoing radiation in the null direction: $\langle T^{\mu\nu} \rangle_{\mathscr{I}^+} \neq 0$, we interpret this phenomenon as the creation of massless particles from the vacuum, i.e. the Hawking effect.

Let us introduce null coordinates in two-dimensional space–time:

$$ds^2 = 2\psi \, du \, dv \qquad \psi > 0. \tag{5.7}$$

There exists a system of null coordinates (u_-, v_-) such that the corresponding ψ satisfies the condition

$$\psi_- |_{\mathscr{I}^-} = 1. \tag{5.8}$$

There also exists a system of null coordinates (u_+, v_+) such that

$$\psi_+ |_{\mathscr{I}^+} = 1. \tag{5.9}$$

However, this does not yet mean that these two coordinate systems coincide. The definition of null coordinates is not unique but admits two arbitrary functions, each of a single variable:

$$u_+ = u_+(u_-) \qquad v_+ = v_+(v_-) \tag{5.10}$$

$$\psi_- = \frac{du_+ \, dv_+}{du_- \, dv_-} \psi_+ . \tag{5.11}$$

The Hawking effect, *if present*, is entirely due to the difference between coordinate systems (5.8) and (5.9).

The main technical difficulty in the calculation of non-local quantities like (5.6) is to find the Green's function in an external field. However, in the two-dimensional case, this problem can be solved in a general form and in a remarkably simple way. In the metric (5.7) one finds

$$\Box = \frac{2}{\psi} \partial_{uv}^2 \qquad \Box \ln \psi = R. \tag{5.12}$$

Thus, $\ln \psi$ is exactly the solution of the inhomogeneous D'Alembert equation (5.12) and the arbitrariness of the solution is just the arbitrariness in the definition of ψ. The retarded solution is $\ln \psi_-$:

$$\hat{G}_{\text{ret}} R = \ln \psi_- . \tag{5.13}$$

If $\mathscr{I}^+$ is $v = $ constant, then the energy flux through $\mathscr{I}^+$ is given by the expression

$$\langle T^{\mu\nu} \rangle l_\mu l_\nu \big|_{\mathscr{I}^+} \tag{5.14}$$

where $l_\mu = \nabla_\mu v_+$. Using (5.6) and (5.13) we find:

$$\langle T^{\mu\nu} \rangle l_\mu l_\nu \big|_{\mathscr{I}^+} = b \left[2 \frac{\partial^2 \ln \psi_-}{\partial u_+^2} + \left(\frac{\partial \ln \psi_-}{\partial u_+} \right)^2 \right] \Bigg|_{\mathscr{I}^+} . \tag{5.15}$$

By virtue of (5.11)

$$\ln \psi_- \big|_{\mathscr{I}^+} = \ln |du_+/du_-| + \text{constant}. \tag{5.16}$$

Therefore the whole effect depends on the form of the function $u_+(u_-)$ which characterizes the dynamics of a given gravitational field.

Until now our considerations have been completely general. Now we may apply the above formulae to the gravitational field of a spherically symmetric collapsing body. As a model of the collapsing body we shall choose a thin massive null shell. Its gravitational field is described in the classical theory by the Schwarzschild metric in which the mass M is replaced by $M\theta(v\text{-constant})$, where $v = $ constant is the worldline of the shell. We shall consider a two-dimensional slice of this space–time. The corresponding Penrose diagram is shown in Figure 1.

Outside the shell (the unshaded region in figure 1) and outside the event horizon, the metric is of the form:

$$\psi_+(u_+, v_+) = 1 - 2M/r \tag{5.17}$$

$$v_+ - u_+ = \sqrt{2}(r + 2M \ln |(r/2M) - 1|) \equiv \sqrt{2} r^*. \tag{5.18}$$

Inside the shell the space–time is flat. In order to obtain the function $u_+(u_-)$ one may consider the worldline of the shell, denoted by $\mathscr{N}$ in figure 1. The equation of $\mathscr{N}$ when approached from the external region is

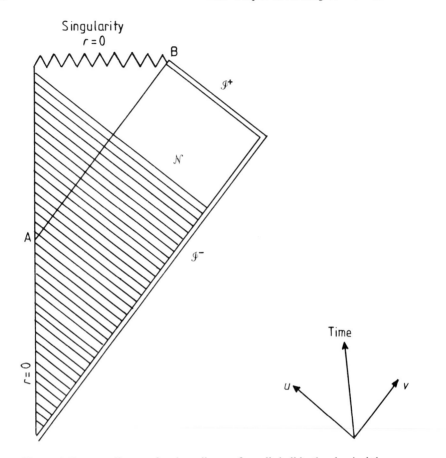

Figure 1 Penrose diagram for the collapse of a null shell in the classical theory. $\mathcal{N}$ is the worldline of the shell. The shaded region of space–time is flat. The line AB is the event horizon.

$$\mathcal{N}: \quad v_+ \equiv \sqrt{2}r^* + u_+ = \text{constant}. \tag{5.19}$$

The equation for $\mathcal{N}$ when approached from the internal region is

$$\mathcal{N}: \quad v_- \equiv \sqrt{2}r + u_- = \text{constant}. \tag{5.20}$$

Equating $r(u_+)$ defined by (5.19) and $r(u_-)$ defined by (5.20), we can find the connection between u_+ and u_-. The result is

$$\frac{du_+}{du_-} = 1 - \frac{2\sqrt{2}M}{u_- - u_-^0} \qquad u_-^0 = \text{constant} \tag{5.21}$$

where $u_- = u_-^0$ or $u_+ = \infty$ is the equation of the event horizon (see figure 1).

It remains now to use the above result in expressions (5.16) and (5.15). Asymptotically at late times ($u_+ \to \infty$):

$$\partial \ln \psi_- / \partial u_+ = 1/2 \sqrt{2} M \tag{5.22}$$

and the energy flux through $\mathscr{I}^+$ is

$$\langle T^{\mu\nu} \rangle l_\mu l_\nu \Big|_{\substack{\mathscr{I}^+ \\ u_+ \to \infty}} = b/8M^2. \tag{5.23}$$

According to (4.43) and (4.16), the contribution of the quantized scalar field to the coefficient b in the effective action (5.3) is:

$$b = 1/48\pi. \tag{5.24}$$

Thus the energy loss with respect to the proper time of the observer at $r = \infty$ equals†

$$\frac{\mathrm{d}E}{\mathrm{d}t} = \langle T_{rt} \rangle |_{\mathscr{I}^+} = -\tfrac{1}{2} \langle T^{\mu\nu} \rangle l_\mu l_\nu |_{\mathscr{I}^+} = -\tfrac{1}{12}\pi \frac{1}{(8\pi M)^2}. \tag{5.25}$$

This is exactly the flux of the blackbody radiation of massless bosons at a temperature $1/8\pi M$.

The calculation of $\langle T^{\mu\nu} \rangle$ for Hawking radiation in four dimensions is one of the important sections of the 'Gospel According to DeWitt' (1975). It remains a problem for the future to derive his results by the effective-action technique. But even this will only be the beginning of the task.

The true purpose of the effective action is in the solution of the self-consistent equations

$$\frac{\delta W}{\delta g} \equiv \frac{\delta W_{\text{classical}}}{\delta g} + \frac{\delta W_{\text{loop}}}{\delta g} = 0.$$

But the self-consistent equations produce still other problems. First of all, it is clear that one must not solve effective equations by expanding in powers of $\hbar$ beginning from the classical solution. If the classical solution contains a singularity, then quantum corrections may not be small at all. On the contrary, we hope that the proper account of the vacuum polarization will radically change the solution at small scales. But if the solution is not to be expanded in powers of $\hbar$, then:

(1) does the one-loop approximation make any sense?

(2) how will we handle the divergences of W even in the single-loop approximation? Only in the iterative solution do the one-loop divergences vanish ('t Hooft and Veltman 1974, Fradkin and Vilkovisky 1977b);

(3) how are the boundary conditions to the effective equations to be formulated?

We shall not discuss these problems here but refer the reader to Frolov and Vilkovisky (1981, 1982, 1983) where some of them are considered.

---

† The final result agrees with other calculations of the Hawking effect in two dimensions (Davies *et al* 1976, Unruh 1976, Davies 1976), and in particular with Christensen and Fulling (1977). The latter authors established the correspondence between the Hawking effect and the trace anomaly, though their approach was not based on the effective action.

It is a pity that a classical Lagrangian for the gravitational field does not exist in two dimensions. We should then have a solvable model for the self-consistent picture of the collapse.

ACKNOWLEDGMENTS

It is a pleasure for me to thank in this volume my colleagues A O Barvinsky, I A Batalin, E S Fradkin and V P Frolov, with whom much of the above work was carried out. Special thanks to Tony Rothman for the English edition of the text, help in preparation of the manuscript, and the title.

REFERENCES

Adler S L, Lieberman J and Ng Y J 1977 *Ann. Phys., NY* **106** 279
Barvinsky A O and Vilkovisky G A 1983a *Generalized Schwinger–DeWitt Technique* in *Gauge Theory and Quantum Gravity* preprint of the University of Texas at Austin
—— 1983b work in progress
Batalin I A and Vilkovisky G A 1977 *Phys. Lett.* **69B** 309
—— 1981 *Phys. Lett.* **102B** 27
—— 1983a *Quantization of Gauge Theories with Linearly Dependent Generators* preprint of the University of Texas at Austin
—— 1983b in *Quantum Gravity* ed M A Markov (London: Plenum)
—— 1983c *Closure of The Gauge Algebra, Generalized Lie Equations and Feynman Rules, Nucl. Phy. B.* submitted
Becchi C, Rouet A and Stora A 1975 *Commun. Math. Phys.* **42** 127
Brown L S and Cassidy J P 1977 *Phys. Rev.* D **15** 2810
Buchbinder J L, Fradkin E S and Gitman D M 1981a *Fortsch. Phys.* **29** 187
—— 1981b Preprint N138 of the Lebedev Physical Institute (Moscow)
Christensen S M 1976 *Phys. Rev.* D **14** 2490
Christensen S M and Fulling S A 1977 *Phys. Rev.* D **15** 2088
Davies P C W 1976 *Proc. R Soc.* A **351** 129
Davies P C W, Fulling S A and Unruh W G 1976 *Phys. Rev.* D **13** 2720
Deser S, Duff M J and Isham C J 1976 *Nucl. Phys.* B **111** 45
The following 12 references constitute The Gospel According To DeWitt
DeWitt B S 1963 in *Relativity, Groups and Topology* (New York: Gordon and Breach)
—— 1964 *Phys. Rev. Lett.* **13** 114
—— 1965 *Dynamical Theory of Groups and Fields* (New York: Gordon and Breach)
—— 1967a *Phys. Rev.* **160** 1113
—— 1967b *Phys. Rev.* **162** 1195
—— 1967c *Phys. Rev.* **162** 1239
—— 1975 *Phys. Rep.* **19** 295
—— 1979 in *General Relativity, an Einstein Centenary Survey* ed S Hawking and W Israel (Cambridge: Cambridge University Press)
—— 1981a in *Quantum Gravity 2* ed C Isham, R Penrose and D Sciama (Oxford: Oxford University Press)

—— 1981b *Phys. Rev. Lett.* **47** 1647

DeWitt B S and van Nieuwenhuizen P 1982 *J. Math. Phys.* **23** 1953

DeWitt B S *Supermanifolds and Supersymmetry*, draft of contribution to the book in collaboration with P van Nieuwenhuizen and P West

Dowker J S and Critchley R 1977 *Phys. Rev. D* **16** 3390

Faddeev L D 1969 *Theor. Math. Phys. (USSR)* **1** 3

Faddeev L D and Popov V N 1967 *Phys. Lett.* **25B** 29

Fradkin E S and Fradkina T E 1978 *Phys. Lett.* **72B** 343

Fradkin E S and Vasiliev M A 1977 *Phys. Lett.* **72B** 70

Fradkin E S and Vilkovisky G A 1973 *Phys. Rev. D* **8** 4241

—— 1975a *Lett. Nuovo Cimento* **13** 187

—— 1975b *Phys. Lett.* **55B** 224

—— 1976 *On Renormalization of Quantum Field Theory in Curved Spacetime* Preprint of Institute for Theoretical Physics, University of Berne

—— 1977a Preprint TH. 2332–CERN

—— 1977b *Lett. Nuovo Cimento* **19** 47

—— 1978 *Phys. Lett.* **73B** 209

Frolov V P and Vilkovisky G A 1981 *Phys. Lett.* **106B** 307

—— 1982 *Spherically Symmetric Collapse in Quantum Gravity* Preprint of the University of Texas at Austin

—— 1983 in *Quantum Gravity* ed M A Markov (London: Plenum)

Gilkey P B 1975 *J. Diff. Geom.* **10** 601

Hawking S W 1977 *Commun. Math. Phys.* **55** 133

't Hooft G and Veltman M 1974 *Ann. Inst. Henri Poincaré* **20** 69

Kallosh R E 1977 *Pis'ma JETP (USSR)* **26** 573

Penrose R 1968 in *Battelle Recontres, 1967 Lectures in Mathematics and Physics* ed C M DeWitt and J A Wheeler (New York: Benjamin Inc)

Rothman T R 1979 *The Magician and The Fool* 3-act Drama

—— 1983 *The Magician and the Fool* Revised edition

Schwinger J S 1951 *Phys. Rev.* **82** 664

Unruh W G 1976 *Phys. Rev. D* **14** 870

Wald R M 1977 *Commun. Math. Phys.* **54** 1

de Wit B and van Holten J W 1979 *Phys. Lett B* **79** 389

Is there a Quantum Theory of Gravity?

ANDREW STROMINGER

1. INTRODUCTION

Can quantum mechanics and general relativity be reconciled within the framework of quantum field theory, or do we need new laws of physics? If one takes the (totally unjustifiable!) point of view that this problem can be solved without recourse to new principles, then one is led to a well defined avenue of research. In this essay we review and discuss attempts to construct a unitary, renormalizable quantum field theory of gravity. Particular attention is paid to 'non-perturbative' expansions that probe above the Planck energy. We will conclude that there is a consistent flat space perturbation theory capable of describing quantum gravitational processes below the Planck energy. It does, however, have certain disturbing, although not inconsistent, features that arise near the Planck scale. The ultimate acceptability of this theory remains an open question.

In § 2 we discuss the quantization of the pure R theory and the renormalizability problem. We also discuss its solution by the addition of R^2 terms and the ensuing difficulties with unitarity and instabilities. In § 3 it is shown how both of these latter problems disappear when the theory is expanded in a parameter that is small at all length scales: $1/N$ = the inverse number of matter fields. The price we pay is that S matrix elements must be computed with the Lee–Wick prescription. The physical acceptability of this prescription, particularly within the context of gravity, remains unclear. In § 4 we discuss several gauge invariant 'non-perturbative' expansions of pure quantum gravity (with no matter fields). The first such expansion is in the inverse number of dimensions. The theory is found to be either unitary *or* renormalizable order by order in this expansion, depending on parameters, but not both. In the second expansion we let the metric become a $D \times D$ matrix, but space–time remains a four-dimensional manifold (as in dimensional reduction). Below the Planck energy, where the Einstein term (R) dominates, the $1/D$ expansion is equivalent to a version of the $1/N$ expansion. It may thus simultaneously solve, with the Lee–Wick prescription, unitarity and renormalizability. At high energies, however, where R^2 terms are important, the set of graphs that contribute to the large

D limit becomes much more complicated (the cactus graphs). The method for summing these graphs is presented, and calculation of coefficients is in progress. In §5 we discuss the implications of these results for particle physics and model building.

2. RENORMALIZABILITY AND UNITARITY IN QUANTUM GRAVITY

The problem of unifying quantum field theory and general relativity was formally solved in 1967 by DeWitt in a series of classic papers (DeWitt 1967a,b,c). Using the standard principles of quantum mechanics, he constructed a unitary set of Feynman rules describing the quantum dynamics of the gravitational field. Unfortunately, the perturbation series in Newton's constant is not renormalizable. This is evident from power counting: the naive degree of divergence of an L loop diagram is $2(L+1)$. Barring miracles, we therefore expect on dimensional grounds that a counterterm constructed from L powers of the Riemann tensor will be necessary to renormalize the $(L+1)$ th order of perturbation theory. Since a new coupling constant is induced at each order, the theory loses its predictive power. This difficulty arises from the fact that the coupling constant has dimension of inverse mass.

A first response to this problem was to conjecture that the full theory is in fact finite (Pauli 1967, Deser 1970). After all, the perturbation expansion is really an expansion in the dimensionless parameter $\kappa^2 E$ (E is an energy scale, $\kappa^2 = 32\pi G$). This expansion parameter is large at large energies. One cannot expect such an expansion to provide a systematic method for computing the effects of virtual gravitons of arbitrarily large energies. The non-renormalizability of the weak coupling expansion may just be due to the bad expansion of a good theory.

The finiteness conjecture has both intuitive and calculational motivations. On the intuitive side, there is a vague notion that one should not be able to propagate at energies where wavelengths are less than the Schwarzschild radius. The Planck energy should thus provide a natural cutoff. This notion has also received some support from explicit calculations. Non-perturbative summations of ladder (DeWitt 1964) and cocoon graphs (Isham *et al* 1971) have in fact produced finite results.

Unfortunately, these calculations are not gauge invariant and do not represent a systematic expansion in some small parameter. Furthermore, in recent years, systematic expansions have been developed in parameters that are small at all energies (e.g. the inverse number of matter fields (Tomboulis 1977, 1980) or dimensions (Strominger 1981, 1982)). Quantum gravity is *not* finite order by order in these expansions.

The prospects for a sensible quantization of the Einstein action thus do not appear terribly bright. The next logical step is to alter the action by adding higher derivative terms such as R^2 (Utiyama and DeWitt 1962). In any case one is forced to add these terms for renormalization.

This step is taken rather hesitantly because actions with four time derivatives generally describe theories that are pathological even at the classical level. In fact it can be shown that classical higher derivative gravity theories either have tachyons or negative energies for small, long wavelength fluctuations† (see formulae in Hartle and Horowitz 1981). Thus they have pathologies that are evident on macroscopic length scales. How, then, can the quantum version of these theories possibly be reasonable? The miracle is that some, if not all, of these instabilities can be systematically eliminated from the quantum theory. This will be discussed further in the next section.

There are three fourth-order invariants that one can form from the metric:

$$R^2$$
$$C^2 \equiv C_{\mu\nu\alpha\beta} C^{\mu\nu\alpha\beta}$$
$$R_{\mu\nu} R^{\mu\nu} \tag{2.1}$$

One linear combination of these is a total derivative and may be ignored at the level of Feynman diagrams. R^2 gives rise to spin zero particles while the conformally invariant C^2 gives rise to spin two particles. We thus take:

$$S = \int d^4x \sqrt{-g} (\Lambda + \frac{1}{\kappa^2} R + \frac{1}{\beta} R^2 - \frac{1}{\alpha} C^2) \tag{2.2}$$

as our fundamental action.

The addition of these higher derivative terms changes the particle content of the linearized theory. In addition to the usual massless graviton, we now have a spin zero and another spin two particle, both with masses near the Planck mass. This increase in the number of degrees of freedom also occurs classically—additional Cauchy data must be specified in higher derivative theories. The causal propagation of these particles fixes α and β to be positive.

If (2.2) is expanded in powers of κ, non-renormalizable counterterms are again encountered. In fact the p^4 vertices make things worse. However, another option is available to us—a loop expansion around the quadratic part of the action (DeWitt 1967a,b,c). The propagator then has the form (we work in Euclidean space throughout)

$$\Delta(q^2) = \frac{P_2}{q^2 + \frac{2\kappa^2}{\alpha}q^4} - \frac{P_0}{q^2 + \frac{6\kappa^2}{\beta}q^4} \tag{2.3}$$

where P_0 and P_2 are the spin zero and spin two projection operators. Due to the $1/q^4$ falloff of this propagator above the Planck energy, the loop expansion is power counting renormalizable. A detailed proof of renormalizability has been given by Stelle (1977). All four terms of (2.2) are needed for renormalization. Speculations have been made that one can begin with the conformally invariant and asymptoti-

---

† Recent work has shown that, for the particular case in which the coefficient of the Einstein term vanishes, the small fluctuations analysis is irrelevant due to linearization instability. Exact solutions in fact have zero energy. The subsequent comments thus do not apply to this case. See Boulware *et al* (1983).

cally free C^2 term alone and induce the Einstein term as a long range effective action with a finite coefficient (e.g. Hasslacher and Mottola 1981). (Adler (1982) has given the conditions for which the contribution to this coefficient from quantum matter fields is finite.) In practice, however, no one has devised a calculational scheme in which this is possible (although see §4).

The price we pay for this renormalizability is very high. The theory is not unitary. The propagator (2.2) may be written:

$$\Delta = \frac{P_2}{q^2} - \frac{P_2}{q^2 + \alpha/2\kappa^2} - \frac{P_0}{q^2} + \frac{P_0}{q^2 + \beta/6\kappa^2} \tag{2.4}$$

from which we clearly see that there is a spin two ghost pole near the Planck mass. This lack of unitarity is a general property of any propagator that has a real spectral representation and goes as $1/q^4$ asymptotically. (The spin zero ghost at $q^2 = 0$ is a gauge artifact and can be shown not to contribute to any physical process.)

The trade-off between renormalizability and unitarity is an old story. The massive vector boson theories of weak interactions could also be made renormalizable at the price of unitarity. Ordinarily when a theory is found to be non-unitary, it is time to give up and look for a new theory. The massive vector boson theories were replaced by the Weinberg–Salam model which is both renormalizable and unitary.

Unfortunately, no analog of the Weinberg–Salam model has been found for gravity. However, all is not lost for higher derivative gravity theories. Unitarity is a statement about the full theory, and does not necessarily hold order by order in perturbation theory. In quantum-electrodynamics, for example, perturbation theory produces the famous Landau ghost. This problem is not considered fatal to the theory since the pole is at high energies, outside the reach of perturbation theory. It is thus expected that higher order corrections will move the pole around. The same is true for our case: the ghost pole is near the Planck mass, where quantum corrections become large.

It should be mentioned that it is not at all clear that a physically sensible theory of gravity should be unitary. Calculations such as the Hawking evaporation of a black hole or more general processes involving non-trivial topology, do not give unitary results. It is an interesting question whether one should still expect a flat space perturbation theory that is unitary at all energies.

It should also be mentioned that we are treading on very thin ice when we begin to discuss particles with masses above the Planck mass, and Compton wavelengths below the Planck length. It is unlikely that such particles can in any sense exist as asymptotic scattering states. This is because their Schwarzschild radius is above the Planck length where classical relativity is still a good approximation. We therefore expect an event horizon to form around these particles (possibly followed by Hawking evaporation).

We do not know how to incorporate these comments into a practical calculational scheme. For the present we continue in pursuit of a renormalizable, unitary field theory. For this purpose what is needed is an expansion parameter that is small at all energies. We can then calculate to leading order and hope that qualitative features are unchanged by higher order corrections. In the rest of the paper we discuss several such expansions.

3. THE $1/N$ EXPANSION[†]

The basic idea of the $1/N$ expansion is very simple: consider quantum gravity (with higher derivative terms) coupled to N matter fields and rescale $\kappa \to \kappa/\sqrt{N}$. S matrix elements may now be computed in a power series in $1/N$. The first term turns out to be simply the iterated one-loop matter corrections. Since $1/N$ is small at all energy scales, this provides a tool for investigating the high energy behavior of quantum gravitational systems.

The amazing thing is that this simple trick solves the unitarity problem! To leading order in $1/N$, for minimally coupled scalar matter, the propagator becomes:

$$\Delta(q^2) = P_2\{q^{-2} - [q^2 + (2\kappa^2/\alpha + \kappa^2/1920\pi^2 \ln q^2/\mu^2)^{-1}]^{-1}\}$$
$$+ P_0\{-q^{-2} + [q^2 + (6\kappa^2/\beta + \kappa^2/384\pi^2 \ln q^2/\mu^2)^{-1}]^{-1}\} \quad (3.1)$$

where μ^2 is a renormalization parameter which is naturally taken to be near the Planck mass.

The effect of quantum corrections on the spin zero and spin two propagation is different. The previously well behaved spin zero particle now acquires a tachyon, since there is a pole for real positive q^2. Ordinarily this is a signal that the $1/N$ expansion is unreliable or that the theory is sick, since $1/N$ is small at any length scale. However in this case, due to a very small combinatoric factor, the tachyon occurs *near* $M_p \exp(1152\pi^2/\beta)$ (M_p is the Planck mass). This makes it difficult to worry!

This tachyon can be avoided to leading order in $1/N$ by coupling conformally invariant matter fields. This is the case usually considered in the literature. However, it seems likely that this tachyon will still appear at higher orders where the counterterms are not conformally invariant. The presence of this tachyon is related to the fact that the R^2 coupling, unlike the C^2 coupling, is *not* asymptotically free. (This is one reason why the possibility of a pure C^2 theory is attractive.)

The situation with the massive spin two ghost is improved. For $\kappa^2\mu^2 < 190\pi^2 \exp(3840\pi^2/\alpha)$ (certainly a reasonable choice!) there is no longer a pole on the real axis. The unitarity violating ghost pole has split into a pair of complex conjugate poles.

Such theories *are* unitary with appropriate boundary conditions, as has been shown in investigations by Lee and Wick (1970) (see also Coleman 1969, Lee 1970, Kay 1981). The existence of complex conjugate poles on the physical sheet indicates

---

[†] For a more detailed discussion of the $1/N$ expansion see Tomboulis (1977, 1980), this volume or Smolin (1982).

that the large N effective Hamiltonian is not Hermitian and its eigenvalues are not real. It can be shown, however, that for an appropriate (indefinite) metric η on the Hilbert space:

$$H = \eta^{-1} H^+ \eta. \tag{3.2}$$

The real energy eigenstates have positive norm. The complex energy eigenstates grow exponentially in time. They must be excluded as asymptotic states in order to have a well defined S matrix. In general one cannot exclude states without violating unitarity. However, since complex energy is conserved, a real energy-in state cannot scatter into a complex energy-out state. This means that the restriction of the S matrix to the real energy subspace is unitary.

It should be emphasized that exclusion of the complex energy states is an alteration that drastically affects the physics of all length scales. All of the states (in the linearized theory) corresponding to massive spin two excitations become states with complex energies in the $1/N$ expansion of the interacting theory. Their exclusion thus wipes out more than half the states of the in out Fock spaces.

Among other things, this has the consequence that the $\hbar \to 0$ limit of the Lee–Wick quantum theory of $R + C^2$ does not resemble classical or even semi-classical $R + C^2$. The classical and semiclassical theories have small fluctuations with negative energies—even at long wavelengths. (Horowitz 1981, Hartle and Horowitz 1981, Yamagishi 1982). The corresponding states are, however, excluded from Lee–Wick quantum gravity by the boundary condition. Lee–Wick quantum gravity has positive energy for all small fluctuations. Thus the quantum theory is better behaved than the classical theory!

Implementing the exclusion of these states in terms of Feynman diagrams involves a complicated prescription for contour integration. This has been worked out for the first few orders by Lee and Wick. Tomboulis (private communication) has an argument that a unitarity preserving prescription does exist to all orders. The prescription is *not* uniquely dictated by unitarity. One must impose an additional principle of 'maximal symmetrization.' (Lee 1970).

This contour prescription does not lead to a theory which is causal in the usual sense of the word. This is because we have imposed the noncausal boundary condition that no runaway modes occur in the infinite future. The resultant acausality is very similar to the pre-acceleration of a Lorentz–Dirac electron. For example, the Lee–Wick ghost can decay before it is produced. The time scale of this causality violation is about a $10^5/N$ times the Planck time—near grand unified time scales.

One might worry that a device could be constructed which detects the decay of the Lee–Wick ghost and then prevents it from being produced! Since the S matrix is well defined, however, no such logical contradiction is possible. The physical construction of such a device is impossible in a world described by a Lee–Wick field theory.

This situation leaves many, myself included, feeling a little queasy. On the one hand, in the fourteen years since the first paper of Lee and Wick, no one has found any logical inconsistency (although efforts continue). We have built a self-consistent quantum S matrix for gravity, even if half of the Hilbert space was butchered in the process. Yet the measures taken seem unnatural, and the resultant acausality distasteful. In the case of Lorentz–Dirac electron theory, the acausality was a signal that the physical model was breaking down at short length scales. Could that not be the case here as well?

Another objection to Lee–Wick quantum gravity is that it seems permanently wed to an *S* matrix expansion in free particle states (as opposed to the more geometric background field formulations). This is particularly unappealing in general relativity where a particle is not a coordinate independent concept. It is not even known if there is a unique Lee–Wick prescription in a general curved space.

Finally there is the question of whether all the instabilities of higher derivative gravity really can be eliminated with the Lee–Wick mechanism. Although they are absent in perturbation theory, it is generically true that these theories have non-perturbative instabilities (Horowitz 1981).

The question of course remains whether or not the $1/N$ approximation is a good approximation to the physical world. The acausality, for example, could just be artifact of the $1/N$ approximation. The net effect of adding many matter fields to the theory is to swamp out the quantum gravitational fluctuations. It is thus unclear if we learn anything about quantized *gravity*. We know that both QED and QCD (quantum-chromodynamics) undergo drastic transformations if we add enough matter fields to them. The $1/N$ expansion is not reliable for these theories.

In light of these considerations, it would be nice to re-sum gravity itself. Tomboulis (1980) and others have argued that this would produce results similar to the $1/N$ expansion.

Such a resummation is in fact possible. Like Yang–Mills, gravity contains a hidden expansion parameter—the inverse dimensionality of the invariance group. There are several inequivalent ways of defining quantum gravity for $D \neq 4$. These lead to inequivalent expansions, which will be the subjects of the next section.

4. $1/D$ EXPANSIONS

A. *D-Dimensional Metric on D-Dimensional Space–time†*
The most obvious form of the $1/D$ expansion begins with a *D*-dimensional metric on a *D*-dimensional manifold. One can then rescale κ and expand quantum gravity in powers of $1/D$. This leads to an *S* matrix which is unitary or renormalizable, but *not* both. This subsection is consequently brief.

Feynman diagrams acquire factors of *D* from two sources. The first source is combinatoric factors arising from the fact that the metric is a $D \times D$ matrix. These factors can be counted with standard techniques. At this point it is seen that ghost loops (like fermion loops in QCD) drop out of the large *D* limit. This is because there are only *D* of them as opposed to the D^2 components of the metric. This has the pleasant consequence that the large *D* limit is manifestly coordinate invariant. The Green's functions obey the 'naive' Slavnov–Taylor identities following from general coordinate invariance, as opposed to the much less restrictive identities that follow from BRS invariance.

† See Strominger (1981).

The second source of factors of D is the dimensionality of the space–time manifold. Feynman diagrams acquire D-dependent phase space factors. It has been shown on a lattice that these factors lead to a mean field expansion (Fisher and Gaunt 1964). Higher loop one particle irreducible diagrams are suppressed.

In our case, the tensor structure of the leading D term, including all factors of D, can be deduced from general arguments and the Slavnov–Taylor identities. The result is that the large D limit consists of tree diagrams plus quantum corrections to spin zero propagation and interactions. In particular, there is no effect on spin two propagation. It thus remains either non-unitary or non-renormalizable.

This is not the case for a different version of the $1/D$ expansion, which we now discuss.

B. *D-Dimensional Metric on a Four-Dimensional Space–time*†

In order to separate the two sources of factors of D discussed in A, one may consider the case where g^{AB} is a $D \times D$ matrix but the dimensionality of space–time is taken to be 4 (or $4 + \varepsilon$ for the purpose of regularization). This is equivalent to a Kaluza–Klein theory in which the extra dimensions are a flat hypertorus and the metric depends only on the four space–time dimensions. The only factors of D are now combinatoric.

For the sake of simplicity, we first consider the case of pure Einstein action. Later on we will find that higher derivative terms are necessary for renormalization, and should therefore be included in the original expansion. Unlike in the $1/N$ expansion, such terms greatly complicate the graphical expansion (although, as will be seen, it is still soluble). The $1/D$ expansion of the pure Einstein action is thus only relevant below the Planck energy, where the higher derivative terms have negligible effects.

We begin by writing the Einstein action in matrix notation:

$$S = \frac{1}{2\kappa^2} \int d^4 x \det^{-1/2}[G] \{\tfrac{1}{2} g^{\alpha\beta} \operatorname{Tr}[G_{,\beta} G^{-1}{}_{,\alpha}]$$

$$+ g^{\alpha\beta} \operatorname{Tr}[G_{,\beta} G^{-1}] \operatorname{Tr}[G_{,\alpha} G^{-1}]$$

$$+ g^{\mu\nu}{}_{,\nu} \operatorname{Tr}[G^{-1}{}_{,\mu} G] + \tfrac{1}{2} g^{A\beta}{}_{,\nu} g^{B\nu}{}_{,\beta} g_{AB}\} \tag{4.1}$$

where $A, B = 1 \ldots D$, $\alpha, \beta, \mu, \nu = 1 \ldots 4$, $(G)^{AB} = g^{AB}$ and x here and hereafter is a four-dimensional space–time coordinate. This action is not invariant under the full group of D dimensional coordinate transformations. It is, however, invariant under

$$g^{AB}(x) \to g^{AB}(x) + \varepsilon^{A;B}(x) + \varepsilon^{B;A}(x) \tag{4.2}$$

with $\varepsilon^A(x)$ an arbitrary D-dimensional vector that depends on only the four space–time dimensions. D gauge conditions and ghost fields are necessary to fix this invariance. Since there are only D of them, they do not affect the large D limit. We again find the nice feature that the large D limit is manifestly general coordinate invariant.

---

† See Strominger (1983).

Equation (4.1) is not the most general action invariant under (4.2). Other invariants, such as $R[g^{\mu\nu}]$, are higher order in $1/D$ and may be neglected.

The large N afficionado will quickly realize that, because of the double trace term, the large D limit of (4.1) is not described by an simple set of diagrams. To circumvent this difficulty, we make the field redefinition:

$$g^{AB} = \exp 2\kappa\Omega/D \; (\exp 2\kappa\phi)^{AB}$$

$$\delta_{AB}\phi^{AB} = 0 \tag{4.3}$$

in terms of which

$$\frac{-1}{\kappa^2}\sqrt{g}\,R = \exp(-\kappa\Omega)(\exp 2\kappa\phi)^{\mu\nu}\mathrm{Tr}[\phi_{,\mu}\phi_{,\nu}] + \dots$$

$$= \sqrt{g}g^{\mu\nu}\mathrm{Tr}[\phi_{,\mu}\phi_{,\nu}] + \dots \tag{4.4}$$

where ... indicates terms that contain at most D fields and may thus be neglected. The large D limit is now evident from inspection of (4.4).

The large D limit of quantum Einstein gravity is equal to the large N limit of gravity coupled to N minimally coupled massless scalar fields, where $N = D^2$.

Below the Planck energy the $1/D$ expansion does give results qualitatively similar to the $1/N$ expansion, albeit the non-conformally invariant version. Above the Planck energy, the situation is different.

We now show how to compute the leading term of the $1/D$ expansion of:

$$S = \int d^4x\sqrt{-g}\left(\Lambda + \frac{1}{\kappa^2}R + \frac{1}{\beta}R^2 - \frac{1}{\alpha}\mathcal{R}_{CD}\,\mathcal{R}^{CD}\right) \tag{4.5}$$

where $\quad \mathcal{R}_{AB} = R_{AB} - \dfrac{1}{D}g_{AB}R$ and $g^{AB}\mathcal{R}_{AB} = 0$ for convenience. (4.6)

The set of dominant diagrams for large D is much less trivial than in the previous case. They are the 'cactus' diagrams. For example, for $\kappa \to \kappa/D$, the graph in figure 1 is order one. (Only a particular index structure is allowed at the vertices.) Because of complicated momentum structure at the vertices, direct solution by integral equations is prohibitively difficult. A simpler method is to use auxiliary fields. Two such fields are added to equation (4.5) to eliminate the curvature squared terms:

$$S = \int d^4x\sqrt{-g}\left(\Lambda + \frac{1}{\kappa^2}R - \frac{2}{\kappa}rR - \frac{\beta}{\kappa^2}r^2 + \frac{2}{\kappa}f_{AB}\mathcal{R}^{AB} + \frac{\alpha}{\kappa^2}f_{AB}f^{AB}\right)$$

$$g^{AB}f_{AB} = 0. \tag{4.7}$$

For $\Lambda = 0$, the quadratic part of S reads:

$$S_2 = \int d^4x\{-\delta^{\mu\nu}(\mathrm{Tr}[\phi_{,\mu}\phi_{,\nu}] - \Omega_{,\mu}\Omega_{,\nu} - 2r_{,\mu}\Omega_{,\nu} + 2\mathrm{Tr}[f_{,\mu}\phi_{,\nu}])$$

$$+ \frac{\alpha}{\kappa^2}\mathrm{Tr}[f^2] - \frac{\beta}{\kappa^2}r^2 + \dots\} \tag{4.8}$$

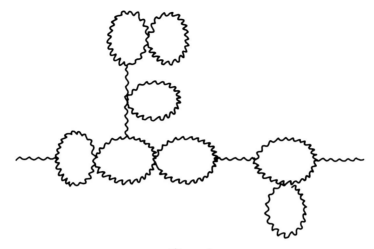

Figure 1

from which we see that we must shift the fields

$$\phi = \tilde{\phi} - \not{r}$$
$$\Omega = \tilde{\Omega} - r \tag{4.9}$$

in order to diagonalize the propagators.
We now find:

$$S = \int d^4 x \exp\left[-\kappa(\tilde{\Omega} - r)\right] \left\{ -\left[\exp 2\kappa(\tilde{\phi} - \not{r})\right]^{\mu\nu} (\text{Tr}[\tilde{\phi}_{,\mu}\tilde{\phi}_{,\nu}] - \Omega_{,\mu}\Omega_{,\nu}\right.$$

$$- \text{Tr}[\not{r}_{,\mu}\not{r}_{,\nu}] + r_{,\mu}r_{,\nu})$$

$$- 2\kappa \not{r}^{\mu\nu} (\text{Tr}[\tilde{\phi}_{,\mu}\tilde{\phi}_{,\nu}] + \text{Tr}[\not{r}_{,\mu}\not{r}_{,\nu}] - 2\text{Tr}[\tilde{\phi}_{,\mu}\not{r}_{,\nu}])$$

$$+ \frac{\alpha}{\kappa^2} \text{Tr}[\not{r}^2] - \frac{\beta}{\kappa^2} r^2 + \Lambda + \dots \}. \tag{4.10}$$

The fourth-order action (4.5) has been reduced to the second-order action (4.10) with twice as many fields. The auxiliary fields have become dynamical because the original action contained more derivatives. The equivalence of the two actions can be checked by comparing propagators:

$$\langle \phi\phi \rangle = \frac{1}{q^2 + \dfrac{\kappa^2}{\alpha}q^4}$$

$$= \langle (\tilde{\phi} - \not{r})(\tilde{\phi} - \not{r}) \rangle$$

$$= \frac{1}{q^2} - \frac{1}{q^2 + \alpha/k^2}. \tag{4.11}$$

The diagrams of this theory are quartically divergent at one loop. But we know from looking at (4.5) that the theory can be renormalized by adjusting Λ, κ and the masses of the r and f fields. This means that there are many cancellations of divergences.

The large D Feynman rules of (4.10) are straightforward to determine. A typical graph is

Figure 2

where

$\sim\!\sim\!\sim = \tilde{\phi}$ propagator

$\text{———} = f$ propagator

These graphs are much easier to sum because the touching bubble graphs have been eliminated. The physical masses of the f and r fields are found by integral equations of the type

$$m^2 \;=\; \text{———}\!\!\!\times + \text{———}\!\bigcirc \qquad\qquad (4.13)$$

With these masses, the two-point function is given by the infinite series

$$\langle \phi\phi \rangle = \langle (\tilde{\phi}-f)(\tilde{\phi}-f) \rangle$$

$$= \text{———} + \sim\!\sim\!\sim + \text{—}\!\bigcirc\!\!\sim\!\sim + \cdots$$

$$+ \cdots + \cdots + \cdots \qquad\qquad (4.14)$$

which can be summed by inverting a two by two matrix.

Explicit evaluation of this propagator is in progress and will be published elsewhere. The calculation is simplified due to the manifest general coordinate invariance. This reduces the number of Lorentz invariant tensor structures from five down to two.

The integral equation (4.13) for the physical mass is similar to the mass gap equation in the $1/N$ expansion of, for example, the sigma model. As such, it can describe dynamical mass generation. Since the mass here is essentially the Planck mass, this raises the interesting possibility of inducing the Einstein action from a bare C^2 action.

A complete discussion of the physical relevance of the $1/D$ expansion awaits evaluation of (4.14).

5. QUANTUM GRAVITY AND PARTICLE PHYSICS

Now that we have a consistent theory of quantum gravity, so what? Even for those who are not concerned with questions of principle, it now appears that quantum gravity is becoming phenomenologically relevant. The relentless charge upward in energy scales is bringing particle physicists dangerously close to the Planck energy. It has been amply demonstrated that gravitational effects are very relevant for grand unified model building (Ovrut and Wess 81, 82, Weinberg 82, Arnowitt *et al* 82, Ovrut and Raby 82, Nilles, to appear, Nilles, Schrednicki and Wyler, to appear, Hall, Lykken and Weinberg, to appear).

It is important that these effects be included in a systematic way, and that all effects of the same order be included. At present, the $1/N$ and $1/D$ expansions are possibly the only systematic methods of computing gravitational effects within the context of a unitary, renormalizable quantum field theory. Even if this theory ultimately proves to be incorrect, it is still plausible that it would provide a good approximation below the Planck energy.

Since the graviton propagator may be written (neglecting logarithms):

$$\frac{1}{q^2 + (2\kappa^2/\alpha)q^4} = \frac{1}{q^2} - \frac{1}{q^2 + \alpha/2\kappa^2} \tag{5.1}$$

the effect of adding R^2 and C^2 is roughly equivalent to cutting the theory off (à la Pauli–Villars) at the Planck scale. This enables us to quickly estimate the magnitude of Feynman diagrams. Consider, for example, the scalar self-energy diagram:

$$\propto \frac{1}{M_P^2} \times M_P^4 = M_P^2$$

Figure 3

where M_P is the Planck mass. Below the Planck energy, where the graviton propagator goes like $1/q^2$, this diagram diverges quartically. This contributes a factor M_P^4. Including the factor of $1/M_P$ from each vertex, we find that this graph gives the scalar a mass of order M_P. (There is also wavefunction renormalization of order one.) It is thus difficult to have naturally light Higgs in quantum theories including gravity.

Similarly, the graph

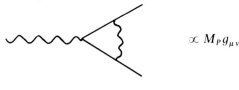

$$\propto M_P g_{\mu\nu}$$

Figure 4

induces an effective vertex $M_P\, g_{\mu\nu}$, as opposed to the tree graph:

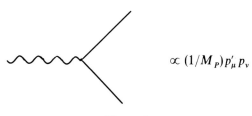

$$\propto (1/M_P)\, p'_\mu\, p_\nu$$

Figure 5

which is of the form $(1/M_P)\, p'_\mu p_\nu$ (p and p' are some lower energy external momenta). Higher order graphs are a power series in α and β. Clearly graviton loops can not be *a priori* neglected in a low energy expansion in p^2/M_P^2.

For fermions and gauge bosons, the situation is different. The former is protected from getting a Planck mass by chiral invariance, the latter by gauge invariance. All gravitational effects are down by at least $1/M_P$, and the theory of gravity discussed here is not in (direct) contradiction with phenomenology. However, it is still true that, in a low energy expansion in p^2/M_P^2, graviton loops can not be neglected relative to graviton tree diagrams.

These general conclusions apply to a very wide class of gravitational theories. The only assumption is that the graviton propagator effectively goes like $1/q^2$ until somewhere near the Planck mass, after which it begins to converge more rapidly. This alteration of the effective propagator could also follow from infinite resummations, the excitation of new dimensions or restoration of supersymmetry. If the alteration occurs at a higher (lower) energy, gravitational effects will be more (less) important.

To make more precise statements about which graphs are relevant, one must consider a specific theory, process and expansion parameter. For example, graphs of figures 3 and 4 are suppressed in both the $1/N$ and $1/D$ expansions. The point we wish to make here is that effects from both tree and virtual gravitons can be large at energies well below the Planck energy. The $1/N$ and $1/D$ expansions provide systematic means of calculating these effects.

ACKNOWLEDGMENTS

The essay is dedicated to Bryce DeWitt on his sixtieth birthday, and I would also like to thank him for interesting discussions. In addition I would like to thank J Dell, G Horowitz, J Lykken, B Ovrut, L Smolin and E Tomboulis for helpful discussions. This work was supported in part by Grant DE-AC02-76ERO220.

REFERENCES

Adler S 1982 *Rev. Mod. Phys.* **54** 729
Arnowitt R, Chamseddine A H and Nath P 1982 *Phys. Rev. Lett.* **14** 49
Boulware D, Horowitz G and Strominger A 1983 *Phys. Rev. Lett.* **50** 1726
Coleman S 1969 *Subnuclear Phenomena* (New York: Academic) p 282
Deser S 1970 'Talk given at the Austin Conference on Particle Physics' (unpublished)
DeWitt B S 1964 *Phys. Rev. Lett.* **13** 114
—— 1967a *Phys. Rev.* **160** 1113
—— 1967b *Phys. Rev.* **162** 1195
—— 1967c, *Phys. Rev.* **162** 1239
Fisher M and Gaunt D S 1964 *Phys. Rev.* **133** 225
Hartle J B and Horowitz G T 1981 *Phys. Rev.* D **24** 257
Horowitz G T 1981 in *Quantum Gravity II* (Oxford: Oxford University Press)
Hasslacher B and Mottola E 1980 *Phys. Lett.* **95B** 237
Isham C J, Salam A and Strathdee J 1971 *Phys. Rev.* D **3** 1805
Kay B S 1981 *Phys. Lett.* **101B** 241
Lee T D 1970 in *Elementary Processes at High Energy* (New York: Academic) p 62
Lee T D and Wick G C 1970 *Phys. Rev.* D **2** 1033
Ovrut B A and Wess J 1982a *Phys. Lett.* **112B** 347
—— 1982b IAS preprint
Ovrut B A and Raby S 1982 IAS preprint
Pauli W 1967 *Theory of Relativity* (London: Pergamon)
Smolin L 1982 *Nucl. Phys.* B. **208** 439
Stelle K S 1977 *Phys. Rev.* D **16** 953
Strominger A 1981 *Phys. Rev.* D **24** 3082
—— 1983 *Lecture Notes in Physics; Gauge Theory and Gravitation* (New York: Springer-Verlag)
Tomboulis E 1977 *Phys. Lett.* **70B** 361
—— 1980 *Phys. Lett.* **97B** 77
Utiyama R and DeWitt B S 1962 *J. Math. Phys.* **3** 608
Weinberg S 1982 Harvard preprint
Yamagishi H 1982 *Phys. Lett.* **114B** 27

Quantum Gravity: The Question of Measurement

JOHN ARCHIBALD WHEELER

THE FORMALISM OF QUANTUM GEOMETRODYNAMICS AS GUIDE TO THE APPROPRIATE QUESTION

No question about quantum gravity is more difficult than the question, 'What is the question?' What does theory tell us to hope to measure? Devices to do the measuring, we should imagine—but sensitive enough to measure what?

This often-raised issue is in principle and at first sight no issue at all. Any theory that is a theory already defines in and by itself, we know, what quantities in it are subject to determination (see for example Bohr and Rosenfeld 1933, DeWitt 1962). The familiar example is the non-relativistic quantum mechanics of the electron. There we have learned which quantities are subject to simultaneous measurement and which are not amongst a list of individually determinable quantities that go on without limit: coordinate, momentum, energy, angular momentum, dipole moments, quadrupole moments, octupole moments and higher order quantities.

Quantum electrodynamics so far exploits nowhere nearly so fully a far greater wealth of options open for what shall be determined. The reason is well known. The most primitive concepts are already by themselves incredibly numerous. The options include

(1) the number of quanta in the k-th 'Fourier' mode,
(2) the phase of that mode,
(3) the real part, ξ, of its amplitude as epitomized for example in the ground state

$$\psi(\xi_1, \xi_2, \ldots) = N \exp\left[-(\xi_1^2 + \xi_2^2 + \ldots)\right].$$

(4) the imaginary part of the amplitude of that mode,
(5) a 'wave packet' or 'coherent wave' description of that mode,

(6) the average of a field component such as E_x or B_y over a region of space and time such as $V_I T_I$ or $V_{II} T_{II}$,

(7) one of these—or any other 'representative part' of the electric or magnetic field—as it appears in one or another *spacelike representation* of the electromagnetic field, such as

$$\psi[B(x,y,z)] = \mathcal{N} \exp\left(-\int\int \frac{B(x_1)\cdot B(x_2)}{16\pi^3\hbar c r_{12}{}^2} d^3x_1 d^3x_2\right).$$

By now the ways available to go back and forth between one of these descriptions and another are understood reasonably well, and we can agree with Feynman (1967) that 'every theoretical physicist who is any good knows six or seven different theoretical representations for exactly the same physics. He knows that they are all equivalent, and that nobody is ever going to be able to decide which one is right at that level, but he keeps them in his head, hoping that they will give him different ideas for guessing.'

THE TWO SPACELIKE REPRESENTATIONS OF QUANTUM GEOMETRODYNAMICS

In gravitation, or 'geometrodynamics,' the representations of which we have the beginning of an understanding are nowhere near so numerous. There are at least two, but it is not clear to this writer that there are any more that meet the requirement, 'the beginning of an understanding.' Both are spacelike. In one the quantum state functional is considered in its dependence on the three-geometry,

$$\Psi = \Psi(^{(3)}\mathcal{G})$$

the momentary configuration reached by space geometry in the course of its dynamic development; and in the other the state functional is considered in its dependence on the conformal part of the three-geometry and the time rate of contraction of the three-geometry; that is, the 'trace of the extrinsic curvature' or 'extrinsic time' or 'York time.'

These two spacelike representations of the state function differ in the concepts of time that they use. In the first, time is intrinsic. 'A three-geometry,' in the words of Baierlein, Sharp and Wheeler (1962), 'is the carrier of information about time.' The information, however, is coded and only to be read out in conjunction with other information. How come? To specify a three-geometry is to specify on each point of a three-dimensional manifold not only the conformal metric but the scale of that metric. In contrast, the dynamical conjugate of that scale, its fractional rate of change, is the extrinsic time of the second representation. Both ways of describing time, both expressions for the state functional, in principle view time as 'many fingered.' In other words, both descriptions admit of our pushing forward this spacelike hypersurface—for which one evaluates the probability amplitude—at different rates in different places. The extrinsic time representation offers a convenient and attractive way for us to limit the work of the exploration by confining our attention to a sequence of hypersurfaces, distinguished one from another by successive fixed values of the extrinsic time, Tr**K**. No such way to limit attention to a natural foliation of configurations immediately offers itself in the intrinsic-time description. The reason is evident, even at the classical level. There it makes sense to speak of a definite 'space–time,' a deterministic history of space geometry evolving in time. One spacelike slice through this four-geometry is as good as another. If we dream up an arbitrary three-geometry, we will not ordinarily find it represented at all in the given four-geometry, in the given deterministic history of space geometry developing in time. Thus, determinism makes the well known sharp division—for a given classical history—between the YES-three-geometries that will be found in that history and the infinitely more numerous NO-three-geometries that will not. The same point lets itself be expressed in the language of superspace (Wheeler 1963, 1968, 1970, DeWitt 1967a,b,c) the familiar infinite dimensional manifold in which each point describes a three-geometry complete with all its warts, bumps and ripples, the three-geometry associated with one point differing from that associated with another point as a Ford fender differs from a Fiat fender. How to transcribe a deterministic four-geometry into the language of superspace is well known. We imagine it unraveled, after the fashion of a Chinese puzzle, into the totality of all the three-dimensional spacelike slices that can be made through it. Each slice, each three-geometry, is then marked as a YES point in superspace. The NO points in superspace are infinitely more numerous. Specifically, the YES points fill out a 'leaf' in superspace, the dimensionality of which, while infinite, is only one-third as great as the dimensionality of the enveloping superspace.

This factor of one-third follows from elementary geometrodynamics, from the fact that there are two gravitational wave degrees of freedom per space point. The two 'coordinates' associated with these two degrees of freedom per space point are already fixed by the specification of the three-geometry in the first place. The two 'momenta' associated with these two degrees of freedom would be free for us to choose if the three-geometry alone were specified. They would give us—per point of the *three*-geometry—two independent directions of 'travel' away from the given point in *superspace*. Pushing the spacelike hypersurface ahead a little gives us a third kind of change available in the three-geometry at that space point, and therefore a third direction of travel in superspace. However, we kill two out of these three independent directions of travel when we specify the four-geometry, the classical history, the 'leaf of history in superspace'—because in so doing we nail down the momenta at each stage in the dynamic evolution. Thus we deprive ourselves of the two directions of travel off the leaf of history. We have left only the one direction of travel in the leaf that goes with our freedom of choice for many-fingered time. There being ∞^3 points in the three-geometry, these considerations repeat themselves for each of them. This is the sense in which the leaf, infinite dimensional though it is, has only one-third the dimensionality of the enveloping superspace.

THE QUANTUM VERSION OF THE INTRINSIC-TIME
SPACELIKE DESCRIPTION

The intrinsic-time version has the advantage over the extrinsic-time description that the equivalent of a Schrödinger equation is local. Symbolically, it reads

$$\delta^2 \Psi / (\delta^{(3)} \mathscr{G})^2 + {}^{(3)}R\Psi = 0$$

or more specifically,

$$\left(G_{ijkl} \frac{\delta}{\delta \gamma_{ij}} \frac{\delta}{\delta \gamma_{kl}} + \gamma^{1/2\,(3)}R \right) \Psi[{}^{(3)}\mathscr{G}] = 0$$

where

$$G_{ijkl} \equiv \tfrac{1}{2} \gamma^{-1/2} (\gamma_{ik}\gamma_{jl} + \gamma_{il}\gamma_{jk} - \gamma_{ij}\gamma_{kl}).$$

In no way does it permit the state functional to vanish everywhere off a classical leaf of history, as would correspond to the classical concept of 'space–time.' On the contrary, the three-geometries that occur with appreciable probability amplitude are far more numerous than can be accommodated in any definite four-geometry. In this sense, space–time is a shattered concept. Even the very terms, 'before' and 'after' lose all meaning at the Planck scale of distances. Moreover—we remind our colleagues in other fields—it takes no deep mathematics to see why space–time is 'no-go.' After all, classical space–time is nothing but the history of space geometry evolving deterministically. We give a configuration of the three-geometry as part of the specification of the initial data for this dynamics. However, that is not enough. We must also specify the time rate of change of this geometry or, in other words, the 'geometrodynamic field momentum' conjugate to the geometrodynamic field coordinate, that is, conjugate to the three-geometry itself. But Heisenberg's principle of indeterminism has taught us that it is absolutely impossible, even in principle, to specify simultaneously a coordinate and its conjugate momentum. That is why the concept of a worldline for a particle fails, and has to be replaced by a wave packet. That is why the concept of space–time, or leaf of history in superspace, fails. What takes its place is a wave packet, or more general functional, in superspace.

So much for the rationale of the intrinsic-time spacelike representation—or 'superspace version' of quantum gravity. But what does it mean and how does it work?

ONE WAY IN: A COHERENT-STATE DESCRIPTION

No development would make the content of quantum gravity more readily accessible to understanding than the development of a *coherent-state, spacelike representation*. Such a description, like the coherent-state representation of electrodynamics (see for example Bohr and Rosenfeld 1933, Klauder and Sudarshan 1968), would have a close affinity to the description of a harmonic oscillator by a wave packet. In this formulation neither coordinate nor momentum is predictable with perfect accuracy, but both are predictable with an accuracy adequate for many purposes.

Current analyses of gravitational radiation implicitly assume the existence of such a coherent representation, and rightly so, because the effective number of quanta in the output from any of the usually considered astrophysical sources is so stupendous. Bohr and Rosenfeld (1933) in their pioneering study of the measurability of the electromagnetic field, consider among other cases such a coherent wave though it was not so titled at that time. In particular, they showed that the familiar Poisson expression,

$$P_n = \bar{n}^n e^{-\bar{n}}/n!$$

gives the probability that a measurement of the number of quanta in such a wave will yield the result n when the average number is $\bar{n}$. It is well known that the motion of the center of gravity for the harmonic oscillator wave packet follows the predictions of classical mechanics. There is no reason to doubt that the corresponding statement will be true in such a future coherent-state representation of quantum gravity when relevant lengthscales are large compared to the Planck length. Exactly the respects in which this classical description fails are what we are looking for in quantum geometrodynamics, and nowhere more so than in the realm of measurement theory.

HOW TO MEASURE THE THREE-GEOMETRY: OPEN QUESTION

If Bohr and Rosenfeld's analysis (1933) of the measureability of the electromagnetic field is the deepest paper ever written in all of theoretical—as distinguished from mathematical—physics, Bryce DeWitt's treatment (1962) also remains the most thorough analysis, in the same spirit, of the measurability of the geometrodynamic field quantities. That all issues of measurement theory in this field have been cleared up, DeWitt himself is nevertheless the first to deny. In particular, Salecker and Wigner (1958), considering an idealized clock of simple construction, find themselves very far from being able to determine space–time geometry with the accuracy that theory promises. This discrepancy creates a lively challenge. However it also reminds one that Bohr and Rosenfeld themselves, using the best adapted idealized apparatus that they could conceive, also found themselves unable in the beginning to reproduce the expectations of measurement theory.

As Rosenfeld (1955) tells us, 'the investigation, however, was very soon brought to an apparent deadlock by difficulty of a most baffling nature . . . It was child's play, with our arsenal of test-bodies and springs, to reduce the product of the indeterminacies [of field quantities] to the *sum* [of two expressions]. But how could one ever hope to get down to the [expected] *difference* [of those two expressions]? At this juncture Bohr did not hesitate to challenge the commutation rules themselves . . . Eventually, we noticed . . . that the 'messages' which could thus be automatically conveyed by the test-bodies contained just enough information about their respective displacements to conjure up the difference predicted by the theory . . . [and that the] self-reactions and mutual perturbations of the test-bodies can only be compensated by spring mechanisms to the extent of their classically evaluated average magnitudes.' These words serve as a reminder that it is easy to devise a measuring device that will not work!

What is it that one would like to measure to flesh out the formalism of the intrinsic-time or superspace formalism of quantum gravity? How is one to give the term 'three-geometry' a meaning that rises above any coordinate ambiguities, or even any use of coordinates whatsoever? In no way more simply than by a Regge (1961) simplicial decomposition of the three-geometry into tetrahedrons. The vertices are most naturally identified as the momentary location of test-masses. The record of the spacelike separations between each vertex and its neighbors constitutes the entire specification of the three-geometry, skeleton three-geometry though it is. Out of these distances the curvature of the three-geometry lets itself be determined by the 'angle of rattle' between the blocks, as it would be if they were embedded in Euclidean three-space, according to the prescription so beautifully spelled out by Regge. This is a three-geometry in its full tangibility!

How to get those distances from vertex to vertex: that is the question. They are spacelike. Signals, however, travel on light paths. How those signals can be used to measure the distance by an appropriate zig-zag pattern of light rays connecting worldlines of appropriately chosen masses has been described in the context of classical theory by Marzke (1959), by Marzke and Wheeler (1964), by Ehlers, Pirani and Schild (1972), by Misner, Thorne and Wheeler (1974) and by Pfister (1983) and, less elaborately but earlier and at the quantum level, by Salecker and Wigner (1958).

Reaching beyond the issues considered in any of these papers is the question how to compensate out an effect not considered in any of them, the complete indeterminism of the geometrodynamic field momentum conjugate to the 'field coordinate,' the conformal part of the three-geometry. The very fixation of the three-geometry that is aimed at implies an enormous indeterminacy in the conjugate momentum and therefore in the gravitational waves which will unpredictably perturb the light rays and test masses. How are we to redesign the network of light rays and test particles so that it will automatically compensate out the effects of these waves? That is the question. The stating of it is perhaps one small step toward the solution of the question. Certainly the solution is not now in hand. Until it is, some of the central meaning of quantum gravity will continue to escape our grasp.

A SECOND WAY IN: AN INITIAL-VALUE ANALYSIS OF THE SUPERSPACE WAVE EQUATION

If the formalism itself is a guide, and perhaps even the best of all guides, as to what to measure, and what to hope to predict, then the wave equation in superspace, as written above, would seem a central source of guidance. The equation itself should supply its own prescription for what should be specified and what should be predicted. In brief, what is the proper initial-value analysis for this wave equation?

Some illuminating remarks relevant to the mathematics of this initial-value problem appear in the first paper in Bryce DeWitt's famous trio (1967a,b,c). Moreover, on the physical side, one derives immediate guidance from the comparison with quantum electrodynamics in its spacelike representation. There the state functional, given at an initial instant in its dependence on the two independent numbers per space point that describe the magnetic field, is predicted in its dependence on the magnetic field at every later time by the equivalent of the Tomonaga (1946) equation. If that is the case in electrodynamics, it would seem reasonable to expect no less in geometrodynamics. What does this conjecture mean, if true? That is easier to say in the extrinsic-time version than in the intrinsic-time description. Specify and specify arbitrarily, that formulation says, the state functional in its dependence on the conformal part of the three-geometry at one value of the extrinsic time. Then, it claims, we can in principle find the state functional in its dependence on the conformal part of the three-geometry for all other values of the extrinsic time. Moreover, this predictive power is not limited to this, that, or the other spacelike hypersurface of fixed extrinsic time. The hypersurface can weave as it will through space–time so long as it remains spacelike.

To have so much information about the state functional in its dependence on the full range of variables in the intrinsic-time representation is enough in principle to give fully the state functional in the intrinsic-time representation; that is, to give

$$\Psi(^{(3)}\mathcal{G})$$

in its entirety.

If we can transform the 'answer' from one representation to the other, can we not transform the initial data from the one representation to the other? No easy way to do so is evident. The very word, 'initial,' is the first of the stumbling blocks in the way of such an enterprise. Initial what: initial intrinsic-time? What does that mean in any understandable terms? No such difficulty of interpretation appeared in electrodynamics. There,

(1) the field evolved in a given space–time manifold

(2) no choice had to be made between two dynamically-conjugate concepts of time.

In geometrodynamics, however, one 'time' is connected with the conformal scale factor; the other, essentially, its conjugate, with its fractional rate of change with time. What is well connected in superspace in one representation may even conceivably be wildly scattered about in the other representation. If it is morally wrong to do a calculation before one knows the nature of the answer, then we would seem to have something to learn before we even start the mathematical analysis of the initial-value problem!

If we are not sure how to formulate 'the initial-value problem in superspace' in the most general context, we may still have some idea how to do it in a special context. For the harmonic oscillator problem, one might not know that $\psi(x)$ can be assigned arbitrarily at the initial time; but one might have less doubt about the appropriateness of the specifying of

$$\psi(x) = \delta^{1/2}(x - 17)$$

as a starting condition for a particle definitely at the point $x = 17$ at the time $t = 0$. However, it would be in absolute contradiction with any reasonable physics to add to the foregoing expression a multiplicative factor $\delta(t-0)$. On the contrary, the equations themselves give the time evolution and indeed in such a way that at every moment, the space integral of $|\psi|^2$ is conserved. Naturally, in the Klein–Gordon equation, the situation is somewhat modified (Pauli and Weisskopf 1934). However, the essential idea is the same: the probability amplitude, initially localized here in space at one time, spreads out at later times. The corresponding lesson has yet to be fully read in the case of quantum gravity.

It is a partial reading of the lesson to say that $\Psi(^{(3)}\mathcal{G})$ cannot differ from zero for only a single three-geometry, just as $\psi(x, t)$ cannot differ from zero for only the single point $x = 17$, $t = 0$. What one can freely specify and what one cannot in non-relativistic quantum mechanics we learn from the Schrödinger equation itself. Nowhere better can one expect to learn the corresponding lesson for quantum gravity than from the superspace wave equation. Nowhere does one see more deeply into the analogy between this equation and the Klein–Gordon equation than in DeWitt (1967a). His analysis will surely someday open the door to the formulation of the initial-value problem for the propagation of waves in superspace, with all of the insights that are bound to follow.

CONCLUSION

The formalism of quantum gravity, in its best developed form, makes three-geometry a central concept; consequently, the finding of a proper way to measure this three-geometry—against the background of the indeterminancy of the conjugate geometrodynamic field momentum—is a central issue.

ACKNOWLEDGMENT

The preparation for publication was assisted by the Center for Theoretical Physics of the University of Texas and by NSF grant PHY 8205717

REFERENCES

Baierlein R F, Sharp D H and Wheeler J A 1962 *Phys. Rev.* **126** 1864–5
Bohr N and Rosenfeld L 1933 *Kgl. Danske Videnskab. Sels. Mat.-fys. Medd.* **12** no. 8 (Engl. Transl. pp 357–400 in Cohen R S and Stachel J J 1979 *Selected Papers of Léon Rosenfeld* (Dordrecht: Reidel)
DeWitt B S 1962 in Witten L ed *Gravitation: An Introduction to Current Research* (New York: Wiley) pp 266–381

—— 1967a *Phys. Rev.* **160** 1113–48
—— 1967b *Phys. Rev.* **162** 1195–1239
—— 1967c *Phys. Rev.* **162** 1239–56
Ehlers J, Pirani F A E and Schild A 1972 in O'Raifeartaigh L ed *General Relativity, Papers in Honor of J L Synge* (London: Oxford University Press) pp 63–84
Feynman R P 1967 *The Character of Physical Law* (Cambridge, Massachusetts: M I T Press)
Klauder J R and Sudarshan E C G 1968 *Fundamentals of Quantum Optics* (New York: Benjamin)
Marzke R 1959 "The theory of measurement in general relativity," *Princeton University A B senior thesis*, unpublished, on file in Fine Library, Princeton
Marzke R F and Wheeler J A 1964 in H Y Chiu and W F Hoffmann eds *Gravitation and Relativity* (New York: Benjamin) pp 40–64
Misner C W, Thorne K S and Wheeler J A 1974 *Gravitation* (San Francisco: Freeman)
Pauli W and Weisskopf V 1934 *Helv. Phys. Acta* **7** 709–31
Pfister H 1983 'Grundlegung der allgemeinen Relativitätstheorie mittels Lichtstrahlen und freien Teilchen' *University of Tübingen preprint*
Regge T 1961 *Nuovo Cimento* **19** 558–71
Rosenfeld L 1955 in W Pauli ed *Niels Bohr and the Development of Physics* (New York: McGraw-Hill) pp 70–95
Salecker H and Wigner E P 1958 *Phys. Rev.* **109** 571–7
Tomonaga S 1946 *Prog. Theor. Phys.* **1** 27–42
Wheeler J A 1963 in DeWitt C and DeWitt B eds *Relativity, Groups and Topology* (New York and London: Gordon and Breach) pp 317–520
—— 1968 in DeWitt C M and Wheeler J A ed *Battelle Recontres: 1967 Lectures in Mathematics and Physics* (New York: Benjamin) pp 242–307
—— 1970 in Gilbert R P and Newton R G ed *Analytic Methods in Mathematical Physics* (New York: Gordon and Breach) pp 335–78

The following is a toast presented by Professor Wheeler at Bryce DeWitt's birthday dinner held at Winnie Schild's home in Austin, Texas on December 11 1982.

A toast!
First to our host and hostess
And to Winnie's farsighted, warmhearted Alfred
Who had the inspiration
To make Austin a world center of relativity
And did it most of all by bringing Bryce here;

And then to *him.*
Who is our 'push-it-to-the-limit man?'
Who is the man unsurpassed in all the world
At pushing solid theory to unexpected limits,
Who by his own work
Opens the door to advances by everyone?
Who explores far corners of the world?
Who makes narrow escapes
    from drowning, disaster and death?
Are those only his setting-up exercises?
Is that how he learns to escape divergences,
    non-unitarity, and infinities?
And who's the girl that sings out that
    song with him, Rigor, Rigor, Rigor?
Who encourages him to let the action act?
To just *do* that calculation?
We toast Cecile
And Bryce, the 'push-it-to-the-limit man.'
Great work, Cecile!
Happy birthday, Bryce!

Steps towards a Quantum Theory of Gravity

W G UNRUH

Bryce DeWitt has contributed immensely to our ideas and approaches to quantum gravity. In spite of his work one still often hears statements like 'Do we really have to quantize gravity?' This paper will attempt to answer a series of such questions by means of some simple gedanken experiments. However any reader is also urged to go back to reread DeWitt's papers on the subject (see references).

Gravity has resisted all attempts at finding a natural way to quantize it. One of the key difficulties has been the lack of any experiments to guide us in finding a theory of quantum gravity. In such a situation a theorist's fallback is often to make up his own gedanken experiments. These serve not to test nature but rather to present the *a priori* prejudices of the theorist in their simplest physical guise. They highlight the beliefs and prejudices the theorist has about the physical world–beliefs which could well be proven wrong by true experiments, but which seem necessary to limit the infinite range of possible theories in the absence of experiment.

One question which is often raised about gravity is 'Does it really need to be quantized?' Can we not regard the arena of all physics, namely space–time, as classical with the rest of quantum physics being played in this arena? This is of course an attitude which has served physics extremely well in the past. The development of modern field theory is unimaginable without the powerful organizational structure which the Poincaré group, the group of isometries of a flat background space–time provides. Unfortunately, if we are to take General Relativity at all seriously, this position of a classical gravity field loses its appeal. After all, the fundamental equation of General Relativity is

$$G_{\mu\nu} = 8\pi T_{\mu\nu} \tag{1}$$

(where I have chosen units such that $G = \hbar = c = 1$). The left-hand side represents the geometry of space–time, while the right is dependent on the rest of the matter in the universe. The left-hand side is 'classical', is an ordinary function of the space–time points while the right-hand side is a quantity which depends on quantum operators.

The two sides of the equation are thus different mathematical structures, and cannot be set equal to each other.

One can try some stratagems to get around this problem. Since the left-hand side is a *c*-number field, any classical theory of gravity must convert the right-hand side into a *c*-number. The simplest such approach would be to convert the right-hand side into a *c*-number field by replacing it by the expectation value of the energy–momentum tensor (Moller 1962, Rosenfeld 1963)

$$G_{\mu\nu} = 8\pi \langle T_{\mu\nu} \rangle. \tag{2}$$

The problem with this approach is twofold. The first is epitomized by the gedanken experiment of figure 1, an elaboration of the 'Schrödinger's Cat' experiment. A perfectly sealed box contains two masses held apart by a rod with a stretched spring between the masses. An explosive is connected to the rod and to a Geiger counter in such a way that when the counter clicks, the explosive will break the stick and allow the two masses to be pulled together by the spring. The box is set up with a radioactive source of such strength that the counter would click on average, say, once a day. Now we place a Cavendish balance outside the box in such a way that

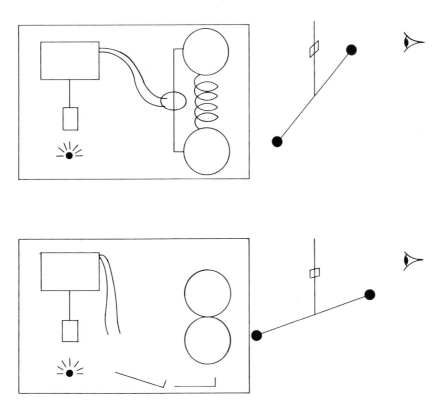

Figure 1 A Schrödinger–Cavendish experiment to test semiclassical quantum gravity.

the equilibrium position of the balance is different when the masses are held apart than when they are drawn together by the spring.

By hypothesis the gravitational field outside the box is given by equation (2), where the state is the state of the whole system inside the box. We can divide the states of the box into two classes—those for which the two masses are apart $|\Updownarrow\rangle$, and those for which the two are together $|\maltese\rangle$. The energy–momentum tensor of the matter making up the box is expected to be dominated by the position of the two masses. Initially, the expectation value will be given by

$$\langle T_{\mu\nu}\rangle_0 = \langle\Updownarrow|T_{\mu\nu}|\Updownarrow\rangle. \tag{3}$$

As time progresses, the state will evolve into a superposition of states with the masses separated and those with the masses together.

$$|\phi\rangle = \alpha(t)|\Updownarrow\rangle + \beta(t)|\maltese\rangle \tag{4}$$

where we know experimentally that

$$|\alpha|^2 \approx (e^{-\lambda t})$$
$$|\beta|^2 \approx (1 - e^{-\lambda t}). \tag{5}$$

The expectation value becomes

$$\begin{aligned}\langle\phi|T_{\mu\nu}|\phi\rangle = {}& |\alpha|^2 \langle\Updownarrow|T_{\mu\nu}|\Updownarrow\rangle \\ &+ |\beta|^2 \langle\maltese|T_{\mu\nu}|\maltese\rangle \\ &+ 2\mathrm{Re}\,(\alpha^*\beta\,\langle\Updownarrow|T_{\mu\nu}|\Updownarrow\rangle).\end{aligned} \tag{6}$$

The last term is expected to be very small (i.e. the states are approximately eigenvectors of the total $T_{\mu\nu}$) leaving us with

$$\begin{aligned}\langle t|T_{\mu\nu}|t\rangle = {}& e^{-\lambda t}\langle\Updownarrow|T_{\mu\nu}|\Updownarrow\rangle \\ &+ (1 - e^{-\lambda t})\langle\maltese|T_{\mu\nu}|\maltese\rangle.\end{aligned} \tag{7}$$

The balance will see a slow change (i.e. over the course of a day in our case) in the gravitational mass distribution, and will slowly swing from one equilibrium to the other.

Of course we do not expect this to happen. We expect the balance to swing suddenly at the 'time the counter clicked'. But if it does move suddenly, equation (2) cannot be a correct description of gravity.

There is a further problem with this theory. In particular, what happens if we look into the box part way through the experiment? The standard interpretation of quantum mechanics would have the state suddenly change from $\alpha|\Updownarrow\rangle + \beta|\maltese\rangle$ to one with a definite position for the masses. This sudden discontinuity would produce a discontinuity in $\langle T_{\mu\nu}\rangle$ which would, in general, not obey the conservation law

$$\langle T_\mu^{\ \nu}\rangle_{;\nu} = 0 \tag{8}$$

which the left-hand side of equation (2) must obey. Such a change in state of the system could be used to transmit signals faster than light by means of an

Einstein–Podolsky–Rosen type experiment as has been pointed out by Eppley and Hannah (1977).

The above gedanken experiment could almost be performed. A variant, in which the geiger counter and dynamite was replaced by the experimentalist who moved the larger masses of a Cavendish balance depending on the outcome of a quantum decay process has been performed by Page and Geilker (1981). Although their experiment makes sense only if one regards them as part of the apparatus in the box and does not believe that they reduce the wavefunction by their observation, it does illustrate the ease with which such a gedanken experiment could be made real.

Another possibility for classical gravity would be if one could somehow regard $G_{\mu\nu}$ as the eigenvalue of the tensor operator $T_{\mu\nu}$ Einstein's equation would then become

$$8\pi T_{\mu\nu}|\phi\rangle = G_{\mu\nu}|\phi\rangle. \tag{9}$$

Unfortunately this equation does not seem to make any sense. Firstly, the 10 operator components of $T_{\mu\nu}(x)$ do not commute with each other. Simultaneous eigenstates will not therefore exist. A more difficult problem is that $T_{\mu\nu}$ is a function of the classical field $g_{\mu\nu}$ and thus of $G_{\mu\nu}$. We thus have an eigenvalue equation in which the operator is a function of its own eigenvalue. This is certainly a non-standard 'eigenvalue' equation, and one which I strongly suspect has no solution.

Having disposed of classical gravity, one can ask whether there is an easy way out. In Newtonian gravity one could claim that the reason the 'Cavendish' experiment does not work is because the gravitational field measured by the Cavendish apparatus is really just a function of its sources. One is thus only measuring the quantum properties of the masses by means of the Cavendish balance. One can write the force on the Cavendish balance as

$$F = \sum_i \frac{G M_i m [x - X_i(t)]}{|x - X_i(t)|^3} \tag{10}$$

where $X_i(t)$ is the quantum operator representing the position of the ith mass, M_i, within the box. The gravitational field is thus just the intermediary by which one measures the quantum location of the large masses.

This position is one which could be held also for the electromagnetic field. Writing the electromagnetic field as

$$F_{\mu\nu}(x) = \int G_{\mu\nu}{}^{\alpha}(x, x') J_{\alpha}(x') \, d^4 x' \tag{11}$$

where J_α is the quantum operator for the source of the electromagnetic field and $G_{[\mu\nu]}{}^{\alpha}$ is the classical retarded Green's function for Maxwell's equations, the electromagnetic field now becomes simply a functional of the quantum current operators. One could argue that the electromagnetic field does not exist as a separate quantum entity. The quantum nature of the electromagnetic field is the result of the quantum nature of its sources alone.

Although this position may be tenable for the electromagnetic field, it suffers some difficulties even there. Consider a charged mass attached to a spring. When set into oscillation, the charge will 'radiate' and be damped in its motion. In particular, this damping will damp away the motion of the oscillator completely unless the oscillator is re-excited because of its interaction with the electromagnetic field or, in this view, because of its non-local electromagnetic interaction with the other current sources in the universe. In particular, the emission of electromagnetic radiation would reduce the commutator $[q, p]$ of the oscillator to zero if both q and p are not driven by other quantum systems. This implies that there must be a sufficient number of such other sources in the universe to maintain the quantum commutation relations of our oscillator.

This theory has an obvious affinity with the action at a distance theories of the electromagnetic interaction, as exemplified by, say, the Wheeler–Feynman (1945, 1948) formulation. In their formulation, one needed a universe which was everywhere sufficiently absorptive that every future-directed wave would be absorbed somewhere in the future. In this formulation, one needs a sufficient supply of sources in the past so that the quantum mechanics of our oscillator does not decay.

Although such a theory could probably be developed in more detail for electromagnetism, it seems highly unlikely to ever be a viable theory for gravitation (see Unruh 1976). Such a source theory for electromagnetism would eventually be entirely equivalent to conventional QED. The self-interaction of the gravitational field, and the importance of the metric in determining the kinematics of the rest of physics make such a quantum action-at-a-distance gravitational theory extremely difficult to formulate.

Let us now grant that if quantum matter is the source of the gravitational field, one must quantize gravity. Does it matter? Is there any way in which the quantum nature of gravity would ever manifest itself? After all, while the charge on the electron, the coupling to the electromagnetic field is of order 10^{-2}, its mass, which is its coupling to gravity is of order 10^{-22}. Does one not need to look at interactions where the interactions are of order unity (or in more conventional units 10^{19} GeV or 10^{-5} g) to see quantum gravity effects? To see for example the wave–particle duality of a graviton with atomic physics experiments would seem to be essentially impossible. And if it is impossible to see the effects of quantum gravity, can't it simply be ignored?

The answer seems to me to be 'No'. Although the coupling of an electron or an atom to gravity is indeed small, the same is not true for large aggregates of matter. The coupling to gravity of many atoms is cumulative. The coupling constant (i.e. the mass) of a neutron star to the gravitational field for example, is of the order of 10^{40} rather than 10^{-20}. One can propose an experiment to see the wave–particle duality of gravitons by use of neutron stars rather than atoms or electrons.

A traditional demonstration of the wave–particle duality for photons is given by the single slit diffraction experiment. On examining a photographic plate placed behind the slit, one finds individual localized grains of silver developed. Alternatively, if a photomultiplier is used, the photomultiplier emits distinct pulses of light, illustrating the particle nature of the light. On the other hand the density of silver grains, or the rate of firing of the phototube varies with position and shows interference phenomenon, demonstrating the wave-like nature of the light.

In our case, let us set up a neutron star vibrating in its fundamental quadrupole mode as a source of the gravitational waves. Instead of a single slit, we will use its

Kirkoff inverse, a spherical absorber of the gravity waves. In particular, we will place a black hole between the emitter and our detector of the gravitational waves (see figure 2).

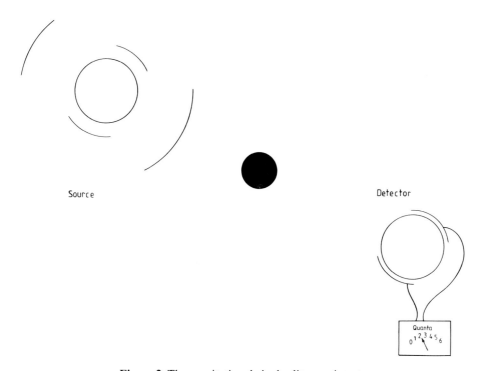

Figure 2 The gravitational single slit experiments.

Finally, we use another cold (0 K) neutron star as the detector. We initialize this detector so that the lowest quadrupole mode is in its ground state, and instrument the neutron star so as to be able to detect the number of quanta of vibrational energy in this lowest quadrupole mode. The technology is identical in principle with that suggested to perform quantum and non-demolition monitoring of current resonant bar gravity wave detectors, although the technology would need to make some advances in practice.

The reason for the use of a neutron star is that the quantum efficiency for capturing a 'graviton' is very high for a neutron star as compared with a Weber bar, or phrased another way, its cross section is of the order of its physical size. If we now watch this fundamental mode, we would expect the number of quanta in the star to change discontinuously at certain times. These discontinuous changes in the number of quanta would be interpreted as the absorption of a gravitational quantum by the star. On the other hand, one would expect that the rate at which these transitions would occur would be proportional to the intensity of the gravitational wave, and would demonstrate interference effects.

In addition to using the neutron star as a detector of gravitational radiation, we can also use it to see the vacuum fluctuations of the quantum gravitational field. The oscillation of the neutron star in its lowest quadrupole mode are rapidly damped by the emission of gravitational radiation (see for example Misner *et al* 1974, p 984). The Heisenberg equations of motion for the mode will have the form

$$M\ddot{Q} + M\Omega^2 Q = -\alpha\dot{Q} + F \tag{12}$$

where Q is the amplitude of oscillation of that lowest mode, M is its effective mass, and Ω is frequency. The damping by gravitational radiation contained in the factor α is large, giving a damping time of the order of a second. The homogeneous solution to this equation will thus have the initial Q_0 decay away on timescales of the order of a second. Unless F is a quantum force, due to the quantum 'vacuum fluctuations' of the gravitational field, the mode will lose its quantum nature, i.e. $[Q, P]$ will decay to zero on timescales of the order of a second. The commutator remains unity only because of the commutation relations of the forcing term $F(t)$.

The above arguments indicate that gravity must be quantized, and that furthermore the quantum nature of gravity can have effects even on macroscopic scales. It doesn't really give much of a clue as to how one must quantize the gravitational field. However, all of the above arguments would suggest that the radiative modes of gravity should be quantized in exactly the same way as for any other field. In particular, the radiative degrees of freedom are carried by the Weyl tensor $C_{\mu\nu\rho\sigma}$. The arguments therefore suggest that Weyl tensor should behave like any other radiative field. Unfortunately we know that all attempts at quantizing gravity in any standard fashion have failed. I will spend the rest of the paper describing a rather handwaving argument that the quantization may be very different from that for other fields.

Let us regard the metric $g_{\mu\nu}$ as a quantum field which we wish to measure. I will argue that the metric and the Einstein curvature obey an uncertainty principle. The usual approach regards the three-metric on a hypersurface and the extrinsic curvature of that surface as canonically conjugate variables. In arguing for my result, I will assume that the Einstein equations (equation (1)) are true as operator equation for $G_{\mu\nu}$.

The first question we must ask is 'Does the metric have any measurable meaning?'. After all, the metric is a tensor, the values of whose components depend on the particular coordinate system one is using. On the other hand, the metric does have an invariant definition as the object which gives the lengths between various points in the space–time.

I will take the attitude that the points of our space–time are not a given structure independent of the rest of physics. The points within the space–time must be specified in terms of the physical objects which make up the universe. In other words, space and time points must ultimately be defined intrinsically to the theory of matter, rather than as additional extrinsic structures.

We can now ask the question of how we can measure, or determine the distance between any two nearby points of our space–time. Let us use as our example, the attempt to measure g_{00}, the time-like metric component. This measurement corresponds to a comparison between the coordinate time difference δt and the proper time difference $\delta\tau$ between the same two events. The coordinate time difference δt is defined by the labeling of the matter field which defines our space–time points (such as the elastic medium used by DeWitt (1962a)). The proper time difference, $\delta\tau$, between these two points must be measured by means of some clock. We have

$$\delta\tau = \sqrt{g_{00}\delta t^2}. \tag{13}$$

The uncertainty, Δg_{00}, in the inferred value of g_{00} is related to the uncertainty, $\Delta\tau$, in the measured time $\delta\tau$ by

$$\Delta\tau \approx \Delta g_{00}\delta\tau \tag{14}$$

where I have assumed that I have chosen my time coordinate $\delta\tau$ such that the expected value of g_{00} is unity.

For any clock, however, the uncertainty in the time measured by that clock is related to the uncertainty in the energy of the clock by

$$\Delta E\Delta\tau \gtrsim 1 \tag{15}$$

(in units where $\hbar = c = G = 1$).

But the total energy E of the clock is given by

$$E \approx T^{00} V^{(3)} \tag{16}$$

where $V^{(3)}$ is the three-volume occupied by the clock. The uncertainty in the energy is therefore given by

$$\Delta E \approx \Delta T^{00} V^{(3)}. \tag{17}$$

We thus have from equation (15)

$$\Delta g_{00}\Delta T^{00} V^{(3)}\delta\tau \gtrsim 1. \tag{18}$$

or

$$\Delta g_{00}\Delta T^{00} V^{(4)} \gtrsim 1.$$

where $V^{(4)}$ is the four-volume occupied by the clock during the experiment. Using Einstein's equation, we obtain

$$\Delta g_{00}\Delta G^{00} \gtrsim 1. \tag{19}$$

A similar argument can be applied to an attempt to measure the spatial distance between two points. In that case, the attempt to measure the distance between two objects defining the coordinate system leads to uncertainties in their relative momenta. The momentum uncertainties come about because of the uncertainties in the stresses within the measuring apparatus. One finds in this case an uncertainty relation of the form

$$\Delta g_{ii}\Delta G^{ii} V^{(4)} \gtrsim 1. \tag{20}$$

In general, uncertainty relations arise from commutation relations between the operators of the theory. The problem is that the above commutation relations are not sufficiently general to give a unique set of commentators. However, if we demand coordinate invariance, and a c-number commutation relation, we have

$$[g_{\mu\nu}(x), G^{\rho\sigma}(x')] = \alpha\delta_\mu{}^\rho\delta_\nu{}^\sigma\delta^{(4)}(x, x').\tag{21}$$

These uncertainty relations are rather unusual. In the ordinary Hamiltonian procedure, one finds commutation relations between the dynamical variable (such as the spatial metric) and its first time derivative. Our relations arose out of the demand for internal consistency, that the lengths as defined by the metric be related to the lengths as defined by physical clocks and measuring rods. In all other quantized theories, one can assume that the space–time manifold structure can be defined by some matter which does not interact with the field of interest. It is precisely the universality of gravity which prevents us from doing this for the gravitational field.

Although we are at this point left far from a theory of quantum gravity, the above considerations do suggest that such a theory may have a structure radically different from that for any other theory.

ACKNOWLEDGMENTS

I would like to thank the NSERC of Canada for partial support of this research. I would also like to thank the Center for Theoretical Physics and J A Wheeler at the University of Texas at Austin where some of the ideas were first developed.

REFERENCES

De Witt B S 1962a in *Gravitation: An Introduction to Current Research* ed L Witten (New York: Wiley) p 266
—— 1962b *J. Math. Phys.* **3** 619
—— 1963 in *Proceedings of the Eastern Theoretical Conference* ed M E Rose (New York: Gordon and Breach) p 355
—— 1967a *Phys. Rev.* **160** 1113
—— 1967b *Phys. Rev.* **162** 1195, 1239
Eppley K and Hannah E 1977 *Found. Phys.* **7** 51
Misner C W, Thorne K S and Wheeler J A 1973 *Gravitation* (San Francisco: W H Freeman)
Moller C 1962 in *Les Theories Relativistes de la Gravitation* ed A Lichnerowicz and M A Tonnelat (Paris: CNRS)
Page D N and Geilker C D 1981 *Phys. Rev. Lett.* **47** 979
Rosenfeld L 1963 *Nucl. Phys.* **40** 353.
Unruh W G 1976 *Proc. R. Soc.* A **348** 447
Wheeler J A and Feynman R P 1945 *Rev. Mod. Phys.* **17** 157
—— 1948 *Rev. Mod. Phys.* **21** 425

Quantum Gravity at Small Distances

M R BROWN

It is a function, perhaps the most important function, of a quantum theory of gravitation to provide a description of the small distance behavior of space–time. In classical, non-quantum, physics this description is provided by General Relativity: Einstein's equations may be solved to obtain the metric tensor, $g_{ab}(x)$, at all points x of the space–time. For nearby points, x and x_1, coordinates can be chosen so that

$$g_{ab}(x) = \eta_{ab}(x_1) + \tfrac{1}{4} R_{abcd}(x_1)(x - x_1)^c (x - x_1)^d + 0\left(|x - x_1|^3\right) \tag{1}$$

where η_{ab} is the usual flat Minkowski metric and R_{abcd} is the curvature tensor. In other words, space–time is locally flat and, moreover, this limit is approached smoothly. In this article we wish to ask how this description might be different when the gravitational field is quantized.

In DeWitt's (1967) effective action formulation of quantum gravity the role of Einstein's equations is taken over by the effective action field equations; these can, it is supposed, be solved to obtain an effective metric $\langle \hat{g}_{ab}(x) \rangle$. There must be a correspondence with a given classical solution to Einstein's equations, $g_{ab}(x)$, in the sense that

$$\langle \hat{g}_{ab}(x) \rangle \rightarrow g_{ab}(x) \qquad \text{as } \hbar \rightarrow 0.$$

A solution to the effective action field equations can be thought of as providing an interpolation for the metric between prescribed values on initial and final hypersurfaces. This endpoint characterization of the boundary conditions is both necessary and natural for the quantum theory: it avoids the difficulty of the non-commutativity of the field operator and its time derivative on a prescribed initial hypersurface. Of course, a solution to the classical Einstein equations can similarly be described and it is most useful for the interpretation of the quantum theory to maintain this parallel discussion.

A precise correspondence with a classical solution, $g_{ab}(x)$, is set up by identifying the metric operator eigenvalues on the initial and final hypersurfaces with the values of the classical solution on the same surfaces, $g_{ab}(x_1)$ and $g_{ab}(x_2)$. This identification is realized by the solution

$$\langle \hat{g}_{ab}(x) \rangle = \frac{\langle g_{ab}(x_2), x_2 | \hat{g}_{ab}(x) | g_{ab}(x_1), x_1 \rangle}{\langle g_{ab}(x_2), x_2 | g_{ab}(x_1), x_1 \rangle} \tag{2}$$

where $|g_{ab}(x_1), x_1 \rangle$ is the eigenstate of $\hat{g}_{ab}(x_1)$ with eigenvalue $g_{ab}(x_1)$. (Strictly speaking we should phrase this discussion in terms of the dynamical three-metric, but for our immediate purposes this would be an irrelevant nicety.)

It is common, when dealing with the effective action theory, to require the initial and final hypersurfaces to be located in the remote past and future respectively (supposing such regions to exist); and, rather than work with field eigenstates, to work with energy eigenstates, most notably the vacuum. By so doing one is either trying to formulate a scattering theory or one might hope to obtain an effective metric that reflected these global properties of the background, classical space–time. Neither of these alternatives is our intention in this article. Our problem is nothing if it is not local. When we say that we wish to understand the small distance behavior of space–time we have in mind the following:

(a) At any one point, x_1, in the space–time we may suppose the metric to be known: the metric state is an eigenstate of the metric operator with eigenvalue $g_{ab}(x_1)$.

(b) We are interested in comparing solutions to the effective action field equations with the corresponding solutions to the classical Einstein equations. Thus we must calculate

$$\langle \hat{g}_{ab}(x) \rangle = \frac{\langle g_{ab}(x_2), x_2 | \hat{g}_{ab}(x) | g_{ab}(x_1), x_1 \rangle}{\langle g_{ab}(x_2), x_2 | g_{ab}(x_1), x_1 \rangle}$$

for some $g_{ab}(x_2)$ of interest.

(c) Finally we ask how $\langle \hat{g}_{ab}(x) \rangle$ behaves for x close to x_1.

We might reasonably expect there to be two distinguishable features of the solution: classical, space–time curvature effects as represented, for example, by equation (1); and quantum-mechanical effects ($\hbar \neq 0$). The latter may manifest themselves as additional curvature effects, indeed they would do so in a semiclassical approximation, but we must allow for the eventuality that it might not be possible so to describe them.

Of particular interest for quantum field theories other than gravity is the solution where the classical space–time curvature effects are absent or, at least, negligible. The ultraviolet divergences of these quantum theories are particularly sensitive to the small distance behavior of space–time (Brown 1981). If the often expressed hope that quantum gravity will provide an effective ultraviolet cut-off (Isham *et al* 1972) is to be realised then, in terms of the formulation presented here, it must be that $\langle \hat{g}_{ab}(x) \rangle$ cannot be represented as a Taylor series in a neighborhood of the point x_1; it must exhibit some sort of non-analytic behavior (Brown 1981). Is this what happens?

Of course, it is difficult to answer this question for the gravitational field and we certainly do not do so here. Instead, we argue, by analogy with quantum mechanics, that the answer is likely to be yes. If this is so then it has important consequences for the approximation of the exact theory of quantum gravity: not least that the familiar multi-loop expansion of the path integral is very likely to be incapable of ever yielding the desired non-analytic behavior, however many loops are calculated. Indeed, this loop expansion seems altogether ill suited to a determination of the small distance behavior of space–time: at each level of approximation it assumes a smooth form for the metric; it is a semiclassical expansion, its purpose is to average over many fluctuations. The crucial point is that as any quantum field theory appears to be ultraviolet divergent because one supposes space–time to be locally, smoothly flat, so it is with gravity itself. Thus it may be that quantum gravity appears to be infinite only because we (unnecessarily) force it to be so.

To see how these non-analytic effects might come about and, in any event, to improve one's feeling for effective field equations we now describe an exactly soluble, quantum-mechanical example. We begin with a quick resumé of the standard construction of the effective action and, in passing, prove what we have hitherto assumed, that $\langle \hat{g}_{ab}(x) \rangle$ (equation (2)) is a solution to its field equations.

$\Gamma\{J\}$ is the generating functional for the Green's functions of the quantum theory that is derived from a classical action $S\{x\}$; it is given by the formal expression

$$\exp\left(\mathrm{i}/\hbar\right)\Gamma\{J\} = \langle x_2, t_2 | x_1, t_1 \rangle_J$$

$$= \int_{x(t_1) = x_1}^{x(t_2) = x_2} \mathrm{D}\{x(t)\} \exp\left(\mathrm{i}/\hbar\right)(S\{x\} + J \cdot x) \tag{3}$$

where $J \cdot x$ is a convenient shorthand for the more explicit

$$\int_{t_1}^{t_2} J(t)x(t) \, \mathrm{d}t.$$

Functional differentiation of equation (3) allows the definition of the field $X\{J\}$:

$$X\{J\} \equiv \frac{\delta\Gamma}{\delta J} = \frac{\langle x_2, t_2 | \hat{x}(t) | x_1, t_1 \rangle_J}{\langle x_2, t_2 | x_1, t_1 \rangle_J}. \tag{4}$$

The effective action $W\{X\}$ is defined by the equation

$$W\{X\} \equiv \Gamma\{J\} - J \cdot X \tag{5}$$

where J is now considered to be a functional of X by equation (4), namely

$$X\{J\} = \frac{\delta\Gamma\{J\}}{\delta J} \Leftrightarrow J = J\{X\}.$$

This definition of W has the important property that

$$\frac{\delta W}{\delta X} = \left(\frac{\delta\Gamma}{\delta J} - X\right) \cdot \frac{\delta J}{\delta X} - J = -J. \tag{6}$$

Thus the effective action field equations, $W'\{X\} = 0$, imply that $J = 0$ and have the solution

$$X\{0\} = \frac{\langle x_2, t_2 | \hat{x}(t) | x_1, t_1 \rangle}{\langle x_2, t_2 | x_1, t_1 \rangle}. \tag{7}$$

The relation of $W\{X\}$ to $S\{X\}$ can be made more apparent by translating the variable of integration, $x(t)$, in equation (3). If we let

$$x(t) = X\{J\} + y(t) \tag{8}$$

then

$$\exp(i/\hbar)\,(W\{X\} - S\{X\}) = \int_{y(t_1)=0}^{y(t_2)=0} D\{y(t)\}\exp(i/\hbar)\,((S'\{X\} + J)y$$

$$+ \tfrac{1}{2}yS''\{X\}y + \ldots). \tag{9}$$

Notice that, if $S\{X\}$ is simply a quadratic functional of its argument then, up to an inessential constant, $W\{X\}$ is the same as $S\{X\}$. In this case $J = -S'\{X\}$.

For more general actions one can generate the loop expansion for W by supposing a solution of the form

$$W = S\{X\} + \hbar W_1\{X\} + 0(\hbar^2). \tag{10}$$

Equation (6) implies that

$$-J = S'\{X\} + \hbar W'_1\{X\} + 0(\hbar^2)$$

and W_1 is determined by equation (9) in terms of the distribution $S''\{X\}$. Here, notice that one is forcing onto W the form of equation (10); this form is not required by the implicit definition of W through equations (4) and (5).

These remarks concerning W apply equally well to any field theory; from now on we deal exclusively with quantum mechanics. Quantum mechanics has the massive advantage over quantum field theory in that the amplitude $\langle x_2, t_2 | x_1, t_1 \rangle$ satisfies a partial, and not functional, differential equation, namely the inhomogeneous Schrödinger equation,

$$\left\{ i\hbar\frac{\partial}{\partial t_2} + \frac{\hbar^2}{2m} \times \frac{\partial^2}{\partial x_2^2} - V(x_2) \right\} \langle x_2, t_2 | x_1, t_1 \rangle = i\hbar\delta(x_2 - x_1)\delta(t_2 - t_1). \tag{11}$$

The identification of equation (11) with quantities appearing in equation (3) is made precise by specifying the action functional $S\{x\}$:

$$S\{x\} = \int_{t_1}^{t_2} dt\,(\tfrac{1}{2}m\dot{x}^2 - V(x)) \tag{12}$$

and the field $x(t)$ is constrained to satisfy $x(t_2) = x_2$, $x(t_1) = x_1$. The existence of equation (11) and the relative ease of its solution probably explains why it is not usual to formulate quantum mechanics in terms of an effective action: information is more easily extracted directly from the Schrödinger equation. In particular, once equation (11) has been solved, it is an easy matter to construct the solution to the effective action equations of motion. Let us, following Feynman and Hibbs (1965), denote the amplitude $\langle x_2, t_2 | x_1, t_1 \rangle$ by $K(x_2, t_2; x_1, t_1)$; then we have

$$\langle x_2,t_2|\hat{x}(t)|x_1,t_1\rangle = \int_{-\infty}^{+\infty} dx \,\langle x_2,t_2|\hat{x}(t)|x,t\rangle\langle x,t|x_1,t_1\rangle$$

$$= \int_{-\infty}^{+\infty} dx \,K(x_2,t_2;x,t)xK(x,t;x_1,t_1). \tag{12}$$

The validity of this equation rests upon the representation of the unit operator in terms of a complete set of position eigenstates at an arbitrary time t:

$$\hat{1} = \int_{-\infty}^{+\infty} dx|x,t\rangle\langle x,t|. \tag{13}$$

In what follows we shall primarily concern ourselves with $\langle\hat{x}(t)\rangle$ as given by equation (12); the effective action itself remains difficult to compute exactly.

At this point it is worth noting that the leading order WKB approximation to K is given by

$$K(x_2,t_2;x_1,t_1) \sim \left(\frac{im}{2\pi\hbar} \times \frac{\partial^2 S}{\partial x_1 \partial x_2}\right)^{1/2} \exp\,(i/\hbar)S(x_2,t_2;x_1,t_1) \tag{14}$$

where $S(x_2,t_2;x_1,t_1)$ is the action of equation (12), now evaluated along the trajectory, $X(t)$, that is the solution to the classical equations of motion $S'(X) = 0$, subject to the boundary conditions: $X(t_1) = x_1, X(t_2) = x_2$.

If there is more than one path satisfying these requirements then by $X(t)$ we mean the one with the least action. It is a pretty and instructive exercise to verify that the substitution of equation (14) into equation (12) does indeed yield, to zero'th order in Planck's constant, the classical solution $X(t)$. This provides a rigorous justification of the formal manipulations that led to equation (10). It is also the analogue of Ehrenfest's theorem (Ehrenfest 1927) for the function $\langle\hat{x}(t)\rangle$; Ehrenfest's theorem applied originally to expectation values or, equivalently, wave packets.

At last we come to our example. We must choose a potential, other than a quadratic one, so that equation (11) is exactly soluble. Our choice is $V = \alpha x^{-2}$, where α is a positive constant. The classical trajectory passing through (x_1, t_1) and (x_2, t_2) is the hyperbola

$$mEx^2 - 2E^2(t - \bar{t})^2 = m\alpha \tag{15}$$

where E, the energy, is given by

$$2E(t_2 - t_1)^2 = m(x_2{}^2 + x_1{}^2) - 2\sqrt{[m^2 x_2{}^2 x_1{}^2 - 2m\alpha(t_2 - t_1)^2]} \tag{16}$$

and

$$\bar{t} = \tfrac{1}{2}(t_2 + t_1) - \frac{m(x_2{}^2 - x_1{}^2)}{4E(t_2 - t_1)}. \tag{17}$$

The classical action $S(x_2, t_2; x_1, t_1)$ is given by

$$S(x_2,t_2;x_1,t_1) = \tfrac{1}{2}\, m(x_2{}^2 + x_1{}^2)(t_2 - t_1)^{-1} + (2m\alpha)^{1/2}(\tan^{-1}\beta - \beta - \tfrac{1}{2}\pi)$$

where

$$\beta\sqrt{(2\alpha)}(t_2 - t_1) = \{mx_2{}^2 x_1{}^2 - 2\alpha(t_2 - t_1)^2\}^{1/2}.$$

Equation (11) can be solved to give K in terms of the Hankel function H_ν^2, namely,

$$K = \exp -\tfrac{1}{2}i\pi(\nu + 1)\,\frac{m(x_1 x_2)^{1/2}}{2\hbar(t_2 - t_1)} \times \exp\frac{im(x_1{}^2 + x_2{}^2)}{2\hbar(t_2 - t_1)}$$

$$\times H_\nu^2\left(\frac{mx_1 x_2}{\hbar(t_2 - t_1)}\right) \times \theta(t_2 - t_1) \tag{18}$$

where

$$\nu^2 = \tfrac{1}{4}(1 + 8m\alpha/\hbar^2) \qquad \nu > 0.$$

We note that switching off the potential (setting $\alpha = 0$) correctly reproduces the free propagator K_0,

$$K_0 = \{m^{-1} 2\pi i\hbar(t_2 - t_1)\}^{-1/2} \exp\frac{im(x_2 - x_1)^2}{2\hbar(t_2 - t_1)} \times \theta(t_2 - t_1). \tag{19}$$

One could now proceed with the general expression, equation (18), but the necessary integrations of products of Hankel functions are, at best, cumbersome; it is better to choose a special value for α. We choose the simplest, non-zero value, $\alpha = \hbar^2 m^{-1}$. This implies that $\nu = \tfrac{3}{2}$ and that, for this value,

$$K_{3/2} = \{1 - i\hbar(t_2 - t_1)(mx_1 x_2)^{-1}\} \times K_0. \tag{20}$$

It can easily be verified that $K_{3/2}$ satisfies the reproductive property

$$\int_{-\infty}^{+\infty} dx \times K_{3/2}(x_2,t_2;x,t)K_{3/2}(x,t;x_1,t_1) = K_{3/2}(x_2,t_2;x_1,t_1). \tag{21}$$

Despite all attempts at simplification, the effective trajectory, $X\{0\}$ (equation (7)), remains a little complicated and is given by

$$X\{0\} = X(t) = \frac{\langle x_2,t_2|\hat{x}(t)|x_1,t_1\rangle}{\langle x_2,t_2|x_1,t_1\rangle}$$

$$= \bar{x}(t) + i\pi \times \frac{\hbar^2\{m(t_2 - t_1)(t_2 - t)(t - t_1)\}^{1/2}}{(2i\pi\hbar)^{1/2} m\{mx_1 x_2 - i\hbar(t_2 - t_1)\}} \times \exp i\lambda\bar{x}^2 \times \Phi\{\bar{x}(i\lambda)^{1/2}\} \tag{22}$$

where

$$\bar{x} \equiv \frac{x_2(t - t_1) + x_1(t_2 - t)}{(t_2 - t_1)}$$

$$\lambda \equiv \frac{m(t_2 - t_1)}{2\hbar(t_2 - t)(t - t_1)}$$

and

$$\Phi\{\bar{x}(i\lambda)^{1/2}\} = (2i)^{1/2}\{C(\bar{x}\lambda^{1/2}) - iS(\bar{x}\lambda^{1/2})\}. \tag{23}$$

The function Φ is known as 'the probability integral' (Gradshteyn 1965). Equation (23) represents it in terms of Fresnel integrals, C and S.

It is easy to verify that this solution for $X(t)$ satisfies the correct boundary conditions at the endpoints and that, in the limit $\hbar = 0$, it correctly reduces to the free trajectory $\bar{x}(t)$, remember that V has been chosen to be proportional to $\hbar^2$. Beyond these simple observations, the interpretation of equation (22) remains obscure. It is more revealing to study the small $\hbar$ behavior ($\hbar(t_2 - t_1) \ll mx_1x_2$). This may easily be obtained from known properties of Φ and is given by the equation,

$$X(t) = \bar{x}(t) - \hbar^2 \frac{(t_2 - t)(t - t_1)}{m^2 x_1 x_2} \left[\bar{x}(t)^{-1} - \sqrt{(\tfrac{1}{2}i\pi\lambda)} \times \exp i\lambda\bar{x}(t)^2\right] + \ldots \tag{24}$$

To the same order in Planck's constant the classical trajectory (equation (15)) is given by

$$x(t) = \bar{x}(t) - \frac{\hbar^2(t_2 - t)(t - t_1)}{m^2 x_1 x_2 \bar{x}(t)} + 0(\hbar^3). \tag{25}$$

In the limit as $\hbar \to 0$ equation (24) becomes

$$X(t) = \bar{x}(t) - \frac{\hbar^2(t_2 - t)(t - t_1)}{m^2 x_1 x_2} \left[\bar{x}(t)^{-1} - i\pi\delta(\bar{x}(t))\right]. \tag{26}$$

This follows from equation (24) by identifying the final term in that equation as a delta convergent series (Gel'fand and Shilov 1964). It is of interest to rewrite equation (26) in the form

$$X(t) = \bar{x}(t) - \frac{\hbar^2(t_2 - t)(t - t_1)}{m^2 x_1 x_2 (\bar{x} + i\varepsilon)} \tag{27}$$

where we have made use of the distributional equation

$$\lim_{\varepsilon \to 0} (\bar{x} + i\varepsilon)^{-1} = \bar{x}^{-1} - i\pi\delta(\bar{x}). \tag{28}$$

Now one can clearly see the effect that the introduction of quantum mechanics has had on the classical trajectory: by equation (24) it is apparent that the interacting part of the trajectory is 'smeared'; moreover that $X(t)$ is not an analytic function of $\hbar$ (anywhere) and of t (at the endpoints). There does not exist a Taylor series expansion in $\hbar$ or in $(t - t_1)$ and it does not make sense to talk of the curvature of the effective trajectory at t_1. By equation (27) it is clear that the averaged effect is to soften the potential. Indeed, although it is not strictly legitimate under the conditions of the approximation that led to equation (24) it is nonetheless interesting that equation (27) does allow trajectories to pass through the origin, $\bar{x} = 0$, by dipping into the complex.

One should probably have expected the non-analytic behavior of the solution at the endpoints, in that the position operator is in an eigenstate there. By the Uncertainty Principle the value of the momentum is undetermined at that instant. However it is interesting that this uncertainty only manifests itself in the interacting (non-quadratic) theory—the gravitational action, one should remember, is not quadratic in the field operators.

The parallels with quantum gravity are self-evident: one might reasonably hope for a similar picture with the effective metric being distributed about its classical value, with the oscillations having characteristic Planck length dimensions. The correspondence principle ensures that our classical description of space–time is appropriate to such relatively large bodies as ourselves—but quantum field theory must be allowed to do its own averaging over this Planck length structure. Gravity may, in this way, provide an ultraviolet cut-off. However, it should be stressed that one's problems do not necessarily end there: all that will have happened, if it happens at all, is that a previously divergent quantum field theory is now regulated, albeit physically, by the introduction of the Planck length. This, in itself, does not ensure that the theory provides a good description of nature; the criterion of renormalizability (now finite renormalizations) may still be needed to sift the physical from the unphysical theories.

An important advance that needs to be made in quantum gravity is the development of an approximation that correctly, and consistently, predicts the small distance behavior of the theory. We feel that it could be instructive first to develop this approximation for quantum mechanics. We hope that the exactly soluble example given here may prove useful in this respect; if not, then at least enjoyable.

REFERENCES

Brown M R 1981 *Quantum Gravity* 2 ed C J Isham, R Penrose and D W Sciama (Oxford: Oxford University Press)
DeWitt B 1967 *Phys. Rev.* **162** 1195
Ehrenfest P 1927 *Z. Phys.* **45** 455
Feynman R P and Hibbs A R 1965 *Quantum Mechanics and Path Integrals* (New York: McGraw-Hill)
Gel'fand I M and Shilov G E 1964 *Generalized Functions* vol 1 (New York: Academic)
Gradshteyn I S and Ryshik I M 1965 *Table of Integrals, Series and Products* (New York: Academic)
Isham C J, Salam A and Strathdee J 1971 *Phys. Rev.* D **3** 967

Renormalization and Asymptotic Freedom in Quantum Gravity

E T TOMBOULIS[†]

1. INTRODUCTION

Classical general relativity accounts very well for all macroscopic gravitational phenomena. Attempts to quantize the theory, however, encounter serious difficulties. The Einstein–Hilbert action defines a perturbatively non-renormalizable theory. This is indeed verified by many explicit calculations of Feynman graphs. These calculations originate in the work of Feynman (1963), and DeWitt (1965, 1967a,b,c) in the sixties. DeWitt's series of seminar papers established quantum gravity as a modern area of research in quantum field theory. In fact, the importance of this work extends well beyond any narrowly defined area of 'quantum gravity'. It involved significant contributions to the quantization of general non-Abelian fields, which, unfortunately, were not fully appreciated at that time.

In this article we will review some recent attempts to construct satisfactory theories of quantum gravity within the framework of local, continuum field theory. These attempts are in part motivated by the deeper understanding of renormalization afforded by the modern formulation of the 'renormalization group' (for a review and references see Wilson and Kogut 1974). We will be led to consider renormalizable theories with actions that involve terms quadratic in the Riemann tensor. We will argue that some class among these are perhaps the only actions which, by virtue of their ultraviolet stability, may give well defined, continuous, i.e. truly cut-off independent, field theories of quantum gravity. It should be stressed that our discussion will be based on the assumption that, upon quantization of any local field-theoretic formulation of gravity, we will encounter infinities. There would obviously be no problem if we had some *completely* finite quantum theory.[‡] It just might be that

[†] A P Sloan Foundation fellow.

[‡] Recently, there has been progress in the theory of relativistic extended objects, i.e. (super)strings. This approach, which at the present time is still at a rather formal stage of development, might eventually produce finite and sensible theories of quantum gravity. (For a review see Schwarz 1982.)

some version of a supergravity theory with a finite S-matrix exists. However, at the present time, this appears an unlikely prospect. We will not, therefore, explicitly include supersymmetry in our considerations. The theories we will examine, however, do admit supersymmetric extensions.

2. THE TROUBLES OF QUANTUM GRAVITY

To the quantum field theorist it is immediately apparent that quantization of the Einstein–Hilbert action:

$$S_E = \int d^4 x \sqrt{-g} \, (K^{-2}R - \Lambda) \tag{2.1}$$

will present problems. Indeed equation (2.1) is a perturbatively non-renormalizable Lagrangian by naive power counting. This is because K, the coupling constant of the theory, has dimensions[†] of $[\text{mass}]^{(2-d)/2}$ at space–time dimensionality d. As it is well known, theories with couplings with dimensions of a negative power of mass are non-renormalizable. In the case of equation (2.1) at $d = 4$, one finds that diagrams with L loops have a degree of divergence equal to $2(L+1)$, and hence divergent parts proportional to $(L+1)$ powers of the Riemann tensor. To cancel divergences to arbitrary loop order one, therefore, needs an infinite number of counterterms, and hence an infinite number of arbitrary parameters.

There is another feature of equation (2.1) which is also immediately apparent: it is not the most general fourth-order Lagrangian allowed by the symmetries of the theory. It is a well known fact that quantum corrections will induce all the terms that occur in the most general allowed Lagrangian. These corrections are, in general, infinite, and have to be renormalized. Since renormalizations are always arbitrary by finite amounts, it is clear that not to include all the allowed terms in the 'bare' Lagrangian is not a meaningful procedure in quantum field theory. If we include all possible fourth-order terms, we obtain an action of the form:[‡]

$$S_{\text{grav.}} = \int d^4 x \sqrt{-g} \, [-\Lambda + K^{-2}R - a(R_{\mu\nu}^2 - \tfrac{1}{3}R^2) + bR^2] \tag{2.2}$$

This action is renormalizable by power counting (DeWitt 1965), and perturbative renormalizability can actually be proven to all orders (Stelle 1977). Equation (2.2) contains fourth-order derivatives and, therefore, has traditionally been rejected, since such theories are generally believed to be sick. Indeed, classically equation (2.2)

[†] We use natural units, $\hbar = c = 1$, so the only dimensional quantity is mass $= (\text{length})^{-1}$.

[‡] Such actions have a long history. They were introduced soon after Einstein's formulation of general relativity as possible generalizations. (See Pauli (1919), Weyl (1921), Eddington (1924).) Other possible invariants quadratic in the Riemann tensor can be expressed, by the Gauss Bonnet theorem, as linear combinations of the ones included in equation (2.2) plus total derivatives.

contains negative energies. In the quantum theory negative energies can be traded off for positive energies but indefinite metric, and, as a consequence, possible loss of physical unitarity. This is already evident in the tree-level propagators in the linearized version of equation (2.2): there is a massive spin-2 ghost pole near the Planck mass. However, this may be a hasty conclusion. The spectrum, and/or the consistent isolation of physical subspaces, in the exact, interacting quantum theory are dynamical questions, the answers to which can be quite different from those given by the linearized theory. Nevertheless, we clearly should expect some problems associated with the higher derivatives. What we wish to emphasize at this point is that, at *a non-perturbative level*, Einstein gravity (2.1) must have similar, and in fact much more severe problems of physical interpretation: at short distances it becomes strongly coupled and full of unphysical singularities and terrible acausalities.[†]

The dimensional coupling and the resulting non-renormalizability of the Einstein–Hilbert action (2.1), the occurrence of fourth-order derivatives in the general fourth-order renormalizable action (2.2), and the related unitarity and causality problems alluded to in the last paragraph, can all be traced to the simple fact that the gravitational field $g_{\mu\nu}$ is dimensionless. This is because $g_{\mu\nu}$ serves as the space–time metric, and, therefore, appears as a special field having features not shared by any other physical field currently employed in physics. As a consequence, the usual connection between Minkowski and Euclidean field theory, as formulated in modern constructive field theory, seems to be disrupted. Related difficulties become evident when one attempts to devise some *non-perturbative*, gauge-invariance preserving regularization scheme for (Euclidean) gravity, such as the lattice formulation of ordinary gauge theories. Such a scheme is, of course, a necessary prerequisite for any really rigorous considerations in field theory.[‡] Attempts to construct lattice formulations of gravity have not been particularly fruitful. However, one thing is apparent. Whether one uses a 'passive' flat lattice (Smolin 1979a, Das *et al* 1979), or a physically more appropriate construction with a dynamical lattice, such as Regge's calculus (Regge 1961), one ends up with complicated, rather ugly looking actions which do not possess any of the nice features of ordinary lattice gauge theories. In particular, such actions generally are not reflection positive. Among those of the Osterwalder–Shrader axioms that pertain to the physical interpretation of field theory, reflection positivity is the most important one. It is this property that guarantees the existence of a positive tranfer matrix in Euclidean space, and, equivalently, a positive Hilbert space in Minkowski space (see for example Glimm and Jaffe 1981). It should come as no surprise then that, at a non-perturbative level, unitarity and causality problems appear to be generic to all local field theory formulations of quantum gravity. It may well be that no such formulation can be fully unitary and causal in the conventional sense. This must be kept in mind when one attempts, as in this article, to discuss quantum gravity

[†] This can be seen, e.g., in 'space–time foam' calculations (Hawking *et al* 1979).
[‡] Also from a physical point of view, the indispensible role played by the lattice formulation in elucidating non-perturbative phenomena such as confinement, or, more generally, the rich phase structure of various gauge theories, need hardly be pointed out.

while staying, as far as possible, within the framework of ordinary relativistic field theory.

In recent years there has been considerable progress towards a deeper understanding of field theory. Central to this development is the notion of the renormalization group as the general language for describing the construction of effective theories at different energy scales, and the emergence of critical behavior. In selecting possibly viable theories of quantum gravity, we will be guided in the following by renormalization group considerations concerning the existence of truly cut-off independent field theories.

3.   THE RENORMALIZATION GROUP (RG) AND ITS FIXED POINTS

Let us recall some of the elementary physical ideas underlying the RG (Wilson and Kogut 1974, Brezin *et al* 1975). Suppose that we consider a system characterized by some fundamental length a, or, equivalently, momentum cut-off Λ. For example, in solid state physics a could be the spacing of an actual physical lattice. Suppose the system at this length scale is described by an action or 'Hamiltonian' $H(a)$. The RG is the group of transformations that relate $H(a)$ to the effective Hamiltonian $H(t)$ describing the theory at length scale $e^t a$. The basic physical idea behind this view of the RG, as formulated by Wilson, is Kadanoff's 'thinning out' of the degrees of freedom of the system. This is achieved by 'integrating out' the fluctuations of the dynamical variables with wavenumber larger than e^{-t}/a. The new effective $H(t)$ will, in general, be more complicated than the original $H(a)$, and contain more interaction terms. However, its parameters will be determined by the parameters of $H(a)$. The evolution of the system under successive such transformations may then be viewed as a trajectory in the space S of all (cut-off) interactions given in dimensionless form. The elements in S can be partitioned into subspaces consisting of all H with given dimensionless correlation length ξ. The critical subspace S_∞ is the one with $\xi = \infty$. A fixed point defines a trivial trajectory $H(t) = H^*$, all t, and hence must have a correlation length $\xi^* = 0$ or $\xi^* = \infty$. It clearly represents critical behavior of the system. Note that growing t corresponds to flow towards the infrared (IR) regime along the RG trajectory. Running the RG 'backwards' gives the ultraviolet (UV) flow. The simplest behavior that one can obtain, and the one almost exclusively dealt with in physical applications, as $t \rightarrow (\pm)\infty$, is that a trajectory hits an IR (UV) fixed point. Associated with each fixed point H^* there is a domain $D(H^*)$ of interactions in S. $D(H^*)$ is the set of all initial interactions H_0 such that trajectories $H(t)$ starting from H_0 eventually hit H^*. In this way S is partitioned into domains, and universality holds separately within each domain. This is because the same critical behavior, determined by the corresponding fixed point, is obtained for all initial interactions within the given domain.

This is then the general framework of the modern theory of critical phenomena. We now turn to continuum field theory. In this case we want to totally remove any fundamental cut-off, i.e. take $a = 1/\Lambda \rightarrow 0$. This defines the problem of renormalization in continuum field theory which is seen to have two parts. The first part is to ensure that there is a finite theory in the limit of finite cut-off. The second part is to obtain a *non-trivial*, i.e. an interacting theory in this limit. It is this second part which presents a rather difficult problem. Conventional perturbative renormalization methods generally cannot say anything about the question of triviality.

A more explicit way of stating this is as follows. Renormalization conventionally means giving the 'bare' masses and coupling constants a cut-off dependence, so that the theory expressed in terms of a number of 'renormalized' parameters is cut-off independent. Let us denote the set of bare parameters by g_{0i}, with g_{0i} having dimensions of $[\text{mass}]^{\delta_i}$. It is then convenient to introduce dimensionless bare couplings $\bar{g}_{0i} \equiv \Lambda^{-\delta_i} g_{0i}$. Suppose the dimensionless correlation length of the theory is given by $\xi(\bar{g}_{0i})$. Introducing a renormalization mass μ that will define our physical units, we may choose our cut-off Λ so that

$$\mu\xi(\bar{g}_{0i}) = \Lambda.$$

The continuum limit is taken by letting $\Lambda \to \infty$ keeping some physical mass, in particular, μ, fixed. To achieve this we let $\bar{g}_{0i}$ be a function of Λ in such a way that $\Lambda/\xi[\bar{g}_{0i}]$ is independent of Λ for large Λ. We write

$$\bar{g}_{0i} = \bar{g}_{0i}\left(\frac{\Lambda}{\mu}, \{\bar{g}_j\}\right)$$

where g_j denotes the set of renormalized couplings, and $\bar{g}_j = \mu^{-\delta_j} g_j$ the corresponding dimensionless quantities. Then, as $\Lambda \to \infty$, $\xi[\bar{g}_{0i}] \to \infty$, i.e. in taking the continuum limit the theory is put on the critical subspace S_∞. The further evolution of the system is then determined by the fixed points H_∞^* on S_∞. In particular, the number of *free* renormalized parameters on which, in addition to μ, the theory depends, is determined by the nature of the fixed point on S_∞ in whose domain the system finds itself. In the neighborhood of a given fixed point, one can define a set of local axes that span the directions along which the system can evolve. The number of free renormalized parameters is equal to the number of directions of IR instability (number of IR relevant operators) which, of course, is the same as the number of directions of UV stability (number of UV irrelevant operators). All other values of $\bar{g}_i$, that are not free renormalized parameters, are determined by the given H_∞^*, *regardless* of the values of the corresponding $\bar{g}_{0i}$—as long as these $\bar{g}_{0i}$ are, of course, in $D(H_\infty^*)$. This is the field theoretic version of universality. The renormalized theory can be trivial if the only free parameter is μ, and all other renormalized couplings fixed to zero.

It is generally believed that the only theories that exist as continuum, interacting theories are those whose RG trajectories can reach S_∞, and which possess UV fixed points at finite couplings. Those that do not satisfy these conditions are believed to be trivial, or to exhibit unphysical singularities and non-localities. ϕ_d^4, $d \geq 4$, is an example of this general statement: the theory does possess a well defined continuum limit, but it is trivial, i.e. a generalized free field, in this limit. This became evident in some numerical work with the RG (Wilson and Kogut 1974), and is also seen in the

$1/N$ expansion (Coleman *et al* 1974). It has recently been proven rigorously (Aizenman 1981, Frohlich 1982). This is a modern restatement of the old argument of Landau and collaborators (Landau 1955). What these proofs show is that Landau's argument, though totally unjustified in its naive extrapolation of RG-improved weak coupling perturbation theory to all couplings, is qualitatively correct. This is because there is no UV fixed point in ϕ_4^4. In particular, Landau's formula for the renormalized coupling, in terms of the cut-off Λ and the bare charge, qualitatively reproduces the behavior of the exact theory: as $\Lambda \to \infty$, the renormalized coupling vanishes no matter what the value of the bare charge, or its dependence on Λ is chosen to be. Landau, of course, was more interested in QED than ϕ_4^4. There is little reason to doubt that ordinary, non-compact QED shows the same behavior. It also is a theory with an IR fixed point at the origin, and it is unlikely that it possesses a UV fixed point at some finite coupling. Note, however, that theories like ϕ_4^4 and QED, being *perturbatively renormalizable*, present problems[†] symptomatic of triviality only at exponentially high energies.[‡] Therefore, as long as they are cut-off at such energies, i.e. $\Lambda \gg 1$ but the actual limit $\Lambda \to \infty$ is not taken, they will behave as weakly coupled, interacting theories for which perturbative methods can be used. However, they will be very weakly cut-off dependent, the cut-off representing the exponentially high energy at which they have to be embedded in a more fundamental theory that solves the short distance problem.

4. FIXED POINTS AND DIMENSIONAL CONTINUATION IN GRAVITY

When we come to consider quantum gravity such a 'phenomenological' attitude towards renormalizability is physically unacceptable. This point seems to have been totally missed in most standard investigations of renormalization in quantum gravity and supergravity. A general statement of the requirement of UV fixed points in gravity has been made by Weinberg (1979), and also by Smolin (1979b). These authors, however, seem to be concentrating on the possible existence of UV fixed points away from the origin of the coupling constant space. Unfortunately, there does not seem to be any evidence for this in four dimensions. On the other hand, equation (2.2), or some conformal or supersymmetric versions of equation (2.2), are not only perturbatively renormalizable, but, as we will see, also possess an UV fixed point at the origin. This should be a compelling reason for considering such theories seriously— according to our considerations in the last section, we may then have a chance of producing an interacting, cut-off independent field theory of gravity. Of course, we will have to deal with problems created by the presence of fourth-order derivatives. However, as we argued before, these problems are generic to all field theories of quantum gravity, and can ultimately be traced to the basic fact that the quantized field serves as the space–time metric and is dimensionless.

---

† e.g. Landau ghosts.
‡ This would not be true in perturbatively non-renormalizable theories.

As it is well known, critical behavior, and hence the location and number of RG fixed points in S, depend sensitively on the space dimensionality d. It is therefore instructive to examine the behavior of gravitational actions (2.1), (2.2) as a function of d. K is dimensionless at $d = 2$, and hence equation (2.1) is renormalizable by power counting. Gravity is of course trivial at $d = 2$, but we may examine the theory† at $d = 2 + \varepsilon$, for small $\varepsilon > 0$ (Gastman *et al* 1978, Christensen and Duff 1978). It is found that the origin is an IR stable fixed point, and that there is an UV fixed point of $O(\varepsilon)$ with one direction of IR instability. This situation is closely analogous to that of the two-dimensional four-fermion interaction theory (Wilson 1973, Parisi 1975, Gross and Neveu 1974). Let us, therefore, review this simpler case. The Lagrangian is given by

$$L = \tfrac{1}{2}\bar{\psi}_i \not{\partial} \psi_i - \frac{g^2}{8}(\bar{\psi}_i\psi_i)^2 \tag{4.1}$$

where ψ is an N component spin-$\tfrac{1}{2}$ field, and g has dimensions of $[\text{mass}]^{(2-d)/2}$. At $d = 4$ the theory is non-renormalizable (the Fermi interaction), but at $d = 2$ it is renormalizable, and also asymptotically free. At $d = 2 + \varepsilon$, the UV fixed point moves a distance $O(\varepsilon)$ away from the origin, which becomes an IR fixed point. All this is then as in the gravitational case. For $d > 2$, equation (4.1) is no longer renormalizable within the ordinary loop expansion. However, the UV asymptotic behavior of all Green's functions is computable since it is controlled by the fixed point. Indeed, if we expand the theory in powers of $1/N$, then it appears renormalizable‡ within this expansion for $d < 4$. This is because the $1/N$ propagators are 'improved' by the summation of infinite sets of diagrams (the fermion self-energy bubbles to leading $1/N$ order) implied by the rearrangement in powers of $1/N$. The improved propagators show precisely the asymptotic scaling behavior implied by the existence of the UV fixed point. At $d = 4$ this breaks down—new divergences appear which cannot be absorbed by the renormalizations that were sufficient for $d < 4$. The UV fixed point does not survive at $d = 4$. It is not known whether *another* non-trivial UV fixed point exists at $d = 4$. Such a fixed point would help us make sense of the Fermi theory at $d = 4$, but there is no evidence that one exists.

In the explicit calculations it is convenient to rewrite equation (4.1) in terms of an auxiliary scalar field σ:

$$L = \tfrac{1}{2}\bar{\psi}_i \not{\partial} \psi_i - \tfrac{1}{2}\sigma\bar{\psi}_i\psi_i + \frac{1}{2g^2}\sigma^2. \tag{4.2}$$

Equation (4.2) is the Lagrangian of the Yukawa theory but without a kinetic energy and self-interaction terms for σ. Such terms will be induced in the effective quantum action, all with computable, finite coefficients for $d < 4$. At $d = 4$, one needs new,

---

† The dimensionally regularized quantum theory must be properly renormalized even for $\varepsilon \to 0$ because of the occurrence of poles.
‡ Wavefunction, coupling constant, and, if a bare mass is included, also mass renormalizations are needed.

infinite counterterms that correspond to precisely such bare terms; to obtain a renormalizable theory in $d = 4$, one must then extend equation (4.2) to the full Yukawa theory. This theory is superrenormalizable at $d < 4$, and has a finite σ-field renormalization constant Z_σ. It can be shown (Shizuya 1980) that imposing the 'compositeness' condition $Z_\sigma = 0$ makes the renormalized Green's functions of this Yukawa theory identical to those of the four-fermion theory (4.1)–(4.2)—the superrenormalizable Yukawa theory with the extra constraint turns into a renormalizable theory. In $d = 4$ one may also set up this correspondence in the presence of a cut-off Λ. However, the unconstrained Yukawa theory is now only renormalizable. If the constraint is imposed, and $\Lambda \to \infty$, one finds that the remaining renormalized parameters acquire cut-off dependence. This is a reflection of the non-renormalizability of the Fermi theory, and the disappearance of the UV fixed point present in lower dimensions. Note that at $d = 4$ the unconstrained, perturbatively renormalizable Yukawa theory has an IR fixed point at the origin. There is no guarantee that it exists as an interacting theory, and actually it is probably trivial.

Gravity is found to behave in a closely analogous manner, though the available calculations are not as detailed as in the four-fermion case (see for example Smolin 1982). One may use an $1/N$ expansion (see next section), with N the number of matter fields, so that ε need not be small. Again, we find that for $d < 4$ the Einstein theory equation (2.1) defines a renormalizable theory due to the presence of the UV fixed point. The analog of the Yukawa theory with its bare kinetic energy term, is now the Lagrangian equation (2.2), which contains the bare terms with the extra derivatives and is superrenormalizable for $d < 4$. At $d = 4$ the situation again changes drastically, and equation (2.4) is the only renormalizable theory. However, there is now a crucial difference from the four-fermion Yukawa case: the fixed point at the origin at $d = 4$ is UV stable.

5.   QUANTUM GRAVITY AT $d = 4$—THE $1/N$ EXPANSION—ASYMPTOTIC FREEDOM

We now have enough motivation to consider equation (2.2) in $d = 4$ as our fundamental gravitational Lagrangians. We will also add couplings to matter which we take in the form of N fermion and/or vector boson fields:

$$\mathscr{L} = \sqrt{-g}\,[-\Lambda + K^{-2}R - a(R_{\mu\nu}{}^2 - \tfrac{1}{3}R^2) + bR^2] + \mathscr{L}_{\text{matter}}(\psi, A, g) \quad (5.1)$$

As already mentioned, equation (5.1) is perturbatively renormalizable within the ordinary loop expansion. However, within this expansion, physical unitarity is absent: the tree-level propagators contain a massive spin-2 excitation which is either a ghost ($a > 0$), or a tachyon ($a < 0$). There is also the constraint of a good Newtonian limit. Classically, this is obtained only if $a > 0$, $b > 0$.

Now unitarity preservation can be a matter of dynamics. In theories where quantum effects are important, the spectrum of excitations at tree level may bear no resemblance to the actual excitations present in the full theory. As we will see, this appears to be the case with equation (5.1). To get a glimpse at such effects, we will use an expansion in powers of $1/N$ (Tomboulis 1977), keeping $k^2 \equiv K^2 N$, $\alpha^2 \equiv N/a$, $\eta^2 \equiv N/b$, $\lambda = \Lambda/N$ fixed. This is an expansion in a gauge invariant parameter, and independent of the strength of the couplings at all scales.[†]

The $1/N$ expansion is related to 'semiclassical gravity' calculations, where gravity is treated classically, but quantum effects are included by coupling to some sort of vacuum expectation value of a quantum energy–momentum tensor $T_{\mu\nu}$. From work on the regularization of $T_{\mu\nu}$, it follows that, in the semiclassical equations, terms with higher derivatives that follow from variation of the action (5.1) must be included. The issue of the boundary conditions to be used for the states used in defining the expectation of $T_{\mu\nu}$ is an important one. With appropriate such conditions the in-out semiclassical theory may be viewed as the lowest term of $1/N$ expansion. (See Horowitz (1981), Kay (1981) for discussion and references.)

To exhibit the new features revealed in the $1/N$ expansion, it suffices to consider N massless fermion fields and ignore all non-gravitational interactions. The resulting rules for the $1/N$ expansion are then very simple. After integrating out the fermions completely, and performing some simple rescalings, we obtain an effective non-local action:

$$\sqrt{-g}\,[\,k^{-2}R - \alpha^{-2}(R_{\mu\nu}{}^2 - \tfrac{1}{3}R^2) + \eta^{-2}R - \lambda\,]$$
$$+\, \mathrm{i}(-1)\ln\det(\mathrm{i}\,S^{-1}[g]) + \Delta\mathscr{L}^{(0)} + \text{(gauge and ghost terms)}. \quad (5.2)$$

Here $S[g]$ denotes the fermion propagator in the presence of the external field $g_{\mu\nu}$, and $\Delta\mathscr{L}^{(0)}$ is the counterterm needed to render the determinant finite. The $1/N$ expansion of the original equation (5.1) is simply the loop expansion of the non-local action (5.2). The graviton propagator in equation (5.2) differs from that in equation (5.1) by the summation of the fermion self-energy bubbles. The scalar and longitudinal parts of this modified propagator again present no unitarity problems. The new crucial feature is in the spin-2 part: with $k^2 \ll 1$, there are, except for the expected graviton pole at $p^2 = 0$, no other poles on the entire real p-axis. Instead, a pair of complex conjugate poles appears on the physical sheet at $\operatorname{Re} p^2 \sim k^{-2}$, $\operatorname{Im} p^2 \sim 10^{-3}k^{-2}$. The interactions made the original unphysical pole 'unstable'. It remains on the physical sheet, but, at the same time, it split into a conjugate pair. It is known that in this case a unitary S-matrix between asymptotic states containing only physical (stable) particles can be defined (Lee and Wick 1969a,b). This is rather similar to defining an S-matrix in field theory in the presence of unstable particles— which are on the second sheet—so that asymptotic states connected by it contain only the stable articles (Veltman 1963). From a Lehmann representation of the modified propagator, it is obvious that the complex pair will not contribute to the absorptive part in the Cutkosky cuts over intermediate states in lowest order in $1/N$. This will remain true in higher orders if, in calculating diagrams, contours are

---

[†] Another possible expansion of this type is the $1/d$ expansion, where d is the space–time dimensionality (Strominger 1981).

deformed according to the Lee–Wick prescription. This prescription may be viewed as essentially a generalization of the iε-prescription, where the infinitesimal ε is replaced by the finite imaginary part of the complex poles. Because of the presence of these poles, and associated contour pinching, the usual analytical structure of the *S*-matrix is modified.

The asymptotic UV behavior of the propagator in equation (5.2) is $[k^4 \ln k^2]^{-1}$, whereas that of all vertices—which in general include the contribution of fermion loops and are non-local—is $k^4 \ln k^2$. The $1/N$ expansion is then renormalizable by power counting. Renormalizability is indeed shown by detailed analysis (Stelle 1977, Tomboulis 1977). We have chosen a simple form for matter in the above discussion, but general matter Lagrangians involving masses and matter self-interactions, do not change the result: all infinities can be absorbed by counterterms of the form

$$\Delta \mathscr{L}_{\text{grav.}} = \left[c_0 + c_1 R + c_2 (R_{\mu\nu}{}^2 - \tfrac{1}{3} R^2) + c_3 R^2 \right] \sqrt{-g} \tag{5.3}$$

plus, of course, the appropriate matter counterterms.

If we return to the simple case of massless fermions, we find that, with dimensional regularization, the necessary counterterm in equation (5.2) is:

$$\Delta \mathscr{L}^{(0)} = \frac{1}{(d-4)} \frac{1}{10(4\pi)^2} (R_{\mu\nu}{}^2 - \tfrac{1}{3} R^2) \sqrt{-g}. \tag{5.4}$$

This has a Weyl-invariant form, and renormalizes the α coupling. The Einstein part, and the remaining terms in equation (5.2) do not get renormalized. This is because matter is conformally invariant, and dimensional regularization was used. We then have very simple RG equations for the running coupling:

$$\tau d\bar{\alpha}(\tau)/d\tau = -b_0 \bar{\alpha}(\tau)^3 \qquad b_0 = \tfrac{1}{20} (4\pi)^{-2}. \tag{5.5}$$

We see that $\bar{\alpha} = 0$ is an UV fixed point, i.e. the theory is asymptotically free (Tomboulis 1980). The sign in equation (5.5) is a consequence of the choice $a > 0$ in equation (5.1). This choice then leads to controllable short-distance behavior which, as discussed in the previous sections, was one of the essential requirements made of the theory. It also leads to a sensible IR limit. At the classical level this was already noted—indeed it is the customary reason for it. In the quantum theory, the propagator and all *n*-point vertices reduce to those of the Einstein theory for small momenta, since then the p^2 terms dominate over p^4 terms. To see this in detail in the context of the $1/N$ expansion, let us examine the theory in the large N limit in which only the leading $1/N$ approximation is retained. In this limit equation (5.5) becomes exact, and its solution holds for all $\tau : \bar{\alpha}(\tau)$ vanishes logarithmically for large Euclidean momenta, whereas it diverges for small momenta. The solutions of the corresponding RG equations for the 1-particle irreducible vertices show then that for momenta small compared to k^{-1}, one has the Einstein theory. For large momenta, however, it is the $R_{\mu\nu}{}^2$ terms that dominate; the dimensional coupling k becomes irrelevant. It is unlikely that these features will be qualitatively altered by higher $1/N$ corrections.

The simple form of equation (5.5) was obtained using conformal matter and dimensional regularization. If we use non-conformal matter and/or dimensional cut-offs, or if we consider higher $1/N$ corrections (even with dimensional regularization), we will have charge renormalization of the other couplings, k, η, and λ also. We would, of course, like to know how the respective running couplings behave. Furthermore, notice that in the leading $1/N$ approximation, the entire contribution to b_0 in equation (5.5) comes from matter loops. It is of interest to inquire whether the one-loop gravitational approximation (which would be *part of* the next to leading $1/N$ order) gives a contribution of the same sign to b_0. Though not all the rather tedious calculations have yet been examined in sufficient detail, there is general agreement that the purely gravitational one-loop contribution to the renormaliz-ation of α is also asymptotically free (Julve and Tonin 1978, Fradkin and Tsytlin 1982). The same seems to be true for the renormalization of the (dimensionless) $\mu^2 K^2$, as well as for the cosmological term Λ/μ^4.[†] The sign of the renormalization of η, however, will be asymptotically free if the bare R^2 in equation (5.1) is chosen with $b < 0$. This may create a problem with the Newtonian limit of the scalar excitation in the theory. If, on the other hand we insist on $b > 0$, then the full propagator may develop a tachyon in its scalar piece (in this connection see Hartle and Horowitz 1981, and Yamagishi 1982). This problem of the scalar mode has a resolution according to an observation by Zee (1982) within the class of theories that we now proceed to introduce.

With conformally invariant matter Lagrangian, we found that, at least to leading order in $1/N$ and with appropriate regularization, we have no infinite renormaliz-ations of η and λ. A theory in accord with experiment is then obtained by taking $\Lambda = 0$, $b = 0$ in equation (5.1). This gives an action which is conformally invariant except for the Einstein term:

$$\mathscr{L} = \sqrt{-g}\,[K^{-2}R - a(R_{\mu\nu}{}^2 - \tfrac{1}{3}R^2) + \mathscr{L}_{\substack{\text{conf.}\\ \text{matter}}}\,(g,\psi,A)]. \qquad (5.6)$$

Now, on general physical grounds, we like to have theories defined by actions containing no dimensional parameters. As we saw in § III, however, even in such theories, renormalization always introduces a mass scale which replaces one of the dimensionless parameters of the bare Lagrangian—a phenomenon known as dimensional transmutation. Scale and conformal invariance is 'intrinsically' broken by the renormalization procedure. Ideally, we would like the theory to contain no remaining free dimensionless renormalized couplings. In such a case the theory would have a single parameter, a (RG invariant) mass which sets the scale of our units. All physical quantities, e.g. all masses, will be computable multiples of powers of this scale.[‡] It would certainly be very attractive to have a theory of this type also in the case of quantum gravity. Indeed, the idea of generating the dimensional parameters of classical relativity, k and λ, dynamically is not new. Usually this is presented in the context of conformal gravity involving scalar fields, and a scalar potential which gives a stable broken-symmetry phase; the scalar field acquires a vacuum expectation value

---

† Assuming $\Lambda > 0$ in equation (5.1).
‡ QCD with light fermions is a theory of precisely this type.

and this induces a Newtonian constant and a cosmological term (see Englert *et al* 1976, Smolin 1979b, Zee 1979, and references given therein). However, in such models because of the presence of scalar fields and dimensional parameters in the scalar potential, k and λ are not really computable. More recently, Adler observed that in theories with no scalar fields, and no bare masses or massive regulators, no term linear in R, and no cosmological term can appear in the action; but such terms may appear as induced, calculable contributions in the effective quantum action (Adler 1980, 1982, Zee 1981). From general physical arguments (Gross and Neveu 1974, Adler 1982), we expect such dynamical mass generation to occur in asymptotically free theories, and the associated breaking of scale and conformal invariance to be 'soft', i.e. all dynamically induced masses have soft, indeed vanishing, UV behavior. The asymptotic freedom property of the theories we have been examining is, therefore, important in this context also. Now, the form of the fundamental matter action that presumably couples to gravity, is not known. However, a model that satisfies the conditions for possible dynamic generation of the Einstein–Hilbert action is the fully conformal version of (5.6):

$$\mathcal{L} = [-a(R_{\mu\nu}{}^2 - \tfrac{1}{3}R^2) + \mathcal{L}_{\substack{\text{conf.} \\ \text{matter}}} (\psi, A, g)]\sqrt{-g}. \tag{5.7}$$

In addition to induced Einstein gravity terms, equation (5.7) will generally also induce R^2-terms in the effective action. Due to the 'soft' breaking of conformal invariance alluded to above, this induced term was shown by Zee (1982) to be finite and computable, and also of the right sign to ensure the right (tachyon-free) Newtonian limit of the scalar sector in the effective action.

There have been some model calculations of induced Einstein gravity from equation (5.7).[†] These calculations indicate that the sign of K is sensitive to the IR details of the mass generation mechanism. Here, we would like to point out that it is perhaps possible to compute this effect within some simple $1/N$ expansion of equation (5.7). Indeed, with non-self-interacting fermions, the leading $1/N$ propagator from equation (5.7) exhibits a tachyon at $p^2 \simeq \mu^2 \exp(-1/b_0\alpha^2)$, which is proportional to the RG invariant mass in this approximation. This tachyon has nothing to do with poles from the higher derivatives. From our previous discussion, we know that this tachyon will disappear when we add the Einstein term by hand, as in equation (5.6). This would indicate that the theory is unstable to dynamical mass generation.[‡] The correct vacuum should be one in which, due to gravitational and other interactions, $\bar{\psi}_i\psi_i$ develops an expectation, and hence a soft fermion mass. This should automatically induce terms linear in R, and also a cosmological term, in the quantum effective action. Though explicit calculations have not yet been

---

[†] These calculations utilize either explicit regulators as a mock-up of the UV damping of soft symmetry breaking (Zee 1981), or instanton approximations which, unfortunately, break down in the IR regime where the effect is supposed to occur (Hasslacher and Mottola 1980).
[‡] This is how mass generation is signaled also in the $1/N$ expansion in the Gross–Neveu model (1974).

attempted, they appear feasible within the formalism of the effective action for composite operators (Cornwall *et al* 1974).

If the induced $1/K \sim M_{\text{Planck}}$, then, in general, the induced $\Lambda \sim M_{\text{Planck}}^4$, in violent contradiction with the observed $|\Lambda| \leq (10^{-42}\,\text{GeV})^2$. A possible way out would be to postulate some symmetry that forbids the appearance of a cosmological term. However, for this to work, the assumed symmetry must remain unbroken. In this connection note that this problem is present at all energy scales—even if the effective gravity theory at Planck length has $\Lambda \simeq 0$, huge values of Λ may be subsequently induced in the phase transitions of elementary particle physics (Veltman 1975). Therefore, perhaps a better resolution of the problem is to start with large Λ, and invoke non-equilibrium mechanisms, such as a Planck era inflationary senario, for the transition to a small effective cosmological constant (Brout *et al* 1979, Guth 1981, Linde 1982, Gott 1981).

6. CONCLUSIONS AND SPECULATIONS

We have argued that only theories whose short-distance behavior is controlled by UV fixed points of the RG may be expected to yield well defined, interacting continuum limits. Applying this premise to quantum gravity, we were led to examine actions quadratic in the Riemann tensor as the only gravitational actions known to possess UV fixed points in $d = 4$. In fact, these theories are asymptotically free.[†] Among them, the conformal versions appear particularly attractive. At a microscopic level, they describe a scale and conformally invariant world, with the Einstein theory emerging as the long-distance effective description of the soft breaking of these symmetries. However, a convincing model calculation of this effect has yet to be performed.

These higher derivative theories appear non-unitary within ordinary perturbation theory. We argued that this indication may be misleading. Indeed, we found that in an approximation scheme, such as the $1/N$ expansion, where the effects of interactions are included even 'to zeroth order', the unitarity picture is substantially modified. In fact, it is modified in such a way that a physical subspace, on which a unitary S-matrix is defined, can be consistently isolated. To do this we had to evoke the Lee–Wick prescription. It must be pointed out, however, that this came about in a natural way[‡]—the absence of unphysical poles on the real axis, and the emergence of complex conjugate pairs was automatically produced by the radiative corrections. The Lee–Wick prescription must be stated in Minkowski space. One may rotate to Euclidean space only below particle thresholds. This is perhaps one more indication

---

[†] Incidentally, the signs in the Lagrangian yielding asymptotic freedom, are also those that result in a well defined, damped Euclidean path integral.

[‡] We did not, for example introduce any imaginary masses by hand, as it was done in earlier model calculations using the prescription.

that gravity theory can only be formulated in Minkowski space. At any rate, the relation, if any, between Euclidean and Minkowski formulations is, to put it mildly, obscure.

The presence of complex poles on the physical sheet modifies the usual analytical structure of the S-matrix. Since this structure is related to the causal properties of wave packets, the modification introduces acausal effects in their motion. These effects, involving Planck times, are obviously very small. It can be shown (Lee and Wick 1969) that they do not affect any classical limit. Moreover, since a well defined unitary S-matrix between in-out physical states exists, no logical paradox could ever arise in a scattering experiment. In fact, it is not clear that any real physical effects can be associated with these 'acausalities'.

It is conceivable that methods incorporating more of the non-perturbative structure of the theory than the low orders of the $1/N$ expansion do, will further transform the unitarity picture, so that eventually the Lee–Wick prescription need not be evoked at all. In this connection note that in the pure conformal theory (5.7), the classical (tree-level) gravitational potential between two external sources is linear, i.e. 'confining'. As expected, this theory is strongly interacting in the infrared, even at tree-level. It is not clear that ordinary perturbation theory is at all meaningful in this case. In canonical quantization one generally encounters constraints which, when imposed on states, define the physical subspace. Kaku (1982) has speculated that the constraints in the conformal gravity theory may have non-perturbative solutions such as string-like objects. Supersymmetry may be important here. It is not even clear that the conformal supergravity version of equation (5.7) does not have some definition of 'energy' which is positive even in the classical theory.

In the renormalizable higher derivative theories, there is a variety of possible, interesting physical effects that have not yet been investigated in any detail. The whole question of the occurrence of space–time singularities must clearly be re-examined. It is easy to see that the assumption of the usual singularity theorems of general relativity are violated. In view of the asymptotic freedom of these theories, it is quite likely that, starting from physically realizable situations, no singularities are formed. Indeed, this controllable short distance behavior should make the final stages of collapse, and of black hole evaporation amenable to meaningful computation. Applications to Planck era cosmology, and the possible relation of dynamical breaking generating the dimensional parameters K and Λ to the unification scale of particle physics are other unexplored areas.

REFERENCES

Adler S L 1980 *Phys. Rev. Lett.* **44** 1567
—— 1982 *Rev. Mod. Phys.* **54** 729
Aizenman M 1981 *Phys. Rev. Lett.* **47** 1
Brezin E, Le Guillon J C and Zinn-Justin J 1975 in *Phase Transitions and Critical Phenomena* ed C Domb and M S Green (New York: Academic)

Brout R, Englert F, and Spindel P 1979 *Phys. Rev. Lett.* **43** 417
Christensen S M and Duff M J 1978 *Phys. Lett.* **79B** 213
Coleman S, Jackiw R and Politzer H D 1974 *Phys. Rev.* D **10** 2491
Cornwall J M, Jackiw R and Tomboulis E 1974 *Phys. Rev.* D **10** 2428
Das A, Kaku M and Townsend P 1979 *Phys. Lett.* **81B** 11
DeWitt B S 1965 *Dynamical Theory of Groups and Fields* (New York: Gordon and Breach)
—— 1967a *Phys. Rev.* **160** 1113
—— 1967b *Phys. Rev* **162** 1195
—— 1967c *Phys. Rev.* **162** 1239
Eddington A 1924 *The mathematical theory of relativity* (Cambridge; Cambridge University Press)
Englert F, Truffin C and Gastmans R 1976 *Nucl. Phys.* B **117** 407
Feynman R P 1963 *Acta Phys. Pol.* **24** 697
Fradkin F S and Tseytlin A A 1982 *Nucl. Phys.* B **201** 469
Frohlich J 1982 *Nucl. Phys.* B **200** 281
Gastmans R, Kallosh R and Truffin C 1978 *Nucl. Phys.* B **133** 417
Glimm J and Jaffe A 1981 *Quantum Physics* (New York; Springer)
Gott J R III 1982 *Nature* **295** 304
Gross D J and Neveu A 1974 *Phys. Rev.* D **10** 3235
Guth A H 1981 *Phys. Rev.* D **23** 347
Hartle J B and Horowitz G T 1981 *Phys. Rev.* D **24** 257
Hässlacher B and Mottola E 1980 *Phys. Lett.* **95B** 237
Hawking S W, Page D N and Pope C N 1979 *Phys. Lett.* **86B** 175
Horowitz G T 1981 in *Quantum Gravity 2* ed C J Isham, R Penrose and D W Sciama (Oxford: Oxford University Press)
Julve J and Tonin M 1978 *Nuovo Cimento* B **46** 137
Kaku M 1982 *Nucl. Phys.* B **203** 285
Kay B S 1981 *Phys. Lett.* B **101** 241
Landau L D 1955 in *Niels Bohr and the development of Physics* (London: Pergamon Press)
Lee T D and Wick G C 1969a *Nucl. Phys.* B **9** 209
——1969b *Nucl. Phys.* B **10** 1
Linde A D 1982 *Phys. Lett.* **108B** 389
Parisi G 1975 *Nucl. Phys.* B **100** 368
Pauli W 1919 *Phys. Z.* **20** 457
Regge T 1961 *Nuovo Cimento* **19** 558
Schwarz J 1982 *Caltech preprint* Calt-68-911
Shizuya K 1980 *Phys. Rev.* D **21** 2327
Smolin L 1979a *Nucl. Phys.* B **148** 333
—— 1979b *Nucl. Phys.* B **160** 253
—— 1982 *Nucl. Phys.* B **208** 439
Stelle K S 1977 *Phys. Rev.* D **16** 953
Strominger A 1981 *Phys. Rev.* D **24** 3082
Tomboulis E 1977 *Phys. Lett.* **70B** 361
—— 1980 *Phys. Lett.* **97B** 77
Veltman M 1963 *Physica* **29** 186
—— 1975 *Phys. Rev. Lett.* **34** 777
Weinberg S 1979 in *General Relativity* ed S Hawking and W Israel (Cambridge: Cambridge University Press)
Weyl H 1921 *Phys. Z.* **22** 473

Wilson K G 1973 *Phys. Rev.* D **7** 2911
Wilson K G and Kogut J 1974 *Phys. Rep.* **12** 75
Yamagishi H 1982 *Phys. Lett.* **114B** 27
Zee A 1979 *Phys. Rev. Lett.* **42** 417
—— 1981 *Phys. Rev.* D **23** 858
—— 1982 *Phys. Lett.* **109B** 183

Quantization of Higher Derivative Theories of Gravity

DAVID G BOULWARE

1. INTRODUCTION

The quantum field theory of gravity, with the metric taken to be a fundamental field and with the usual Einstein action, is a non-renormalizable theory. The coupling to matter and the self-coupling of the gravitational field both give rise to divergences which require counterterms proportional to the square of the curvature tensor (Stelle 1977, 1978, Julve and Tonin 1978, Kawasaki *et al* 1981, Martinelli 1981, Nakanishi 1978a, b). There are three known models for gravity in which such counterterms do not appear in the fundamental Lagrangian:

1) Gravity is embedded in some larger theory such as supergravity which has fundamentally different short distance properties so that it is either renormalizable or finite.

2) Gravity is a phenomenological theory valid for distances large compared to the Planck scale so that the short distance behavior need not be taken seriously, i.e., the graviton is an excitation of the theory analogous to the phonon with no fundamental underlying field (such a theory is consistent with observed gravitational phenomena (Weinberg 1964, 1965, Boulware and Deser 1975) but cannot be regarded as complete at this time).

3) The basic quantum field theory structure fails in some general way at extremely short distances to be replaced by, e.g., a lattice theory or some modification of quantum mechanics.

Although each of these possibilities has attractive features and has been proposed in various forms as the solution to the problems of quantum gravity, none is fully compelling. This paper is concerned with a fourth, perhaps simpler, possibility. When the required counterterms involving the square of the curvature tensor are added to the action, it describes a satisfactory, well defined quantum field theory in the sense that the perturbation theory is well behaved and formally renormalizable. This possibility has been discussed from various points of view (Stelle 1977, Julve and

Tonin 1978, Martinelli 1981, Nakanishi 1978a, b); however, the relationship of the formal covariant perturbation theory to the canonical quantization of the theory has not been discussed in full (see, however, the work of Kaku (1982)).

The coupling of the gravitational field to matter involves second derivatives of the matter field; as a consequence, the effective action of the matter field, viewed as a function of the metric, contains divergent terms proportional to the square of the curvature tensor as well as the Einstein action and the cosmological term. Further, if the graviton propagator has the usual $1/p^2$ behavior for high energy then the graviton exchange contribution to the matter fields is not renormalizable because, in a scalar self-energy graph for example, there are four powers of momentum in the denominator and four in the numerator due to the vertices leading to a quartic divergence. For small perturbations around flat space, the counter terms proportional to the two possible squares of the curvature tensor,

$$\delta W = \int dx \sqrt{-g} [-\alpha C^{\mu\nu\lambda\sigma} C_{\mu\nu\lambda\sigma}/4 + \beta R^2/8] \qquad (1.1)$$

then lead to an inverse graviton propagator proportional to p^4. Then the scalar self-energy graph has only the usual quadratic divergence and the theory is renormalizable. As was first analyzed in detail by K Stelle (1977), the same arguments apply to the self-interactions of the gravitational field, which is itself renormalizable in perturbation theory. The fact that the propagator behaves as $1/p^4$ immediately implies that there are other problems with the theory: in a Lorentz invariant theory with positive semi-definite energies and a positive definite metric in the Hilbert space, the high energy behavior of the propagator must behave as $1/p^2$ (Smolin 1982), thus the renormalizable theory must involve either negative energies, indefinite Hilbert space metric or both.

The equivalence of the functional integral and operator formulations of quantum systems follows from a careful discussion of the canonical quantization of the systems. A given classical system is written in terms of an appropriate set of canonical variables, $\{q, p\}$, and a general matrix element of the form

$$\langle q''t | q'0 \rangle \qquad (1.2)$$

may then be rewritten as the functional integral,

$$\langle q''t | q'0 \rangle = \int [dq][dp] \exp i \int \{ p\dot{q} - H \} \qquad (1.3)$$

by repeatedly inserting a complete set of states at intermediate times. The functional integral is over all paths in phase space such that,

$$q(t) = q'' \qquad q(0) = q'$$

and the momenta are unconstrained at the initial and final times (if the initial and final states are not taken to be eigenstates of q, then the space of paths over which the integration is done will be different). As long as the Hamiltonian is quadratic in the momenta and as long as the variables are unconstrained, the momenta may be integrated out and the matrix element becomes an integral over the configuration space paths which satisfy the boundary conditions. In the case in which there are constraints, the procedure must be modified; this has been done by Faddeev and Slavnov (1980) in the case of Yang–Mills fields.

In order to establish the functional integral formulation of the R^2 theories of gravity, the theory must first be cast in canonical form and the constraints identified. Then the correct functional integral for the theory may be written down. After that the momenta and any auxiliary variables may be integrated out, provided that the form of the action allows it.

In the second section, the general R^2 action for a metric theory in four space–time dimensions,

$$W = \int dx \sqrt{-g} [\Lambda + R/16\pi G - \alpha C^{\mu\nu\lambda\sigma} C_{\mu\nu\lambda\sigma}/4 + \beta R^2/8] \qquad (1.4)$$

is recast in first (rather than fourth) order form and the constraints are found. These constraints are analogous to those of the Einstein theory. In the case of the Weyl theory,

$$W_c = -\int dx \sqrt{-g} \; \alpha C^{\mu\nu\lambda\sigma} C_{\mu\nu\lambda\sigma}/4 \qquad (1.5)$$

there are two additional constraints which follow from the conformal invariance. These constraints are discussed in some detail in the third section where it is shown that they form a complete set of constraints and that the Poisson brackets of the constraints among themselves are proportional to the constraints, thus the theory of generalized Hamiltonian dynamics discussed by Dirac (1959) and by Faddeev and Slavnov (1980) may be applied. The major difficulty in analyzing the constraints arises from the fact that, just as in the case of the Einstein theory, one of them is nominally the Hamiltonian. Care must be taken in the integrations by parts and in what surface terms are included in the action. If this is done, it is possible to formally eliminate the constraints and proceed with the canonical quantization, thereby obtaining the quantum-mechanical matrix element between states of definite three-geometry as a functional integral over paths in the space of three-geometries.

The process of expressing the matrix element as a functional integral works as long as the integral coverages. If one works in space–times with Minkowski signature, the integrand of the functional integral is of the usual form,

$$\exp i \int (p\dot{q} - H) \qquad (1.6)$$

and the integral is marginally convergent regardless of the sign of the action. In the higher order theories considered here, the energy is not bounded from below, hence the system is not stable. This is known in the classical theory (Yamagishi unpublished, Horowitz and Wald 1978, Horowitz 1981) where the addition of the R^2 terms to the action causes the flat space solution to be unstable and the cosmological solutions to the field equations possess runaway modes as well as the modified Robertson–Walker solutions. This causes no formal difficulty other than the absence of a ground state, but the physics of the theory as such is unacceptable.

Lee and Wick (1971) showed how to eliminate such instabilities in some cases: quantize the theory in an indefinite metric Hilbert space. Then, the variables for which the energy becomes negative as the variables become large are effectively multiplied by a factor of i, thereby rendering the energy positive, but the Hilbert space metric is indefinite. As a result the runaway modes are still present because the amplitude for a state of zero norm can grow without bound. These modes are eliminated as a boundary condition in the future, leaving (by assumption) a physical Hilbert space with positive definite norm and positive definite energies. The theory exhibits acausal behavior because the growing modes are eliminated as a boundary condition; this behavior is strange but does not appear to conflict with any fundamental principles. In addition, in order to preserve the unitarity of the theory, the virtual growing modes must also be eliminated. The theory which results after all this has been done is known as the Lee–Wick theory and it is generally assumed that R^2 gravity theories incorporate the Lee–Wick mechanism, as indeed they must if the theory is not to be energetically unstable.

Because of the complex values of the fields, the functional integral does not exist in space–times with Minkowski signature; the integral must be done in imaginary time. In § 5, the stability of the theory and the problem of doing the indefinite metric quantization are discussed. There it is shown that, although the Euclidean space functional integral may be written down and converges, it does not follow from the canonical quantization. Further, the functional integral for a general indefinite metric theory does not incorporate the Lee–Wick prescription for eliminating the virtual growing modes. Thus, although the Euclidean functional integral may be well defined, it is not simply the indefinite metric field theory incorporating the Lee–Wick prescription.

2. FIRST-ORDER FORM

The general action for R^2 theories of gravity in four space–time dimensions is

$$W = \int dx \sqrt{-g} [\Lambda + (R/16\pi G) - \alpha C^{\mu\nu\lambda\sigma} C_{\mu\nu\lambda\sigma}/4 + \beta R^2/8]. \tag{2.1}$$

The Riemann curvature tensor is given by

$$R^{\mu}{}_{\nu\lambda\sigma} = \Gamma^{\mu}{}_{\nu\sigma,\lambda} - \Gamma^{\mu}{}_{\nu\lambda,\sigma} - \Gamma^{\tau}{}_{\nu\lambda}\Gamma^{\mu}{}_{\tau\sigma} + \Gamma^{\tau}{}_{\nu\sigma}\Gamma^{\mu}{}_{\tau\lambda} \tag{2.2}$$

and the traceless Weyl tensor in n-dimensions is

$$C^{\mu}{}_{\nu\lambda\sigma} = R^{\mu}{}_{\nu\lambda\sigma} - [\delta_{\lambda}{}^{\mu}R_{\nu\sigma} - \delta_{\sigma}{}^{\mu}R_{\nu\lambda} + g_{\nu\sigma}R_{\lambda}{}^{\mu} + g_{\nu\lambda}R_{\sigma}{}^{\mu}]/(n-2)$$
$$+ [\delta_{\lambda}{}^{\mu}g_{\nu\sigma} - \delta_{\sigma}{}^{\mu}g_{\nu\lambda}]R/(n-1)(n-2). \tag{2.3}$$

In three dimensions the Weyl tensor vanishes identically because it has the symmetries of the Riemann tensor, hence it has as many independent components as its (vanishing) trace.

In order to define a canonical theory, a foliation of spacelike surfaces of the space–time region under consideration must exist. The canonical variables are then defined relative to these surfaces. The canonical decomposition of the standard theory (Arnowitt *et al* 1962, O'Murchadha and York 1974) is most simply done in terms of the intrinsic metric of the three-dimensional surface, g, and the extrinsic curvature, or second fundamental form,

$$K_{kl} = \tfrac{1}{2}\mathscr{L}_n g_{kl} = (\partial g_{kl}/\partial t - N_{k|l} - N_{l|k})/2N \tag{2.4}$$

where the normal to the surface is given by

$$n^\mu = (1, -N^m)/N \tag{2.5}$$

$$ds^2 = -(N^2 - N^k N_k)dt^2 + 2N_k dx^k dt + dx^k g_{kl} dx^l \tag{2.6}$$

and

$$N_k = g_{kl} N^l. \tag{2.7}$$

The $|$ indicates covariant differentiation in the surface. The Gauss–Codazzi equations give the Riemann tensor in terms of the intrinsic geometry of the surface, g_{kl}, the extrinsic curvature, K_{kl}, and the lapse function, N, which determines the separation of the space-like surfaces. The curvature components are given by,

$$
\begin{aligned}
R^{\hat{0}}{}_{l\hat{0}m} &= -n_\mu R^\mu{}_{l\sigma m} n^\sigma \\
&= \mathscr{L}_n K_{lm} - K_{lm} K_n{}^n - (N_{|lm}/N) \\
R^{\hat{0}}{}_{lmn} &= -n_\mu R^\mu{}_{mn} = K_{ln\,|m} - K_{lm\,|n}
\end{aligned} \tag{2.8}
$$

and

$$R_{klmn} = {}^3R_{klmn} + K_{km}K_{ln} - K_{kn}K_{lm}.$$

The term $\mathscr{L}_n K$ is the Lie derivative of the extrinsic curvature along the normal to the surface,

$$\mathscr{L}_n K_{kl} = (\partial K_{kl}/\partial t - N^m K_{kl\,|m} - N^m{}_{|k}K_{ml} - N^m{}_{|l}K_{km})/N \tag{2.9}$$

which is a covariant second rank tensor under changes of coordinates in the surface. The Ricci tensor is then given by

$$
\begin{aligned}
R^{\hat{0}\hat{0}} &= -g^{mn}R^{\hat{0}}{}_{m\hat{0}n} = K_{mn}K^{mn} - g^{mn}\mathscr{L}_n K_{mn} + (\nabla^2 N)/N \\
R^{\hat{0}}{}_k &= K_{,k} - K^l{}_{k\,|l}
\end{aligned} \tag{2.10}
$$

and

$$R_{kl} = {}^3R_{kl} + K_{kl}K - 2K_{km}K_k{}^m + \mathscr{L}_n K_{kl} - (N_{|kl})/N$$

while the scalar curvature is

$$R = g^{kl}R_{kl} - R^{\hat{0}\hat{0}} = {}^3R + K^2 - 3K_{mn}K^{mn} + 2g^{mn}\mathscr{L}_n K_{mn} - 2(\nabla^2 N)/N. \tag{2.11}$$

From these results it is straightforward to compute the Einstein and Weyl tensors,

$$G^{\hat{0}\hat{0}} = \tfrac{1}{2}({}^3R + K^2 - K_{mn}K^{mn})$$

$$G_k{}^{\hat{0}} = R_k{}^{\hat{0}} = K_{|k} - K^l{}_{k|l}$$

$$\begin{aligned}G_{kl} ={}& {}^3G_{kl} + K_{kl}K - 2K_{km}K^{ml} - \tfrac{1}{2}g_{kl}(K^2 - 3K_{mn}K^{mn}) \\ &+ (\delta_k{}^m\delta_l{}^n - g_{kl}g^{mn})\mathcal{L}_nK_{mn} - [(\nabla_k\nabla_l - g_{kl}\nabla^2)N]/N \end{aligned} \tag{2.12}$$

$$C^{\hat{0}}{}_{l\hat{0}n} = (\delta_l{}^s\delta_n{}^t - g_{ln}g^{st}/3)(\mathcal{L}_nK_{st} - KK_{st} - {}^3R_{st} - N_{|st}/N)/2$$

$$C^{\hat{0}}{}_{lmn} = [\delta_l{}^r\delta_m{}^s\delta_n{}^t - (g_{nl}\delta_m{}^s - g_{ml}\delta_n{}^s)g^{rt}/2](K_{rt|s} - K_{rs|t})$$

and, since ${}^3C^k{}_{lmn} = 0$

$$C_{klmn} = -(g_{km}C^{\hat{0}}{}_{l\hat{0}n} - g_{kn}C^{\hat{0}}{}_{l\hat{0}m} + g_{lm}C^{\hat{0}}{}_{k\hat{0}m} - g_{lm}C^{\hat{0}}{}_{k\hat{0}n}).$$

The Weyl tensor squared may be written as

$$C^{\mu\nu\lambda\sigma}C_{\mu\nu\lambda\sigma} = 8C^{\hat{0}k}{}_{\hat{0}l}C^{\hat{0}}{}_{k\hat{0}l} - 4C^{\hat{0}klm}C^{\hat{0}}{}_{klm} \tag{2.13}$$

where all spatial indices are raised with g^{kl}, the three-dimensional inverse of g_{kl}. The action may now be written in terms of g and K,

$$\begin{aligned}W = \int dx\, N\sqrt{g}[&\Lambda + (R/16\pi G) + \beta R^2/8 - 2\alpha C^{\hat{0}k\hat{0}l}C^{\hat{0}}{}_{k\hat{0}l} \\ &+ \alpha C^{\hat{0}klm}C^{\hat{0}}{}_{klm}] \end{aligned} \tag{2.14}$$

where the various quantities are defined above.

The action is now in fourth-order $(\mathcal{L}_nK)^2$ form. In order to cast the theory in canonical form, the action must be varied with respect to the various explicit time derivatives which appear, the results set equal to new independent variables, and Legendre transformed so as to cast the theory in $p\dot{q} - H$ form. The highest derivative appearing is $\mathcal{L}_nK$, hence define Q by

$$\delta_{\mathcal{L}_nK}W \equiv -\int dx[Q^{kl} - (g^{kl}\sqrt{g}/8\pi G)]\delta\mathcal{L}_nK_{kl} \tag{2.15}$$

where Q^{kl} is a density of weight one. The variable Q is then

$$Q^{kl} = \sqrt{g}(2\alpha C^{\hat{0}k}{}_{\hat{0}}{}^l - g^{kl}\beta R/2) \tag{2.16}$$

with the additive term, $g^{kl}\sqrt{g}/8\pi G$, having been chosen so that Q would vanish at flat space.

The Weyl tensor is traceless on kl, hence

$$C^{\hat{0}k}{}_{\hat{0}}{}^l = Q^{Tkl}/2\alpha\sqrt{g}$$

and

$$R = -2Q/3\beta\sqrt{g} \tag{2.17}$$

where Q^{Tkl} is the traceless part of Q^{kl},

$$Q^{Tkl} = (\delta_m{}^k\delta_n{}^l - g^{kl}g_{mn}/3)Q^{mn} \tag{2.18}$$

and Q is the trace,

$$Q = g_{mn} Q^{mn}.$$

For $\beta \neq 0$, this equation may be solved for

$$\mathcal{L}_n K_{kl} = K K_{kl} + {}^3 R_{kl} + N_{|kl}/N + Q_{kl}{}^T/\alpha\sqrt{g}$$
$$- g_{kl} Q/9\beta\sqrt{g} + g_{kl}[-{}^3 R + (K_{mn} K^{mn} - K^2)]/2 \qquad (2.19)$$

The Legendre transformation then yields, with the aid of (see equation (A.7),

$$\int dx \, N(-Q^{kl} \mathcal{L}_n K_{kl}) = \int dx \, N(\mathcal{L}_n Q^{kl}) K_{kl} - \int d^3x \, N \, K_{kl} Q^{kl} |_-^+$$
$$W = \int dx \, N(K_{kl} \mathcal{L}_n Q^{kl} - \mathcal{H}(g, Q, K)] + \int d\sigma_k (N_{|l} Q^{kl} - N Q^{kl}{}_{|l}$$
$$- \sqrt{g} \, N^{|k}/8\pi G) \qquad (2.20)$$

where

$$\mathcal{H} = -\alpha\sqrt{g} \, C^{\hat{o}klm} C^{\hat{o}}{}_{klm} - Q^{Tkl} Q^T{}_{kl}/2\alpha\sqrt{g} - \sqrt{g}\Lambda + Q^2/18\beta\sqrt{g}$$
$$- (Q^{kl}{}_{|kl} + Q^{kl}{}^3 R_{kl} - Q \, {}^3 R/2)$$
$$- \sqrt{g}({}^3 R + K^{kl} K_{kl} - K^2)/16\pi G - Q^{kl} K_{kl} K - Q(K^{kl} K_{kl} - K^2)/2.$$

The surface term at spatial infinity arises from integrating the

$$\int dx \, Q^{kl} \, N_{|kl} \quad \text{and} \quad \int \nabla^2 N$$

terms by parts; one may require that Q^{kl} drop off sufficiently fast so that the integration by parts of the $Q^{kl} N_{|kl}$ term yields no surface term. The surface term coming from the time integration by parts depends only on the initial and final configurations and may be dropped.

If $\beta = 0$, but $G^{-1} \neq 0 \neq \Lambda$, the theory is not conformally invariant because of the Newton and cosmological constants; when the theory is quantized an R^2 term is induced. As a result, the $\beta = 0$ case is of interest only when $G^{-1} = 0 = \Lambda$, thereby yielding a conformally invariant theory. In that case, $Q = 0$ and equation (2.16) cannot be solved for the trace of $\mathcal{L}_n K$ which remains undetermined. The expression for $\mathcal{L}_n K$ becomes

$$\mathcal{L}_n K_{kl} = (K \, K_{kl} + {}^3 R_{kl} + N^{-1} N_{|kl} + Q_{kl}/\alpha\sqrt{g})^T$$
$$+ g_{kl} g^{mn} \mathcal{L}_n K_{mn}/3 \qquad (2.19')$$

and the action becomes

$$W_c = \int dx \, N(K_{kl} \mathcal{L}_n Q^{kl} - \mathcal{H}_c(g, Q, K))$$

where

$$\mathcal{H}_c = -\alpha\sqrt{g} \, C^{\hat{o}klm} C^{\hat{o}}{}_{klm} - Q^{kl} Q_{kl}/2\alpha\sqrt{g}$$
$$- (Q^{kl}{}_{|kl} + {}^3 R_{kl} Q^{kl}) - Q^{kl} K_{kl} K \qquad (2.20')$$

and

$$Q = g_{kl}Q^{kl} = 0.$$

The action is now a function of g_{kl} and Q^{kl}; time derivatives of each appear and the action is quadratic in explicit time derivatives. The momenta conjugate to Q and g are defined by

$$\delta_{Q,g}W = \int dx N (P_{kl} \delta \mathcal{L}_n Q^{kl} + p^{kl} \delta \mathcal{L}_n g_{kl})/2 \tag{2.21}$$

or

$$P_{kl} = 2 K_{kl}$$

and

$$p^{kl} = \mathcal{L}_n Q^{kl} - \delta \bar{\mathcal{H}}/\delta K_{kl}. \tag{2.22}$$

The Legendre transformation then yields

$$W = \int dx \ N [(P_{kl} \mathcal{L}_n Q^{kl} + p^{kl} \mathcal{L}_n g_{kl})/2 - \mathcal{H}(g, Q, p, P)] \tag{2.23}$$

where

$$\begin{aligned}
\mathcal{H}(g, Q, p, P) = {}& p^{kl} P_{kl}/2 - \alpha\sqrt{g} C^{\hat{\delta}klm} C^{\hat{\delta}}{}_{klm} - Q^{Tkl} Q^T{}_{kl}/2\alpha\sqrt{g} \\
& - \sqrt{g}\Lambda + Q^2/18\beta\sqrt{g} - (Q^{kl}{}_{|kl} + Q^{kl\,3}R_{kl} - Q^3R/2) \\
& - \sqrt{g}[^3R + (P^{kl}P_{kl} - P^2)/4]/16\pi G - Q^{kl}P_{kl}P/4 \\
& - Q(P^{kl}P_{kl} - P^2)/8.
\end{aligned} \tag{2.24}$$

and

$$C^{\hat{\delta}}{}_{klm} = [\delta_k{}^r \delta_l{}^s \delta_m{}^t - \tfrac{1}{2}(g_{km}\delta_l{}^s - g_{kl}\delta_m{}^s)g^{rt}][P_{rt|s} - P_{rs|t}]/2.$$

Since the action is linear in N and $\mathcal{H}$ is its coefficient, $\mathcal{H}$ must vanish and the variables are constrained. Since $\mathcal{H}$ is nominally the Hamiltonian, the Hamiltonian appears to vanish, just as in the Einstein theory. There, the Hamiltonian is introduced as the extrinsic curvature of the bounding surface at infinity integrated over that surface (Hawking 1979, Gibbons and Hawking 1977, York 1972); the same may be done here. The constraint is regarded as a constraint on the conformal factor of the three-metric, which factor determines the extrinsic curvature of the bounding surface. The Hamiltonian can also be obtained by integrating the second derivative term in R by parts.

In the conformally invariant case, $Q = 0$, hence the variations of $\mathcal{L}_n g$ and $\mathcal{L}_n Q$ are not independent. Rather than using a Lagrange multiplier to enforce the constraint, thereby allowing the free variation of the variables, it is useful to impose the constraint on Q and its derivatives and variations; in this way the second constraint which appears may be more easily understood. The Lie derivative of the constraint reads,

$$0 = \mathcal{L}_n(g_{kl}Q^{kl}) = g_{kl}\mathcal{L}_n Q^{kl} + Q^{kl}\mathcal{L}_n g_{kl} \tag{2.25}$$

hence,

$$g_{kl}\delta\mathscr{L}_n Q^{kl} + Q^{kl}\delta\mathscr{L}_n g_{kl} = 0 \tag{2.26}$$

and the variation of $\mathscr{L}_n Q^{kl}$ must be taken as

$$\delta\mathscr{L}_n Q^{kl} = (\delta\mathscr{L}_n Q^{kl})^T - g^{kl}Q^{mn}\delta\mathscr{L}_n g_{mn}/3 \tag{2.27}$$

where $\mathscr{L}_n Q^T$ represents the traceless part of $\mathscr{L}_n Q$. The momenta conjugate to Q and g are then defined by

$$\delta_{Q,g}W_C = \int dx\, N[P_{kl}(\delta\mathscr{L}_n Q^{kl})^T + p^{kl}\delta\mathscr{L}_n g_{kl}]/2 \tag{2.28}$$

where P_{kl} is traceless (the trace would not contribute in any case). This variation yields

$$P_{kl} = 2K^T{}_{kl} \tag{2.29}$$

and

$$P^{kl} = \mathscr{L}_n Q^{kl} - Q^{kl}2K/3 - \delta\bar{\mathscr{H}}_c/\delta K_{kl}.$$

Just as the trace of $\mathscr{L}_n K$ could not be solved for in the equation for Q, now the trace of $\mathscr{L}_n g = 2K$ cannot be solved for. Both quantities transform non-trivially under conformal transformations, hence they do not appear in a conformally invariant theory. By the same token, the canonical variables which would be determined by K and its Lie derivative in a conformally non-invariant theory must be constrained in the invariant theory. The trace free part of K_{kl} and Q^{kl} may be solved for, yielding,

$$(\mathscr{L}_n Q^{kl})^T = (p^{kl} + 2Q^{kl}K/3 + \delta\bar{\mathscr{H}}_c/\delta K_{kl})^T \tag{2.30}$$

and

$$K_{kl} = P_{kl}/2 + g_{kl}K/3.$$

In addition, the trace of the equation for p yields the further constraint

$$p = -Q^{kl}P_{kl} - g_{kl}\delta\bar{\mathscr{H}}_c/\delta K_{kl} = Q^{kl}P_{kl}/2 \tag{2.31}$$

which, as will be shown below, is the generator of conformal transformations. The Legendre transformation, if P is taken to be zero, then yields the conformally invariant first-order action,

$$W_c = \int dx\, N[(P_{kl}\mathscr{L}_n Q^{kl} + p^{kl}\mathscr{L}_n g_{kl})/2 - \mathscr{H}_c(g, Q, p, P)] \tag{2.32}$$

where

$$\mathscr{H}_c(g, Q, p, P) = p^{kl}P_{kl}/2 - \alpha\sqrt{g}\, C^{\hat{\delta}klm}C^{\hat{\delta}}{}_{klm}$$
$$- Q^{kl}Q_{kl}/2\alpha\sqrt{g} - (Q^{kl}{}_{|kl} + Q^{kl\,3}R_{kl}).$$

Note that the trace of K does not appear, Q and P are traceless, and it will be shown in the next section that $\mathscr{H}_c$ transforms multiplicatively under conformal transformations.

Just as in the general case, the density, $\mathscr{H}_c$, is constrained; now, however, there is no leading, $^3R/16\pi G$, term which may be solved for the conformal factor. This is only to be expected since the conformal factor may be chosen freely. The constraint must now be solved for the longitudinal component of Q. Now, the surface terms,

$$\int dS_k (N Q^{kl}_{\ |l} - N_{|l} Q^{kl})$$

cannot be taken to vanish; they are determined by the constraints and, after the contraints are imposed, constitute the Hamiltonian.

The consistency of the constraints will be discussed in the following section.

3. THE CONSTRAINTS

In both the conformally invariant case and the general case, the action has been cast into first-order form; in both cases there are constraints, the explicit constraints of the C^2 theory and the usual constraints due to coordinate invariance which arise in general relativity from the variation of N and N^k. The latter constraints are

$$\mathcal{H} = 0 \tag{3.1}$$

and

$$\mathcal{H}_m = Q^{kl} P_{kl|m}/2 - (P_{mk}Q^{kl})_{|l} + \overset{3}{g}_{mk}p^{kl}_{\ |l} = 0$$

in both cases. In the C^2 theory, $\mathcal{H}$ is replaced by $\mathcal{H}_c$ and there are the additional trace constraints,

$$g_{kl}Q^{kl} = 0 \tag{3.2}$$

and

$$g_{kl}p^{kl} - Q_{kl}P^{kl}/2 = 0.$$

The constraints are local functions of the field quantities at a particular point of space, hence the Poisson bracket of two such quantities is a sum of delta functions and derivatives of delta functions. It is simpler to integrate the constraints over the surface with a testing function, so that the constraint $C(x) = 0$ becomes

$$C[\psi] = \int d^3x \, \psi(x)C(x) = 0. \tag{3.3}$$

Then, the Poisson bracket of two such integrated quantities is

$$[C^1[\psi], C^2[\psi']] = 2 \int d^3x \left(\frac{\delta C^1[\psi]}{\delta Q^{kl}(x)} \frac{\delta C^2[\psi']}{\delta P_{kl}(x)} \right.$$
$$\left. + \frac{\delta C^1[\psi]}{\delta g_{kl}(x)} \frac{\delta C^2[\psi']}{\delta p^{kl}(x)} - C^1 \leftrightarrow C^2 \right) \tag{3.4}$$

where the canonical variables are varied freely, independent of the constraints.

In order to use the usual process of quantization by a functional integral formalism (Faddeev and Slavnov 1980), the constraints must be first-class constraints such that their classical Poisson brackets are linear in the constraints,

$$[C^i[\psi], C^j[\psi']] = \int d^3y \, C_k^{\,ij}[\psi, \psi', y] C^k(y) \tag{3.5}$$

where the C^i are constraints.

It is straightforward but tedious to evaluate the Poisson brackets between the various constraints and show that they are indeed first class constraints. The Poisson brackets involivng $\mathscr{H}_m$ are easily calculated and understood since $\mathscr{H}_m$ is the generator of coordinate transformations in the surface. Let

$$G[\psi] = \int d^3x \, \mathscr{H}_m(x) \psi^m(x) \tag{3.6}$$

then,

$$[G[\psi], g_{kl}(x)] = \psi_{k|l}(x) + \psi_{l|k}(x)$$
$$[G[\psi], p^{kl}(x)] = (\psi^m p^{kl})_{|m}(x) - \psi^k{}_{|m} p^{ml}(x) - \psi^l{}_{|m} p^{km}(x) \tag{3.7}$$
$$[G[\psi], Q^{kl}(x)] = (\psi^m Q^{kl})_{|m}(x) - \psi^k{}_{|m} Q^{ml}(x) + \psi^l{}_{|m} Q^{kl}(x)$$

and

$$[G[\psi], P_{kl}(x)] = \psi^m P_{kl|m}(x) + P_{km}\psi_{|l}{}^m(x) + P_{ml}\psi^m{}_{|k}(x).$$

Since $\mathscr{H}_k$ is a vector density, its Poisson bracket with G is

$$[G[\psi], \mathscr{H}_k(x)] = (\psi^m \mathscr{H}_k)_{|m}(x) + \psi^m{}_{|k}\mathscr{H}_m(x) \tag{3.8}$$

the remaining constraints are all scalar densities, hence their Poisson brackets with G are

$$[G[\psi], C(x)] = (\psi^m C)_{|m}(x). \tag{3.9}$$

The most difficult bracket is that of $\mathscr{H}(x)$ with $\mathscr{H}(x')$; the calculation presented in Appendix B yields the result

$$(\mathscr{H}[\xi], \mathscr{H}[\xi']) = -\int d^3x (\xi\xi'_{|m} - \xi'\xi_{|m}) g^{mn} \mathscr{H}_n(x) \tag{3.10}$$

where

$$\mathscr{H}[\xi] = \int d^3x \, \xi(x) \mathscr{H}(x).$$

The brackets calculated above are qualitatively similar to those in the Einstein theory, modified only by the difference in the interactions. In the conformally invariant theory there are two new constraints related to the conformal invariance. First consider the Poisson brackets of Q,

$$[G[\psi], Q(x)] = (\psi^m Q(x))_{|m} \tag{3.11}$$

since Q is a scalar density. The Poisson bracket with $\mathscr{H}_c$ is

$$[\mathscr{H}_c[\xi], Q(x)] = -\xi J(x) + \xi(x) Q \, P/2 \tag{3.12}$$

where

$$J(x) = g_{kl} p^{kl} - P_{kl} Q^{kl}/2.$$

All the Poisson brackets to date have only involved the constraints $\mathscr{H}$, $\mathscr{H}_m$, and Q, hence they all vanish if the constraints are satisfied. This Poisson bracket involves the new quantity, $J = g_{kl}P^{kl} - Q^{kl}P_{kl}/2$, the vanishing of which is not guaranteed by the imposition of the previous constraints. The system is not consistent unless J is also constrained to vanish. The Poisson brackets of J with the other constraints are then

$$[G[\psi], J(x)] = (\psi^m J)_{|m}(x)$$

$$[\mathscr{H}_c[\xi], J(x)] = -\xi(x)\mathscr{H}_c(x)$$

and

$$[Q[\xi], J(x)] = \xi(x)Q(x). \tag{3.13}$$

Now, the Poisson brackets of the constraints among themselves close and the constraints are first class constraints. The new constraint, J, is the generator of conformal transformations on the variables of the conformally invariant theory.

In the next section the quantization of the theory is discussed using the formal properties of these constraints and Poisson brackets.

4. FORMAL QUANTIZATION

The R^2 theories have now been cast into the standard first-order form with constraints. These constraints have been shown to be of the first class, i.e., the Poisson brackets of the constraints among themselves is linear in the constraints,

$$[C^i(x), C^j(x')] = \int dy\, f^{ij}{}_k(x, x', y)C^k(y)$$

where the action explicitly includes the constraints, C, with Lagrange multipliers. Whenever such constraints appear, the theory is undetermined in that there are variables which are not determined by the equations of motion; these variables depend upon the choice of the Lagrange multipliers, or, conversely, the variables may be specified freely, thereby determining the Lagrange multipliers. The general procedure for quantizing such systems has been discussed by Faddeev and Slavnov (1980).

In the case of general relativity, these non-dynamical variables are determined by the coordinate conditions. The constraint quantities $\mathscr{H}$ and $\mathscr{H}_m$ generate infinitesimal coordinate transformations, the coordinates must be specified either by doing so directly or by specifying the Lagrange multipliers, N and N^m. If the Lagrange multipliers are specified, then the coordinates on any one spacelike surface are still arbitrary but the coordinates on all other surfaces are determined; in general, however, the specification of the lapse and shift functions will lead to coordinate singularities, hence coordinate conditions on the canonical variables will be used here.

The same is true of the R^2 theories under consideration. In the general case, the constraints are just generalizations of the general relativity constraints; as a result coordinate conditions must be specified. In the conformally invariant theory, there are two additional constraints and, concomitantly, two additional 'coordinate conditions'. The coordinate condition corresponding to J is the choice of conformal gauge: one may choose the conformal factor freely. The coordinate condition corresponding to Q has no direct physical significance: in the general theory, P is related to the trace of K_{kl}, the extrinsic curvature, but here K is determined by the choice of conformal factor and P may be chosen freely.

In summary, the full action with the constraints enforced by Lagrange multipliers must be supplemented by a set of coordinate conditions which complete the determination of the variables. In the general case, these determine the space–time coordinates while in the conformally invariant theory they determine both the space–time coordinates, the conformal factor, and the unphysical variable, P. These coordinate conditions may be written as

$$\chi^\alpha(g, Q, p, P, x) = 0 \qquad (4.2)$$

where

$$[\chi^\alpha(r, t), \chi^\beta(r', t)] = 0$$

and

$$[\chi^\alpha(r), C^i(r')] = M^{\alpha i}(r, r')$$

where M is a non-singular matrix for all field configurations. If this matrix were singular for some particular field configuration, then, for that configuration, the constraints would generate a variation which would still satisfy the coordinate condition, i.e., the coordinates would not be completely determined. Thus, this condition is necessary for the coordinate condition to be complete.

It is now possible to make a change of variables such that the coordinate conditions, χ^α, are themselves canonical variables,

$$q^\alpha(x) = \chi^\alpha(g, Q, p, P, x) \qquad (4.3)$$

while the remaining canonical qs are defined so that their Poisson brackets with the q^α vanish,

$$[q^A(x), q^B(x')] = 0. \qquad (4.4)$$

The variables, p_A, conjugate to the qs are then given implicitly by the requirement that they be canonically conjugate to the qs,

$$[q^A(x), p_B(x')] = \delta_B{}^A(x - x') \qquad (4.5)$$

Since the Poisson bracket of the constraints with the coordinate conditions is non-singular,

$$M^{\alpha i}(x, x') = [\chi^\alpha(x), C^i(x')] = [q^\alpha(x), C^i(x')]$$
$$= \delta C^i(x') / \delta p_\alpha(x) \qquad (4.6)$$

and the constraints may be solved for the p_αs: the coordinates q^α vanish and their conjugate momenta are determined by the constraints.

The quantum theory is defined by the functional integral over the remaining, unconstrained, variables,

$$\langle q't|q''0 \rangle = \int d[q]d[p]\exp i \int \{\mathscr{L}\} \tag{4.7}$$

where

$$\int dx\, \mathscr{L} = \int dx \{p^a \dot{q}_a - h(q,p)\}$$

and the Hamiltonian arises from the surface term given by the constraint. In the case of a compact space, there is of course no surface term but the volume integral,

$$\int d^3x\, Q^{kl}{}_{|kl} = \int (\mathscr{H} + Q^{kl}{}_{|kl})d^3x = \int d\sigma_k Q^{kl}{}_{|l}$$

vanishes identically on the left, hence the remaining variables are not unconstrained. In this paper only the case of an open space will be considered.

The constraints and the coordinates q^α may then be reintroduced using the identity,

$$1 = \int [dq^\alpha][dp_\alpha]\delta[q^\alpha]\delta[C_\beta(p,q)]\det(\delta C/\delta p) \tag{4.8}$$

whereupon, the functional integral becomes

$$\int [dq^A][dp^A]\delta[q^\alpha]\delta[C^\beta(p,q)]\det(\delta C/\delta p)\exp i \int \mathscr{L} \tag{4.9}$$

which, after changing back to the original variables, becomes

$$\int [dq][dQ][dp][dP]\delta[\chi]\delta[C]\det[[\chi,C]]\exp i \int \mathscr{L} \tag{4.10}$$

The delta-functionals involving the constraints may be eliminated by reintroducing the Lagrange multipliers with the result,

$$\int [dq][dQ][dp][dP][d\lambda]\delta[\chi]\det[\chi,C]\exp i \int (p\dot{q} - \lambda^\alpha C^\alpha - h) \tag{4.11}$$

which is precisely the usual result in the gauge ('coordinates') specified by

$$\chi^\alpha = 0$$

and the functional determinant is recognized as the usual Faddeev–Popov determinant. Again the choice of gauge is arbitrary. If another choice of gauge had been made using a different set of coordinate conditions, χ'; then a gauge transformation, ξ, would exist such that if the fields $\{q,p\}$ satisfy the original condition, then the fields $\{q(\xi), p(\xi)\}$, satisfy the new coordinate condition. Since the constraints generate the coordinate transformations,

$$\frac{\delta F}{\delta \xi^\alpha(x)} = [F, C^\alpha(x)], \tag{4.12}$$

and the identity

$$\int [d\xi]\delta[\chi'(q(\xi), p(\xi))]\det[\chi', C] \equiv 1$$

may be invoked to cast the theory into an arbitrary gauge,

$$
\begin{aligned}
\int [dq][dQ][dp]&[dP][d\lambda]\delta[\chi]\det[\chi, C]\exp i \int (p\dot{q}' - \lambda C - h) \\
&= \int d\xi[dq][dQ][dp][dP][d\lambda]\delta[\chi(g)]\det[\chi, C]\exp i \int (p\dot{q}' - \lambda C - h) \\
&\quad \times \delta\{\chi'[g(\xi)]\}\det[\chi', C][g(\xi)] \\
&= \int [dg][dQ][dp][dP][d\lambda]\delta[\chi']\det[\chi', C]\exp i \int (p\dot{q} - \lambda C - h)
\end{aligned} \tag{4.13}
$$

If the coordinate condition, χ, contains an arbitrary additive function, a, independent of the canonical variables, then the integral which is independent of χ, and therefore a, may be integrated with a Gaussian functional of a; the value of the integral is unchanged and becomes,

$$
\begin{aligned}
\langle q't|q''0 \rangle = \int [dq][dQ][dp]&[dP][d\lambda]\det[\chi, C] \\
&\times \exp i \int (p\dot{q}' - \lambda^\alpha C^\alpha - \tfrac{1}{2}\chi^i\sigma_{ij}\chi^j - h)
\end{aligned} \tag{4.14}
$$

the standard result which has now been derived from the canonical quantization.

In the original formulation, the dependence of the coordinate conditions on the canonical variables was taken to be local in time. As long as the dependence is taken to be local, the Faddeev–Popov determinants contain dependence upon Q which prevents one from performing the Q integration to obtain a fourth-order action depending only on g. However, the freedom to make general coordinate transformations may be used to impose conditions of the form,

$$
\chi^\alpha[q] = 0
$$

where the χ^α depend only on g. In that case, the dependence must involve time derivatives of g (or dependence on the lapse and shift functions N, N^m) in order to fully determine the gauge. The function, $\delta\chi/\delta\xi$, which appears in the Faddeev–Popov determinant will then also depend only on g, not p, Q, or P, and the latter integrations may all be done explicitly.

The general result may be written in terms of the standard variables,

$$
\begin{aligned}
\langle g'', Q'', t''|g', Q', t \rangle \\
= \int [dg][dQ][dp][dP][dN][dN^m]\delta[\chi](\det \delta\chi/\delta\xi)\exp i W
\end{aligned} \tag{4.15}
$$

where W is given in equation (2.23) and is linear in p. The coordinate conditions are taken to depend only on g, hence the p integral may be done explicitly, yielding,

$$
\begin{aligned}
\langle | \rangle = \int [dg][dQ][dP][dN^m]\delta[\tfrac{1}{2} P - K]/\det N^6 \\
\times \delta[\chi](\det \delta\chi/\delta\xi)\exp i W
\end{aligned} \tag{4.16}
$$

where,

$$
W = \int dx \, N[\tfrac{1}{2} P_{kl} \mathcal{L}_n Q^{kl} - \mathcal{H}(g, Q, K)] - h.
$$

The delta functional may be used to do the P integration with the result,

$$
\langle | \rangle = \int [dg][dQ][dN^\mu]\delta[\chi][\det(\delta\chi/\delta\xi)/\det N^6]\exp i W \tag{4.17}
$$

which is Gaussian in Q. Since the coordinate conditions are independent of Q, that integral may be done yielding the functional integral,

$$\langle g'',Q'',t''|g',Q',t'\rangle = \int [dg_{\mu\nu}] \exp iW[g]\delta[\chi(g)]$$
$$\times (\det \delta\chi/\delta\xi) \det(N^9\sqrt{g}) \qquad (4.18)$$

where W is the original fourth-order action given in equation (2.1) and the integral is over all paths in g subject to the condition that

$$g_{kl}(r, t') = g'_{kl}(r) \qquad\qquad g_{kl}(r, t'') = g''_{kl}(r)$$
$$Q^{kl}(r, t') = Q^{kl'}(r) \qquad\qquad Q^{kl}(r, t'') = Q^{kl''}(r)$$

with Q^{kl} defined as in equation (2.16).

In the case of the conformally invariant theory, the argument is more complicated because of the conformal constraints. In addition to the usual coordinate conditions, χ^α, one must specify the trace of P. One cannot choose the new coordinate conditions to depend only on g because the variable P has no direct physical significance. The new conditions may be taken to be of the form

$$\chi^{\alpha'}[g'] = 0 \qquad\qquad \alpha = 0, 1, 2, 3, \quad \text{or} \quad c \qquad (4.19)$$

and

$$g'^{kl} P'_{kl} = 0.$$

The metric, g, transforms in the usual way under gauge transformations,

$$\delta g'_{\mu\nu} = g'_{\mu\nu}\delta\xi^\lambda_{;\nu} + g'_{\sigma\nu}\delta\xi^\sigma_{;\mu} + \delta\phi\lambda g'_{\mu\nu} \qquad (4.20)$$

and is invariant under the transformations generated by Q. The variation of P under the gauge transformations is given by,

$$\delta(g'^{kl} P'_{kl}) = 6\delta\psi + \delta\xi, \delta\phi \text{ terms.} \qquad (4.21)$$

The new Faddeev–Popov determinant is given by the variation of the new coordinate conditions with respect to a gauge transformation,

$$\delta\chi'^\alpha/\delta\xi^\lambda, \delta\chi'^\alpha/\delta\phi, \delta\chi'^\alpha/\delta\psi = 0$$

$$\delta(g'^{kl} P'_{kl})/\delta\xi^\lambda = ? = \delta(g'^{kl} P'_{kl})/\delta\phi \qquad\qquad \delta(g'^{kl} P'_{kl})/\delta\psi = 6.$$

Since $\delta\chi^\alpha/\delta\psi = 0$ the det is independent of $\delta P/\delta\xi$ and $\delta P/\delta\phi$ terms, hence the Faddeev–Popov determinant becomes,

$$\det(\delta\chi^\alpha/\delta\xi) \cdot 1$$

which depends only on g.

The dependence of P on the coordinate transformation is complicated by the fact that P is essentially the extrinsic curvature of the $t = $ constant surfaces, hence it depends implicitly on the normal to those surfaces. Under the coordinate transformation, the normal changes since the surface to which the curvature refers changes. Fortunately, the dependence does not enter the final result, hence we will not present it.

The variables, g and P, which appear in the new coordinate conditions may be expressed as gauge transforms of the old variables which appeared in the original coordinate conditions,

$$g'_{\lambda'\sigma'}(y) = \left(\frac{\partial x^\mu}{\partial y^{\lambda'}} g_{\mu\nu} \frac{\partial x^\nu}{\partial y^{\sigma'}}\right)[x(y)]\exp\phi[x(y)] \tag{4.22}$$

then the identity,

$$1 = \int d[y][d\phi][d\psi]\delta\{\chi'^\alpha[g'(y)]\}\delta[P' + 6\psi] \times \det[\delta\chi/\delta(y,\phi)] \tag{4.23}$$

may be used to rewrite the functional integral in the form

$$\langle\,|\,\rangle = \int [dg][dQ][dp][dP][dN^\mu][dy][d\phi][d\psi]$$
$$\times \delta[\chi]\det(\delta\chi/\delta\xi)\exp iW\,\delta\{\chi'[g(y)]\}\delta[P' + 6\psi]\det\delta[\chi'/\delta(y,\phi)]$$
$$W = \int dx\, N[\tfrac{1}{2}p^{kl}\mathcal{L}_ng_{kl} + \tfrac{1}{2}P_{kl}\mathcal{L}_nQ^{kl} - \mathcal{H}_c - \lambda g_{kl}Q^{kl}$$
$$+ \sigma(g_{kl}p^{kl} - \tfrac{1}{2}Q^{kl}P_{kl})]. \tag{4.24}$$

The old variables may then be expressed as coordinate transforms of the new variables; since the coordinate transformation is canonical, the transformed integral is simply the same integral in terms of the new variables,

$$\langle\,|\,\rangle = \int [dg][dQ][dp][dP][dN^\mu][dy][d\phi][d\psi][d\sigma][d\lambda]$$
$$\times \delta[\chi'](\det\delta\chi'/\delta\xi)\exp iW\,\delta\{\chi[g(y,\phi)]\}$$
$$\times \delta[P + 6\psi]\det[\delta\chi/\delta(y,\phi)] \tag{4.25}$$

and the gauge transformation may be integrated over, eliminating the old Faddeev–Popov determinant and yielding,

$$\langle\,|\,\rangle = \int [dg][dQ][dp][dP][dN^\mu][d\lambda][d\sigma]\,\delta[\chi']\delta[P']\det\delta\chi'/\delta\xi$$
$$\times \exp iW \tag{4.26}$$

Now, the p integration may be done yielding a delta-functional which determines P in terms of the extrinsic curvature, K, and the Lagrange multiplier, σ,

$$\langle\,|\,\rangle = \int [dg][dP][dQ][d\sigma][d\lambda]\,\delta[\chi'(g)]\delta[2K + 6\sigma]$$
$$\times \delta(P_{kl} - 2K_{kl} - 2\sigma g_{kl})\det[\delta\chi'/\delta(\xi,\phi)](\exp iW^1)/(\det N^6)$$
$$W^1 = \int dx\, N[(K_{kl} + \sigma g_{kl})\mathcal{L}_nQ^{kl} + (Q^{kl}Q_{kl}/2\alpha\sqrt{g}) \tag{4.27}$$
$$+ \alpha\sqrt{g}\, C^{\hat{0}klm}C^{\hat{0}}{}_{klm} + Q^{kl}{}_{|kl} + {}^3R_{kl}Q^{kl}$$
$$- (K + 3\sigma)(K_{kl} + \sigma g_{kl})Q^{kl} - \lambda g_{kl}Q^{kl} - \sigma Q^{kl}(K_{kl} + g_{kl}\sigma)]$$

where the delta-functional has been used to write the P dependence of W in terms of K and σ. The coordinate condition delta-functional determining P may then be used to eliminate the σ dependence. The action W^1 appearing in the integrand then becomes,

$$W^2 = \int dx \, N[K^T_{kl} \, \mathscr{L}_n Q^{kl} + (Q^{kl} Q_{kl} \, 2\alpha \sqrt{g}) + \alpha \sqrt{g} \, C^{\hat{o}klm} \, C^{\hat{o}}_{klm}$$

$$+ Q^{kl}(N^{-1} N_{|kl} + {}^3 R_{kl}) - \lambda g_{kl} Q^{kl} + \frac{K}{3} Q^{kl} K_{kl}{}^T] \tag{4.28}$$

where the spatial derivative terms involving Q have been integrated by parts, eliminating the surface terms which appeared in the original reduction to first-order form. With the aid of the relation,

$$C^{\hat{o}}_{k\hat{o}l} = \tfrac{1}{2}(\mathscr{L}_n K_{kl} - K K_{kl} - {}^3 R_{kl} - N^{-1} N_{|kl})^T$$

$$= \tfrac{1}{2}(\mathscr{L}_n K^T_{kl} - \tfrac{1}{3} K K_{kl} - {}^3 R_{kl} - N^{-1} N_{|kl})^T$$

the coefficient of Q may then be re-expressed in terms of the Weyl tensor components, $C^{\hat{o}}_{k\hat{o}l}$, and λ. The Gaussian integral over Q may then be done, yielding,

$$\langle \, | \, \rangle = \int [dg][d\lambda][dN^\mu] \, \delta[\chi'(g)] \, (\det \delta\chi/\delta(\xi, \phi))/(\det N^9 \sqrt{g}) \tag{4.29}$$

$$\times \exp i \int dx \, N\sqrt{g}[-\alpha C^{\mu\nu\lambda\sigma} C_{\mu\nu\lambda\sigma}/4 - (\lambda - \zeta)^2/2]$$

which is Gaussian in λ; the result of doing that leaves the final, and expected result,

$$\langle g'', Q'', t'' | g', Q', t' \rangle = \int [d^4 g] \, \delta[\chi'(g)][\det \delta\chi'/\delta(\xi, \phi)] \tag{4.30}$$

$$\times \exp i \int dx \, N\sqrt{g}(-\alpha C^{\mu\nu\lambda\sigma} C_{\mu\nu\lambda\sigma})$$

where the coordinate condition functions are arbitrary, so long as they in fact determine both the space–time coordinates and the conformal factor. The Faddeev–Popov determinant is given above, where Q^{kl} is defined in equation 2.19'.

A local factor of $\det N(\sqrt{g})$ has been suppressed in the conversion to an integral over g. This factor has been the object of a long controversy (Fradkin and Vilkovisky 1977). In any case, being local, it is divergent and, if present, must be renormalized. Also, if the theory is dimensionally regulated, those terms vanish.

5. STABILITY

In the preceding, the space–time has been taken to have the Minkowski signature so that the theory was a real-time theory. The Hamiltonian and constraint functions are all real, and as a result the integrals are all marginally convergent. One cannot improve the convergence by rotating to Euclidean time. This is true even after the constraints have been eliminated and the coordinate conditions imposed.

In the case of the general theory with $G \neq 0$, the Hamiltonian is defined just as it is in the Einstein theory (Hawking 1979, Gibbons and Hawking 1977, York 1972). The super-Hamiltonian constraint,

$$\mathscr{H} = 0$$

may be written as

$$^3R = -(P^{kl}P_{kl} - P^2)/4 + \frac{16\pi G}{\sqrt{g}} \, [p^{kl}P_{kl}/2 - \alpha C^{\hat{0}klm}C^{\hat{0}}_{klm} \tag{5.1}$$

$$- Q^{Tkl}Q^T_{kl}/2\alpha\sqrt{g} + Q^2/18\beta\sqrt{g} - \Lambda\sqrt{g} - Q^{kl}_{\ |kl} - Q^{kl\,3}G_{kl}$$

$$- Q^{kl}P_{kl}P/4 - Q(P^{kl}P_{kl} - P^2)/8 + \varepsilon]$$

and solved for the conformal factor of the intrinsic metric of the three-dimensional surface. The energy density, ε, has been added to represent possible matter energy density. The time at spatial infinity is defined by the requirement that the lapse function have the asymptotic form,

$$N \sim 1 + O(1/r). \tag{5.2}$$

The (non-vanishing) Hamiltonian is then added to the action by including the integral over the bounding surface at infinity of the difference between the extrinsic curvature of the bounding surface and the extrinsic curvature of the bounding surface in a flat three-dimensional space,

$$h = -\int dS(K_S - K_S^0). \tag{5.3}$$

Einstein theory is obtained by finding the solution to the equations of motion for the new variables, Q and P, in the limit in which α and β vanish. The result is

$$^3R = (8\pi G)^2(p^{kl}p_{kl} - \tfrac{1}{4}p^2) + 16\pi G\varepsilon$$

which yields a positive-definite energy (Schoen and Yao 1981, Witten 1981), provided that ε is positive-definite. In the case of the higher derivative theories, the additional terms in the constraint equation may be regarded as additional contributions to ε which are certainly not positive-definite. As a result, the positive energy theorems fail and the energy need not be positive. The integrand on the right side of the equation for ^3R has no positive-definiteness properties whatsoever, nor are any expected. The induced variables, Q and P, appear in this expression with the wrong sign, and the cross term, $pP/2$, is clearly indefinite. The indefiniteness of the energy for small perturbations around flat space is explicitly exhibited in Appendix C. This indefiniteness is consistent with the instability of the solutions to the field equations: runaway cosmological solutions exist (Yamagishi, unpublished, Horowitz and Wald 1978, Horowitz 1981) and small perturbations around flat space are not stable. There is a paradox here: although the energy is manifestly indefinite in the purely gravitational theory, the higher derivative theories have been embedded in a larger supersymmetric theory (van Nieuwenhuizen 1981, van Nieuwenhuizen et al 1977, 1978) for which a formal positive energy theorem holds. In the case of scale invariant theories with $\alpha\beta \leqslant 0$, the total energy as defined by the surface term,

$$E = \int d\sigma_k Q^{kl}_{\ |l}$$

must vanish (Boulware et al 1983b). The formal argument based on the conformal supersymmetric theory fails because the fermionic sector must be quantized using an indefinite metric (Boulware et al 1983a), hence the anticommutator of the supersymmetry charges, Q, has no positivity properties and

$$E = \mathrm{Tr}\{Q, Q\} \neq 0.$$

In the case of the conformally invariant theory, the analysis is quite different although the conclusions are the same. The constraint cannot determine the conformal factor because the theory is conformally invariant (the lapse and shift functions may be chosen by the choice of coordinates and the conformal freedom may be used to choose the conformal factor of the three metric). Now, the constraint reads

$$Q^{kl}{}_{|kl} = p^{kl}P_{kl}/2 - \alpha\sqrt{g}\,C^{\hat{o}klm}C^{\hat{o}}{}_{klm} \tag{5.4}$$
$$- Q^{kl}Q_{kl}/2\alpha\sqrt{g} - Q^{kl\,3}R_{kl}$$

which determines the longitudinal part of Q^{kl}. Using the standard transverse trace decomposition of Q,

$$Q^{kl} = Q^{kl\pi} + Q^{k\,|l} + Q^{l\,|k} - 2g^{kl}Q^{m}{}_{|m}$$

the longitudinal part of Q^k is determined by the constraint equation. Then, the Hamiltonian is defined by the surface integral,

$$\int \mathrm{d}S_k(N\,Q^{kl}{}_{|l} - Q^{kl}N_{|l}) = h \tag{5.5}$$

which is determined by the asymptotic behavior of Q^k. The parameter, h, is then the total energy. Again, since the right side of the constraint is not positive-definite, the energy is not positive-definite. Also the subtleties which arise in the definition of the energy in the Einstein theory are less severe because the constraint equation is simpler.

The standard response to this situation is to invoke the Lee–Wick mechanism (Lee and Wick 1969) in which the fields are quantized using an indefinite metric Hilbert space. The energy (in the case discussed by them) then becomes positive semi-definite, thereby insuring energy stability; however, states of zero norm then exist. In the presence of interactions, these zero norm states can grow or decay exponentially with time. Because the growing states have zero norm, any given state will have bounded norm because the growing portion has vanishing inner product with any state which is not decaying. The norm of an arbitrary state is then bounded, but not positive semi-definite. Lee and Wick define the physical Hilbert space to consist of the states of positive norm which also have real, positive energies. The zero norm growing modes are required to be absent as a boundary condition at I^+. As a result of imposing this boundary condition, the theory exhibits the same kind of acausal behavior as does the classical electron theory: the system must 'pre-respond' to an external force so that the growing modes are not excited.

In the case of quantum electrodynamics, Lee and Wick showed that the theory, although very strange, was physically acceptable in perturbation theory. However, because the growing modes were eliminated as a boundary condition, the theory was not unitary without further modification: the boundary condition prevents the real creation of the growing modes but not their virtual creation. The S-matrix includes contributions from the virtual creation as a result of which it is not unitary without further modification. The modification required to produce an S-matrix which is unitary in the physical sector was then discussed by Cutkosky *et al.* (1969) who showed how to extend it to all orders in perturbation theory.

Recently, Gross and the author discussed these theories further (Boulware and Gross, unpublished), showing that the indefinite metric theories could be written as a functional integral but that the resultant functional integral could not incorporate the Lee–Wick, Cutkosky, Landshoff, Olive, and Polkinghorne prescription for a unitary *S*-matrix. In the presence of interactions (the only interesting case) a given indefinite energy theory can either be quantized in Minkowski space with indefinite energy and a positive-definite Hilbert space metric, or in Euclidean space with positive-definite energy and an indefinite Hilbert space metric.

In the previous section, it was shown how to formulate the R^2 theories of gravity as a functional integral, using space–times with Minkowski signature. The energy is indefinite and the classical instability associated therewith is present.

If the same quantization procedure is attempted in Euclidean space, the energy cannot be taken to be indefinite because the integral will not converge: the argument of the exponential is real, except for the kinetic, $p\dot{q}$, term, but it is not positive-definite. The only possibility is the indefinite metric quantization in which one integrates over purely imaginary values of the indefinite metric fields (Boulware and Gross, unpublished). In that case the constraints become complex, e.g.,

$$\mathcal{H} = -\mathrm{i}p^{kl}P_{kl}/2 - \mathrm{i}Q^{kl}{}_{|kl} - \mathrm{i}Q^{kl\ 3}G_{kl} + \alpha C^{\delta klm}C^{\delta}{}_{klm} + \ldots \tag{5.6}$$

and cannot be implemented by means of a functional delta-functional of the form that was used in § 4,

$$\delta[\mathcal{H}] = \int [\mathrm{d}N]\exp -\mathrm{i}\int N\mathcal{H}.$$

Thus, the Euclidean indefinite metric theory cannot be represented as the functional integral over unconstrained variables with a Wiener–Kac measure. However, one may simply ignore this problem and proceed to write down the integral corresponding to equation (4.12), the functional integral after the constraint delta-functionals have been written as integrals over the Lagrange multipliers. The result is obtained by analytically continuing the real-time result to Euclidean time; the transformation is,

$$t \to -\mathrm{i}\tau \qquad N^m \to -\mathrm{i}N^m \qquad N \to N \tag{5.7}$$

and the action becomes,

$$\mathrm{i}\int \mathcal{L} \to \int \mathrm{d}x\, N[\mathrm{i}(p^{kl}\mathcal{L}_n g_{kl} + P_{kl}\mathcal{L}_n Q^{kl})/2 - N\mathcal{H} - \mathrm{i}N^m\mathcal{H}_m] = W^E$$

The functional integral then becomes,

$$\langle\,|\,\rangle = \int [\mathrm{d}g][\mathrm{d}p][\mathrm{d}Q][\mathrm{d}P][\mathrm{d}N][\mathrm{d}N^m]\,\delta[\chi)\det(\delta\chi/\delta\xi)\exp W^E$$

which does not converge when integrated over real values of P and Q. The indefinite metric quantization using integration over imaginary values must be used, thus,

$$Q^{klT} \to \mathrm{i}Q^{klT} \qquad P_{kl}{}^T \to \mathrm{i}P_{kl}{}^T$$

and the quantity, $\mathcal{H}$, becomes,

$$\begin{aligned}
\mathcal{H} = &-\mathrm{i}p^{kl}P_{kl}/2 - \mathrm{i}Q^{kl}{}_{|kl} - \mathrm{i}Q^{kl\ 3}G_{kl} + \alpha C^{\delta klm}C^{\delta}{}_{klm} \\
&+ Q^{klT}Q^T{}_{kl}/2\alpha\sqrt{g} + Q^2/18\beta\sqrt{g} - \Lambda\sqrt{g} - Q^{kl}P_{kl}P/4 \tag{5.8} \\
&+ Q(p^{kl}P_{kl} + P^2/3)/8.
\end{aligned}$$

Note that only the trace free components of Q and P are rotated to the imaginary axis. This is because there are no dynamical scalar modes in the Einstein theory; in some gauges there are negative metric scalar modes, but the physical massive scalar mode of the R^2 theory is free to be, and is, a positive energy and metric mode.

The coordinate conditions are taken to restrict components of $g_{\mu\nu}$, and, in the case of the conformally invariant theory, P (the part conjugate to $Q^{kl}g_{\mu\nu}$). Then, just as in the pseudo-Riemannian case, the p integral is done first producing a delta-functional relating P to K which can be used to do the P integration. The resultant expression is Gaussian in Q, hence the Q integral may be done, thereby reproducing the original action,

$$\langle\,|\,\rangle = \int [dg]\exp\int\left(-\frac{\alpha}{4}C^{\mu\nu\lambda\sigma}C_{\mu\nu\lambda\sigma} - \beta R^2 + \ldots\right). \tag{5.9}$$

It should be emphasized that this does not constitute a derivation of the Euclidean positive-definite form for the functional integral of the canonical R^2 theory. The constraint delta-functional cannot be written as the functional Fourier integral and, even if that problem could be circumvented, the resultant integral would still not incorporate the Lee–Wick prescription for a unitary S-matrix.

ACKNOWLEDGMENTS

It is a pleasure to thank David Gross for asking the critical questions which led to the work. I am also indebted to Stephen Adler and Gary Horowitz for helpful comments and encouragement. This work was begun at the Institute for Advanced Study, whose support and hospitality made this work possible. This work was supported in part by the US Department of Energy under contract DE-AC06-81ER40048.

APPENDIX A

In this appendix some results for covariant differentiation and differentiation in a surface are summarized. A spacelike slicing of the space–time with slices $t = $ constant and a normal n^μ is assumed.

In general, the Lie derivative of a covariant vector along a vector, n, is given by,

$$\mathcal{L}_n A_\mu = n^\nu A_{\mu,\,\nu} + n^\nu{}_{,\,\mu} A_\nu \tag{A.1}$$

$$= n^\nu A_{\mu;\nu} + n^\nu{}_{;\mu} A_\nu.$$

If the vector A_μ is orthogonal to the surface, $n \cdot A = 0$, then,

$$n^\mu \mathcal{L}_n A_\mu = (n \cdot A)_{,\,\mu} n^\mu = 0 \tag{A.2}$$

and $\mathscr{L}_n A$ is also orthogonal to the surface. Since,

$$n_\mu = (-N, 0), n^\mu = (1, -N^m)/N$$

$$A_\mu = (N^l, \delta_m{}^l) A_l$$

and the space components completely determine the vector.

The extrinsic curvature of the surface is defined by

$$K_{kl} = \mathscr{L}_n g_{kl}/2$$

$$= (\partial g_{kl}/\partial t - N_{k|l} - N_{l|k})/2N \tag{A.3}$$

where the $|$ indicates covariant derivatives within the surface. This equation is a special case of the general relation: if T_{kl} is a covariant second rank tensor orthogonal to the normal to the surface, then the Lie derivative along the normal is,

$$\mathscr{L}_n T_{kl} = n^\mu T_{kl,\,\mu} - (N^m{}_{,k} T_{ml} + N^m{}_{,l} T_{km})/N$$

$$= (\partial T_{kl}/\partial t - N^m T_{kl\,|m} - N^m{}_{|k} T_{ml} - N^m{}_{|l} T_{km})/N. \tag{A.4}$$

The Lie derivative of a contravariant vector is defined as

$$\mathscr{L}_n V^\mu = V^\mu{}_{,\lambda} n^\lambda - n^\mu{}_{,\lambda} V^\lambda{}_. \tag{A.5}$$

and, if A_μ is an arbitrary vector in the surface and $n \cdot V = 0$,

$$A_\mu \mathscr{L}_n V^\mu = A_m(V^m{}_{,\lambda} n^\lambda + N^m{}_{,l} V^l/N)$$

$$= A_m(\partial V^m/\partial t - N^l V^m{}_{|l} + {}^+ N^m{}_{|l} V^l)/N$$

thus, $\mathscr{L}_n V$, projected into the surface, is given by

$$(\mathscr{L}_n V)^m = (\partial V^m/\partial t - N^l V^m{}_{|l} + N^m{}_{|l} V^l)/N. \tag{A.6}$$

Note that the product rule holds,

$$\mathscr{L}_n(V^\mu A_\mu) = n^\lambda \partial_\lambda(A_m V^m)$$

$$= V^l \mathscr{L}_n A_l + A_l \mathscr{L}_n V^l$$

and that the Lie derivative of a scalar is just the ordinary derivative along n but that the space–time volume integral of

$$\mathscr{L}_n(V^\mu A_\mu)$$

is not zero but,

$$\int dx\, N\sqrt{{}^3g}\, \mathscr{L}_n(V^\mu A_\mu) = -\int dx\, NK A_\mu V^\mu + \text{surface terms.} \tag{A.7}$$

If V is taken to be a contravariant vector density under transformations in the surface, its Lie derivative is,

$$(\mathscr{L}_n v)^m = (\partial v^m/\partial t - (N^l v^m){}_{,l} + N^m{}_{,l} v^l)/N \tag{A.8}$$

$$= (\partial v^m/\partial t - (N^l v^m){}_{|l} + N^m{}_{|l} v^l)/N.$$

The space–time volume integral may then be integrated by parts with no additional terms,

$$\int dx\, N(A_m \mathcal{L}_n v^m + v^m \mathcal{L}_n A_m) = \int dx\, \partial_\lambda(Nn^\lambda A_\mu v^\mu) = \text{surface terms.} \quad \text{(A.9)}$$

APPENDIX B

In this appendix, the Poisson bracket of the Hamiltonian constraint with itself is calculated. The constraint contains second derivatives of the canonical variables, hence the Poisson bracket involves third derivatives of the testing functions (the fourth derivative terms cancel).

In order to calculate the Poisson bracket, the integrated constraint,

$$\mathcal{H}[\xi] = \int d^3x\, \xi(x)\mathcal{H}(x) \quad \text{(B.1)}$$

must be functionally differentiated with respect to the canonical variables; these functional derivatives may be expanded as follows according to the number of ordinary (covariant) derivatives which appear,

$$\delta\mathcal{H}[\xi]/\delta g_{kl}(x) = \xi(x)(\partial\mathcal{H}/\partial g_{kl})(x) - (\xi(x)\partial\mathcal{H}/\partial g_{kl|m})|_m$$
$$+ (\xi(x)\partial\mathcal{H}/\partial g_{kl|mn})|_{nm'}$$
$$\partial\mathcal{H}[\xi]/\delta p^{kl}(x) = \xi(x)(\partial\mathcal{H}/\partial p^{kl})(x) \quad \text{(B.2)}$$
$$\partial\mathcal{H}[\xi]/\delta Q^{kl}(x) = \xi(x)(\partial\mathcal{H}/\partial Q^{kl})(x) - (\xi(x)\partial\mathcal{H}/\partial Q^{kl}{}_{|m})$$
$$+ (\xi(x)\partial\mathcal{H}/\partial Q^{kl}{}_{|mn})$$

and

$$\partial\mathcal{H}[\xi]/\delta P_{kl}(x) = \xi(x)(\partial\mathcal{H}/\partial P_{kl})(x) - (\xi(x)\partial\mathcal{H}/\partial P_{kl|m})|_m$$

where the derivatives of $\mathcal{H}$ are ordinary derivatives with the covariant derivatives of the various canonical variables being taken as independent. The derivatives which appear above are explicitly given by

$$\partial\mathcal{H}/\partial Q^{kl}{}_{|mn} = -(\delta_m{}^k\delta_n{}^l + \delta_n{}^k\delta_m{}^l)/2$$
$$\partial\mathcal{H}/\partial g_{kl|mn} = -[Q^{km}g^{ln} + Q^{lm}g^{kn} - Q^{kl}g^{mn} - \tfrac{1}{2}Q^{mn}g^{kl}$$
$$- Q(g^{km}g^{ln} + g^{kn}g^{lm} - 2g^{kl}g^{mn})/4$$
$$+ \sqrt{g}(g^{km}g^{ln} + g^{kn}g^{lm} - 2g^{kl}g^{mn})/16\pi G]$$
$$\partial\mathcal{H}/\partial g_{kl|m} = -(Q^{km|l} + Q^{lm|k} - Q^{kl|m})/2$$
$$+ \alpha g(S^{kul}P^m{}_u - S^{kum}P^l{}_u - S^{lum}P^k{}_u)/2 \quad \text{(B.3)}$$

where

$$S^{kul} \equiv \sqrt{g}(C^{\hat{o}kul} + C^{\hat{o}luk})$$

$$\partial \mathcal{H}/\partial P_{kl} = p^{kl}/2 - \sqrt{g}(P^{kl} - g^{kl}P)32\pi G - Q^{kl}P/4$$
$$- g^{kl}Q^{mn}P_{mn}/4 - Q(P^{kl} - g^{kl}P)/4$$

$$\partial \mathcal{H}/\partial p^{kl} = P_{kl}/2$$

and

$$\partial \mathcal{H}/\partial Q^{kl} = -Q_{kl}{}^{T}/\alpha\sqrt{g} + g_{kl}Q/9\beta\sqrt{g}$$
$$- {}^{3}G_{kl} - P_{kl}P/4 - g_{kl}(P^{mn}P_{mn} - P^{2})/8.$$

The Poisson bracket, equation (3.4), may then be written in the following form,

$$[\mathcal{H}[\xi], \mathcal{H}[\xi']] = \int dx\{[\xi\partial\mathcal{H}/\partial g_{kl} - (\xi\partial\mathcal{H}/\partial g_{kl|m})_{|m} + (\xi\partial\mathcal{H}/\partial g_{kl|mn})_{|mn}]P_{kl}\xi'$$
$$+ 2[\xi\partial\mathcal{H}/\partial Q^{kl} - \xi_{|kl}][\xi'\partial\mathcal{H}/\partial P_{kl} - (\xi'\partial\mathcal{H}/\partial P_{kl|m})_{|m}]$$
$$- (\xi \to \xi')\} \tag{B.4}$$

from which the terms involving various numbers of derivatives may be easily read off. There are no fourth derivative terms. The third derivative terms are

$$C^{3} = -\alpha\int d^{3}x\sqrt{g}[\xi_{|kl}(\xi'S^{kml})_{|m} - (\xi \leftrightarrow \xi')]$$
$$= -\alpha\int d^{3}x\sqrt{g}\, S^{kml}\, {}^{3}R_{kl}(\xi_{|m}\xi' - \xi'_{|m}\xi) \tag{B.5}$$

which cancel, leaving a first derivative residue as a result of the non-commutativity of the covariant derivatives. The Riemann tensor may be eliminated in favor of the Ricci tensor by virtue of the vanishing of the Weyl tensor in three dimensions.

The second derivative terms may also be written in terms of first derivatives of the testing functions by virtue of the identity,

$$\xi_{|mn}\xi' - \xi'_{|mn}\xi = -[(\xi\overleftrightarrow{\nabla}_{m}\xi')_{|n} + (\xi\overleftrightarrow{\nabla}_{n}\xi')_{|m}]/2$$

whence they become,

$$C^{2} = \int d^{3}x\sqrt{g}[(P_{kl}\overleftrightarrow{\nabla}_{n}\partial\mathcal{H}/\partial g_{kl|\underline{mn}}) + (\partial\mathcal{H}/\partial P_{mn})_{|n}]$$
$$\times (\xi'\xi_{|m} - \xi\xi'_{|m}). \tag{B.6}$$

Note that the part of $\partial\mathcal{H}/2g_{kl|mn}$ symmetric in mn is taken.

The explicit first derivative term is,

$$C^{1} = \int d^{3}x\sqrt{g}[-(P_{kl}\partial\mathcal{H}/\partial g_{kl|m})/2 - \alpha\sqrt{g}\, S^{kml}\partial\mathcal{H}/\partial Q^{kl}]$$
$$\times (\xi'\xi_{|m} - \xi\xi'_{|m}) \tag{B.7}$$

and the non-derivative terms cancel identically.

All terms involve only first derivatives of the testing functions; they may be collected and written in the form,

$$(\mathcal{H}[\xi], \mathcal{H}[\xi']) = \int d^3x (\xi' \xi_{|m} - \xi \xi'_{|m}) t^m \tag{B.8}$$

where,

$$t^m = [-(P_{kl}\partial\mathcal{H}/\partial g_{kl|m})/2 + (P_{kl}\overleftrightarrow{\nabla}_n\partial\mathcal{H}/\partial g_{kl|mn})/2$$
$$+ (\partial\mathcal{H}/\partial P_{mn})|_n - \alpha\sqrt{g}\,S^{kml}(^3R_{kl} + \partial\mathcal{H}/\partial Q^{kl})].$$

The terms in t^m explicitly involving the Ricci tensor cancel. There are several terms involving P which may be collected and written in the following form,

$$\alpha\sqrt{g}\{P^{mk}P_{kl}P^{ln} - (P^{mk}P_k{}^n)P_l^l - \tfrac{1}{2}P^{mn}(P^{kl}P_{kl} - P_k{}^k P_l^l)$$
$$- g^{mn}[(P_k{}^k)^3 - 3(P_k{}^k)P^{st}P_{st} + 2P^{kl}P_{lr}P_k{}^r]/6\}_{|n}. \tag{B.9}$$

The tensor in curly brackets is the characteristic equation for the tensor P in three dimensions; it vanishes identically.

The remaining terms may be combined to find,

$$t^m = P^{mn}{}_{|n} + Q^{kl}P_{kl}{}^{|m} - (P_k{}^m Q^{kn})_{|n} = g^{mn}\mathcal{H}_n. \tag{B.10}$$

APPENDIX C

In this Appendix, we exhibit a set of initial value data which satisfies the constraint equations through linear order. As in Einstein gravity, the quadratic terms become the Hamiltonian of the linearized theory; the resultant energy may, according to the particular choice of initial value data, be either positive or negative, thereby giving an explicit demonstration, in the linearized theory, of the lack of positivity of the energy.

The linearized constraints are,

$$-Q^{kl}{}_{,kl} - (h_{kl,kl} - \nabla^2 h)/16\pi G = 0$$

and

$$p^{kl}{}_{,l} = 0$$

which, for small perturbations around flat space, may be trivially satisfied by taking the momenta, p and P, to vanish, and the variables h and Q to be transverse and traceless. If the linearized theory is written in the fourth-order form, the transverse traceless components of $h = g - \eta$ then satisfy the equation

$$(-\partial^2)(\mu^2 - \partial^2)h_{kl}{}^\pi = 0 \qquad \mu^2 = 1/8\pi G$$

which has both the usual graviton, zero mass, solution and a massive solution. In the two cases,

$$\ddot{h}_{kl} = \nabla^2 h_{kl}$$

and

$$\ddot{h}_{kl} = -(\mu^2 - \nabla^2)h_{kl}.$$

Then, the variable, Q^{kl}, may be written, in the linear approximation, as

$$Q^{kl} \simeq -\alpha(\mu^2 - 2\nabla^2)h_{kl}.$$

The quadratic term in the super-Hamiltonian, $\mathscr{H}$, then becomes the Hamiltonian for the linearized system, which, with vanishing momenta and transverse traceless h and Q, becomes,

$$\int d^3r\,\mathscr{H}^Q = \int d^3r[-Q^{kl}Q_{kl}/2\alpha + Q^{kl}\tfrac{1}{2}(\nabla^2)h_{kl} + (16\pi G)^{-1}\tfrac{1}{2}h_{kl,\,m}h_{kl,\,m}]$$

which is positive in the graviton, $\mu^2 = 0$, case,

$$\int d^3r\,\mathscr{H}^Q = \frac{1}{\pi G}\int d^3r\, h_{kl,\,m}h_{kl,\,m}$$

and negative in the case of the massive excitation,

$$\int d^3r\,\mathscr{H}^Q = -\tfrac{1}{2}\int d^3r\left(\frac{1}{8\pi G\alpha}\,h_{kl}h_{kl} + h_{kl,\,m}h_{kl,\,m}\right).$$

In the linearized theory, the energy may be made arbitrarily large of either sign by an appropriate choice of initial conditions.

REFERENCES

Arnowitt R, Deser S and Misner C 1962 in *Gravitation: An Introduction to Current Research* ed L Witten (New York: Wiley)
Boulware D G and Deser S 1975 *Ann. Phys., NY* **89** 193
Boulware D G, Deser S, Gibbons G and Stelle K 1983a unpublished
Boulware D G and Gross D unpublished
Boulware D G, Horowitz G and Strominger A 1983b *Phys. Rev. Lett.* **50** 1726
Cutkosky R E, Landhoff P V, Olive D I and Polkinghorne J C 1969 *Nucl. Phys.* B **12** 281
Dirac P A M 1959 *Phys. Rev.* **114** 924
Faddeev L and Slavnov A A 1980 *Gauge Fields: Introduction to Quantum Theory* (New York: Benjamin/Cummings) p 72ff
Fradkin E S and Vilkovisky G A 1977 *Lett. Nuovo Cimento* **19** 47
Gibbons G W and Hawking S W 1977 *Phys. Rev.* D **15** 2752
Hawking S W 1979 in *General Relativity: An Einstein Centenary Survey* ed S W Hawking and W Israel (Cambridge: Cambridge University Press)
Horowitz G 1981 in *Quantum Gravity 2: A Second Oxford Symposium* ed C J Isham, R Penrose and D W Sciama (Oxford: Oxford University Press) p 107
Horowitz G and Wald R 1978 *Phys. Rev.* D **17** 414
Julve J and Tonin M 1978 *Nuovo Cimento* **46B** 137
Kaku M 1982 'Quantization of conformal gravity: another approach to the renormalization of gravity' *CUNY preprint*
Kawasaki S, Kimura T and Kitago K 1981 *Prog. Theor. Phys.* **66** 2085

Lee T D and Wick G C 1969 *Nucl. Phys.* B **9** 209
—— 1971 *Phys. Rev.* D **2** 1033
Martinelli M 1981 *Nuovo Cimento* **64** B 137
Nakanishi N 1978a *Prog. Theor. Phys.* **59** 972
—— 1978b *Prog. Theor. Phys.* **60** 1190
van Nieuwenhuizen P 1981 *Phys. Rep.* **68** 191
van Nieuwenhuizen P, Ferrara, Kaku M and Townsend P K 1977 *Nucl. Phys.* B **129** 125
van Nieuwenhuizen P, Kaku M and Townsend P K 1978 *Phys. Rev.* D **17** 3179
O'Murchadha N and York J W 1974 *Phys. Rev.* D **10** 428, 437
Schoen R and Yao S T 1981 *Commun. Math. Phys.* **79** 231
Smolin L 1982 'A fixed point for quantum gravity' *Institute for Advanced Study preprint*
Stelle K 1977 *Phys. Rev.* D **16** 953
—— 1978 *Gen. Relativ. Grav.* **9** 353
Weinberg S 1964 *Phys. Rev.* B **135** 1049
—— 1965 *Phys. Rev.* B **138** 988
Witten E 1981 *Commun. Math. Phys.* **79** 231
Yamagishi Y 'Instability of flat space' *Princeton preprint*
York J W 1972 *Phys. Rev. Lett.* **28** 1082

Donaldson's Moduli Space: a 'Model' for Quantum Gravity?

ROGER PENROSE

I fear that I must apologize to Bryce DeWitt for the thoughts which follow, since though dedicated to him, they are of the nature merely of vaguest speculation rather than being a contribution to the thorough-going, well founded type of approach to quantum gravity of which he has been a pioneer and major exponent. These thoughts do, however, owe to him the helpful suggestion that insights may be gained by studying the simpler types of non-linearity which occur in Yang–Mills theory as a preliminary to tackling those more complicated and elusive ones of general relativity (cf DeWitt 1964).

For various reasons, which I do not propose to go into here (cf Penrose 1976, 1979, 1981, 1982, also Karolyhazy 1966, Mielnick 1974, Komar 1969, Kibble 1981), I have found myself driven to take a decidedly unorthodox viewpoint with regard to the features I would expect (or hope) to be present in the 'true' quantum gravity theory. Among such features are:

    (a) time-asymmetry (including effective CPT-violation),

    (b) a non-linear modified quantum theory,

    (c) an objective theory to take over the role of subjective 'wavefunction reduction',

    (d) space–time singularities not to be replaced by non-singular 'bounces'; instead, insofar as a space–time manifold remains a good approximation, the Weyl curvature should approach zero at initial singular boundary points.

It seems highly unlikely that any too 'orthodox' approach to quantum gravity could produce anything much resembling (a), . . . , (d)—which, I suspect, would be reason enough for most people simply to reject these desiderata and go on their own merry ways. Nevertheless, I strongly feel that the evidence from other sources (primarily the existence of a second law of thermodynamics, coupled with cosmological evidence constraining the structure of the big bang singularity) is sufficiently persuasive to indicate that something of the nature of (a), . . . , (d) ought to be striven for.

This suggests to me that while 'orthodox' quantization may supply necessary clues, we should look beyond these standard procedures for the eventual formulation of a quantum gravity theory. Rather than a theory obtainable by 'correct' application of standard quantization to standard general relativity, I would anticipate something more in the nature of a 'grand synthesis', which would be a theory having standard quantum theory and standard general relativity as two different limits (i.e. probably the limits of small energies and of large dimensions). Since this means we are concerned with a theory whose structure is different both from that of a Hilbert space and that of a (pseudo-) Riemannian manifold, any insights into possible mathematical objects which might, even if only very superficially, possess the right kind of limiting behaviors would be something worthy of serious examination. It is in this spirit that I put forward the following highly tentative 'model'. It is not at all to be thought of as a proposal for a quantum gravity theory (which it certainly is not), but as something possibly analogous, in some limited but possibly suggestive ways, to such a theory.

The 'model' I am bringing forward is a construction (due to Donaldson 1983) which has, over the past year, caused a certain amount of excitement in pure mathematical circles, since it has led to the proof of an outstanding (and not altogether anticipated) result concerning the topology of compact differentiable 4-manifolds. When combined with other results one obtains, as a corollary, the remarkable fact that $\mathbb{R}^4$ possesses more than one distinct differentiable structure—a property not shared by $\mathbb{R}^n$ for any other value of n (cf Donaldson 1983, and Atiyah 1983 for an overall view).

Let $\mathscr{M}$ be a compact differentiable 4-manifold which is topologically 'positive' in the sense that the intersection matrix of its 2-cycles is positive-definite. Supply it with a C^∞ Riemannian (positive-definite) metric g. We look for solutions of SU(2)–Yang–Mills theory on $\mathscr{M}$ which are self-dual and have instanton number one. Let $\mathscr{D}$ be the moduli-space of these solutions (i.e. the points of $\mathscr{D}$ represent solutions under gauge equivalence; so distinct points of $\mathscr{D}$ represent Yang–Mills instantons not related to one another by a gauge transformation). Now $\mathscr{D}$ may possess singular points, of one or other of two types A or B, but apart from that it is a smooth 5-manifold-with-boundary. The B-singularities arise only for 'special' choices of the metric g. If we are concerned only with topological (or differential-topological) questions concerning $\mathscr{M}$, then the particular choice of metric g that has been made is of no consequence. We need only ensure that g is suitably 'generic' and all B-singularities will disappear.

The A-singularities remain, however. These consist of a certain number, N, of isolated points within $\hat{\mathscr{D}}$, where N is the second Betti number of $\mathscr{M}$. Locally, each A-point is the vertex of a cone whose cross-section is $\mathbb{C}\mathbb{P}^2$ (complex projective two-space). The boundary $\partial\mathscr{D}$ of $\mathscr{D}$ turns out to be diffeomorphic to $\mathscr{M}$. We may envisage a truncated manifold $\check{\mathscr{D}}$, obtained from $\mathscr{D}$ by 'cutting off' all the A-points, so that $\check{\mathscr{D}}$ is, indeed, a smooth manifold-with-boundary, the boundary being diffeomorphic to a disjoint union:

$$\underbrace{\mathbb{C}\mathbb{P}^2 \cup \ldots \cup \mathbb{C}\mathbb{P}^2}_{N \text{ copies}} \cup \mathscr{M}$$

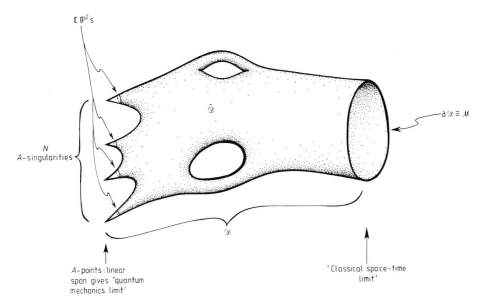

Figure 1 Donaldson's moduli space for SU(2)-instantons, taken as a tentative 'model' for a quantum gravity theory.

This supplies, in a concrete way, a *cobordism* between $\mathscr{M}$ and n copies of $\mathbb{CP}^2$ (see figure 1). Moreover, each $\mathbb{CP}^2$ turns out to count with the same orientation, and from this the required topological theorem can be derived (that the intersection matrix of two-cycles is diagonal).

The significance (if any) for a quantum gravity program lies in the interpretation of the A-points and the identity of $\mathscr{M}$ with $\partial \mathscr{D}$. The A-points arise from those particular SU(2)-instantons which 'degenerate' to the *linear* Maxwell case (with constant 'Yang–Mills direction'). The relation to the second Betti number comes from de Rham–Hodge theory. The boundary $\partial \mathscr{D}$ arises from those limiting instantons which are concentrated at points of $\mathscr{M}$. I would think of the A-points as the analogs of 'physically realized quantum states' or a 'natural basis' for the quantum Hilbert space, where the full Hilbert space is given as their linear span. This Hilbert space corresponds, in some sense, to the linear limit of the Yang–Mills theory. On the other hand the boundary $\partial \mathscr{D}$ is viewed as corresponding to the classical space–time (general relativity) limit of the theory, with individual instantons concentrated at points.

The idea would be that the 'correct' quantum gravity theory would in some sense correspond to the *interior* points of $\hat{\mathscr{D}}$, so that neither the purely quantum-mechanical nor the classical general-relativistic picture would ever be strictly accurate (A-singularities and boundary points being removed) but could be approached arbitrarily closely in a limiting sense.

One can raise many objections to this as a serious model for a quantum gravity picture (and only (b) of my list is in any way incorporated), but it might be worth thinking about further.

ACKNOWLEDGMENT

I am grateful to Michael Atiyah for a beautifully clear exposition of the Donaldson scheme.

REFERENCES

Atiyah M F 1983 Report to 1982 Durham conference
DeWitt B S 1964 in *Relativity, Groups and Topology: The* 1963 *Les Houches Lectures* ed
    B DeWitt and C DeWitt (New York: Gordon and Breach)
Donaldson S K 1983 *Bull. Am. Math. Soc.* **8** 81
Karolyhazy F 1966 *Nuovo Cimento* A **42** 390
Komar R 1969 *Int. J. Theor. Phys.* **2** 157
Kibble T W 1981 in *Quantum Gravity 2* ed C J Isham, R Penrose and D W Sciama (Oxford:
    Oxford University Press)
Mielnick B 1974 *Commun. Math. Phys.* **37** 221
Penrose R 1976 *Gen. Relativ. Gravity* **7** 31
—— 1979 in *General Relativity, An Einstein Centenary Survey* ed S W Hawking and W Israel
    (Cambridge: Cambridge University Press)
—— 1981 in *Quantum Gravity 2* ed C J Isham, R Penrose and D W Sciama (Oxford: Oxford
    University Press)
—— 1982 in *Quantum Structure of Space and Time* ed M J Duff and C J Isham (Cambridge:
    Cambridge University Press)

Quantum Geometry

C J ISHAM

1. INTRODUCTION

This article is dedicated with respect and admiration to Bryce DeWitt on the happy occasion of his sixtieth birthday. There is scarcely a significant idea in quantum gravity to which Bryce has not made a major contribution and it is a fair reflection of the enduring quality of his work that so many of his early papers are still compulsive reading. However, the subject with which his name is most often, and rightly, linked is the background-field method: a technique which has wide-ranging applications to all gauge theories and which played a crucial role in the astonishing renaissance of quantum field theory that we have witnessed in the last decade. Like so many of his ideas, much of this material has been absorbed in the professional collective consciousness with the original source being, regrettably, often forgotten.

In preparing my contribution to this volume I must confess to employing canonical rather than covariant methods and it is true that on occasions Bryce has been known to express 'reservations' about such approaches to quantum gravity. My excuse lies in the connection of the work with the material in DeWitt (1967a): the first of the famous trilogy of papers (DeWitt 1967 a, b, c) which still form such an important keystone in our understanding of quantum gravity.

All workers in quantum gravity, canonical or covariant, will join me in wishing Bryce the happiest of birthdays and in expressing the hope that he will continue to inspire us for many years to come.

The severe difficulties that are encountered in all attempts to quantize the gravitational field have resulted in a fragmented and complex evolution; numerous schemes have been enthusiastically proposed, intensely investigated and then quietly dropped as the inevitable obstacles to progress have appeared. Diversity of ideas involves a diversity of opinion which has, not infrequently, occasioned a noticeable 'warmth' in both private and public discussions of the best path to follow. We may recall in particular the antithetical requirements of covariant and canonical quantization, of perturbative and non-perturbative methods and the still unresolved question of whether *any* conventional quantum field theory scheme is applicable.

Rather, should we radically revise our most fundamental ideas on quantum theory and the nature of space and time?

Fortunately, the wide-ranging nature of the methodology and philosophy of quantum gravity is occasionally relieved by the appearance of certain questions that have a universal appeal and hence transcend the horizons of any particular approach. One such is surely 'what is the nature of quantum geometry?' Or in a more cumbersome paraphrase, 'to what extent are the geometric concepts that underlie classical general relativity still applicable in the quantum domain?' This is indeed an old problem but its interest has not waned with the passage of time and is shared by all votaries of the subject. In what follows I hope to shed a little light on this intriguing question using methods that recover many of the recent important results of Pilati but in a way that hopefully places them in a broader context and suggests new paths for future development.

The first step is to phrase the problem within a framework that is amenable to technical investigation. For example, in a Hilbert-space approach we could enquire about the nature of $\langle \psi | \hat{g}_{ij}(x) | \psi \rangle$ for various states $| \psi \rangle$, in a canonical quantization we might investigate the type of mathematical object that appears as the argument of a state functional $\Psi[g_{ij}(\cdot)]$ and in a functional integral scheme the problem would be to delineate the support of the measure. Is there any reason to suppose that these mathematical entities bear any relation to the infinitely differentiable metric tensor of classical general relativity? True, Wheeler in his epic investigations of canonical quantization presented a picture of superspace—the domain space of the state functionals—based on the space of *smooth* metrics but all our experience with conventional quantum field theory would suggest otherwise†. Indeed it is the singular nature of the quantized metric that is responsible for the notorious non-renormalizability of (weak-field) quantum gravity.

Any deviation from the classical idea of a metric tensor is likely to have dramatic effects. If, for example, we find that the quantized 'metric' is forced to vanish in certain regions how are we to interpret the defining equation for the inverse? The conventional perturbative approach simply ignores this problem by positing a weak-field expression for the metric‡ and then expanding the inverse $g^{ij} = \eta^{ij} - h_{ij} + \ldots$ in the appropriate infinite series. Evidently the nature of quantum geometry is, like most Planck length effects, best investigated within a framework that does not involve such expressions. It would indeed be unwise to impose any *a priori* conditions on physics at these minuscule scales of length and time and, in particular, we should bear in mind the relations often conjectured between quantum topology change and the singular and/or degenerate form of a quantized metric§. In such circumstances

† The possibility of singular geometries was clear to DeWitt when he wrote 'we shall be willing to admit any sort of pathology for γ (the three-metric) which we can get away with, i.e. for which some sort of physical interpretation exists, however idealized, which permits γ to be handled in a consistent fashion' (DeWitt 1967a).

‡ i.e. $g_{ij} = \eta_{ij} + h_{ij}$ $|h_{ij}| \ll 1$.

§ Apologies to the dedicatee who has recently affirmed that ' . . . I believe that to talk of space–time foam and changing topology is nonsense' DeWitt (1982).

the conventional view of space or space–time as a differentiable manifold will require a radical revision.

I have opted to study the problem of quantum geometry within a canonical, Hilbert-space-based approach but many of the general ideas are immediately applicable in either canonical or covariant path integral contexts. Attention is centered on the unitary representations of the affine commutation relations and of the diffeomorphism group of the compact three-manifold; the use of the former in quantum gravity was first suggested in Klauder (1970) and is widely applied in Pilati (1982a, b).

It can be said fairly that the ensuing picture of quantum geometry is as unlike the classical version as one could ever wish for (or dread). The state functionals are concentrated on 'distribution' metrics which vanish outside spatial points or other suitably pathological sets. The possibility that distributions would arise was given some serious thought in Isham (1976) but it was not anticipated that the results would be quite so bizarre!

The treatment that follows is intentionally pedagogical with the main mathematical steps being only sketched; some of the details are quite complicated and will be discussed at greater length in future papers.

2. THE STRATEGY

In the canonical quantization of gravity attention is focused on the metric tensor $g_{ij}(x)$ of three-space Σ (here assumed compact) and the canonical conjugate $\pi^{ij}(x)$ which is related to the extrinsic curvature of Σ embedded in four-dimensional space–time (see Kuchař (1982) for a convenient review of the canonical formalism). An elegant starting point is that of Teitelboim (1980 a, b) who expresses the transition probability between three-geometries g and g', as the (formal) path integral†

$$\langle g, t; g', t' \rangle = \int dg \, d\pi \exp\left(i \int_t^{t'} (\pi^{ij}\dot{g}_{ij} - N\mathcal{H}_\perp - N_i\mathcal{H}^i)d^4x \right) \qquad (2.1)$$

where N and N_i are the lapse and shift functions and $\mathcal{H}^i$ is the infinitesimal generator of spatial coordinate transformations; the dynamical information is essentially carried by the 'Hamiltonian':

$$\mathcal{H}_\perp(g,\pi) = G_{ijkl}\pi^{ij}\pi^{kl} - \kappa^{-1}g^{1/2}\,^{(3)}R. \qquad (2.2)$$

Note that $g = (\det g)^{1/2}$, $\kappa = 8\pi G/c^2$ in terms of the usual constants, and

$$G_{ijkl} = \tfrac{1}{2}g^{-1/2}(g_{ik}g_{jl} + g_{jk}g_{il} - g_{ij}g_{kl}). \qquad (2.3)$$

---

† Ghost terms are also included but these are not relevant to the present discussion.

The (N-dependent) overlap between two state functionals Ψ and Φ is

$$\langle \Psi, \Phi \rangle^N = \int d\mu(g) d\mu(g') \Psi^*[g] \langle g,t; g',t' \rangle \Phi[g'] \tag{2.4}$$

and the final inner product is obtained by functionally integrating over N:

$$\langle \Psi, \Phi \rangle = \int dN \langle \Psi, \Phi \rangle^N. \tag{2.5}$$

Within this framework, the central problem is to find the support of the measure μ in (2.4) on the space of 3-metrics or, to echo Wheeler's famous question (Wheeler 1964, 1968) 'what is the domain space of the quantum gravity state functional?' i.e. 'what is quantum geometry?' Note in particular that any reply that involves singular fields will have a striking effect on our concept of *time*, which for compact Σ, needs to be defined intrinsically in terms of g_{ij} and/or π^{ij} (t and t' drop out of the final answer (2.5)).

We can rephrase our question by noticing that, formally, equation (2.1) implies an imaginary time diffusion equation for $\langle g,t; g',t' \rangle$.

$$i\frac{\partial}{\partial t} \langle g,t; g',t' \rangle + \int_\Sigma \left\{ N \mathcal{H}_\perp \left(g, -i\hbar\frac{\delta}{\delta g} \right) + N_i \mathcal{H}^i \left(g, i\hbar\frac{\delta}{\delta g} \right) \right\} \langle g,t; g',t' \rangle = 0 \tag{2.6}$$

whose exact structure can be investigated only when the significance of the substitution $\hat{\pi}^{ij} = -i\hbar\delta/\delta g_{ij}$ is clarified. This arises of course as a formal representation of the canonical commutation relations (CCR)

$$[\hat{g}_{ij}(x), \hat{\pi}^{kl}(y)] = i\hbar\delta_{(i}{}^k\delta_{j)}{}^l\delta^{(3)}(x,y) \tag{2.7}$$

on functionals of $g_{ij}(\cdot)$ and was much explored in the past in the form of the Wheeler–DeWitt equation $\mathcal{H}_\perp(g,-i\hbar\delta/\delta g)\Psi[g..(\cdot)] = 0$. However, this substitution is mathematically ambiguous and from this point of view our task is to study the representations of (2.7) and hence derive information on the measure μ. It is important to note that the state functionals also carry a representation of the diffeomorphism† group Diff Σ of Σ, and the theory of such representations must be intertwined with that of the CCR. Eventually, of course, we need to factorize out the Diff Σ action and pass to Wheeler's superspace but the details of this procedure will depend crucially on the support of μ; for example, if the 'metrics' are forced to vanish in some region then the action of Diff Σ is exceedingly non-free and the usual theory of superspace or extended superspace is inapplicable.

Let us consider now the problem of imposing the quantum analogue of det $g > 0$ with the signature of g_{ij} being $+ + +$. Of course, in order to discuss the important issue of gravitational singularities we may wish to admit metrics with det $g = 0$ and in any event it seems desirable to have some control over the value of the determinant; we shall follow the lead of Klauder (1970) and Pilati (1982 a, b) and

---

† We shall only consider orientation preserving diffeomorphisms.

replace the CCR (2.6) with the affine commutation relations (AF CR)

$$[\hat{g}_{ij}(x), \hat{\pi}^k{}_l(y)] = i\hbar\delta_{(i}{}^k g_{j)l}(x)\delta^{(3)}(x, y) \tag{2.8}$$

where, heuristically, $\hat{\pi}^k{}_l(y) = \hat{\pi}^{km}(y)\hat{g}_{ml}(y)$. The use of the AF CR is motivated by Klauder's observation that the relations $[\hat{x}, \hat{\pi}] = \hat{x}$ have representations in which the spectrum of $\hat{x}$ is strictly greater than zero, zero, or less than zero and it is supposed that the quantized metric analogue is det $g > 0, = 0, < 0$, respectively. Actually there are various ways in which det g can vanish, for example g_{ij} can have rank 2, 1 or 0 and, as shown below, this is reflected in the details of the representation theory.

Anticipating *a posteriori* the possibility of topological complexity in the support of the metric distributions, it is highly desirable to employ a *global* quantum field theory formalism that does not involve local coordinate charts (as do (2.7) and 2.8)) on Σ. To this end we note that Σ is a compact and orientable three-manifold and is hence necessarily parallelizable; thus there exists a global frame $(\theta^1, \theta^2, \theta^3)$ of nowhere vanishing, linearly independent, one-forms to which all tensors can be referred; for example the metric tensor has a global expression $g_x = g_{AB}(x)\theta_x{}^A \otimes \theta_x{}^B$ where $g_{AB}(x)$ is a set of six scalar functions with det $g_{AB}(x)$ greater than zero everywhere. For various reasons it is useful to employ a triadic, rather than metric, formalism (for example the future inclusion of spinors, the removal of unwanted degrees of freedom in (2.8) (see Pilati 1982a) and the existence of certain topological properties of Σ that may be probed with frames) and we introduce the triad $e_x{}^{(a)} = e_A{}^a(x)\theta_x{}^A, a = 1, 2, 3$; with $g_{AB}(x) = e_A{}^a(x)e_B{}^b(x)\delta_{ab}$. The specific choice of the frame $\{\theta^A\}$ is not too important and drops out of most of the results; indeed there is a manifest frame-free version of the formalism which is not employed in the present paper simply because the symbolism is more abstract and consequently some of the basic ideas might be obscured.

It is important that the quantum geometry effects which we seek should be related in some way to the characteristic Planck length $L_P = (G\hbar/c^3)^{1/2} \approx 10^{-33}$ cm; indeed on macroscopic scales we expect geometry to be perfectly smooth and presumably there exist quantum states which realize this expectation. One natural way of exploring the Planck length is to use a perturbation theory in *inverse* powers of G (Liang 1972, Isham 1976, Teitelboim 1980b); the expression (2.2) for $\mathcal{H}_\perp$ is well adapted to this idea and gives the lowest-order approximation:

$$\mathcal{H}_\perp{}^{(0)} = g^{-1/2}(g_{ik}g_{jl} + g_{jk}g_{il} - g_{ij}g_{kl})\pi^{ij}\pi^{kl}. \tag{2.9}$$

The crucial feature of $\mathcal{H}_\perp{}^{(0)}$ is the absence of spatial derivatives with a corresponding implication that a form of ultra-local quantum field theory is called for (Klauder 1973a); in particular we need a 'renormalization' technique for defining the products of potentially singular field operators but, to this order of $1/G$, we do not need to construct products involving spatial *derivatives* of the quantum fields. This observation leads naturally to the class of representations of the AF CR and Diff Σ considered below and is also compatible with the earlier remarks on loss of differentiability: it is an attractive conjecture that to leading order in $1/G$ only continuous or even measurable functions are relevant with the differential structure reappearing in higher orders of perturbation theory.

A final cautionary remark: one should not necessarily assume that the physically relevant Af CR representations are irreducible; such an assumption presupposes an essentially one-to-one correspondence between classical and quantum observables but there is no *a priori* reason for believing this to be true in quantum gravity; indeed one might well expect new observables that relate to specific Planck length features and, as such, are not simply functions of the canonical variables. It has also been suggested in Komar (1979) that the constraints $\mathscr{H}^i$ and $\mathscr{H}_\perp$ may not be represented by *Hermitian* operators and the possibility has frequently been raised (e.g. DeWitt 1967a) that the parts of g_{ij} and π^{kl} related to intrinsic time are not Hermitian. In the present context this would correspond to non-unitary representations for the Af CR and/or Diff Σ; for the sake of simplicity I shall not consider this possibility here but merely remark that, in any event, the proofs in Komar (1979) are inapplicable if, as seems quite likely, the spectra of the constraint operators are continuous rather than discrete.

3.   THE AFFINE COMMUTATION RELATIONS

In our global formalism, the full set of affine commutation relations is

$$\left[\hat{e}_A{}^a(x), \hat{e}_B{}^b(y)\right] = 0 \tag{3.1}$$

$$\left[\hat{\pi}^a{}_b(x), \hat{\pi}^c{}_d(y)\right] = i\hbar\{\delta^a{}_d\hat{\pi}^c{}_b(x) - \delta^c{}_b\hat{\pi}^a{}_d(x)\}\delta^{(3)}(x,y) \tag{3.2}$$

$$\left[\hat{e}_A{}^a(x), \hat{\pi}^b{}_d(y)\right] = i\hbar\hat{e}_A{}^b(x)\delta_d{}^a\delta^{(3)}(x,y) \tag{3.3}$$

with the δ-function defined so that, formally,

$$f(x) = \int_\Sigma f(y)\delta^{(3)}(x,y)\,d\theta_y$$

where $d\theta$ denotes the measure on Σ associated with the volume element $\theta \equiv (3!)^{-1}\varepsilon_{ABC}\theta^A \wedge \theta^B \wedge \theta^C$. Pilati (1982a) noted that (3.2) defines a $GL(3,\mathbb{R})$ 'current algebra' and this feature plays a key role in the ensuing development.

When handling commutation relations of unbounded operators it is always useful to exponentiate them and re-express the relations in 'Weyl' form. Thus, smearing $\hat{e}_A{}^a$ and $\hat{\pi}^a{}_b$ with respectively triplets of smooth vector fields† $v_a{}^A$ and nonets $\varepsilon_b{}^a$ of functions, the unitary operators $V(v)$ and $U(\varepsilon)$ are defined formally by

$$V(v) = \exp\left(i\int_\Sigma \hat{e}_A{}^a(x)v_a{}^A(x)d\theta_x\right)$$

$$U(\varepsilon) = \exp\left(i\int_\Sigma \hat{\pi}^a{}_b(y)\varepsilon_b{}^a(y)d\theta_y\right). \tag{3.4}$$

† $v_a{}^A$ is not subject to any condition such as det $v > 0$.

Now the $U(\varepsilon)$ afford a representation of that part of the 'gauge group' $C^\infty(\Sigma, GL^+(3, \mathbb{R}))$ (smooth functions from Σ into the group of real 3×3 matrices with positive determinant) that can be reached by exponentiation of the gauged Lie algebra of $GL(3, \mathbb{R})$. It is natural to look for representations that extend to the full group and, in this case, equations (3.1)–(3.3) are replaced by

$$V(v_1)V(v_2) = V(v_1 + v_2) \tag{3.5}$$

$$U(\Lambda_1)U(\Lambda_2) = U(\Lambda_1 \cdot \Lambda_2) \tag{3.6}$$

$$U(\Lambda)V(v)U(\Lambda)^{-1} = V(\Lambda v) \tag{3.7}$$

where $(\Lambda v)_a{}^A \equiv \Lambda_a{}^b v_b{}^A$ and $\Lambda \varepsilon\, C^\infty(\Sigma, GL^+(3, \mathbb{R}))$. Equations (3.5)–(3.7) reveal that, in fact, we must seek unitary representations of the gauge group $C^\infty(\Sigma, M(3, \mathbb{R})) \,\circledS\, GL^+(3, \mathbb{R}))$ where the semidirect product of $GL^+(3, \mathbb{R})$ with the additive group $M(3, \mathbb{R})$ of all 3×3 real matrices is defined by the group law $(M, g)(M', g') = (M + gM', gg')$; it is the replacement of the conventional Weyl commutation relations with a *gauge* group of functions on Σ that leads to many of the characteristic features of the theory developed below.

We must also find a unitary representation $D(\phi)$ of diffeomorphisms ϕ of Σ (so that $D(\phi_1)D(\phi_2) = D(\phi_1 \circ \phi_2)$) which intertwines correctly with the U and V operators. Detailed calculation gives the complete set of relations

$$D(\phi_1)D(\phi_2) = D(\phi_1 \circ \phi_2) \tag{3.8}$$

$$D(\phi)U(\Lambda)D(\phi)^{-1} = U(\Lambda \circ \phi^{-1}) \tag{3.9}$$

$$D(\phi)V(v)D(\phi)^{-1} = V(v \circ \phi^{-1} \cdot L_\phi). \tag{3.10}$$

In deriving these expressions the Jacobian $\mathscr{J}_{\phi A}{}^B$ of the diffeomorphism ϕ with respect to the frame $\{\theta^A\}$ is defined by $(\phi^*\theta^A)_x = \theta_x{}^B \mathscr{J}_{\phi B}{}^A(x)$ giving the basic Diff Σ action on the space of triads: $e_A{}^a \longrightarrow \mathscr{J}_{\phi A}{}^B e_B{}^a \cdot \phi$. Then $(v \circ \phi^{-1} L_\phi)_a A = v_a{}^B \circ \phi^{-1} L_{\phi B}{}^A$ where $L_{\phi B}{}^A \equiv \det \mathscr{J}_{\phi^{-1}} \mathscr{J}_{\phi B}{}^A \circ \phi^{-1}$ and it can be shown that

$$L_{\phi_1 \circ \phi_2} = L_{\phi_2} \cdot L \circ \phi_1^{-1} L_{\phi_1}. \tag{3.11}$$

Thus L_ϕ defines a type of $GL^+(3, \mathbb{R})$ valued one-cocycle and the multiplication law (3.11) plays an important role in that part of the representation theory of Diff Σ that is relevant to quantum gravity.

Note from equation (3.5) that the V operators provide a unitary representation of the Abelian group $C^\infty(\Sigma, M(3, \mathbb{R}))$. The usual Von Neumann algebra spectral theory shows that this representation is unitarily equivalent to one defined on a Hilbert space $L^2(E', d\mu)$† by

$$(V(v)\Psi)[\chi] = e^{i\chi(v)} \Psi[\chi] \tag{3.12}$$

where E' is the space of distributions on the test fields $v_a{}^A$ and μ is some measure that is quasi-invariant under the actions of $C^\infty(\Sigma, GL^+(3, \mathbb{R}))$ and Diff Σ whose

---

† The state vectors may themselves be vector space valued if the representation is not cyclic.

representations are now of the form

$$(U(\Lambda)\Psi)[\chi] = \sqrt{RN}\,\Psi[\chi\cdot\Lambda] \tag{3.13}$$

$$(D(\phi)\Psi)[\chi] = \sqrt{RN}\,\Psi[\mathscr{J}_\phi\chi\circ\phi]. \tag{3.14}$$

Here RN denotes the appropriate Radon–Nikodym derivatives and $\chi\cdot\Lambda$ (resp. $\mathscr{J}_\phi\chi\circ\phi$) are the distributions whose components are respectively $\chi_A{}^b\Lambda_b{}^a$ and $\mathscr{J}_{\phi A}{}^B\chi_B{}^a\circ{}^a\phi$. Note that equation (3.13) is the rigorous form of the affine version of the Wheeler–DeWitt representation of $\hat{\pi}^a{}_b(x) = -i\hbar\,e_A{}^a(x)[\delta/\delta e_A{}^b(x)]$ and that our search for the nature of quantum theory now resolves into discovering the support properties of μ; in particular is there any sense in which χ satisfies $\det\chi > 0$?

We wish to choose representations of $C^\infty(\Sigma, M(3,\mathbb{R})\circledS GL^+(3,\mathbb{R}))$ and Diff Σ that are compatible with the posited ultralocal structure of Planck length quantum gravity. Gauge group representations of this type were discussed in the past within the context of representations of the current algebra commutation relations; a useful general reference is Guichardet (1972). Suitably adapting these techniques, our programme has the following steps.

(i) Study representations $\tilde{T}$ of $M(3,\mathbb{R})\circledS GL^+(3,\mathbb{R})$ on some Hilbert space $\mathscr{V}$.

(ii) Construct the direct integral $h = \int_\Sigma^\oplus \mathscr{V}^x \, d\theta_x$ and define a representation T on h of the gauge group

$$C^\infty(\Sigma, M(3,\mathbb{R})\circledS GL^+(3,\mathbb{R})) \text{ by } (T(g)\Psi)(x) := \tilde{T}(g(x))\Psi(x)$$

where $\qquad\qquad g\in C^\infty(\Sigma, M(3,\mathbb{R})\circledS GL^+(3,\mathbb{R}))$

and $\qquad\qquad \Psi\in L^2(\Sigma,\mathscr{V};d\theta)\approx h. \tag{3.15}$

(iii) Interwine the natural Diff Σ action on Σ with a unitary representation on $\mathscr{V}$ of the $GL^+(3,\mathbb{R})$ cocycle L_ϕ (equation (3.11)).

(iv) At this point we have sets of unitary operators $V(v), U(\Lambda)$ and $D(\phi)$ satisfying the group laws (3.5)–(3.10). Now rewrite this representation in the $L^2(E',d\mu)$ form and read off the support of μ; the result?—*Quantum Geometry*.

(v) If appropriate, exponentiate the representation on h onto the boson Fock space $\mathscr{H} = \exp h$ (possibly involving a cocycle) and repeat the analysis of μ.

(vi) Look for interesting generalizations of this construction.

Let us now proceed to step (i) and find the unitary representations of $M(3,\mathbb{R})\circledS GL^+(3,\mathbb{R})$.

4.   REPRESENTATIONS OF $M(3,\mathbb{R})\circledS GL^+(3,\mathbb{R})$

Since this group is a semidirect product it is natural to employ Mackey's induced representation theory (for example, Mackey 1978) which, since the semidirect structure given by the group law $(M,g)(M',g') = (M + gM', gg')$ is regular, provides an exhaustive account of the irreducible representations. Recall that induced representations are defined on functions on a $GL^+(3,\mathbb{R})$ orbit (with a quasi-invariant measure) in the character group of $M(3,\mathbb{R})$, with values in a vector space carrying a unitary representation of the little group/isotropy group of the orbit, and hence our first task is to study these orbits.

The character group of $M(3,\mathbb{R})$ is isomorphic to $M(3,\mathbb{R}) \approx \mathbb{R}^9$ via $N \rightsquigarrow \chi_N$ where $\chi_N(M) = \exp i \operatorname{Tr}(MN)$ and $g \in GL(3,\mathbb{R})$ acts by $\chi_N \rightsquigarrow \chi_{Ng^{-1}}$. The matrix space $M(3,\mathbb{R})$ may be decomposed into a disjoint union $M(3,\mathbb{R}) = M^{(3)} \cup M^{(2)} \cup M^{(1)} \cup \{0\}$ where the superscript denotes the rank of the matrix and is *unchanged* by a $GL^+(3,\mathbb{R})$ action. Hence, the orbits divide into three non-trivial categories that may be conveniently considered separately:

I. $M^{(3)}$

The subset $M^{(3)} \subset M(3,\mathbb{R})$ of rank three matrices is the full general linear group $GL(3,\mathbb{R})$ of real, invertible, 3×3 matrices. It comprises two disjoint $GL^+(3,\mathbb{R})$ orbits (labeled by the sign of the determinant) each of which is diffeomorphic to $GL^+(3,\mathbb{R})$; indeed the group action is free and the little group is trivial.

II. $M^{(2)}$

The orbit structure on the subset $M^{(2)}$ of rank two matrices may be obtained by noticing that each such matrix can be expressed in row reduced echelon form. This leads at once to a decomposition of the orbit space of the full $GL(3,\mathbb{R})$ action as a disjoint union $M^{(2)}/GL(3,\mathbb{R}) \approx \mathbb{R}^2 \cup \mathbb{R}^1 \cup \{*\}^{(2)} \approx \mathbb{R}P^2$ (real two-dimensional projective space). An orbit belonging to the $\mathbb{R}^2$ or $\mathbb{R}^1$ classes is labeled by the rank two matrix that lies on it of the row reduced form:

$$\begin{pmatrix} 1 & 0 & a \\ 0 & 1 & b \\ 0 & 0 & 0 \end{pmatrix} \quad (a,b) \in \mathbb{R}^2 \quad \text{and} \quad \begin{pmatrix} 1 & d & 0 \\ 0 & 0 & 1 \\ 0 & 0 & 0 \end{pmatrix} \quad d \in \mathbb{R}$$

respectively while $\{*\}^{(2)}$ denotes the single orbit containing the rank two matrix

$$\begin{pmatrix} 0 & 1 & 0 \\ 0 & 0 & 1 \\ 0 & 0 & 0 \end{pmatrix}.$$

In all cases the little group is the semidirect product $\mathbb{R}^2 \circledS \mathbb{R}_+$ with the group law

$$\left[\binom{a}{b}, r \right] \left[\binom{a'}{b'}, r' \right] = \left[\binom{a + ra'}{b + rb'}, rr' \right].$$

III. $M^{(1)}$

The $GL(3,\mathbb{R})$ orbit structure on the subset $M^{(1)}$ of rank one matrices also decomposes into a disjoint union $\mathbb{R}^2 \cup \mathbb{R}^1 \cup \{*\}^{(1)} \approx \mathbb{R}P^2$ where the $\mathbb{R}^2$ and $\mathbb{R}^1$ orbits are now classified by the matrices

$$\begin{pmatrix} 1 & a & b \\ 0 & 0 & 0 \\ 0 & 0 & 0 \end{pmatrix} \quad (a,b) \in \mathbb{R}^2 \quad \text{and} \quad \begin{pmatrix} 0 & 1 & d \\ 0 & 0 & 0 \\ 0 & 0 & 0 \end{pmatrix} \quad d \in \mathbb{R}$$

respectively while $\{*\}^{(1)}$ denotes the orbit containing the matrix

$$\begin{pmatrix} 0 & 0 & 1 \\ 0 & 0 & 0 \\ 0 & 0 & 0 \end{pmatrix}.$$

In all cases the little group is $\mathbb{R}^2 \circledS GL(2,\mathbb{R})$ with the group law

$$[(^a_b),A][(^{a'}_{b'}),A'] = [(^a_b)+A(^{a'}_{b'}),AA'].$$

Notice that in every case the little group is itself a regular semidirect product and hence the Mackey technique may be iterated to yield the representations of $M(3,\mathbb{R})$ s $GL^+(3,\mathbb{R})$. In this paper we shall discuss Type I only with a single orbit diffeomorphic to $GL^+(3,\mathbb{R})$. The representation space is then the Hilbert space $L^2(GL^+(3,\mathbb{R}); dH)$ of complex functions that are square integrable with respect to Haar measure dH. The group action is

$$(\tilde{T}(M,g)\psi)(e_A{}^a) = \exp(i M_a{}^A e_A{}^a)\psi(e_A{}^b g_b{}^a) \tag{4.1}$$

where the indices on the group elements are written as shown with a view to the next stage of development of the theory. Evidently we have selected the technically simplest case; it also provides the quantum analogue of the classical condition $\det e > 0$.

The remaining cases describe singular geometries and are more complicated since the unitary representations of the little groups are themselves of infinite dimension. This leads to an infinite component state vector for quantum gravity. Notice that the $\det e = 0$ matrices lie on the boundary of $M^{(3)}$ in $M(3,\mathbb{R})$; this is connected with the discussion in DeWitt (1967a) of the geometry of the boundaries of the regular metrics in superspace. The role of the type II and III representations is an intriguing one to which we shall return in a later publication.

5. QUANTUM GEOMETRY AND THE REPRESENTATIONS OF $C^\infty(\Sigma, M(3,\mathbb{R}) \circledS GL(3,\mathbb{R}))$

The construction of the ultralocal representations of the AfCR gauge group follows the general plan outlined in §3; the basic Hilbert space is the direct integral[†]

$$\mathcal{H} = \int_\Sigma^\oplus L^2(GL^+(3,\mathbb{R}); dH)^x d\theta_x \approx L^2(\Sigma, L^2(GL^+(3,\mathbb{R}), dH); d\theta)$$

$$\approx L^2(\Sigma \times GL^+(3,\mathbb{R}), d\theta \otimes dH) \tag{5.1}$$

and the group representation is

$$(V(v)\psi)(x,e) = \exp(iv_a{}^A(x)e_A{}^a)\psi(x,e) \tag{5.2}$$

$$(U(\Lambda)\psi)(x,e) = \psi(x,e\Lambda(x)). \tag{5.3}$$

Note that this representation can be extended to the group $C^\circ(\Sigma, M(3,\mathbb{R}) \circledS GL^+(3,\mathbb{R}))$ of *continuous* functions (including those that are

[†] In the frame-free formalism the state vectors are functions on the principal $GL(3,\mathbb{R})$ bundle of coframes.

non-differentiable) and in this sense, and as anticipated, the differentiable structures on Σ and $GL^+(3,\mathbb{R})$ become irrelevant.

The representation of Diff Σ that interwines correctly with (5.2) and (5.3) is simply

$$(D(\phi)\psi)(x,e) = J_{\phi^{-1}}^{1/2}(x)\psi(\phi^{-1}(x), L_\phi(x)e). \tag{5.4}$$

The group actions in (5.2)–(5.4) are essentially the same as those of Pilati (1982) whose results we rederive within a more general and global context. Notice also that the representation (5.4) is a natural 'index extended' version of the simplest type of diffeomorphism group representation considered in Versik *et al* (1975) and Ismagilov (1972).

To reveal the nature of quantum geometry that is hidden in equations (5.2)–(5.4) we must express these group representations in a L^2 (E', dμ) form. Fortunately the answer is simple: there is a natural injection $\Sigma \times GL^+(3,\mathbb{R}) \to E'$ given by $(x, e_A{}^a) \to (x, e_A{}^a\delta_x^{(3)}(\cdot))$ where $\delta_x^{(3)}(\cdot)$ denotes the Dirac distribution supported by $x \in \Sigma$. Hence the measure μ is concentrated on metric distributions of the type

$$\chi_{x,e}(v) := v_a{}^A(x)e_A{}^a \tag{5.5}$$

and leads to an exotic picture of quantum geometry. *The quantized triad vanishes everywhere except at a single point where it is a δ-function; the quantum analogue of the non-vanishing of the classical determinant lies in the non-vanishing of the matrix coefficient of this singular function.* It is indeed difficult to conceive of a picture of quantum geometry that could be further from our classical notions! In particular the idea of superspace as the space of (equivalence classes under Diff Σ) smooth metrics needs radical revision.

Many implications and developments may be attached to the scheme above but pressure of space permits me to mention briefly only a few. As is usual in this type of quantum field theory, it is possible to define algebraic functions of the quantum fields in a fairly natural way. Note first that (5.2) gives the field operator

$$(\hat{e}_A{}^a(y)\psi)(x,e) = e_A{}^a\delta_x^{(3)}(y)\psi(x,e) \tag{5.6}$$

and hence

$$(\hat{e}_A{}^a(y)\hat{e}_{Ba}(y')\psi)(x,e) = e_A{}^a e_{Ba}\delta^{(3)}(y,y')\delta_x^{(3)}(y)\psi(x,e)$$

in particular the metric tensor $\hat{g}_{AB}(y)`` = ``\hat{e}_A{}^a(y)\hat{e}_{Ba}(y)$ contains an ill defined $\delta^{(3)}(y,y)$ singularity. However, there is a natural *regularized* product (cf Klauder 1973a) obtained by a wavefunction renormalization that results in this singular coefficient being dropped, in particular:

$$(\hat{g}_{AB}^{\text{reg}}(y)\psi)(x,e) := e_A{}^a e_{Ba}\delta_x^{(3)}(y)\psi(x,e). \tag{5.7}$$

Similarly one may define the 'inverse' triad $\hat{e}_a{}^A$:

$$(\hat{e}_a{}^A(y)\psi)(x,e) := e_a{}^A\delta_x^{(3)}(y)\psi(x,e) \tag{5.8}$$

where $e_a{}^A$ denotes the genuine, matrix inverse of $e_A{}^a$. Evidently

$$[\hat{e}_A{}^a(y)\hat{e}_a{}^B(y)]^{\text{reg}} = \delta_A{}^B\hat{\alpha}(y) \tag{5.9}$$

where $\hat{\alpha}(y)$ is itself a δ function supported field:

$$(\hat{\alpha}(y)\psi)(x,e) := \delta_x{}^{(3)}(y)\psi(x,e). \tag{5.10}$$

This is as far as one can go with the construction of a quantized analogue of the classical inverse of the triad (or metric). As anticipated in the introduction, we are evidently a long way from the conventional weak-field perturbative approach in which $\hat{e}_a{}^A(x) = \delta_a{}^A + \hat{h}_a{}^A(x)$!

The field $\hat{\alpha}$ is of considerable interest in its own right; in smeared form† (5.10) becomes

$$(\hat{\alpha}(f)\psi)(x,e) = f(x)\psi(x,e) \tag{5.11}$$

which commutes with the $V(v)$ and $U(\Lambda)$ operations. Hence we have a *reducible* representation of the AfCR; it can be localized in any region $B \subset \Sigma$ by smearing $\hat{\alpha}$ with the characteristic function of B and then using the resulting projection operator to pick out the appropriate subspace h_B of $h = L^2(\Sigma \times \mathrm{GL}^+(3,\mathbb{R}), \mathrm{d}\theta \otimes \mathrm{d}H)$; note however that $D(\phi)$ does *not* commute with $\hat{\alpha}(f)$, rather the diffeomorphisms translate the localized representations from one region to another according to the rule $D(\phi)\hat{\alpha}(f)D^{-1}(\phi) = \hat{\alpha}(f \circ \phi^{-1})$. Note also that $\hat{\alpha}(f)$ obeys the relations

$$\hat{\alpha}(\lambda f + \mu g) = \lambda\hat{\alpha}(f) + \mu\hat{\alpha}(g) \qquad f,g \in C^\circ(\Sigma,\mathbb{R}) \tag{5.12}$$

$$\hat{\alpha}(fg) = \hat{\alpha}(f)\hat{\alpha}(g) \qquad \lambda,\mu \in \mathbb{R} \tag{5.13}$$

so we have a representation of the *ring* of functions on Σ. But the *topological* structure of Σ is determined precisely by the algebraic structure and in this sense *the topology of Σ is coded into the quantum field theory*. Evidently the operator $\hat{\alpha}$ has considerable significance and will be discussed further in future publications.

We note also that the expectation of $\hat{g}_{AB}$ in a quantum state ψ is given by (5.7) as

$$\langle\psi|\hat{g}_{AB}{}^{\text{reg}}(x)|\psi\rangle = \int\limits_{\mathrm{GL}^+(3,\mathbb{R})} e_A{}^a e_{Ba}|\psi(x,e)|^2\,\mathrm{d}H(e). \tag{5.14}$$

Therefore, there exist physically accessible states (i.e. with finite norm) which smear out the singularity over any given region; in particular, nowhere vanishing (in x), C^∞, wavefunctions can give an averaged $\hat{g}_{AB}{}^{\text{reg}}(x)$ that is a conventional metric and hence correspond to classical states of the system.

A natural extension of the theory employs distributions $\Sigma_{n=1}^N e_A^{(n)a}\delta_{x^{(n)}}{}^{(3)}(\cdot)$ concentrated on a finite set of points $x^{(1)}, x^{(2)}, \ldots, x^{(N)}$.

The Hilbert space‡ is $L^2(\Pi_{i=1}^N \Sigma \times \mathrm{GL}^+(3,\mathbb{R}), \otimes_{i=1}^N(\mathrm{d}\theta \otimes \mathrm{d}H))$ with the group

---

† Note that we can use test functions that are merely continuous or, even (worse) measurable.
‡ Strictly speaking, the domain space should be factored by the action of the permutation group acting on the N points; e.g. Versik *et al* (1975).

actions (cf Versik *et al* 1975, Ismagilov 1972).

$$(V(v)\psi)(x^{(1)} \ldots x^{(N)}; e^{(1)} \ldots e^{(N)})$$

$$= \exp\left(i\sum_n v_a{}^A(x^{(n)})e_A^{(n)a}\right)\psi(x^{(1)} \ldots x^{(N)}; e^{(1)} \ldots e^{(N)}) \quad (5.15)$$

$$(U(\Lambda)\psi)(x^{(1)} \ldots x^{(N)}; e^{(1)} \ldots e^{(N)})$$

$$= \psi(x^{(1)} \ldots x^{(N)}; e^{(1)}\Lambda(x^{(1)}) \ldots e^{(N)}\Lambda(x^{(N)})) \quad (5.16)$$

$$(D(\phi)\psi)(x^{(1)} \ldots x^{(N)}; e^{(1)} \ldots e^{(N)})$$

$$= \prod_{n=1}^{N} J_{\phi^{-1}}{}^{1/2}(x^{(n)})\psi(\varphi^{-1}(x^{(1)}) \ldots \varphi^{-1}(x^{(N)}); L_\phi(x^{(1)})e^{(1)} \ldots L_\phi(x^{(N)})e^{(N)}) \quad (5.17)$$

Variable N may be accommodated by employing the boson Fock space $\exp L^2(\Sigma \times GL^+(3, \mathbb{R}), d\theta \otimes dH)$ with various cocycles; it can be shown that the singular support points $x^{(1)} \ldots x^{(N)}$ are distributed on $\Sigma \times GL^+(3, \mathbb{R})$ with Poisson statistics† which in turn suggests the use of more general stochastic point processes.

Finally we note that $\Sigma \times GL^+(3, \mathbb{R})$ is multiply connected with a fundamental group $\pi_1(\Sigma) \times Z_2$ and, as usual in the quantum theory of a system with a non-simply-connected configuration space, one may construct inequivalent quantization schemes that reflect this topological property. These correspond to the various, flat‡, complex line bundles that can be placed over the configuration space and whose smooth cross-sections carry different, essentially self-adjoint, extensions of certain key operators. In particular it can be shown in the present case that a state vector ψ is more properly of the form $\psi(x, e; h)$ where h belongs to the cohomology group $H^1(\Sigma; Z_2)$. The gauge group $C^\circ(\Sigma, GL^+(3, \mathbb{R}))$ now acts, roughly speaking by (cf equation (5.3))

$$(U(\Lambda)\psi)(x, e; h) = \psi(x, e\Lambda(x); h + [\Lambda]) \quad (5.18)$$

where $[\Lambda]$ denotes the cohomology element $\Lambda^*(\iota)$ and ι is the characteristic element of $H^1(GL^+(3, \mathbb{R}); Z_2) \approx Z_2$. Now the elements of $H^1(\Sigma; Z_2)$ label inequivalent spin structures on Σ and it is intriguing to conjecture that the present phenomenon is related to the intrinsic topological spin discussed in Friedman and Sorkin (1980); indeed if such spin does exist then there must be some way in which the different possible spin structures are reflected in the quantum field theory and the construction above is a good candidate.

6. FROM POINTS TO LOOPS TO STOCHASTIC GEOMETRY

From a group theoretic angle the main features of our construction lay in the embedding

$$\Sigma \times GL^+(3, \mathbb{R}) \to E' \qquad \text{i.e.} \ \chi_{x,e}(v) = v_a{}^A(x)e_A^a \quad (6.1)$$

$$(x, e_A{}^a) \rightsquigarrow e_A{}^a \delta_x^{(3)}(\cdot)$$

† Recall that in non-relativistic quantum field theory it is the positions of the *particles* that obey Poisson statistics.

‡ That is, the Ul field strength vanishes.

with the group actions

$$(\Sigma \times GL^+(3,\mathbb{R})) \times C^0(\Sigma, GL^+(3,\mathbb{R})) \to \Sigma \times GL^+(3,\mathbb{R}) \qquad (6.2)$$

$$((x, e_A{}^a), \Lambda) \rightsquigarrow (x, e_A{}^b \Lambda_b{}^a(x))$$

$$(\Sigma \times GL^+(3,\mathbb{R})) \times \mathrm{Diff}\,\Sigma \to \Sigma \times GL^+(3,\mathbb{R}) \qquad (6.3)$$

$$((x, e_A{}^a), \phi) \rightsquigarrow (\phi^{-1}(x),\ L_{\phi A}{}^B(x) e_B{}^a)$$

and the emphasis was placed on the representation theory of the affine commutation relations with the diffeomorphism group playing a relatively secondary role. However, if we consider the theory from the view point of the diffeomorphism group then the central feature is the action of this group on a point, or finite set of points, in Σ. But why should we restrict ourselves to points? Why not consider any other collection of 'geometrical shapes' in Σ upon which there is a diffeomorphism group action?

As a concrete example consider the space of 'circles', or more precisely, continuous loops in $\Sigma \times GL^+(3,\mathbb{R})$, i.e. elements of $C^0(S^1, \Sigma \times GL^+(3,\mathbb{R}))$. The analogue of (6.1) is

$$C^0(S^1, \Sigma \times GL^+(3,\mathbb{R})) \to E' \qquad \chi_{\sigma, e}(v) = \oint v_a{}^A(\sigma(t)) e_A{}^a(t) dt$$

$$(\sigma(\cdot), e_A{}^a(\cdot)) \rightsquigarrow \chi_{\sigma, e} \qquad (6.4)$$

where $\sigma(\cdot)$ and $e_A{}^a(\cdot)$ denote loops in Σ and $GL^+(3,\mathbb{R})$ respectively and the group actions (cf equations (6.2), (6.3)) are

$$C^0(S^1, \Sigma \times GL^+(3,\mathbb{R})) \times C^0(\Sigma, GL^+(3,\mathbb{R})) \to C^0(S^1, \Sigma \times GL^+(3,\mathbb{R}))$$

$$((\sigma(\cdot), e_A{}^a(\cdot)), \Lambda(\cdot)) \rightsquigarrow (\sigma(\cdot), e_A{}^b(\cdot)\Lambda_b{}^a(\cdot)) \qquad (6.5)$$

$$C^0(S^1, \Sigma \times GL^+(3,\mathbb{R})) \times \mathrm{Diff}\,\Sigma \to C^0(S^1, \Sigma \times GL^+(3,\mathbb{R}))$$

$$((\sigma(\cdot), e_A{}^a(\cdot)), \phi) \rightsquigarrow (\phi^{-1} \circ \sigma(\cdot), L_{\phi A}{}^B \circ \sigma(\cdot) e_B{}^a(\cdot)) \qquad (6.6)$$

with for example, the quantum triad representation[†]:

$$(\hat{e}(v)\psi)(\sigma(\cdot), e(\cdot)) = \left\{ \oint v_a{}^A(\sigma(t)) e_A{}^a(t) dt \right\} \psi(\sigma(\cdot), e(\cdot)). \qquad (6.7)$$

Thus in this picture of quantum geometry the metric is concentrated on loops in Σ and there are natural generalizations of the definitions of inverses, regularized products etc. Clearly many features of this construction need to be clarified and expanded, not least being the existence of the desired quasi-invariant measure on the new configuration space $C^0(S^1, \Sigma \times GL^+(3,\mathbb{R}))$.

Brownian motion techniques are possibly applicable and it is noteworthy that the range space of a multi-dimensional Brownian motion may easily have a non-integral

---

[†] The AFCR representations are direct integrals and highly reducible, however the components are mapped into each other under the action of the diffeomorphism group.

Hausdorff dimension. Thus our picture of quantum geometry must be extended to include singular support sets that are fractals!

The use of loops is particularly interesting as it enables the topological structure of Σ to be probed by the quantum field theory in a highly non-trivial way; clearly the loop structure reflects the fundamental group $\pi_1(\Sigma)$ and this group essentially determines the global structure of the compact three-manifold Σ. One is also reminded of the loops and strings which feature prominently in modern accounts of quantum chromodynamics (for example, Polyakov 1981a, b).

We can also contemplate the use of geometrical objects other than circles and replace S^1 in (6.4)–(6.7) with a more general topological space T. Thus we seek a theory of random 'T-sets' in Σ and, fortunately, probability theorists have investigated such objects in depth; a definitive reference is Kendall (1973) which contains a fund of fascinating information, much of which is directly applicable to the quantum gravity situation.

7.   CONCLUSION

Our final picture of quantum geometry is certainly striking: the triad/metrics are concentrated on random T-sets that are very likely to be fractals. Surely here we have a genuine manifestation of the quantum foam ideas that were first suggested in a canonical context by Wheeler (1964, 1968) and then developed by Hawking (1979) in his four-dimensional, Riemannianized program. Clearly many standard concepts (such as intrinsic time), based as they are on a smooth picture of geometry, will require some radical rethinking in the light of the very singular nature of quantum geometry that has been revealed by these investigations.

There is no problem in accommodating matter fields within this framework and supergravity looks particularly promising; since the work of Mathon and Streater (1971) it has been known that non-trivial ultralocal fermion fields must have some unusual features and indeed Klauder (1973b) showed that they were invariably accompanied by an ultralocal *scalar* field. It is an attractive conjecture that, in an appropriate sense, the ensuing structure is *intrinsically supersymmetric*; if true, this would throw together supersymmetry, supergravity and the Planck length in a new and intriguing way. Tony Kakas has been studying the fermion problem and has shown that the quantum states appear naturally as functions on a graded, superspace extension of $\Sigma \times GL^+(3, \mathbb{R})$; this and, other aspects of the theory, will appear in a forthcoming series of papers by Kakas and the author.

Many questions remain to be asked and answered and the work sketched above can only be regarded as the germ of a research program; nevertheless I hope that I have managed to convince the reader that the old problem of the nature of quantum geometry is far from being dead and may have many surprises in store for us.

REFERENCES

DeWitt B S 1967a *Phys. Rev.* **160** 1113–48
——1967b *Phys. Rev.* **162** 1195–239
——1967c *Phys. Rev.* **162** 1239–56
——1982 *Quantum Gravity* 2 ed C J Isham, R Penrose and D Sciama (Oxford: Oxford University Press) pp 439–87
Friedman J L and Sorkin R D 1980 *Phys. Rev. Lett.* **44** 1100–3
Giuchardet A 1972 *Symmetric Hilbert spaces and related topics* in *Lecture Notes in Mathematics* **261** (Berlin: Springer)
Hawking S W 1979 *General Relativity* ed S W Hawking and W Israel (Cambridge: Cambridge University Press) ch 15
Isham C J 1976 *Proc. R. Soc.* A **351** 209–32
Ismagilov R S 1972 *Math USSR Izv.* **6** 181–210
Kendall D G 1973 *Stochastic Geometry* ed D G Kendall (New York: Wiley) pp 322–76
Klauder J R 1970 *Phys. Rev.* D **2** 272–6
——1973a *Lectures in theoretical physics* ed W E Britten (Boulder: Colorado Associated University Press)
——1973b *Ann. Phys., NY* **79** 111–30
Komar A 1979 *Phys. Rev.* D **20** 830–3
Kuchař K 1982 *Quantum Gravity* 2 ed C J Isham, R Penrose and D Sciama (Oxford: Oxford University Press) pp 329–76
Liang E P T 1972 *Phys. Rev.* D **5** 2458–66
Mackey G W 1978 *Unitary Group Representations in Physics, Probability and Number Theory* (London: Benjamin/Cummings)
Mathon D and Streater R F 1971 *Z. Wahr.* **20** 308–16
Pilati M 1982a *Phys. Rev.* D **26** 2645
——1982b *Phys. Rev.* D to appear
Polyakov A M 1981a *Phys. Lett.* **103B** 207–10
——1981b *Phys. Lett.* **103B** 211–3
Teitelboim C 1980a *Phys. Lett.* **96B** 77–82
——1980b in *An Einstein Centenary Volume* ed A Held (New York: Plenum)
Versik A M, Gel'fand I M and Graev M J 1975 *Russ. Math Survey* **30** 1–50
Wheeler J A 1964 in *Relativity Groups and Topology* ed B S DeWitt and C DeWitt (London: Gordon and Breach) pp 317–520
——1968 in *Battelle Recontres* 1967 ed C DeWitt and J A Wheeler (New York: Benjamin)

The Role of Time in Path Integral Formulations of Parametrized Theories

JAMES B HARTLE and KAREL V KUCHAŘ

1. INTRODUCTION

In the history of the efforts to construct a quantum theory of space–time the contrast between the canonical and the covariant approaches to this goal is striking. They differ both in their focus and in their techniques. Consider just a short list of issues of concern for each. For the canonical approach there is the identification of time, the construction of the Hilbert space of states and the realization of dynamics via quantum constraint equations. For the covariant approach one could mention the preservation of manifest coordinate invariance, the construction of perturbation theory, renormalization and the implementation of dynamics by functional integrals. One might almost be describing two different theories rather than two approaches to a common goal.

Bryce DeWitt has long championed the covariant approach to quantum gravity. Yet he has not neglected the canonical theory. His three papers of 1967 (DeWitt 1967a, b, c) are landmarks in both pathways of the subject. On the contrast between them he remarked then '. . . no rigorous mathematical link has thus far been established between the canonical and covariant theories. In the case of infinite worlds it is believed that the two theories are merely two versions of the same theory, expressed in different languages, but no one knows for sure.' Since 1967 the connection between the formalisms of the canonical and covariant approaches *has* been explored (see, e.g., Faddeev and Popov 1973). Despite the links which have been established there is still a lack of congruence between the issues stressed by the two developments. Where, for example, is the issue of the choice of time mirrored in the covariant approach? How does the problem of non-renormalizability appear in the canonical tack?

It is the belief of the present authors that progress could be made towards quantum gravity by exploring relations between the two approaches not only at the level of formalism but also at the level of issues. This paper is a modest step in this direction. It is concerned with how the choice of time, so central to the canonical method, is reflected in the formalism of the covariant version of the theory. We hope that it is a fitting appreciation to Bryce DeWitt, a pioneer in both versions of the theory.

The spacelike surfaces which foliate space–time and define time are central to canonically formulated theories because time plays a privileged role in Hamiltonian mechanics and therefore a privileged role in quantum mechanics. The issue of the appropriate choice of time arises in theories like relativity in which the variables describing these surfaces appear as dynamical variables. Such theories are called parametrized theories. The choice of time becomes important because different choices may lead to inequivalent quantum-mechanical theories. The freedom in the choice of time is reflected as an invariance of the theory formally similar to invariances of gauge theories. Indeed, it is just this formal similarity which is exploited in deriving the connection between the canonical and covariant approaches to quantum gravity: The covariant functional integral is constructed from the canonical theory using the procedure well tested for gauge theories. If the issue of the choice of time is to be reflected in the covariant quantization procedure for quantum gravity it must arise as a difference in the quantization procedure from that used for gauge theories. It is this difference which we shall investigate here in the context of a simple model. A careful distinction between the gauge theories and parametrized theories will thus be important to us and we discuss this distinction in the next section.

2.   GAUGE THEORIES AND PARAMETRIZED THEORIES

The conventional actions for electromagnetism and gravitation are both invariant under transformations of the field variables on which they depend. For electromagnetism, the action functional

$$S = -\frac{1}{16\pi} \int d^4x (\nabla_\mu A_\nu - \nabla_\nu A_\mu)^2 \tag{2.1}$$

is invariant under gauge transformations

$$A'_\mu = A_\mu + \nabla_\mu \Lambda. \tag{2.2}$$

The gravitational action functional

$$S = \frac{1}{16\pi G} \int d^4x (-g)^{1/2} R + \text{(surface terms)} \tag{2.3}$$

is invariant under the action of diffeomorphisms $x^\alpha = x^\alpha(x^{\beta'})$ on the metric

$$g_{\alpha'\beta'} = \frac{\partial x^\alpha}{\partial x^{\alpha'}} \frac{\partial x^\beta}{\partial x^{\beta'}} g_{\alpha\beta}. \tag{2.4}$$

The existence of such types of invariance means that neither the action nor the resulting dynamical equations are expressed entirely in terms of the physical degrees of freedom of the system. They depend on a redundant set of variables which are physically equivalent if connected by a transformation resulting from the invariance. Two vector potentials are physically equivalent if they are connected by a gauge transformation. Two metrics are equivalent (i.e., describe the same space–time) if they are connected by a diffeomorphism.

There is good reason for such redundancy introduced in the conventional descriptions of electromagnetism and gravity. In electromagnetism one could eliminate the redundant degrees of freedom (say, by expressing the action in terms of the transverse components of A_μ in the temporal gauge), but only at the expense of losing the manifest Lorentz invariance. In general relativity the use of redundant variables is almost a necessity. It is not possible to eliminate explicitly the redundant degrees of freedom from the action. For this reason, if for no other, it is important to be able to deal with theories expressed in terms of redundant variables.

From the Lagrangian perspective the gauge transformations of electromagnetism and the diffeomorphisms of general relativity appear structurally similar. An important distinction emerges, however, if one takes the Hamiltonian viewpoint where one must specify the time. In electromagnetism this might be the Minkowski time of a particular inertial frame. In gravity it would be a family of spacelike hypersurfaces. In electromagnetism gauge transformations change the field variables at a particular instant of time, leaving the action unchanged. General relativity also displays this type of invariance in that there are diffeomorphisms which only change field variables on a given spacelike surface. We shall generally call such transformations gauge transformations and the theory gauge invariant.

In addition to gauge transformations which change the dynamical variables at a particular instant of time, general relativity is also invariant under transformations which involve the time. This is because general relativity is an example of a parametrized theory in which the physical time—that intrinsic or extrinsic geometrical quantity which locates a unique member of the family of spacelike surfaces—is one of the redundant field variables describing the theory. The invariance corresponding to this redundance is that of changes in the foliation of space–time by spacelike surfaces. We call transformations having this character reparametrization transformations and the theory re-parametrization invariant.

A perhaps more familiar example of a parametrized theory is that of a free relativistic particle. The action is

$$S = -m \int d\tau \left(-\eta_{\alpha\beta} \frac{dx^\alpha}{d\tau} \frac{dx^\beta}{d\tau} \right)^{1/2} \tag{2.5}$$

where $\eta_{\alpha\beta}$ is the Minkowski metric. The physical time, x^0, enters as a dynamical variable parametrized by a parameter τ. As in general relativity the action is invariant under the action of diffeomorphisms in τ on the variables $x^\alpha(\tau)$. These are the re-parametrization transformations of this theory.

In the Hamiltonian framework a distinction can thus be drawn between gauge invariance and re-parameterization invariance. This distinction would not have the importance it does were it not for the close connection between classical Hamiltonian mechanics and quantum mechanics and the consequent special role played by time in quantum-mechanical theories.

The transition from classical mechanics to quantum mechanics is most simply achieved when the dynamics of the system is expressed in terms of the true physical degrees of freedom. For example, the transition can be accomplished by constructing a Hilbert space of states on which the physical coordinates q^a, $a = 1, \ldots n$ and their conjugate momenta p_a satisfy the commutation relations

$$[q^a, p_b] = i\delta_b{}^a. \tag{2.6}$$

The quantum dynamics is then given by translating the classical Hamiltonian $h(q^a, p_a)$ into a quantum operator and writing down the Schrödinger equation. One can then construct, for example, the transition amplitudes $\langle q''^a t''| q'^a t' \rangle$.

An equivalent prescription for making the passage is the expression for the transition amplitudes as path integrals over paths in the classical phase space

$$\langle q''^a t''| q'^a t' \rangle = \int \frac{\delta^n p \, \delta^n q}{[(2\pi)^n]} \exp\{i\, S[p_a(t), q^a(t)]\}. \tag{2.7}$$

Here $S[p_a(t), q^a(t)]$ is the canonical action and the measure is the natural one induced by the Liouville measure on phase space.

A modified prescription is needed for theories such as electromagnetism and gravity which are formulated in terms of redundant variables. It is desirable to have a prescription for the transition amplitudes as path integrals on the extended phase space including these variables and conjugate momenta. It is on this space that we are given the action.

Such a prescription has been thoroughly worked out for gauge theories. An analogous but more restricted procedure is needed for re-parametrization invariant theories. Our purpose in this article is to emphasize the differences between these two prescriptions by considering a simple model. The model is a non-relativistic system described by a Lagrangian

$$L\left(q^a, \frac{dq^a}{dt}\right) = \tfrac{1}{2}m\delta_{ab}\frac{dq^a}{dt}\frac{dq^b}{dt} + A_a\frac{dq^a}{dt} - V \tag{2.8}$$

where q^a is a position vector in $\mathbb{R}^n$ and A_a and V are prescribed functions of q^a. The flat metric on $\mathbb{R}^n$ is used to construct the first term. We have discussed more general models, in more general language, with more general aims elsewhere (Kuchař 1983, Hartle and Kuchař 1983). Our purpose here is to draw the distinction between the path integral quantization of gauge and re-parametrization invariant theories as simply as possible not as generally as possible, and for this the model decribed by the Lagrangian equation (2.8) will suffice. We can turn it into a gauge theory by adding a redundant dynamical degree of freedom. We can turn it into a parametrized theory by adding the time as a dynamical variable. In each case by following the transition from the basic path integral in equation (2.7) to one on the extended phase space of variables we arrive at a path integral quantization prescription. By tracing the role of the choice of time for re-parametrization invariant theories we shall see how this choice is reflected in the covariant quantization procedure.

3. QUANTUM MECHANICS IN PHYSICAL VARIABLES

To make the transition from classical to quantum mechanics for the non-relativistic system described by the Lagrangian (2.8) we first construct the Hamiltonian

$$h(p_a, q^a) = \frac{1}{2m} \delta^{ab} (p_a - A_a)(p_b - A_b) - V(q^a) \tag{3.1}$$

then the canonical action for a phase space path

$$S[p_a, q^a] = \int_{t'}^{t''} dt \left(p_a \frac{dq^a}{dt} - h(h_a, q^a) \right) \tag{3.2}$$

and finally the transition amplitudes by the path integral in equation (2.7). The meaning of the path integral can be made concrete by first dividing the time interval (t', t'') up into intermediate slices at $t(K)$, $K = 0, 1, \ldots, N$ where $t(0) = t', t(N) = t''$. A path in phase space is then represented by a skeletonized path specified by the values of p_a and q^a on the slices $K = 0, \ldots, N$. The integrand can be skeletonized by using for the action to go from $t(K)$ to $t(K+1)$ the canonical action equation (3.2) evaluated along the classical path for $q^a(t)$ with $p_a(t) = p_a(K)$. The skeletonized measure is the product

$$(2\pi)^{-n} d^n p(0) \prod_{K=1}^{N-1} (2\pi)^{-n} d^n p(K) d^n q(K). \tag{3.3}$$

There is no integral over the initial and final qs because they are fixed by the specification of the transition amplitude. There is a sum over the initial momentum because it is free but none over the final momentum because with the above skeletonization procedure the action does not depend on it. The path integral is the limit of the skeletonized integrals as the division of the time interval (t', t'') is increasingly refined. This skeletonization procedure is spelled out in detail in Kuchař (1983) but for us its existence will be more important than its exact form.

4. A PARAMETRIZED THEORY

Let us now turn our non-relativistic model into a parametrized theory by including the time t as a dynamical variable q^0. The configuration space then becomes $\{q^\alpha\}$, $\alpha = 0, \ldots, n$. By specifying the position along the path with a parameter τ the canonical action of equation (3.2) can be written as

$$S[p_a, q^\alpha] = \int_{\tau'}^{\tau''} d\tau \, (p_a \dot{q}^a - h\dot{q}^0) \tag{4.1}$$

where a dot denotes the derivative with respect to the parameter τ. Variations of this action with respect to p_a and q^a still yield the correct equations of motion. Variation with respect to q^0 yields another correct relation expressing the conservation of energy. If we now introduce p_0 by the equation

$$H = p_0 + h(p_a, q^a) = 0 \tag{4.2}$$

the action can be expressed in canonical form

$$S[p_\alpha, q^\alpha] = \int_{\tau'}^{\tau''} d\tau \, p_\alpha \dot{q}^\alpha. \tag{4.3}$$

Correct equations of motion are obtained from this action if the variation of p_α and q^α are constrained by equation (4.2). This is the 'Hamiltonian constraint' for this system. It can be enforced implicitly by introducing a Lagrange multiplier N so that the action reads

$$S[p_\alpha, q^\alpha, N] = \int_{\tau'}^{\tau''} d\tau \, (p_\alpha \dot{q}^\alpha - NH). \tag{4.4}$$

Variations with respect to N now give the Hamiltonian constraint. Free variation of the p_α, q^α give the correct equations of motion. The theory is now in a fully parametrized form.

The sequence of transformations of the classical action $(3.2) \to (4.1) \to (4.3) \to (4.4)$ can be mirrored in the quantum theory by transformations of the path integral for transition amplitudes. Start with the path integral (2.7) as made concrete by time-slicing. Pick any monotonically increasing function $t(\tau)$ which through inversion assigns a unique value of the parameter τ to each slice. Add an integration over q^0 on each intermediate slice by inserting the relation

$$1 = \int dq^0 \, \delta[q^0 - t(\tau)] \tag{4.5}$$

with τ having the value appropriate to the slice. Extend the action to be a functional of q^0 by equation (4.1) and define $q^0(0) = t'$, $q^0(N) = t''$. The value of the path integral remains unchanged but its form is now

$$\langle q''^a t'' | q'^a t' \rangle = \int \delta^n p \, \delta^{n+1} q [(2\pi)^{-n} \delta(q^0 - t(\tau))] \exp\left(i \int_{\tau'}^{\tau''} d\tau \, (p_a \dot{q}^a - h\dot{q}^0) \right). \tag{4.6}$$

Here, by the square brackets in the measure we indicate a product of factors—one δ function for each intermediate slice and one factor of $(2\pi)^{-n}$ for the initial slice and each intermediate slice.

In a similar way one can introduce the integrations over p_0 by inserting the relation

$$1 = \int dp_0 \, \delta[p_0 + h(p_a, q^a)] = \int dp_0 \, \delta(H) \tag{4.7}$$

on each slice but the last and extending the action to be a functional of p_0 by equation (4.3). The value of the functional integral is unchanged and following equation (4.6) we write

$$\langle q''^{\alpha}|q'^{\alpha}\rangle = \int \delta^{n+1}p\,\delta^{n+1}q\,\{(2\pi)^{-n}\delta[q^0 - t(\tau)]\delta(H)\}\exp\left(i\int_{\tau'}^{\tau''}d\tau p_{\alpha}\dot{q}^{\alpha}\right) \quad (4.8)$$

where the bracket now additionally indicates one $\delta(H)$ on the initial and each intermediate slice. Finally, if one rewrites $\delta(H)$ on each slice as

$$\delta(H) = (2\pi)^{-1}\int_{-\infty}^{+\infty}\varepsilon\,dN\exp(i\varepsilon NH) \quad (4.9)$$

where ε is the separation in τ from the slice of interest to the next one then the transition amplitude can be written

$$\langle q''^{\alpha}|q'^{\alpha}\rangle = \int \delta^{n+1}p\,\delta^{n+1}q\,\delta N\,\{(2\pi)^{-n-1}\varepsilon\delta[q^0 - t(\tau)]\}\times$$
$$\exp\left(i\int_{\tau'}^{\tau''}d\tau\,(p_{\alpha}\dot{q}^{\alpha} - NH)\right). \quad (4.10)$$

The action now has the fully parametrized form on the extended phase space. The path integral is over paths in that phase space and over the Lagrange multiplier function $N(\tau)$. Lest there be any confusion we write out explicitly the measure in its skeletonized form

$$\frac{d^{n+1}p(0)}{(2\pi)^n}\frac{\varepsilon(0)\,dN(0)}{2\pi}\prod_{K=1}^{N-1}\frac{d^{n+1}p(K)d^{n+1}q(K)}{(2\pi)^n}\frac{\varepsilon(K)\,dN(K)}{2\pi}\times$$
$$\delta[q^0(K) - t(K)] \quad (4.11)$$

where $\varepsilon(K) = \tau(K+1) - \tau(K)$.

In making the passage $(2.7) \rightarrow (4.6) \rightarrow (4.8) \rightarrow (4.10)$ we have used the methods developed for gauge theories to transform the path integral in a way which mirrors the parametrization process of the classical action $(3.2) \rightarrow (4.1) \rightarrow (4.3) \rightarrow (4.4)$. We have achieved a path integral prescription for calculating transition amplitudes starting from a fully parametrized action. Although the techniques of gauge theories were used, the result is not the gauge theory result as we shall show in subsequent sections.

Various Lagrangian forms of the path integral for the transition amplitude can be obtained by carrying out the momentum integrations in the phase space forms. For example, carrying out the Gaussian integrals over the momenta in equation (2.7) we arrive at the familiar physical configuration space representation in terms of the physical Lagrangian l:

$$\langle q''^a t''|q'^a t'\rangle = \int \delta^n q\left(\frac{m}{2\pi i\eta}\right)^{1/2}\exp\left(i\int_{t'}^{t''}l(q^a,\dot{q}^a)\,dt\right) \quad (4.12)$$

where η is the spacing in t between one slice and the next.

An analogous form on the extended configuration space can be obtained from equation (4.10) as follows: first, integrate over N to obtain a delta function which can be used to do the p_0 integral. The remaining Gaussian integrals over the momenta p_a can be carried out to give

$$\langle q''^\alpha | q'^\alpha \rangle = \int \delta^{n+1} q \left[\left(\frac{m}{2\pi i \dot{q}_0 \varepsilon} \right)^{1/2} \delta[q^0 - t(\tau)] \right]$$

$$\times \exp\left[i \int\limits_{\tau'}^{\tau''} d\tau \left(\frac{m}{2\dot{q}^0} \delta_{ab} \dot{q}^a \dot{q}^b + \dot{q}^a A_a - \dot{q}^0 V \right) \right]. \tag{4.13}$$

One recognizes the exponent as a valid classical action for the system. Numerically, this action is equal to $\int l \, dt$. Variation with respect to q^a gives the usual equations of motion. Variation with respect to q^0 gives the conservation of energy.

Other configuration space path integrals with multipliers can be constructed (Hartle and Kuchař, 1983) but we shall not display them here.

5.   A GAUGE THEORY

We shall now review the procedure for quantizing a gauge theory using a path integral of a gauge invariant action. We shall be brief because the procedure in its standard form has been set forth lucidly by Faddeev (1969) (see also Faddeev and Popov 1967, 1973, Fradkin and Vilkovisky 1977, DeWitt 1979 and the references contained therein).

Our non-relativistic model can be made into a simple gauge theory by adjoining a single spurious (gauge) degree of freedom, ϕ, and considering the theory on the enlarged space of variables $\{q^a, \phi\}$. (We achieve some formal simplification by considering just a single gauge degree of freedom.) The evolution of ϕ is arbitrary and so the velocity $\dot{\phi}$ may be freely prescribed,

$$\dot{\phi}(t) = \lambda(t). \tag{5.1}$$

(We use a dot in this section for derivative with respect to t.) Equation (5.1) can be reproduced in a canonical framework by starting from the canonical action

$$\sigma[\phi, \pi; \lambda] = \int dt (\pi \dot{\phi} - \lambda \pi). \tag{5.2}$$

Variation with respect to π gives equation (5.1), variation with respect to the multiplier λ gives the constraint

$$\pi = 0 \tag{5.3}$$

and variation with respect to ϕ just gives the same constraint differentiated in time. Adding σ to the original action (3.2) yields a canonical action on the extended space of physical and gauge variables:

$$S[q^a, \phi, p_a, \pi; \lambda] = \int dt \, [p_a \dot{q}^a + \pi \dot{\phi} - h(p, q) - \lambda \pi]. \tag{5.4}$$

In a typical gauge theory the physical and gauge degrees of freedom do not appear decoupled as in equation (5.4) but mixed together. We can achieve this for our model by making an arbitrary transformation of the variables $\{q^a, \phi\}$ to a new set $\{Q^A\}$ $A = 1, \ldots, n+1$. By requiring that $p_a \dot{q}^a + \pi \dot{\phi} = P_A \dot{Q}^A$ we determine the transformation of the momenta necessary to make the transformation canonical. The resulting action is

$$S[Q^A, P_A; \lambda] = \int dt \, [P_A \dot{Q}^A - h(P, Q) - \lambda \pi(P, Q)] \tag{5.5}$$

with π necessarily linear in the P_A. This is the form in which gauge theories usually appear. For example, in electromagnetism the Q^A are the $A_i(x^b)$, the P_A the $E_i(x^b)$. λ is $A_0(x^b)$, π is $\nabla \cdot E(x^b)$ and h is the usual Hamiltonian for the electromagnetic field $(E^2 + B^2)/8\pi$. A sum over spatial points is then necessary to generate the complete action.

We can now trace this evolution of the classical theory from (3.2) to (5.5) in the path integral which gives the transition amplitude. Begin again with the path integral (2.7). Pick any function $\Phi(\phi, \pi, p_a, q^a)$ for which the equation $\Phi = 0$ yields a unique value of ϕ. Add integrations over π and ϕ to the path integral by inserting on each intermediate slice the relation

$$1 = \int d\pi \, d\phi \, \delta(\pi) \delta(\Phi) \left| \frac{\partial \Phi}{\partial \dot{\phi}} \right| \exp(i\pi\eta\dot{\phi}) \tag{5.6}$$

where $\dot{\phi}$ is the differenced form for the velocity. Further, express $\delta(\pi)$ on each slice as

$$\delta(\pi) = (2\pi)^{-1} \int_{-\infty}^{+\infty} \eta \, d\lambda \, \exp(i\eta\lambda\pi). \tag{5.7}$$

One arrives at a path integral expression for the propagator in which the action is (5.4), the integration is over paths in the extended phase space, and the measure is the invariant Liouville measure multiplied by remaining factors from equations (5.6) and (5.7). If we now make the point transformation $\{q^a, \phi, p_a, \pi\} \rightarrow \{Q^A, P_A\}$ on each slice, the Liouville measure remains unchanged and the action becomes (5.5). The result is

$$\langle q''^a t'' | q'^a t' \rangle = \int \delta^{n+1} P \delta^{n+1} Q \, \delta\lambda [(2\pi)^{-n-1}\eta] \left| \{\Phi, \pi\} \right|$$

$$\times \delta(\Phi(P, Q))] \exp\left(i \int_{t'}^{t''} dt \, [P_A \dot{Q}^A - h(P, Q) - \lambda\pi(P, Q)] \right). \tag{5.8}$$

Here, we have written $\partial\Phi/\partial\dot{\phi}$ as the Poisson bracket $\{\Phi, \pi\}$. This is the path integral prescription for quantizing a theory with one gauge degree of freedom. In the usual terminology $\delta(\Phi)$ is the gauge fixing δ function and $|\partial\Phi/\partial\phi|$ is commonly called the Faddeev–Popov determinant.

6.   THE DIFFERENCE BETWEEN GAUGE AND PARAMETRIZED THEORIES

The prescription for quantizing gauge theories in equation (5.8) resembles that for parametrized theories in equation (4.10). In each case there is an integral over paths in an extended phase space of the exponential of an invariant action. In each case there is an integral over a multiplier to enforce the constraint. The difference is in the delta functions in the measure. In gauge theories we are permitted an essentially arbitrary function of the coordinates and momenta to fix the gauge. In parametrized theories the argument of the analogous delta function has a more specific form. It contains an arbitrary function of time and can depend on no other dynamical variables.

Lest the reader believe that there has simply been a lapse in exploiting potential generality in deriving equation (4.10), imagine applying the gauge theory prescription equation (5.8) to the parametrized theory. In the simplest case, this is achieved by putting $t(\tau) = 0$ and identifying t with $\Phi(P, Q)$. Of course, our derivation of equation (5.8) for the quantum propagator is no longer valid because $t(\tau) = 0$ implies $t' = 0 = t''$. When we insist that the expression (5.8) represents the quantum propagator from t' to $t'' > t'$ even for $t(\tau) = 0$, we predictably end with an absurd result. On the other hand, when we put $t(\tau) = 0$ and simultaneously restrict ourselves to $t' = 0 = t''$, the expression (5.8) for the quantum propagator equally predictably yields a correct triviality: it reduces to the delta function because the dynamics are frozen at a single instant of time.

Viewed on the space of possible paths, each point of which represents a particular functional dependence of P_A and Q^A on τ, the condition $\Phi(P, Q) = 0$ is not effective in lifting re-parametrization invariance because it does not assign a unique parametrization to each path. Put differently, the condition $\Phi(P, Q) = 0$ does not represent a surface in this space much less one which intersects the orbits of the re-parametrization group in one and only one point.

To single out a unique parametrization, Φ must depend explicitly on τ. However, one is not permitted an arbitrary function of τ, but only that which assigns the same value of physical time to every phase space point on an intermediate slice. That is, the function $t(\tau)$ in equation (4.10) cannot be generalized to include a dependence on p_α and q^α. The origin of this restriction is easily traced in the derivation of equation (4.10). When using equation (4.5) to obtain equation (4.6), we would not have recovered the correct form of the action in the exponent if we had permitted $t(\tau)$ to depend also on p_α and q^α. Further, it is not possible to introduce a dependence on p_α, q^α by a re-parametrization of the paths and a transformation of the path integral. Such transformations lead from equation (4.10) to identical expressions with different functions $t(\tau)$.

Our conclusion is that, while formally similar, the path integral prescription for quantizing parametrized theories differs essentially from that for gauge theories in the restricted form of the delta function which specifies the time in the path integral for parametrized theories. This function must

  (i) assign the same value of physical time to every phase space point within a slice

  (ii) assign a different value of time to every slice.

Thus enters the privileged physical time into re-parametrization invariant theories.

The reason for the distinction between parametrized theories and gauge theories emerges clearly in the Hamiltonian framework from the different roles played by the associated constraints. In gauge theories the constraint function generates changes in field quantities f due to gauge transformations through the Poisson bracket relation (written here for one gauge degree of freedom)

$$\delta f = \{f, \pi\} \delta \phi. \tag{6.1}$$

Any physical quantity is unchanged by such a gauge transformation.

By contrast the constraint associated with re-parametrization invariance, the super-Hamiltonian H, generates the change in an arbitrary quantity f with time

$$\delta f = \{f, H\} \delta \tau \tag{6.2}$$

that is, it generates the dynamics. Any physical quantity evolves with time.

The distinction between classical gauge theories and parametrized theories, so clear in the Hamiltonian framework, is preserved in their quantum-mechanical versions. It serves yet again to emphasize the special role played by time in quantum mechanics.

ACKNOWLEDGMENT

This work was supported in part by the National Science Foundation under grants PHY 81-06909, PHY 80-26043 and PHY 81-07384.

REFERENCES

DeWitt B 1967a *Phys. Rev.* **160** 1113

—— 1967b *Phys. Rev.* **162** 1195

—— 1967c *Phys. Rev.* **162** 1239

—— 1979 in *General Relativity, an Einstein Centenary Survey* ed S W Hawking and W Israel (Cambridge: Cambridge University Press)

Faddeev L 1969 *Teor. Mat. Fiz.* **1** 3 (1970 *Theor. Math. Phys.* **1** 1)

Faddeev L and Popov V 1967 *Phys. Lett.* **25B** 30

—— 1973 *Usp. Fiz. Nauk* **111** 427

Fradkin E S and Vilkovisky G A 1977 'Quantization of Relativistic Systems with Constraints, Equivalence of Canonical and Covariant Formalisms in the Quantum Theory of the Gravitational Field' *CERN report TH*-2332
Hartle J and Kuchař K 1983 *J. Math. Phys.* to appear
Kuchař K 1983 *J. Math. Phys.* to appear

The Hamiltonian Structure of Two-Dimensional Space–Time and its Relation with the Conformal Anomaly

CLAUDIO TEITELBOIM

1. INTRODUCTION AND SUMMARY

This article is concerned with three different aspects of field dynamics in two-dimensional space–time and their mutual relationship:

(i) the algebra of the Hamiltonian generators of surface deformations and its Schwinger terms

(ii) gravitation theory

(iii) the conformal anomaly.

These are all subjects close to Bryce DeWitt's heart and it seems appropriate to discuss them here, although he might of course not like what is being written in his honor.

The paper is organized as follows. To begin with, the 'integrability' or 'path independence' conditions for a Hamiltonian field theory in which the states are defined on an arbitrary spacelike hypersurface are reviewed in §2. It is shown that when the dimension of space–time is two, special properties, absent in higher dimensions, arise:

(i) it becomes possible to have a Schwinger term in the algebra of surface deformations or—in other words—one may look for 'central extensions' or 'projective (ray) representations' of that algebra,

(ii) The Hamiltonian generators are not constrained to vanish as a consequence of the algebra and path independence.

Moreover, whenever a Schwinger term is present, the generators *cannot* vanish for all times.

It is also shown that if there is a Schwinger term the action is not invariant under space–time re-parametrizations but, nevertheless, the equations of motion are generally covariant.

In §3 an example of a dynamical system with a non-vanishing Schwinger term at the classical level (i.e. in terms of a representation of the surface deformation algebra by Poisson brackets) is given. The system consists of a field ϕ which describes the local space–time scale and its conjugate π. It provides a non-trivial analog of Einstein's theory of gravitation in two space–time dimensions.

A system with a Schwinger term of quantum-mechanical origin (which vanishes classically) is then discussed in § 4. It is just a quantized massless scalar field for which the Hamiltonian generators are normal ordered as in string theory.

Next, it is observed in § 5 that since the magnitude of the gravitational Schwinger term is adjustable (it is inversely proportional to the gravitational constant), whereas that of the scalar field is fixed, it is possible for those terms to mutually cancel provided the gravitational constant is given a special value. If this is done the 'effective action' (the normal ordered operator action of the scalar field plus the c-number action of the gravitational field considered as a fixed background) is coordinate invariant and its associated energy–momentum tensor is divergenceless.

Furthermore the trace of the energy–momentum tensor has the value predicted by the conformal anomaly. Thus a one-to-one relationship emerges between the conformal anomaly and the Schwinger terms in the algebra of the scalar field Hamiltonian generators.

An important property possessed by the effective energy and momentum densities is that they are 'state functions' in the sense that they depend solely on magnitudes defined on a given surface—a property not shared by the densities derived from other (non-local) effective actions sometimes used.

Finally in § 5 the gravitational field is considered as a dynamical quantum field coupled to several scalar fields. It is pointed out that at the quantum level it becomes permissible to impose the constraint equations and hence the analogy with gravitation, which was only partial classically, becomes complete.

All the considerations in this paper can be extended to incorporate local supersymmetry including, in particular, the construction of an analog of super-gravity in two space–time dimensions. That generalization, as well as other further aspects of the work reported here, has been discussed elsewhere (Teitelboim 1983b).

2. HAMILTONIAN STRUCTURE IN TWO SPACE–TIME DIMENSIONS, RAY REPRESENTATIONS OF SURFACE DEFORMATIONS, SCHWINGER TERMS

Hamiltonian Dynamics on Curved Surfaces

When space–time is a curved Riemannian manifold the concept of an instant corresponds to a spacelike hypersurface. Since in general there is no privileged family of surfaces, one must use a dynamical scheme which describes the evolution of the physical state from an arbitrary initial surface to an arbitrary final one. Such a formalism was developed by Dirac (1950, 1951, 1964).

The key feature in Dirac's formalism is that the Hamiltonian which generates the evolution of the system under a deformation of the surface contains arbitrary functions of space and time, the total number of which is—in the absence of internal gauge freedom—equal to the dimension of space–time.

The equations of motion take the form,

$$\dot{F} = \int dx \eta^\mu(x)[F, \mathcal{H}_\mu(x)]. \tag{2.1}$$

Here the bracket on the right-hand side denotes a Poisson bracket in the classical theory or a commutator divided by $i\hbar$ in the quantum case. The xs are coordinates on the surface and t is a timelike coordinate which labels successive surfaces.

The coordinates (t, x) provide an arbitrary parametrization of space–time so that $t = $ constant defines a generic spacelike hypersurface with a generic coordinate system on it. The η^μ are arbitrary functions of x and t which determine the deformation taking the surface $t = $ constant onto the one assigned to $t + \delta t$. More precisely the vector field which joins the points (t, x) and $(t + \delta t, x)$ is given by,

$$\eta = \eta^\perp n + \eta^i e_i \tag{2.2}$$

where n is the normal to the hypersurface and the e_i are the tangent vectors to the x coordinate lines. Thus the index μ runs over $\perp, i = 1, \ldots, d$ where d is the dimension of the surface (the space–time dimension is $d + 1$).

Usually the vector n in (2.2) is taken to be the unit normal so that $n \cdot n = -1$. However it is significantly more convenient to adopt a different normalization, namely,

$$n \cdot n = -g \tag{2.3}$$

where g is the determinant of the spatial metric,

$$g_{ik} = e_i \cdot e_k. \tag{2.4}$$

The space–time metric g_{ab} $(a, b = 0, i)$ can be written in terms of η^μ and g_{ik} as,

$$g_{ab} = g \begin{bmatrix} g^{-1} g_{ij} \eta^i \eta^j - (\eta^\perp)^2 & g^{-1} g_{ij} \eta^j \\ g^{-1} g_{ij} \eta^j & g^{-1} g_{ij} \end{bmatrix} \tag{2.5a}$$

and its inverse as,

$$g^{ab} = g^{-1} \begin{bmatrix} -(\eta^\perp)^{-2} & \eta^i (\eta^\perp)^{-2} \\ \eta^i (\eta^\perp)^{-2} & g g^{ij} - \eta^i \eta^j (\eta^\perp)^{-2} \end{bmatrix}. \tag{2.5b}$$

Path Independence

There exists a fundamental consistency requirement for a theory in which the state is defined on arbitrary surface. It is this: the evolution from a given initial surface Σ_1 to a given final surface Σ_2 must be independent of the sequence of intermediate surfaces employed to deform Σ_1 onto Σ_2 ('path independence,' 'integrability').

It may be shown by a purely geometrical argument (Teitelboim 1973a,b, 1980) that this requirement translates into an equation which involves the Poisson brackets (or the commutators in the quantum case) of the generators of surface deformations, namely

$$C^{\mu}(\eta_1, \eta_2)[F, \mathcal{H}_{\mu}] = \eta_1{}^{\mu}\eta_2{}^{\nu'}[F, [\mathcal{H}_{\mu}, \mathcal{H}_{\nu'}]] \tag{2.6}$$

for an arbitrary dynamical variable F.

In this equation the summation convention has been extended to include the continuous label x besides the discrete index μ, a practice often followed by Bryce DeWitt.

The quantity $C^{\mu}(\eta_1, \eta_2)$ appearing in equation (2.6) is the commutator of the deformations η_1 and η_2. It has the form,

$$C^{\mu''}(\eta_1, \eta_2) = \kappa_{\rho\nu'}{}^{\mu''}\eta_1{}^{\rho}\eta_2{}^{\nu'} \tag{2.7}$$

where the only non-zero κs are given by,

$$\kappa_{\perp\perp}{}^{i''} = -\kappa_{\perp'\perp}{}^{i''} = (gg^{ij})(x'')\delta_{,j}(x, x')\,[\delta(x, x'') + \delta(x', x'')] \tag{2.8a}$$

$$\kappa_{i\perp}{}^{\perp''} = -\kappa_{\perp i}{}^{\perp''} = \delta_{,i}(x, x')[\delta(x, x') + \delta(x', x'')] \tag{2.8b}$$

$$\kappa_{ij}{}^{k''} = -\kappa_{ji}{}^{k''} = \delta_i^k \delta_{,j}(x, x')\delta(x', x'') + \delta^k_j \delta_{,i}(x, x')\delta(x, x''). \tag{2.8c}$$

Equations (2.8a, b) differ from those given in Teitelboim (1973) because of the normalization (2.3) used here.

Now, one may attempt fulfilling (2.6) by demanding

$$[\mathcal{H}_{\mu}, \mathcal{H}_{\nu'}] = \kappa_{\mu\nu'}{}^{\rho''}\mathcal{H}_{\rho''} + \sigma_{\mu\nu'} \tag{2.9}$$

where, in the terminology of the algebras of currents, the 'Schwinger term' $\sigma_{\mu\nu'}$ is a c-number i.e., a quantity which has Poisson brackets (commutators) equal to zero with all the dynamical variables. We may assume without loss of generality that $\sigma_{\mu\nu'}$ cannot be absorbed in a redefinition of $\mathcal{H}_{\rho''}$.

However if one inserts equation (2.9) into equation (2.6) one finds that both expressions are equivalent if and only if,

$$[F, \kappa_{\mu\nu'}{}^{\rho''}]\mathcal{H}_{\rho''} = 0 \tag{2.10}$$

for every F.

These equations would be an identity if the κs were c-numbers. However one sees from equation (2.8a) that if the dimension d of the hypersurface is different from unity (i.e. if the dimension of space–time is different from two) $\kappa_{\perp\perp}{}^{i''}$ involves g_{ij}, which is not a c-number in the present formulation even if one deals with a field in a prescribed (non-dynamical) background. (In that case the spatial metric is brought in through Dirac's (1964) 'surface variables' which describe the location of the hypersurface.)

It thus follows that, if $d > 1$, (2.6) and (2.9) are only compatible if, in addition, one restricts the allowed values of the canonical variables (allowed states in quantum mechanics) by demanding,

$$\mathcal{H}_i = 0. \tag{2.11a}$$

Furthermore this equation must be preserved under surface deformations, and one can easily see from (2.1), (2.8) and (2.9) that this can only happen if one also has,

$$\mathcal{H}_{\perp} = 0 \tag{2.11b}$$

and

$$\sigma_{iv'} = 0. \tag{2.12a}$$

Finally demanding that $(2.11b)$ as well be preserved under surface deformations yields,

$$\sigma_{\perp\perp'} = 0. \tag{2.12b}$$

If one passes to the Hamiltonian from a generally covariant action which is the time integral of a Lagrangian depending on the dynamical coordinates and its first time derivatives—and without 'external' (i.e. not varied) fields—then equations $(2.11a, b)$ follow automatically as the constraint relations in Dirac's method for degenerate Lagrangians. Equations (2.9) and (2.12) are also valid in that case.

However, it is interesting to examine, as was just done, to what extent these equations are necessitated by the consistency of the Hamiltonian formalism itself with the Riemannian structure of space–time. In fact we shall now see, in the light of this analysis, that new possibilities emerge for the one case where the reasoning leading to equations (2.11) and (2.12) does not apply, namely when the dimension of space–time is two. We shall be concerned with that case from now on.

Special Features of Two-Dimensional Space–time
When the spatial section has dimension one, one has $gg^{ij} = 1$ and therefore the dependence on g_{ij} of the structure functions κ drops out. These functions become then the structure constants of a bona fide infinite-dimensional Lie algebra. This happens only with the normalization (2.3) and it is one of the key reasons for adopting it. (The choice (2.3) is also privileged in higher dimensions. For example it simplifies considerably the measure for the path integral of the quantized gravitational field (Teitelboim 1982, 1983a).)

As a consequence equations (2.10) are identically satisfied and hence the constraint relations (2.12) are not implied by the integrability of the Hamiltonian equations. Thus one has the possibility of admitting a non-zero $\sigma_{\mu\nu'}$ in (2.9). This can be done only if one does not impose equations (2.12) from the outside either, since once they are imposed their preservation under surface deformations implies $\sigma_{\mu\nu'} = 0$.

Therefore in a space–time of two dimensions one has the possibility of having path independence without constraints and with Schwinger terms in the surface deformation algebra.

The possible Schwinger terms are severely restricted by the Jacobi identity applied to any three of the generators $\mathcal{H}_\mu$. It turns out in fact that, apart from terms that can be absorbed in redefinitions of the $\mathcal{H}_\mu$, the most general permissible $\sigma_{\mu\nu'}$ has as its only non-vanishing components,

$$\sigma_{1\perp'} = -\sigma_{\perp'1} = Z\delta'''(x, x') \tag{2.13}$$

where Z is an arbitrary constant. The result (2.13) is familiar to those who have worked in string theory, where it is usually exhibited in terms of the Fourier components of $\mathcal{H}_\mu$ (See, for example, Goddard *et al* (1975).)

Using (2.13) and (2.9) one may write explicitly (2.10) as,

$$[\mathscr{H}_\perp(x), \mathscr{H}_\perp(x')] = [\mathscr{H}_1(x) + \mathscr{H}_1(x')]\delta'(x, x') \tag{2.14a}$$

$$[\mathscr{H}_1(x), \mathscr{H}_1(x')] = [\mathscr{H}_1(x) + \mathscr{H}_1(x')]\delta'(x, x') \tag{2.14b}$$

$$[\mathscr{H}_1(x), \mathscr{H}_\perp(x')] = [\mathscr{H}_\perp(x) + \mathscr{H}_\perp(x')]\delta'(x, x') \tag{2.14c}$$

$$+ Z\delta'''(x, x').$$

In the language of group theory, equations (2.14) form a 'ray' or 'projective' representation, also referred to as 'central extension,' of the algebra of surface deformations. The constant Z in equation (2.14) is then the central charge in the extension (if considered as a new generator it has a zero bracket with the $\mathscr{H}_\mu$). Thus we may rephrase the conclusions above as stating that the constraints $\mathscr{H}_\mu = 0$ cannot be imposed when the central charge does not vanish.

To end this paragraph we point out another important special feature of two dimensions which is intimately connected with the possibility of having a non-vanishing central charge. It is the fact that the deformation parameters η^μ are conformally invariant.

This may be seen by specializing equations (2.5) to two dimensions. They read,

$$g_{ab} = \exp(-\phi) \begin{bmatrix} (\eta^1)^2 - (\eta^\perp)^2 & \eta^1 \\ \eta^1 & 1 \end{bmatrix} \tag{2.15a}$$

$$g^{ab} = \exp(-\phi) \begin{bmatrix} -(\eta^\perp)^2 & (\eta^\perp)^{-2}\eta^1 \\ (\eta^\perp)^{-2}\eta^1 & 1 - (\eta^\perp)^{-2}(\eta^1)^2 \end{bmatrix} \tag{2.15b}$$

where we have denoted $g_{11} = \exp(\phi)$. Thus under a conformal transformation $g_{\alpha\beta} \to \exp(\lambda)g_{\alpha\beta}$ one has,

$$\phi \to \phi + \lambda \qquad \eta^\mu \to \eta^\mu. \tag{2.16}$$

Now, since the commutator of two deformations $\eta_1{}^\mu$, $\eta_2{}^\mu$ is itself a deformation it should be also conformally invariant. It then follows from (2.8) that the κs should be conformally invariant which means that ϕ, or what is the same g_{11}, cannot appear in the algebra of the $\mathscr{H}_\mu$. Thus we see that the dropping out of g_{ij} from the algebra—which is what makes the constraints not to follow from (2.11) in the case $d = 1$ and hence permits $Z \neq 0$—may be thought of as a consequence of the conformal properties of two-dimensional space–time.

Action Principle

The equations of motion (2.1) are the conditions for an extremum of the action,

$$S = \int_{t_1}^{t_2} dt\,(\dot{q}p - \eta^\mu \mathscr{H}_\mu) \tag{2.17}$$

under variations of q and p such that q is held fixed at t_1 and t_2.

Note that since the constraints $\mathcal{H}_\mu = 0$ are not to be imposed when $Z \neq 0$, the functions η^μ are not to be varied. Thus the action (4.1) describes a theory in which the conformal space–time metric,

$$\tilde{g}_{ab} = \exp(-\phi)g_{ab} \qquad (2.18)$$

is prescribed as an external field.

An important property of the action (4.1) is its behavior under mappings onto itself of the space–time region included between the surfaces $t = t_1$ and $t = t_2$.

In Hamiltonian language the infinitesimal version of that transformation takes the form $q \to q + \delta q$, $p \to p + \delta p$, $\eta^\mu \to \delta\eta^\mu$, with (Teitelboim 1973b, 1977, Fradkin and Vilkovisky 1977),

$$\delta q = \varepsilon^\mu [q, \mathcal{H}_\mu] \qquad (2.19a)$$

$$\delta p = \varepsilon^\mu [p, \mathcal{H}_\mu] \qquad (2.19b)$$

$$\delta\eta^{\rho''} = \dot{\varepsilon}^{\rho''} - \kappa_{\mu\nu'}{}^{\rho''}\eta^\mu \varepsilon^{\nu'}. \qquad (2.20)$$

Here the ε^μ are the components of the vector field,

$$\varepsilon = \varepsilon^\perp n + \varepsilon^1 e_1 \qquad (2.21)$$

which defines the mapping. The normal component $\varepsilon^\perp$ must vanish at the endpoints,

$$\varepsilon^\perp(t_1) = \varepsilon^\perp(t_2) = 0 \qquad (2.22)$$

in order for the region in consideration to be mapped onto itself.

If one inserts equations (2.19), (2.20) and (2.21) into the action (2.17) one finds for its change,

$$\delta S = - \int_{t_1}^{t_2} dt\, \sigma_{\mu\nu'}\eta^\mu \varepsilon^{\nu'}$$

$$= -Z \int_{t_1}^{t_2} dt \int dx \left(\eta^\perp \frac{\partial^3 \varepsilon^1}{\partial x^3} - \eta^1 \frac{\partial^3 \varepsilon^\perp}{\partial x^3} \right). \qquad (2.23)$$

Thus we see that the action is not invariant under space–time re-parametrizations when $Z \neq 0$. However the change (2.23) in S depends only on the η^μ, which are not varied in the action principle. Therefore the transformation (2.19), (2.20) takes solutions of the equations of motion (2.1) into other solutions.

In other words due to the Schwinger term the action is not invariant under changes of the space–time coordinates but, nevertheless, the equations of motion are generally covariant.

Lastly we should note that although (2.23) was derived above in terms of the Hamiltonian form of the action it is also valid for its Lagrangian form. More precisely, if one expresses p as an a function of $\dot{q}$ by means of the equations of motion,

$$\frac{\delta S}{\delta p} = 0 \qquad (2.24)$$

then the change in the resulting action under the transformation (2.19a), (2.20) is given by (2.23). Furthermore, if (2.24) is inserted in the right side of (2.19a), that equation coincides with the standard tensorial transformation formulas for a given field. Thus if, for example, q is scalar field ψ (t, x) then (2.19a) will read,

$$\delta\psi = \psi_{,a}\varepsilon^a \tag{2.25}$$

where ε^a $(a = 0, 1)$ are the components of the vector field (2.21) along the basis $[(\partial/\partial t), (\partial/\partial x)]$.

3. GRAVITATION IN TWO SPACE–TIME DIMENSIONS. CLASSICAL SCHWINGER TERMS.

Space–time of Constant Curvature as a Hamiltonian System
The possibility of allowing for a non-vanishing Z in (2.14) permits one to construct a non-trivial analog of Einstein's theory of gravitation in two space–time dimensions.

The clue in the construction is the observation that when $Z \neq 0$ the action is not invariant under coordinate transformations but the equations of motion are. Hence we start by writing the only non-trivial local analog of Einstein's equations in two space–time dimensions,

$$R - \Lambda = 0 \tag{3.1}$$

where R is the scalar curvature and Λ is a constant.

Equation (3.1) does not follow from extremization of the Hilbert action $\int \sqrt{-g}$ $(R - \Lambda)$ for a space–time of two dimensions. Indeed the curvature term in that action becomes in that case a topological invariant (the Euler characteristic) and hence extremization of the action yields the statement $\Lambda = 0$, but no restriction on g_{ab}.

There is indeed no invariant action constructed out of the g_{ab} only which is the time integral of a local Lagrangian and gives (3.1) as its equation of motion. However a non-invariant action of the type (2.17) does exist. Its Hamiltonian generators are built from the scale field ϕ appearing in (2.15) and its conjugate momentum π. They read,

$$\mathcal{H}_{\perp}{}^{\text{grav}} = \tfrac{1}{2}(k\pi^2 + k^{-1}\phi'^2) - 2k^{-1}\phi'' - k^{-1}\Lambda \exp(\phi) \tag{3.2a}$$

$$\mathcal{H}_{1}{}^{\text{grav}} = \pi\phi' - 2\pi' = -2\nabla\pi. \tag{3.2b}$$

Here k is an arbitrary non-zero constant and ∇ denotes the intrinsic covariant derivative in the one-dimensional surface.

In terms of the generators (3.2) the following identity holds:

$$\dot{\pi} - \eta^{\mu}[\pi, \mathcal{H}_{\mu}] = k^{-1}\eta^{\perp}\exp(\phi)(R - \Lambda) \tag{3.3a}$$

provided π is expressed in terms of $\dot{\phi}$ through,

$$\dot{\phi} - \eta^{\mu}[\phi, \mathcal{H}_{\mu}] = 0. \tag{3.3b}$$

Relation (3.3a) was arrived at and may be verified by what is essentially an analysis of the Gauss–Codazzi equations of a line embedded in an arbitrary two-dimensional curved space. That analysis will not be given here. We just note that (3.3b) simply relates the momentum π to the extrinsic curvature scalar K (the only curvature the line has) by,

$$\pi = -2k^{-1}\exp(\phi/2)K. \tag{3.4}$$

Equations (3.3) show that (2.1) can indeed be considered a Hamiltonian system with generators given by (3.2). Furthermore, a straightforward calculation shows that the generators (3.2) obey the algebra (2.14) with a central charge,

$$Z^{\text{grav}} = -4k^{-1}.$$

So we see that the analog of the gravitational field in two space–time dimensions possesses a Schwinger term in the surface deformation algebra at the classical (non-quantum) level.

The reader familiar with the Hamiltonian formulation of Einstein's equations will note the great similarity between equations (3.2)–(3.4) and the corresponding ones for Einstein's theory. There is however a key difference, namely that due to (3.5) the η^μ are not to be varied in the present case and hence the generators (3.2) are not constrained to vanish. The system at hand has therefore one dynamical degree of freedom per space point, the field ϕ.

Lagrangian Action

It will be useful for what follows to discuss here a little further the properties of the Lagrangian form of the action constructed from the generators (3.2).

If one inserts in equation (2.17) the expression for π as a function of $\dot{\phi}$ obtained from equation (3.3b) (which is in turn equation (2.24) applied to the present case) one finds,

$$S = k^{-1}\int dt\,dx\,\tfrac{1}{2}\{(\eta^\perp)^{-1}[\dot{\phi} - \phi'\eta^1 - 2(\eta^1)']^2$$
$$-\eta^\perp\phi'^2 + 4\eta^\perp\phi'' + 2\eta^\perp\Lambda\exp(\phi)\}. \tag{3.5}$$

This action is a functional of the space–time metric g_{ab} which is expressed in terms of ϕ and η^μ in equation (2.15). However, in equation (3.5) only the local scale ϕ is supposed to be a dynamical field whereas the conformally invariant piece described by η^μ is given as an external field.

It follows from equation (3.3) that under a conformal variation the change in equation (3.5) is,

$$\frac{\delta S}{\delta\phi} = -k^{-1}\eta^\perp\exp(\phi)(R - \Lambda) \tag{3.6a}$$

which can be rewritten in terms of g_{ab} as

$$g_{ab}\frac{\delta S}{\delta g_{ab}} = -k^{-1}(-{}^{(2)}g)^{1/2}(R - \Lambda) \tag{3.6b}$$

where ${}^{(2)}g$ denotes the determinant of g_{ab}.

On the other hand, under a change of the space–time coordinates,

$$g_{ab} \rightarrow g_{ab} + \nabla_a \varepsilon_b + \nabla_b \varepsilon_a \qquad (3.7)$$

one has, according to (2.23),

$$\delta S = 4k^{-1} \int dt \, dx [\eta^\perp (\varepsilon^1)''' - \eta^1 (\varepsilon^\perp)'''] \neq 0. \qquad (3.8)$$

(As stated at the end of § 2, equations (2.19a) and (2.20) become the tensorial law (3.7) provided one expresses π as a function of $\dot{\phi}$ through (3.3b).)

Now, as was explained before, the reason for the covariance of the variational derivative (3.6b) in spite of the non-invariance (3.8) of the action is the fact that (3.8) is conformally invariant. One may therefore ask whether a conformally invariant piece—i.e. a functional of η^μ only—could be added to equation (3.5) so that its variation under a change of coordinates would exactly cancel the term (3.8). That addition would render the new action invariant under space–time reparametrizations without spoiling equation (3.6).

It turns out that this can indeed be done, but only at the price of sacrificing the locality of (3.5), which means that the system loses its Hamiltonian structure. Indeed, if one adds to (3.5) the functional,

$$\Sigma = (2k)^{-1} \int d^2x \int d^2x' \{ [(-{}^{(2)}\tilde{g})^{1/2} \tilde{R}] (x) K(x, x') [(-{}^{(2)}\tilde{g})^{1/2} \tilde{R}] (x') \} \quad (3.9)$$

where $\tilde{g}_{ab}$ is the conformally invariant metric (2.18) and $K(x, x')$ is a conformally invariant Green's function obeying,

$$\frac{\partial}{\partial x^a} \left[(-{}^{(2)}\tilde{g})^{1/2} \tilde{g}^{ab} \frac{\partial K}{\partial x^b} \right] = \delta^{(2)}(x, x') \qquad (3.10)$$

one obtains the coordinate invariant non-local action (Polyakov 1981),

$$S^{\text{inv}} = (2k)^{-1} \int d^2x \int d^2x' \{ [(-{}^{(2)}g)^{1/2} R] (x) K(x, x') [(-{}^{(2)}g)^{1/2} R] (x') \}$$

$$(3.11)$$

In other words one may regard equation (3.6) as a functional differential equation for $S[g_{ab}]$. Any solution of that equation differs from (3.5) by a 'constant of integration' which is conformally invariant. In order to fix that constant of integration an additional criterion must be used. If that criterion is the demand of coordinate invariance the constant of integration is equation (3.9) and the action turns out to be non-local. On the other hand, if the criterion is locality (and hence Hamiltonian structure) the constant of integration is zero and the action is just (3.5).

However, it so happens that if one considers the quantum dynamics of a massless scalar field, as we will do in the next section, new Schwinger terms of quantum-mechanical nature appear. It becomes then possible to arrange for the classical central charge (3.5) to be exactly cancelled by that coming from the scalar field. In this way the change in equation (3.8) in the gravitational action is cancelled by a similar contribution coming from the (local) scalar action, without the need to resort to equation (3.9). Thus one can have both, locality and coordinate invariance. (This is particularly important in connection with the energy–momentum tensor, as we shall see in §5.) For this reason we shall consider the local action (3.5) to be the appropriate one for the gravitational field, in spite of its lack of coordinate invariance.

4. QUANTIZED MASSLESS SCALAR FIELD. QUANTUM-MECHANICAL SCHWINGER TERMS

As we saw in the previous section Schwinger terms arise already at the classical level in the dynamics of the gravitational field in two space–time dimensions. However a different situation may also arise. For a given system, Schwinger terms may be classically absent but present quantum-mechanically. Indeed if one considers the simplest 'matter' field—the massless scalar field—evolving in a two-dimensional space–time that is precisely the case.

To analyze the scalar field, start from the action,

$$S^{\text{scalar}} = -\tfrac{1}{2} \int (-^{(2)}g)^{1/2} g^{ab} \, \partial_a \psi \partial_b \psi \, d^2 x. \tag{4.1}$$

This integral involves the metric g_{ab} only through the conformally invariant combination,

$$(-^{(2)}g)^{1/2} g^{ab} = (-^{(2)}\tilde{g})^{1/2} \tilde{g}^{ab} \tag{4.2}$$

where $\tilde{g}^{ab}$ is the inverse of $\tilde{g}_{ab}$ given by equation (2.18). That is equation (4.1) depends, besides the scalar field ψ, on η^μ but is independent of the scale ϕ.

If one passes to the Hamiltonian form of equation (4.1) one finds,

$$S^{\text{scalar}} = \int (\dot\psi p - \eta^\mu \mathscr{H}_\mu) \, dx \, dt \tag{4.3}$$

with,

$$\mathscr{H}_\perp{}^{\text{scalar}} = \tfrac{1}{2}(p^2 + \psi'^2) \tag{4.4a}$$

$$\mathscr{H}_1{}^{\text{scalar}} = p\psi'. \tag{4.4b}$$

The generators (4.4) obey classically (i.e. in terms of Poisson brackets) the algebra (2.14) with $Z = 0$. However, in the quantum-mechanical case the situation changes. In fact one may pass to the quantum theory by expanding the spatial dependence of ψ and its conjugate p in terms of a set of instantaneous (time-dependent) creation and annihilation operators and subsequently normally ordering the generators. This has been, in effect, done in detail by investigators working in string theory (see, for example, Rebbi 1974). One finds that if one encloses the system in a spatial box of coordinate length l ($\psi(x, t)$ is expanded for fixed t in terms of $\exp(\text{i}2\pi n x/l)$) the algebra of the normal ordered version $:\mathcal{H}_\mu:$ of the scalar field generators (4.4) is given by,

$$[:\mathcal{H}_\perp(x):,:\mathcal{H}_\perp(x'):] = [:\mathcal{H}_1(x): + :\mathcal{H}_1(x'):]\delta'(x,x') \tag{4.5a}$$

$$[:\mathcal{H}_1(x):,:\mathcal{H}_1(x'):] = [:\mathcal{H}_1(x): + :\mathcal{H}_1(x'):]\delta'(x,x') \tag{4.5b}$$

$$[:\mathcal{H}_1(x):,:\mathcal{H}_\perp(x'):] = [:\mathcal{H}_\perp(x): + :\mathcal{H}_\perp(x'):]\delta'(x,x')$$
$$+ Z^{\text{scalar}}\left[\delta'''(x,x') + \left(\frac{2\pi}{l}\right)^2\delta'(x,x')\right] \tag{4.5c}$$

with,

$$Z^{\text{scalar}} = -(12\pi)^{-1}\hbar. \tag{4.6}$$

The bracket on the left-hand side of equation (4.5) denotes a commutator divided by $i\hbar$.

Technically the reason for the c-number in equation (4.5c) is that the commutator of two normal ordered quadratic operators differs in general from normal ordering by a c-number. The term containing $\delta'(x,x')$ in equation (4.5c) vanishes in the limit $l \to \infty$ and may be eliminated even before that limit (or when the limit cannot be taken as is the case for a closed space) by the redefinition $\mathcal{H}_\perp \to \mathcal{H}_\perp - \hbar\pi/6l^2$. The central charge in equation (4.6) is, on the other hand, independent of l.

It should be emphasized that the normal ordering in terms of instantaneous creation and annihilation operators does not involve a special choice of the space–time coordinates. The functions η^μ remain therefore arbitrary. Nothing is said however about the physical interpretation of the associated Fock space in terms of particles. We plan to return to that question elsewhere taking into account the results of the next section.

5. RELATION WITH THE CONFORMAL ANOMALY. EFFECTIVE ENERGY–MOMENTUM TENSOR.

If one considers the fields appearing in equation (4.1) as operators the action itself becomes an operator. That action may be given a definite sense by normal ordering it in terms of instantaneous creation and annihilation operators. If this is done one finds, according to equation (2.23) and equation (4.6), that the operator action is not invariant under a change of space–time coordinates. This is of no consequence for the equation of motion of the field ψ but it creates a serious difficulty for the energy–momentum tensor. Indeed T_{ab} not only has a non-vanishing divergence,

$$\nabla^a:T_{ab}: \neq 0 \tag{5.1}$$

when the equations of motion of the scalar field hold, but, worse than that, its components do not even transform as those of a tensor under a change of the space–time coordinates.

One may attempt to remedy this problem by adding to the normal ordered scalar field action a piece which would make the sum invariant under general coordinate transformations. However, the extra piece must have an important, and quite restrictive, additional property: it should be such that the total or 'effective' energy and momentum densities on a given spacelike hypersurface depend only on variables defined on that surface. Thus the densities should be independent of how one chooses to deform the surface into the future during the subsequent evolution of the system. As one says in short, they should be 'functions of state'.

In the present context this means that, in the absence of other matter fields, these densities can only depend on the scalar field variables ψ, p and on geometrical properties of the surface $t = $ constant, such as its intrinsic metric and extrinsic curvature. On the other hand they cannot depend on the functions η^μ which describe how the surface is deformed into the next one.

To proceed with the discussion, recall that the energy–momentum tensor is defined by,

$$T_{ab} = -2\frac{\delta S}{\delta g^{ab}}. \tag{5.2}$$

Formula (5.2) yields T_{ab} as a tensor density in space–time. In order to extract from it energy and momentum densities on a hypersurface one must divide equation (5.2) by the square root $(-{}^{(2)}g)^{1/2}$ of the space–time metric, multiply it subsequently by the square root $\exp(\phi/2)$ of the determinant of the (one by one) spatial metric and finally project it along the normal and tangent directions to the surface. In this way one finds, in terms of the decomposition (2.15), that the energy and momentum densities are given respectively by,

$$[\exp(\phi/2)\eta^{\perp}]^{-1} T_{\perp\perp} = \exp(-\phi/2)\frac{\partial S}{\partial\eta^{\perp}} \tag{5.3a}$$

$$[\exp(\phi/2)\eta^{\perp}]^{-1} T_{\perp 1} = \exp(-\phi/2)\frac{\partial S}{\partial\eta^{1}}. \tag{5.3b}$$

(Here the symbol $\perp$ denotes projection onto a normal obeying $n \cdot n = -1$.)

From equation (5.3) one sees that the extra piece that should be added to the normal ordered scalar field action so that the sum is coordinate invariant must be of the form (2.17), with the additional contributions to $\mathscr{H}_\perp$ and $\mathscr{H}_1$ constructed solely from state functions in the sense explained above. This is so because taking S in equation (5.3) to be the normal ordered scalar field action yields the generators $:\mathscr{H}_\mu^{\text{scalar}}:$ obtained by normally ordering equation (4.4), which are state functions by themselves.

At this stage it becomes apparent that the solution is provided by the gravitational action of §3. Indeed if one adds to the normal ordered version of equation (4.1) the functional (3.5) with the constant k chosen so that,

$$Z^{\text{eff}} = Z^{\text{grav}} + Z^{\text{scalar}} = 0 \tag{5.4}$$

i.e., given by,

$$k^{-1} = -(48\pi)^{-1}\hbar \tag{5.5}$$

then the action,

$$S^{\text{eff}} = S^{\text{grav}} + :S^{\text{scalar}}: \tag{5.6}$$

is both local (i.e., of the form equation (2.17)) and invariant under general coordinate transformations. As a consequence the 'effective energy–momentum tensor' derived from equation (5.6) is such that the two required conditions hold, namely:

(i) It is divergence free

(ii) The energy and momentum densities are state functions.

(In order to verify the coordinate invariance of equation (5.6) one may use equations (2.19), (2.20) applied to both the gravitational and scalar field canonical variables (ϕ, π) and (ψ, p) respectively. In the scalar part, the bracket in equation (2.19) is understood as a commutator divided by $i\hbar$. As mentioned before after the momenta are eliminated that transformation coincides with the standard geometrical one

$$\delta\psi = \psi_{,a}\varepsilon^a \qquad \delta g_{ab} = \nabla_a\varepsilon_b + \nabla_b\varepsilon_a.)$$

Now, using equation (3.6) and the fact that equation (4.1) is independent of ϕ, one finds,

$$(-{}^{(2)}g)^{-1/2} g^{ab} T_{ab}^{\text{eff}} = \frac{\hbar}{24\pi}(R - \Lambda) \tag{5.7}$$

which is the well known trace or conformal anomaly (see, for example, DeWitt 1979) for the massless scalar field in curved space. Thus the conformal anomaly reappears as the condition for the gravitational and scalar Schwinger terms to mutually cancel.

Equations (5.3) and (5.7) permit one to write explicitly all the components of the effective energy–momentum tensor. They are given by,

$$[\exp(\phi/2)\eta^{\perp}]^{-1}T_{\perp\perp}^{\text{eff}} = \exp(-\phi/2)[:\tfrac{1}{2}(p^2 + \psi'^2): + \mathscr{H}_{\perp}^{\text{grav}}] \tag{5.8a}$$

$$[\exp(\phi/2)\eta^{\perp}]^{-1}T_{\perp 1}^{\text{eff}} = \exp(-\phi/2)(:p\psi': + \mathscr{H}_{1}^{\text{grav}}) \tag{5.8b}$$

$$[\exp(\phi/2)\eta^{\perp}]^{-1}T_{11}^{\text{eff}} = \exp(-\phi/2)[-2k^{-1}\exp(\phi)(R - \Lambda) + (\eta^{\perp})^{-1}T_{\perp\perp}^{\text{eff}}. \tag{5.8c}$$

In equation (5.8) the space–time geometry is considered as prescribed, with $\mathscr{H}_{\mu}^{\text{grav}}$ given by equation (3.2) but with π being understood as expressed in terms of the extrinsic curvature through equation (3.4). The constant k is given by equation (5.5) and Λ is here simply an arbitrary parameter which must be fixed by additional considerations. (This ambiguity is also present in the more usual treatments, where Λ is usually made to vanish by demanding that the trace (5.7) should be zero in flat space–time.) Some comments on the possibility of treating the gravitational and scalar fields as a coupled dynamical system will be given in §6.

Equations (5.8a) and (5.8b) give respectively the energy and momentum densities. They are clearly functions of state as it was demanded. The component (5.8c), on the other hand, being a momentum flux is not a function of state (R contains the η^{μ} and $\dot{\pi}$—recall equation (3.3a)).

It should be noted here that the contributions to (5.8) from $\mathscr{H}_{\mu}^{\text{grav}}$ differ on a generic surface from those obtained by applying the definition (5.2) to the non-local invariant action (3.11). Indeed the components $[\exp(\phi/2)\eta^{\perp}]^{-1} T_{\perp\mu}$ derived from equation (3.11) are not state functions and hence are not physically appropriate definitions of the densities of energy and momentum. However on special sequences of surfaces—such as those defined by $\eta^\perp = 1$, $\eta^1 = 0$— both expressions coincide. The actions (3.5) and (3.11) also coincide in that case. There is nothing contradictory in this since the gravitational part of $T_{ab}{}^{\text{eff}}$ given by equation (5.8) is not a tensor by itself (and neither is the scalar field part). It is only the total expression (5.8) which, being derived from an invariant action, is a tensor (whose divergence vanishes when the scalar field equations of motion hold). On the other hand since the action (3.11) is invariant, its associated T^{ab} is a tensor constructed solely from the metric and such that its divergence vanishes identically.

6. GRAVITATIONAL AND SCALAR FIELDS AS A COUPLED QUANTUM SYSTEM.

It was shown in §3 that there exists an analog of the gravitational field in two space–time dimensions. In that system the field variable is the local scale ϕ of space–time.

But, at the classical level, the analogy with gravitation is only partial. This is so because the conformally invariant part of the metric, described by the functions η^μ, is treated as a prescribed external field and is not varied in the classical action principle. As a consequence the constraints $\mathscr{H}_\mu = 0$ are not imposed.

However if one passes to the quantum theory the analogy becomes complete as we now briefly describe.

What happens is that in the quantum case a new option appears. Namely, one may decompose the operators $\mathscr{H}_\mu$ into positive and negative frequency parts. Then it turns out that the positive frequency parts $\mathscr{H}_\mu{}^{(+)}$ form a sub-algebra of equation (2.14) *with no Schwinger term* even if $Z \neq 0$ (see, for example Rebbi 1974). Therefore one may impose the constraint equations in the weaker form,

$$\mathscr{H}_\mu{}^{(+)}|\chi\rangle = 0. \tag{6.1}$$

(The splitting of $\mathscr{H}_\mu$ into $\mathscr{H}_\mu{}^{(+)}$ and $\mathscr{H}_\mu{}^{(-)}$ is of course also possible classically. However in that case $\mathscr{H}_\mu{}^{(+)} = 0$ implies $\mathscr{H}_\mu{}^{(-)} = 0$ since $\mathscr{H}_\mu$ is real. Hence the weaker option equation (6.1) is strictly quantum-mechanical.)

Equation (6.1) possesses a most remarkable property which was discovered by dual theorists. It is the fact that it is consistent with the basic requirement,

$$\langle\chi|\chi\rangle \geq 0 \tag{6.2}$$

only if the central charge is given by,

$$Z = -\frac{26}{12\pi}. \tag{6.3}$$

Thus one sees that when ϕ is quantized one should not demand that the total operator action be generally covariant since that would correspond to setting $Z = 0$. On the other hand when the metric is treated as a classical background (as in § 5 above) the correct criterion is $Z = 0$. (A comment about the relation between these two procedures will be made below.)

Now, it has also been shown in the context of dual theory that when equation (6.3) holds the allowed states $|\chi\rangle$ are grouped in equivalence classes. Each class being such that any two of its members differ from each other by a vector orthogonal to all the allowed states (including itself). This effectively reduces the dimensionality of the space of solutions of equation (6.1) to precisely what would result if one imposed two extra conditions per space point on $|\chi\rangle$ in addition to equation (6.1) itself.

In this sense the full gauge freedom (coordinate invariance) of the theory is regained at the quantum level and the analogy with the structure of gravitation in higher space–time dimensions becomes complete.

It should be emphasized here that it is not possible to construct a non-empty quantum theory of gravitation of this kind containing only the field ϕ, as such a model would not have any degrees of freedom left after the Hamiltonian constraints (6.1) are imposed. Thus an analog of quantum gravity in two space–time dimensions must necessarily cont. $\cdot$ matter fields.

The simplest viable model is obtained by admitting, as in string theory, N scalar fields of the type discussed in § 5 in addition to ϕ. The $\mathscr{H}_\mu$ appearing in equation (6.1) are then the sum of the gravitational ones and of N generators of the form of equation (4.4) and one has $N - 1$ independent degrees of freedom per space point.

In that case equation (6.3) translates into the following relation for the constant k appearing in the gravitational action,

$$-4(\hbar k)^{-1} + Z' - \frac{N}{12\pi} = -\frac{26}{12\pi} \tag{6.4}$$

or,

$$(\hbar k)^{-1} = (48\pi)^{-1}(26 - N) + \frac{Z'}{4} \tag{6.5}$$

where Z' is a quantum-mechanical correction to the classical central charge of equation (3.5) of the gravitational Hamiltonian.

If the cosmological constant is zero one can give definite meaning to the operators $\mathscr{H}_\mu^{\text{grav}}$ by instantaneous normal ordering just as for the scalar field. Then one finds that Z' is given simply by the scalar field value equation (4.6) and its effect in equation (6.5) amounts to replacing N by $N + 1$. When $\Lambda \neq 0$, the $\Lambda \exp(\phi)$ term makes instantaneous normal ordering not viable and a different method must be used. That problem will not be dealt with here.

One may regard equation (6.5) as a determination of the gravitational constant in terms of the content of the universe. It is then seen that the quantum effects of the gravitational field itself must be considered on the same footing with those of matter, which is quite satisfactory.

It should be noted that due to the positive sign of the contribution $26/48\pi$ in equation (6.5) one has $k > 0$ (unless N is too large). This is intimately connected with equation (6.2), since if $k < 0$ negative norms would be introduced in the Fock space of the field ϕ. Another way of saying the same thing is that in such a case the kinetic energy term in the action (3.5) would be negative. This is of no concern if ϕ is an external field—as in §5—but is a key point if ϕ is a quantum-dynamical variable.

This last remark brings us back to the possible relationship between the values (5.5) and (6.5) obtained for k in two different circumstances. There is indeed a connection and it appears most transparently in terms of the work of Polyakov (1981) who arrives at equation (6.5) (with Z' omitted) by starting from the path integral over the scalar fields alone.

In his analysis S^{grav} emerges from the regularization of the path integral over the N scalar fields alone (a regularization which has the unconventional feature of bringing in a new field—ϕ—into the problem).

The amount of S^{grav} which is introduced is fixed by the requirement that the effective action of the scalar field be generally covariant. Thus k results to be given, at that stage, by equation (5.5) and the net Schwinger term is zero, as in §5.

The next step is to include in the action a ghost contribution, in preparation for path integrating over the gravitational field. That contribution is also regularized by adding to it an appropriate amount of S^{grav} so as to make the effective ghost action separately generally covariant. That is, the net ghost Schwinger term is also set equal to zero. This step brings in the $26/48\,\pi$ term in equation (6.5).

As a result of this procedure the total action consisting of S^{scalar} plus S^{ghost} plus the total amount of S^{grav} is also re-parametrization invariant and it would seem that, since the total Schwinger term is zero, there is a contradiction with equation (6.3).

However when one goes on-shell (which is just another name for equation (6.1)), the ghosts are no longer present but the amount of S^{grav} which was added to S^{ghost} survives. This extra piece of S^{grav} then destroys the re-parametrization invariance of the regularized S^{scalar} and makes the total on-shell action to have a central charge given by equation (6.3), thus ensuring that the on-shell states have a non-negative norm.

The possibility of linking the Liouville Hamiltonian (3.2) with an analog of gravitation theory in two space–time dimensions has also been considered by Jackiw (1984). However, in his analysis the functions η^μ are fixed as $\eta^\perp = 1$, $\eta^1 = 0$ and therefore the re-parametrization invariance is abandoned from the start.

ACKNOWLEDGMENTS

The author would like to thank Dr Bengt Nilsson for carefully reading the manuscript. He would also like to express his gratitude to his colleagues at the Istituto Nazionale di Fisica Nucleare and at the Istituto di Fisica Teorica dell' Universita for their kind hospitality in Torino. This work has been supported in part by US National Science Foundation Grant No PHY-8216715 to the University of Texas at Austin. The author is a recipient of a John S Guggenheim fellowship.

REFERENCES

Dirac P A M 1950 *Can. J. Math.* **2** 129
—— 1951 *Can. J. Math.* **3** 1
—— 1964 *Lectures on Quantum Mechanics* (New York: Academic)
Fradkin E S and Vilkovisky G A 1977 *CERN report TH*-2332 unpublished
Goddard P, Hanson A J and Ponzano G 1975 *Nucl. Phys.* B **56** 109
Jackiw R 1984 in *Quantum Theory of Gravity* ed S Christensen (Bristol: Adam Hilger)
    pp 403–20
Polyakov A M 1981 *Phys. Lett.* **103B** 207
Rebbi C 1974 *Phys. Rep.* **12** 1
Teitelboim C 1973a *Ann. Phys., NY* **79** 542
—— 1973b *The Hamiltonian Structure of Space–time* Doctoral Dissertation Princeton
    University, unpublished
—— 1977 *Phys. Rev. Lett.* **38** 1106
—— 1980 *General Relativity and Gravitation One Hundred Years after the Birth of Albert
    Einstein* vol 1 ed A Held (New York: Plenum) ch 6
—— 1982 *Phys. Rev.* D **25** 3159
—— 1983a *Phys. Rev.* D **28** 297
—— 1983b *Phys. Lett.* **126B** 41, 46, 49

Supersymmetry, Finite Theories and Quantum Gravity

K S STELLE

1. THE ULTRAVIOLET DILEMMA

In the search for a satisfactory quantum theory of gravity, the ultraviolet problem has always presented an unavoidable stumbling block. Heisenberg suggested long ago (Heisenberg 1938) that theories with coupling constants whose dimensions are of a negative power of mass should not be applicable at energy scales higher than that of the mass associated to the coupling constant. With the development of covariant quantization rules for perturbative quantum gravity (Feynman 1963, DeWitt 1967, Faddeev and Popov 1967), this problem came to be formulated in terms of the ultraviolet divergences of loop graphs, where the naive degree of divergence of an L-loop graph is $D = (d-2)L + 2$, where d is the dimensionality of space–time. This non-renormalizability of quantum gravity implies a loss of predictive power of the theory at energies of the order of the Planck mass $M_P = G^{-1/2} = 1.2 \times 10^{19}$ GeV, and perhaps worse: even if one subtracts all the infinities in quantum amplitudes, the finite parts remaining will still grow with large external momenta like $(Gk^2)^{D/2}$. This non-Froissart boundedness of the theory implies trouble with unitarity unless perturbation theory is giving a completely misleading picture of the true high energy behavior of the theory.

There have been several lines of response to the ultraviolet dilemma. One of these has been to search for non-perturbative mechanisms through which the large momentum/small distance structure might be smoothed out, with gravity acting as its own regulator. A concrete attempt in this direction has been the 'non-polynomial technique' (DeWitt 1964, Khriplovich 1966, Salam and Strathdee 1970, Isham *et al* 1971), summing classes of divergent diagrams to produce a convergent result. A shortcoming of this approach is the arbitrariness inherent in the choice of graphs to be summed. Also, this technique does not aid in making manifest the unitarity of the theory, which would require a further resummation.

Another response has been to include in the graviton propagator contributions due to the higher derivative terms that are needed as counterterms in order to cancel the ordinary perturbative infinities (Utiyama and DeWitt 1962, Weinberg 1974, Deser 1975, Stelle 1976, 1978). In this approach, the terms R^2 and $(C_{\mu\nu\rho\sigma})^2 \sim R_{\mu\nu}^2$ $-\frac{1}{3}R^2 +$ div are included in the tree-level gravitational Lagrangian instead of just being treated as perturbative counterterms. The propagators now fall off like k^{-4} for large momenta, and this has the effect of regulating the ultraviolet divergences so that the maximum naive degree of divergence is now 4, allowing the higher derivative terms' coefficients to be renormalized. Obviously, this helps with the Froissart boundedness, although it falls short of the usual requirement that the high energy behavior not grow except by powers of log (k^2). The real shortcoming of the higher derivative Lagrangians is that they manifestly violate unitarity, since the mechanism of ultraviolet cancellations involves the propagation of massive spin 2 ghost states, as can be seen by separating into partial fractions a typical propagator term

$$m^2 k^{-2} (k^2 + m^2)^{-1} = k^{-2} - (k^2 + m^2)^{-1}.$$

The relative minus sign on the second term forces an interpretation of the corresponding massive states as having either negative energy or negative norm in the state vector space. Indeed, in supersymmetric extensions of the higher derivative gravity model, one's hand is forced and the ghost states can only be taken to have negative norm (Boulware *et al* 1983).

Since the massive ghost states of the higher derivative theory couple universally, as do the ordinary massless graviton states, it is easy to arrange for a Lee–Wick mechanism (Lee and Wick 1969, Coleman 1970, Stelle 1978, Kay 1981) to give an imaginary part to the location of the ghosts' pole and so to remove them from the 'physical sector' of the state vector space. This requires a complicated prescription for contour integration in higher order diagrams. It also causes a violation of causality on timescales of the order of the ghost states' lifetime, and also non-analyticity of the *S*-matrix. Moreover, the classical limit of such a theory is not given by the starting Lagrangian, since the ghost states have been excised.

It has been suggested that renormalization group effects might help in removing the ghost states from the spectrum. For example, if the ratio $M(\lambda)/\lambda$ were to tend to infinity as λ tends to infinity, where $M(\lambda)$ is the renormalized ghost mass with subtractions performed at the scale λ_μ, then the ghosts would not actually occur in the spectrum (Salam and Strathdee 1978, Julve and Tonin 1978). Unfortunately, in ordinary higher derivative gravity precisely the opposite happens, at least as far as one can tell in perturbation theory, where the spin 2 ghost's mass is asymptotically free (Fradkin and Tseytlin 1981). This leaves open the possibility of a non-trivial fixed point with the desired anomalous dimension for $M(\lambda)$, but to date there are no techniques for investigating this possibility.

Another renormalization group approach to defining a sensible quantum gravity is the proposal that, if not renormalizable, quantum gravity might have the property of asymptotic safety (Weinberg 1979). In an asymptotically safe theory, subtractions would be made at all orders to remove the ultraviolet divergences of the theory, but among the infinite number of renormalized coefficients of terms in the full quantum action, only a finite number would be independent freely adjustable parameters. The essential coupling constants among these are those that cannot be changed by field redefinitions. The set of all essential renormalized coupling constants for some reference renormalization scale μ constitutes the initial conditions for the renormalization group equations, determining a trajectory in coupling constant space. If this trajectory hits a fixed point g_i^* as $\lambda \to \infty$, the trajectory is said to be on the ultraviolet critical surface of that fixed point. The condition of asymptotic safety is that the ultraviolet critical surface be of finite non-vanishing dimensionality. For the fixed point at the origin, the condition of asymptotic safety is equivalent to renormalizability plus asymptotic freedom for strictly renormalizable couplings.

In the case of gravity, the theory is formally renormalizable but rather empty in two dimensions, for the number of degrees of freedom of the graviton field in d dimensions is $\frac{1}{2}d(d-3)$, i.e. -1 for $d = 2$! Nonetheless, the behavior of the theory in $2 + \varepsilon$ dimensions could give some information about its behavior in four dimensions by analytic continuation in ε. The counterterm relevant to the determination of the renormalized Newton's constant G is the surface integral of the trace of the extrinsic curvature K over the boundary of $2 + \varepsilon$ dimensional space–time, $\Phi = \int_{\partial M} \mathrm{d}y \sqrt{\gamma} \, K$ where γ_{ij} is the induced metric on the surface (Gastmans *et al* 1978, Christensen and Duff 1978). This term is required in the Einstein–Hilbert Lagrangian in order to write the action in a form which depends only upon first derivatives, as needed for quantization in the functional formalism (Gibbons and Hawking 1977). Christensen and Duff (1978) have given the result for the one-loop renormalization group β function for G in a $2 + \varepsilon$ dimensional theory of gravity coupled to $n_{3/2}, n_1, n_{1/2}$ and n_0 massless matter fields plus $N_{3/2}, N_1, N_{1/2}$ and N_0 massive matter fields of spin $\frac{3}{2}, 1, \frac{1}{2}$ and 0 respectively:

$$\beta(G, \varepsilon) = \varepsilon G - bG^2 + \mathrm{O}(G^3) \tag{1.1}$$

where

$$b = \tfrac{2}{3}(1 - n_{3/2} + n_{1/2} - n_0 - N_1 + N_{1/2} - N_0). \tag{1.2}$$

The result (1.2) shows that the one-loop coefficient b is given by a count of the number of degrees of freedom of the various spins ($-1, 0, 1$ and 1 for massless fields; 0, 1, 1 and 1 for massive fields) with opposite signs for the contributions of bosons and fermions. The condition for asymptotic safety at the critical point $G^* = \varepsilon/b + \mathrm{O}(\varepsilon^2)$ given by equation (1.1) is that b be positive. Obviously, this can be arranged, but the real problem with this whole approach is how to carry out the analytic continuation in ε up to four dimensions, in the course of which all the higher loop effects neglected in (1.1) will come into play.

The most extreme (desperate ?) hope for obtaining a sensible quantum theory of gravity is that despite power counting and all appearances, one particular combination of gravity with matter fields might actually be finite in each order of perturbation theory. This hope has most frequently been expressed in the context of supergravity theories, where some dramatic cancellations do take place. For example, simple supergravity has no essential one- and two-loop counterterms (in the above sense of having coefficients insensitive to field redefinitions, which is equivalent to their not vanishing subject to the equations of motion). The first possible essential counterterm occurs at the three-loop level, and has as gravitational part the square of the Bel–Robinson tensor (Deser *et al* 1977, Ferrara and Zumino 1978, Ferrara and van Nieuwenhuizen 1978). This result holds also for all extended supergravities (Deser and Kay 1978, Kallosh 1981, Howe *et al* 1981).

2. SUPERSYMMETRY AND FINITE THEORIES

The existence of a counterterm does not guarantee that it will actually occur in the quantum corrections to a theory. This is especially the case in supersymmetric theories where there can be non-renormalization theorems that prohibit the occurrence of counterterms that are nonetheless fully supersymmetric (Wess and Zumino 1974, Iliopoulos and Zumino 1974, Ferrara *et al* 1974, Ferrara and Piguet 1975, Grisaru *et al* 1979, Grisaru and Siegel 1982). The most dramatic instance of such cancellations is the maximally extended $N = 4$ supersymmetric Yang–Mills theory for an arbitrary gauge group. (Gliozzi *et al* 1978, Brink *et al* 1977). In this renormalizable theory, the action is obviously an essential supersymmetric counter-term, but explicit calculation through the three-loop order shows it not to occur (Tarasov and Vladimirov 1980, Grisaru *et al* 1980, Caswell and Zanon 1980). A heuristic argument (Ferrara and Zumino (unpublished), Sohnius and West 1981) explains the reason for these cancellations in terms of the link supersymmetry makes between the trace anomaly, which is given by the renormalization of the action, and the anomaly in a certain axial current. In the $N = 4$ theory, the only axial symmetries of the theory are the chiral generators of the SU(4) symmetry that rotates the four supersymmetries. Since SU(4) is expected to be anomaly-free, this would explain the absence of renormalization counterterms proportional to the action.

A more direct argument (Grisaru *et al* 1980, Stelle 1982) for the vanishing of the gauge coupling constant β function, and hence the trace anomaly, uses the fact that when the Lagrangian for the $N = 4$ theory is written in terms of superfields of $N = 1$ supersymmetry, an interaction term is necessary which can only be written as an integral over a chiral superspace, i.e. with half of the anti-commuting coordinates of the full superspace. While such terms are supersymmetric, their occurrence as infinite counterterms is forbidden by the non-renormalization theorem. Moreover, the coefficient of this interaction term is related by gauge symmetry and the internal symmetry (SU(4) or even its non-chiral subgroup SO(4)) to the coefficients of the two other terms in the $N = 4$ theory's $N = 1$ superfield action. Thus, if at least the non-chiral SO(4) symmetry is preserved by the quantum corrections to the theory, then the β function must vanish to all orders in perturbation theory.

The full proof of the vanishing of the β function of $N = 4$ supersymmetric Yang–Mills theory requires the use of a formalism that maximizes the part of the theory's extended supersymmetry that is manifestly and linearly realized. Unfortunately, it appears most unlikely that the full Lorentz covariance and $N = 4$ supersymmetry can be linearly realized in a superfield formulation of the theory

(Roček and Siegel 1981, Rivelles and Taylor 1982). Faced with this difficulty, two orthogonal strategies have been adopted. In the first, Lorentz covariance is kept but only $N = 2$ extended supersymmetry is realized in superfields, with the remaining two supersymmetry transformations being nonlinear (Howe *et al* 1982b, 1983a). The superfield action is now just the sum of two gauge covariant terms, the action for the $N = 2$ supersymmetric Yang–Mills theory, given in its Abelian version by Mezincescu (1979), plus the gauge-coupled superfield action for the $N = 2$ supersymmetric matter 'hypermultiplet' containing only propagating spinors and scalars. Both parts of the superfield action are now covered by $N = 2$ supersymmetric non-renormalization theorems since they must be written as integrals over subsurfaces of $N = 2$ superspace, thus giving $\beta = 0$ to all loop orders. The second strategy keeps the $N = 4$ theory's rigid SU(4) symmetry but abandons the manifest realisation of Lorentz invariance to work in a lightcone gauge (Mandelstam 1982, Brink *et al* 1982), with the same consequence of vanishing to all orders of the β function.

The general lessons that can be drawn from the experience with $N = 4$ supersymmetric Yang–Mills theory are that cancellations of renormalization coefficients are determined by the maximum amount of supersymmetry that can be linearly realized in a superfield formalism. Thus, the cancellations in the $N = 4$ super-Yang–Mills theory are really a consequence of linearly realized $N = 2$ supersymmetry. This is underscored by the existence of other finite linearly realized $N = 2$ supersymmetric theories (Howe *et al* 1983b). These other theories are constructed from $N = 2$ super-Yang–Mills theory coupled to a definite number of $N = 2$ matter hypermultiplets, but with the hypermultiplets in a representation other than the adjoint (which would give the $N = 4$ super-Yang–Mills theory already discussed). For example, if the gauge group is SU (N) and the hypermultiplet matter is written in the fundamental representation, then a finite theory requires $2N$ hypermultiplets.

For supersymmetric gauge theories, the non-renormalization theorem (Grisaru and Siegel 1982) works for β functions like the Adler–Bardeen theorem does for ordinary U(1) axial anomalies (Adler and Bardeen 1969). That is, the full contribution to a β function covered by the theorem comes from the one-loop graphs. This follows from the existence of special ghost graphs (Grisaru and Siegel 1982) that couple only to the background when using the background field method (DeWitt 1967, 't Hooft 1975, Boulware 1981, DeWitt 1981, Abbot 1981) and thus contribute only at the one-loop order. Thus, some theories have an essential infinite counterterm at the one-loop order, but not at higher order. This happens for example in the $N = 2$ super-Yang–Mills theory without hypermultiplet couplings. This situation has also been checked by explicit Feynman diagram calculations to the three-loop order (Avdeev and Tarasov 1982).

For the finite theories built from $N = 2$ super-Yang–Mills theory coupled to $N = 2$ hypermultiplet matter, the finite theories are thus simply constructed by requiring the one-loop β function to vanish, with the non-renormalization theorem covering all higher loops. As long as the strictly renormalizable parts of a Lagrangian can be written in the required extended superfields, it is possible to add super-renormalizable terms of lower symmetry as long as they are protected by their own non-renormalization theorems. For example, it is possible to add $N = 1$ supersymmetric chiral superfield mass terms to the $N = 4$ supersymmetric Yang–Mills theory without violating the vanishing of the β function (Parkes and West 1982). The mass parameter cannot affect the gauge coupling β functions, which would come from strictly renormalizable counterterms, and in addition the mass term is protected by its own $N = 1$ supersymmetry non-renormalization theorem.

Supersymmetric Feynman rules have been constructed in detail for $N = 1$ supergravity (Grisaru and Siegel 1981, 1982), but not yet for extended supergravity theories. Nonetheless, from the experience gained with $N = 2$ supersymmetric gauge theories, we can anticipate the cancellations that can be expected on the basis of non-renormalization theorems. We first must find out what amount of the supersymmetry in a theory can be linearly realized in superfields. For the maximally extended $N = 8$ supergravity theory (Cremmer and Julia 1979), the same counting arguments that almost certainly rule out an $N = 4$ superfield formulation of the $N = 4$ super-Yang–Mills theory indicate that an $N = 8$ superspace formulation of $N = 8$ supergravity will not be possible (Rivelles and Taylor 1982).

However, just as in $N = 4$ super-Yang–Mills theory, the counting arguments do not exclude a formulation with half of the full supersymmetry, i.e. $N = 4$ superfields for $N = 8$ supergravity. Reduced to representations of $N = 4$ supersymmetry, the irreducible $N = 8$ supergravity would have to be described by two $N = 4$ superfield theories: the first containing $N = 4$ supergravity coupled to six $N = 4$ Maxwell gauge multiplets, and the second containing four $N = 4$ multiplets with maximum spin $\frac{3}{2}$. The first of these multiplets has been constructed for the free theory (Howe *et al* 1982a). The second has the same number of physical spin $\frac{3}{2}$ and spin $\frac{1}{2}$ fields as the first, and so satisfies the counting arguments in the same way, so there seems no obstacle other than hard work to developing an $N = 4$ superfield formalism for it.

In terms of $N = 4$ superfields, the prospects for ultraviolet cancellations beyond those predicted just on the basis of the existence of essential counterterms are not encouraging. According to the non-renormalization theorem, a permitted counterterm would have to be built from the superspace vielbeins $E_M{}^A$ and the constrained superfields containing the matter fields, and would have to be integrated over the full $N = 4$ superspace, i.e. with $\int d^{16}\theta$. This integration carries dimension 8 in units of mass, and $\det(E_M{}^A)$ has dimension zero, so the lowest dimension space–time integrand of a permitted counterterm has dimension 8. This is just the dimension of the counterterm containing the square of the Bel–Robinson tensor, which corresponds to the three-loop essential ultraviolet divergence. Thus, the indications are that $N = 8$ supergravity will start to have essential ultraviolet divergences from the three-loop order onwards. This also agrees with the estimate obtained from the $N = 2$ supersymmetric string theory in ten dimensions in the limit of simultaneous dimensional compactification to four dimensions and of letting the slope tend to zero (Green *et al* 1982).

3. PROSPECTS FOR HIGHER DERIVATIVES AND LOWER DIMENSIONS

If the quantum theory of the maximal supergravity theory seems set to diverge at the three-loop order, we should take stock and ask what benefits supersymmetry has given us beyond a two-loop postponement of the ultraviolet dilemma.

The power of extended supersymmetry to make a renormalizable theory finite can be applied to the renormalizable generalizations of general relativity, the higher derivative theories containing R^2 and $(C_{\mu\nu\rho\sigma})^2$. Given a formulation of $N = 8$ supergravity in $N = 4$ superfields, $N = 4$ supersymmetric expressions containing R^2 and $(C_{\mu\nu\rho\sigma})^2$ can be constructed, as well as corresponding higher derivative invariants built from the $N = 4$ 'matter' sector, the four $N = 4$ spin $\frac{3}{2}$ multiplets. The addition of such terms gives rise to a large number of ghost states in the theory, including ghosts coming from what in the original theory were auxiliary fields, now promoted to massive propagating ghosts by the higher derivative term. All of these terms have dimension four (in geometrical dimensions, in which $\dim(g_{\mu\nu}) = 0$). The non-renormalization theorem for $N = 4$ supersymmetry allows only counterterms of dimension 8 and higher, as we have seen. Thus, the $N = 4$ supersymmetric higher derivative extensions of $N = 8$ supergravity will have no essential counterterms from the two-loop level on.

Due to the extra graphs that enter into the background-field method calculations at the one-loop order, a separate check of the ultraviolet divergences needs to be made. For the essential, gauge invariant counterterms, this check can be made using any convenient method of calculation. Until the $N = 8$ theory has been written out in detail in $N = 4$ superfields, it is difficult to make a complete discussion of the one-loop infinities. Nonetheless, we can see that at least the coefficient of $(C_{\mu\nu\rho\sigma})^2$ must be zero.

By explicit calculation, the coefficient of the $(C_{\mu\nu\rho\sigma})^2$ counterterm has been found to vanish in the usual $N = 8$ supergravity without higher derivatives, if one takes the version of the theory obtained from eleven dimensions by dimensional reduction (Fradkin and Tseytlin 1982). A convenient way to calculate the coefficient of $(C_{\mu\nu\rho\sigma})^2$ (Fradkin and Tseytlin 1981) is to calculate the coefficient of $(F_{\mu\nu})^2$ for one of the axial vector fields A_μ, which plays the role of an auxiliary field in the theory without higher derivatives. It arises naturally in supersymmetric theories for this term to be considered as a gauge field, but since it has dimension 2 it does not have the usual kinetic action, but rather the Stueckelberg gauge invariant term $(A_\mu - \partial_\mu B)^2$, where B is a compensating pseudoscalar field. Thus, A_μ couples to a chiral current which is actually one of the currents for SU(4), the manifest internal symmetry of the $N = 4$ superfield formalism. The coefficient of $(F_{\mu\nu})^2$ would have to be the same as the coefficient of the $F_{\mu\nu} * F^{\mu\nu}$ contribution to the anomaly for this current. At the one-loop level, we can see that this anomaly is zero because it is proportional to the sum of the charges that A_μ couples to, which vanishes for any chiral generator of SU(4). The addition of the higher derivative terms to the tree level action does not change this conclusion, because they give rise to multiplets of massive ghost states without violating the SU(4) symmetry, so for every ghost that couples to A_μ, there is one with the opposite charge, so again the sum of the charges is zero, thus ruling out an anomaly and also the coefficients of $(F_{\mu\nu})^2$ and $(C_{\mu\nu\rho\sigma})^2$.

Until more detail is known of the structure of the $N = 8$ theory and of the $N = 4$ supersymmetric higher derivative terms that can be added to it, we cannot make firm conclusions about cancellations in the other possible one-loop infinities. Nonetheless, it seems likely that these other terms can be ruled out by the non-renormalization theorems for matter fields, which remain valid at the one-loop order.

What implications can we draw from the apparent finiteness of the $N = 8$ theory with the $N = 4$ supersymmetric higher derivative terms? This maximal theory appears to saturate the possibilities for ultraviolet cancellations in much the same way as $N = 4$ Yang–Mills theory. Obviously, the ghosts rule out the direct physical applicability of such a theory. As in a merely renormalizable higher derivative theory, any attempt to turn off the coefficients of the higher derivative terms in the fully quantized theory seems doomed to encounter a singularity in the limit of vanishing coupling constants for the higher derivative terms. In a renormalizable theory, this can be seen in the occurrence of $\log (M_{\text{ghost}}/\mu)$ factors in the renormalized amplitudes, resulting from subtractions at the scale μ. In this case, the nature of the singularity might be strikingly modified due to the absence of renormalization effects: if the theory is finite, all coupling constants remain equal to their unrenormalized values, and there is no scaling dependence other than that due to canonical dimensionality.

The power of supersymmetry to control ultraviolet infinities to all loop orders in renormalizable theories is also relevant to the asymptotic safety program, starting with an ε-expansion from two dimensions. As we have seen, general relativity is renormalizable in two dimensions, although the graviton, spin $\frac{3}{2}$ and spin 1 fields don't propagate. All supergravity theories automatically have a vanishing re-normalization of Newton's constant at the one-loop order, since the contribution to the coefficient of the extrinsic curvature surface integral counterterm Φ from each spin is proportional to the number of degrees of freedom for that spin, with a minus sign for fermions. If we have an $N = 4$ supergravity theory in two dimensions (obtained by dimensional reduction from $N = 2$ supergravity in four dimensions), then the non-renormalization theorem is sufficient to rule out the action as a counterterm to all orders, since the full superspace integral $\int d^8\theta$ contributes dimension 4 to any allowed counterterm, while the action is only of dimension 2.

Thus in two dimensions $N = 4$ $(d = 2)$ supergravity theory saturates the possibilities for ultraviolet cancellations like $N = 4$ Yang–Mills theory in four dimensions. In the ε expansion around two dimensions, this would leave the dimensionality of Newton's constant equal to its canonical dimension $(-\varepsilon)$ if supersymmetry were still valid in $2 + \varepsilon$ dimensions. As can be seen from equation (1.1), this would just violate the requirements of asymptotic safety, since $\beta(g)$ keeps rising with slope ε and never turns over.

Is supersymmetry valid in $2 + \varepsilon$ dimensions? The question is the same as whether one can regularize supersymmetric theories by analytic continuation in the number of dimensions. Such a scheme for regularizing directly in superfields has been proposed (Siegel 1979) and later found to be formally inconsistent (Siegel 1980). This inconsistency can be removed by writing a theory out in components, but the consistent scheme then violates supersymmetry (Avdeev *et al* 1981). Thus it seems possible that supersymmetry-violating contributions to the β function for Newton's constant could occur in $N = 4$ ($d = 2$) (or higher N) supergravity theories. Do these restore the condition of asymptotic safety? Is there a way to use this dimensional breaking of supersymmetry to obtain a new kind of perturbation theory encompassing effects from all loop orders in the conventional perturbation expansion? All that is clear is that our understanding of the interface between supersymmetry and the problem of quantum gravity is still at an early stage.

DEDICATION

This article is dedicated to Bryce DeWitt on the occasion of his sixtieth birthday. Bryce's papers are the indispensable roots of the subject of quantum gravity, and there is scarcely an approach or technique that cannot be traced back to his fundamental work. It gives me great pleasure to extend my heartfelt greetings to Bryce on this happy occasion.

REFERENCES

Abbot L 1981 *Nucl. Phys.* B **185** 189
Adler S L and Bardeen W A 1969 *Phys. Rev.* **182** 182
Avdeev L V, Chochia G A and Vladimirov A A 1981 *Phys. Lett.* **105B** 272
Avdeev L V and Tarasov O V 1982 *Phys. Lett.* **112B** 356
Boulware D G 1981 *Phys. Rev.* D **23** 389
Boulware D, Deser S, Gibbons G, and Stelle K S 1983 to appear
Brink L, Lindgren O and Nilsson B E W 1982 *University of Texas preprint* UTTG-1-82
Brink L, Schwarz J and Scherk J 1977 *Nucl. Phys.* B **121** 77
Caswell W and Zanon D 1980 *Phys. Lett.* **100B** 152
Christensen S M and Duff M J 1978 *Phys. Lett.* **79B** 213
Coleman S 1970 in *Subnuclear Phenomena* ed A Zichichi (New York: Academic)
Cremmer E and Julia B 1979 *Nucl. Phys.* B **159** 141
Deser S 1976 in *Proc. Conf. on Gauge Theories and Modern Field Theory* ed R Arnowitt and P Nath (Cambridge, Mass: MIT Press)
Deser S and Kay J H 1978 *Phys. Lett.* **76B** 400
Deser S, Kay J H and Stelle K S 1977 *Phys. Rev. Lett.* **38** 527
DeWitt B S 1964 *Phys. Rev. Lett.* **13** 114
—— 1967 *Phys. Rev.* **162** 1195, 1239 (erratum (1968) *Phys. Rev.* **171** 1834)
—— 1981 in *Quantum Gravity 2* ed C J Isham, R Penrose and D W Sciama (Oxford: Oxford University Press) p 449
Faddeev L D and Popov V N 1967 *Phys. Lett.* **25B** 29
Ferrara S, Iliopoulos J and Zummino B 1974 *Nucl. Phys.* B **77** 413
Ferrara S and van Nieuwenhuizen P 1978 *Phys Lett.* **78B** 578

Ferrara S and Piguet O 1975 *Nucl. Phys.* B **93** 261
Ferrara S and Zumino B 1978 *Nucl. Phys.* B **134** 301
—— unpublished
Feynman R P 1963 *Acta Phys. Pol.* **24** 697
Fradkin E S and Tseytlin A A 1981 *Phys. Lett.* **104B** 377
—— 1982 *Lebedev Inst. preprint* No 114
Gastmans R, Kallosh R and Truffin C 1978 *Nucl. Phys.* B **133** 417
Gibbons G W and Hawking S W 1977 *Phys. Rev.* D **15** 2752
Gliozzi F, Scherk J and Olive D 1978 *Nucl. Phys.* B **133** 253
Green M B, Schwarz J H and Brink L 1982 *Caltech preprint* CALT-68-972
Grisaru M T, Roček M and Siegel W 1980 *Phys. Rev. Lett.* **45** 1063
Grisaru M T, and Siegel W 1981 *Nucl. Phys.* B **187** 149
—— 1982 *Nucl. Phys.* B **201** 292
Grisaru M T, Siegel W and Rocek M 1979 *Nucl. Phys.* B **159** 429
Heisenberg W 1938 *Z. Phys.* **110** 251
't Hooft G 1975 *Acta Universitatis Wratislavensis No 38, XII Winter School of Theoretical Physics in Karpacz*
Howe P S, Nicolai H and Van Proeyen A 1982a *CERN preprint* TH3226
Howe P S, Stelle K S and Townsend P K 1981 *Nucl. Phys.* B **191** 445
—— 1982b *CERN preprint* TH3211 (*Nucl. Phys.* B, in press)
—— 1983a in preparation
Howe P S, Stelle K S and West P C 1983b *ITP Santa Barbara preprint*
Iliopoulos J and Zumino B 1974 *Nucl. Phys.* B **76** 310
Isham C J, Salam A and Strathdee J 1971 *Phys. Rev.* D **3** 867
Julve J and Tonin M 1978 *Nuovo Cimento* B **46** 137
Kallosh R 1981 *Phys. Lett.* **99B** 122
Kay B S 1981 *Phys. Lett.* **101B** 241
Khriplovich I B 1966 *Yad. Fiz.* **3** 575 (*Sov. J.–Nucl. Phys.* **3** 415)
Lee T D and Wick G C 1969a *Nucl. Phys.* B **9** 299
—— 1969b *Nucl. Phys.* B **10** 1
Mandelstam S 1982 in *Proc. 21st Int. Conf. on High Energy Physics* ed P Petiau and M Porneuf, *J. Physique, Colloque* C3 Supp au No 12, p 331
Mezincescu L 1979 *JINR Report* P2-12572
Parkes A J and West P C 1982 *King's College preprint*
Roček M and Siegel W 1981 *Phys. Lett.* **105B** 275
Rivelles V and Taylor J G 1982 *King's College preprint*
Salam A and Strathdee J 1970 *Lett. Nuovo Cimento* **4** 101
—— 1978 *Phys. Rev.* D **18** 4480
Siegel W 1979 *Phys. Lett.* **84B** 193
—— 1980 *Phys. Lett.* **94B** 37
Sohnius M F and West P C 1981 *Phys. Lett.* **100B** 245
Stelle K S 1977 *Phys. Rev.* D **16** 953
—— 1978 *Gen. Relativ. Gravity* **9** 353
—— 1982 in *Quantum Structure of Space and Time* ed M J Duff and C J Isham (Cambridge: Cambridge University Press) p 337
Tarasov O V and Vladimirov A A 1980 *Phys. Lett.* **96B** 94
Utiyama R and DeWitt B S 1962 *J. Math Phys.* **3** 608

Weinberg S 1974 in *Proc XVII Int. Conf. on High Energy Physics* ed J R Smith (Rutherford Laboratory, Chilton, Didcot) III-59
—— 1979 in *General Relativity—An Einstein Centenary Survey* ed S W Hawking and W Israel (Cambridge: Cambridge University Press)
Wess J and Zumino B 1974 *Phys. Lett.* **49B** 52

Properties of Extended Theories of Rigid Supersymmetry

P C WEST

1. INTRODUCTION

We will begin this introduction by making some rather speculative comments regarding the relation between symmetries and consistency in quantum field theory. Demanding consistency has been an important element in the development of quantum field theory. It would seem that field theories, although well behaved at the classical level, can develop diseases at the quantum level. The most common such diseases are uncontrollable ultraviolet behavior and acausality. The classic example of the former is the non-renormalizable four-fermi theory of weak interactions. An example of acausality is thought to occur in the form of the Landau ghosts which appear in the ladder approximation of QED (Landau and Pomeranchuk 1955, Fradkin 1955, Kirzhnits and Linde 1978 and references therein).

The cure of these two diseases, at least in the examples considered above, involves introducing a theory with a deeper symmetry. The non-renormalizability of four-fermi theory is overcome provided the theory is replaced by an anomaly-free non-Abelian gauge theory based on the group SU (2) $\times$ U (1). The Landau ghosts in QED or its extension, the Salam–Weinberg theory, are thought to be absent if these theories are embedded in a theory with a semi-single gauge group such as SU (5) (Georgi and Glashow 1974) SO(10) (Fritzsch and Minkowski 1975) or E_6 (Gursey et al 1976, Achiman and Stech 1978, Sikivie and Gursey 1977).

An interesting recent development is the remarkable fact that the $\lambda\phi^4$ theory, at least when lattice regulated, is a free field theory (Frohlich 1982). It has been speculated that this result may extend to other theories, although not to non-Abelian gauge theories.

It is certainly a remarkable fact that demanding consistency within the framework of the principles of causality, relativity and quantum mechanics alone leads to such a restrictive class of theories which include those found by much experimental input. However, it is far from clear that the demand of consistency has been satisfied. One recent attempt to extend this process is 't Hooft's (1982) requirement that one should be able to uniquely re-sum perturbation theory.

In this context it is enlightening to recall an old argument of Heisenberg (1958). Although very heuristic the argument provides a physical understanding of the relationship between the principles of quantum field theory on the one hand and the emergence of acausality and ultraviolet divergences on the other. The argument is as follows: relativity and causality impose an infinitely sharp boundary between events which could be causally connected and those which are not. The Uncertainty Principle then implies that for this infinite localization there is an infinite probability of finding high momentum states in the region of the lightcone. Study of, for example, the one-loop divergences of quantum field theory bears this out. Although far from complete, the argument is interesting as it suggests that infinities are an inevitable feature of quantum field theory. However, in agreement with known facts, it suggests that if we give up causality we can reduce divergences (Dirac 1942, Lee and Wick 1969).

The above argument could also provide an intuitive understanding of how it is that supersymmetry leads to fewer or no infinities. Despite the fact that supersymmetric theories are causal, the length squared of a space–time displacement $(x_1{}^\mu - x_2{}^\mu)^2$ is not invariant with respect to supersymmetry, rather it is the square of the superspace displacement which is invariant,

i.e. $$(x_1{}^\mu - x_2{}^\mu + i\,\bar{\theta}_1{}^i \gamma^\mu \theta_2)^2.$$

One may speculate that this change in the line element is sufficient to account for the remarkable ultraviolet properties of supersymmetric theories. It is in agreement with detailed calculations of divergences in supersymmetric theories which are seen to be suppressed by the presence of anti-commuting factors in the super-Feynman rules.

It may be worthwhile to attempt to place this old argument on a less heuristic footing. Should this be possible one could examine if the argument makes a distinction between ultraviolet divergences which are renormalizable and those which are not, and also to see how gauge invariance leads to fewer ultraviolet divergences. In this context it is interesting to note that, historically, it was the U (1) theory of Maxwell that led to our present understanding of space and time.

The theories of extended rigid supersymmetry are the most symmetrical consistent theories that we know. They are the most symmetric theories possible within the framework of S-matrix theory (Haag *et al* 1975). They do not involve spin two particles which would seem to lead to non-renormalizable theories and particles of spin $\frac{3}{2}$ are excluded as they lead to acausalities in the absence of spin two particles (Deser and Zumino 1976). From the previous discussion we may expect these most symmetrical theories to be the most consistent. In fact, the extended theories of rigid supersymmetry do have remarkable ultraviolet properties. We will show in §3 and §4 that there are two classes of theories of extended supersymmetry which are finite to all orders of perturbation theory. These theories can incorporate non-Abelian gauge groups and we can expect them to be free from Landau ghosts.

The particle content of the theories of extended supersymmetry is listed in the table below. The $N = 1$ theories have been included for comparison. Theories with more than four supercharges ($N > 4$) involve spins greater than one. The $N = 4$ Yang–Mills theory (Gliozzi *et al* 1977, Brink *et al* 1977) is unique up to the choice of gauge group and the one coupling constant. The $N = 3$ theory has the same particle content as the $N = 4$ theory and is thought to be identical. The $N = 2$ Yang–Mills theory (Salam and Strathdee 1974, Fayet 1976) can be coupled to the $N = 2$ matter (Fayet 1976) (the hypermultiplet) in an arbitrary representation, however, the coupling is given entirely by the gauge coupling constant. In the particular case where the gauge group contains U (1) factors, it is possible also to add terms linear in these U (1) fields.

Spin \ N	1		2		4
1	–	1	–	1	1
$\frac{1}{2}$	1	1	2	2	4
0	2	–	4	2	6

In §2 the arguments for the finiteness of $N = 4$ Yang–Mills are summarized. The ultraviolet properties of theories of rigid $N = 2$ supersymmetry are found in §3. These theories are always finite above one loop and there exists a class of $N = 2$ theories which are finite to all orders. In §4 it is shown that one can add terms which explicitly break supersymmetry to the $N = 4$ Yang–Mills theory and still maintain the finiteness of the theory.

At first sight these extremely restrictive theories do not seem promising candidates for constructing models of particle physics. These theories, by definition, involve more than one supersymmetry charge, and it is a simple consequence of this fact that if the theory has left-handed fermions in a given representation then the theory will also contain right-handed fermions in the same representation (the so-called mirror fermions). Although mirror fermions are not experimentally excluded, the mass splitting between the observed fermions and the mirror fermions breaks SU (2) × U (1) and so is likely to be of the order of the weak interaction energy scale. What mechanism can lead to such a splitting is unknown. Another objection against mirror particles is that they violate the so-called survival hypothesis. That is, there would seem to be no reason why the observed fermions and their mirror partners could not acquire a mass at any energy scale and one cannot use the chirality of SU (2) × U (1) to 'explain' the almost masslessness of the observed particles. This objection, however, is not quite as serious in the extended supersymmetric theories as it is in more usual theories since, as we will see, masses in extended supersymmetric theories are not renormalized and so once fixed in the classical theory they will not be altered by radiative corrections.

Another related problem with extended theories of extended supersymmetry is how does one break the supersymmetry? In the $N = 4$ Yang–Mills theory the six scalar fields, ϕ_{ij}, in the adjoint representation have the following potential

$$V(\phi_{ij}) = -\tfrac{1}{4} \text{Tr}[(\phi_{ij} \phi_{kl})^2].$$

Clearly, this potential has degenerate minima and one can shift the fields so as to break the gauge group, but unfortunately not $N = 4$ supersymmetry. Any resulting massless fermions will belong to the adjoint representation of the unbroken subgroup (Fayet 1979). Radiative corrections within a perturbative framework do not improve this situation; as a consequence of the same argument (West 1976, Capper and Ramon Medrano 1976, Witten 1981) as was used in $N = 1$ supersymmetric theories, namely, if supersymmetry is not broken at the tree level it will not be broken by radiative corrections. This, however, does not rule out non-perturbative mechanisms although no such mechanism is known in the context of four-dimensional field theory.

It has been pointed out (Fayet 1976) that in $N = 2$ rigid supersymmetric theories with $U(1)$ gauge factors, it is possible to add terms linear in fields and as a result break $N = 2$ supersymmetry. The possible implications of such breaking have not been widely explored.

Another feature of extended theories which is extremely restrictive is the fact that the Yukawa coupling constant is the same as the gauge coupling constant. This can present difficulties for finding realistic models in which the quark masses arise by spontaneous symmetry breaking and so are governed by the size gauge coupling constant. In present models some of the Yukawa coupling constants are required to be a factor of 10^{-6} less than the gauge coupling constants. Finding a realistic extended model would involve explaining this 10^{-6} factor.

A final problem concerns the fact that in a finite theory the parameters of the theory are related directly to the Green's functions and so are measurable, and as a result there can be no 'dimensional transmutation'. However, this does not mean that these theories cannot have energy scales. The superconformal invariance can be spontaneously broken as can occur at the tree level, or as we will see in §4 it is possible to introduce masses without destroying finiteness.

Although one could examine the possibilities for non-perturbative breaking, we will in this article consider two other possibilities. One is to examine the effect of adding explicit symmetry breaking terms. It is shown in §4 that provided certain relations hold between these breaking terms then the finiteness of $N = 4$ Yang–Mills is maintained. The other possibility considered in §5 is to use dimensional reduction to break the symmetry of $N = 4$ Yang–Mills theory.

2. FINITENESS OF $N = 4$ YANG–MILLS

There are three known arguments for the finiteness of $N = 4$ Yang–Mills. The oldest argument relies on the fact that in supersymmetric theories the superconformal anomalies lie in a supermultiplet. Another argument uses a generalization of the well known non-renormalization theorems of $N = 1$ supersymmetry to extended supersymmetry. The third argument uses lightcone techniques. We will discuss the first two arguments in turn and refer the reader to reference (Mandelstam 1982, Brink *et al* 1982) for a discussion of the lightcone argument.

The Anomalies Argument (Sohnius and West 1981, *Ferrara and Zumino, unpublished)*
This argument for the finiteness of $N = 4$ Yang–Mills theory makes two assumptions which are as follows:
  (a) in the quantum theory $N = 1$ supersymmetry and SU(4) internal symmetry are preserved;
  (b) when viewed as an $N = 1$ theory the anomalies must belong to a chiral multiplet of the form

$$(\theta_\mu{}^\mu, \partial^\mu j_\mu{}^{(5)}, \gamma^\mu j_{\mu\alpha}, C, D). \tag{2.1}$$

In the above multiplet $\theta_{\mu\nu}$ is the energy–momentum tensor, $j_\mu{}^{(5)}$ the chiral current (the R current), $j_{\mu\alpha}$ is the supercurrent and C and D are objects of dimension 3 for which no interpretation is known. This anomaly multiplet represents the breaking, due to quantum effects, of dilatation, chiral and S supersymmetry invariance.
  The finiteness of $N = 4$ Yang–Mills is given by the following simple argument. Assumption (a) implies that the SU(4) currents are preserved. However, $N = 4$ Yang–Mills is only SU(4) invariant not U(4) invariant and consequently all the nine chiral currents of the theory are preserved. The fact that $N = 4$ Yang–Mills does not possess an additional U(1) invariance follows from the CPT self-conjugate nature of the $N = 4$ Yang–Mills multiplet and the relation between the generator, B of this extra U(1) and the supersymmetry charge $Q_{\alpha i}$, namely

$$[Q_{\alpha i}, B] = \frac{(N-4)}{4} (\gamma_5)_\alpha{}^\beta Q_{\beta i} \tag{2.2}$$

which is zero for $N = 4$. Consequently, in any $N = 1$ decomposition of the theory the R current will be preserved, i.e. $\partial^\mu j_\mu{}^{(5)} = 0$.
  Assumption (b) then implies that $\theta_\mu{}^\mu = 0$, which in turn implies that $\beta(g) = 0$.
  The finiteness of the theory in a background field calculation or possibly with a particular gauge choice then follows from $\beta(g) = 0$.
  The above argument can also be formulated in terms of $N = 2$ supersymmetry. In this case the assumptions are that $N = 2$ supersymmetry and O(4) internal symmetry be preserved in the quantum theory and that the $N = 1$ anomaly multiplet of equation (2.1) is replaced by a corresponding $N = 2$ supermultiplet. We refer the reader to Sohnius and West (1981) and Ferrara and Zumino (unpublished) for the details of this argument.
  The assumptions made above have been found to be correct in all models for which the results are known (Clark *et al* 1977, 1978, 1979). However, these results are restricted to $N = 1$ models, and for the above argument to become complete it is necessary to establish the validity of these assumptions for $N = 4$ Yang–Mills. We will return to this point later. It is interesting to note, however, that the above argument would apply to any theory in which the chiral current is preserved and its divergence sits in the same multiplet as the trace of the energy–momentum tensor.

The Non-Renormalization Argument

This argument is based on the use of extended super-Feynman rules and the associated non-renormalization theorems. Let us consider $N = 4$ Yang–Mills in terms of $N = 2$ superfields; it then consists of $N = 2$ Yang–Mills containing physical component fields $(C, D, \lambda_{\alpha i}, A_\mu)$ represented (Grimm *et al* 1978) by a superfield potential $A_{\alpha i}(x^\mu, \theta_{\beta j})$ and $N = 2$ matter (the hypermultiplet) containing physical component fields $(A_i, B_i, \chi_{\alpha i})$ and represented by the superfield $\phi_i(x^\mu, \theta_{\beta j}, z); i = 1, 2$ (Sohnius 1980). When carrying out an $N = 2$ super-Feynman rule calculation in the background field formalism any contribution to the effective action must be

(a) an integral over all superspace
(b) a gauge invariant local function of $A_{\alpha i}$ and ϕ_i.

In other words any contribution must be of the form

$$\int d^4x \, d^8\theta \, \mathcal{L} \left(A_{\alpha i}, \phi_i, \mathcal{D}_{\alpha j} \phi_i, \ldots \right)$$

where $\mathcal{L}$ is a gauge invariant function. Consequently, any infinite counter-term for the superfield $A_{\alpha i}$ must be of the form

$$\int d^4x \, d^8\theta \, A_{\alpha i} D_{\gamma k} \ldots D_{\delta l} A_{\beta j}. \tag{2.3}$$

However since $A_{\alpha i}$ and $D_{\gamma k}$ have dimension $(\text{mass})^{1/2}$ and the integration measure has dimension zero such a term is impossible. A similar argument applies to counter-terms involving the hypermultiplet ϕ_i which has dimension $(\text{mass})^1$ or indeed any infinite counterterms. In this context it is interesting to note that in the case of $N = 1$, the argument fails as the integration measure has dimension $(\text{mass})^2$ so allowing the gauge invariant term

$$\int d^4x \, d^4\theta \, A_{\alpha i} \bar{D}^2 A^{\alpha i}. \tag{2.4}$$

Examining the background field method for extended supersymmetry, however, requires us to refine the above argument. In extended supersymmetry one finds, having fixed the gauge and found the relevant ghosts, that the ghosts themselves have a gauge invariance. This new gauge invariance requires new ghosts which in turn have a gauge invariance. This process in fact goes on indefinitely requiring an infinite number of ghosts. Fortunately, since these new ghosts only couple to background fields this infinity of ghosts only affects the one-loop contribution to the effective action.

We may conclude from the above argument that $N = 4$ Yang–Mills is finite above one loop. The finiteness of $N = 4$ Yang–Mills then follows, since it is known by explicit calculation to be finite at one loop.

An argument for the finiteness of $N = 4$ Yang–Mills along similar lines to that given above, but involving $N = 4$ superfields can be found in Grisaru and Siegel (1982), Grisaru *et al* (1980) and Stelle (1982a). These references also contain a discussion of non-renormalization theorems and the background field for extended supersymmetry.

Like the anomalies argument the above $N = 2$ argument has an assumption which in this case is the existence of an $N = 2$ superfield background field formalism for $N = 4$ Yang–Mills.

The validity of this assumption was more subject to doubt than the above argument suggests for two reasons. Firstly the field is subject to constraints which must be solved in terms of an unconstrained superfield which can then be used to formulate super-Feynman rules. The solution of this constraint in the Abelian case (Mezincesu 1979) is given in terms of a dimension (mass)$^{-2}$ superfield U^{ij} and more recently it has been shown (Koller 1982, Howe *et al* 1982b, Howe *et al* in preparation) that this solution can be systematically iterated to provide the non-Abelian solution. The second problem concerns the occurrence of the extra bosonic coordinate, z, in the ϕ^i formulation (Sohnius 1980) of the hypermultiplet. This coordinate is not integrated over in the action and represents an off-shell central charge (Sohnius *et al* 1980a,b). It is not known how to construct Feynman rules for such superfields. Fortunately, an alternative description of $N = 2$ matter which does not involve off-shell central charges has been found (Howe *et al* 1982b). It involves superfields S, L^{ij} and L^{ijkl} which are of dimension (mass)^1 and subject to constraints which can be solved.

With the solution of these two problems an $N = 2$ background superfield formalism for $N = 4$ Yang–Mills when expressed in terms of $N = 2$ superfields is guaranteed and the non-renormalization argument for the finiteness of $N = 4$ Yang–Mills given above can be regarded as complete.

The $N = 2$ background field formalism for $N = 4$ Yang–Mills theory can also be used to justify the assumptions made in the anomalies argument. It is evident from the formalism that the $N = 4$ Yang–Mills theory, when viewed as an $N = 2$ theory has a manifest supersymmetry and U (2) internal symmetry. These symmetries will also be manifest at the quantum level provided the theory is regulated in a supersymmetric, and U (2) invariant fashion. Such a regulator is provided above one loop by higher covariant derivatives. In fact, it has also been explicitly checked in the case of $N = 4$ Yang–Mills that the higher derivatives do not introduce any new infinities at one loop (Stelle 1982b). The supermultiplet of anomalies must be a dimension (mass)^3 gauge invariant function of the $N = 2$ superfields. A preliminary analysis confirms that this multiplet is indeed of the correct form in that it contains at least one of the divergences of the U (2) chiral currents as well as the trace of the energy–momentum tensor. An application of the anomalies argument to these facts gives the desired finiteness. Further details of this will be given elsewhere.

The reader may be struck by the fact that the anomaly argument as given just above does not seem to use any of the properties of $N = 4$ Yang–Mills other than one-loop finiteness and the $N = 2$ superspace formulation of the theory. This line of reasoning leads to the results of the next section.

3. A CLASS OF FINITE $N = 2$ SUPERSYMMETRIC THEORIES

The results presented in this section have been found in collaboration with P Howe and K Stelle (Howe *et al* 1983). The anomalies and non-renormalization arguments for the finiteness of $N = 4$ Yang–Mills presented in the previous section, have been given in a form which used an $N = 2$ decomposition of $N = 4$ Yang–Mills. Examination of the arguments shows that they apply not only to $N = 4$ Yang–Mills, but to an arbitrary $N = 2$ theory consisting of $N = 2$ Yang–Mills coupled to $N = 2$ matter (the hypermultiplet), provided there exists a background superfield formalism for an arbitrary $N = 2$ theory. One can, however, demonstrate that it is possible to extend the new formulation of the hypermultiplet to an arbitrary representation provided there are an even number of hypermultiplets. The hypermultiplet is then represented by the complex superfields S, L^{ij}, L^{ijkl} which again are of dimension $(\text{mass})^1$. The reader is referred to Howe *et al* (1983) for more details. This being the case we may conclude that an arbitrary $N = 2$ theory consisting of $N = 2$ Yang–Mills and $N = 2$ matter will be finite above one loop, although infinities can occur at one loop as indeed they do in the case of $N = 2$ Yang–Mills by itself.

The coupling of $N = 2$ matter to $N = 2$ Yang–Mills in the absence of U (1) factors is entirely determined by the gauge coupling constant. As a consequence, (see the next section), the necessary and sufficient condition for finiteness is that the Callan–Symanzik beta function vanish at one loop. The one-loop beta function for a supersymmetric theory consisting of n_i Wess–Zumino multiplets ($N = 1$ chiral superfields) in the representation R_i of a gauge group G and one super Yang–Mills multiplet is given by Ferrara and Zumino (1974)

$$\beta(g) = g^3/16\pi^2 \left(\sum_i n_i \, T(R_i) - 3C_2(G) \right). \tag{3.1}$$

An arbitrary $N = 2$ theory is composed from $N = 2$ Yang–Mills, which consists of: $N = 1$ Yang–Mills and one Wess–Zumino multiplet in the adjoint representation, as well as $N = 2$ matter which consists of two Wess–Zumino multiplets, one in the representation R_i and the other in the complex conjugated representation $\bar{R}_i$. Consequently the β function for such a theory with m_i hypermultiplets in the representation $R_i(\bar{R}_i)$ is given by

$$\beta(g) = 2g^3/16\pi^2 \left(\sum_i m_i \, T(R_i) - C_2(G) \right). \tag{3.2}$$

The equation $\beta(g) = 0$ has many solutions. One example is given by taking G to be SU (N) and R_i to be the fundamental representation; (in this case $C_2(N) = N$ and $T(R_i) = \frac{1}{2}$; the resulting theory is finite provided

$$m = 2N. \tag{3.3}$$

Another example is provided by taking G to be SU (5) where finiteness occurs if the $N = 2$ matter belongs to four $5(\bar{5})$ dimensional and two $10(\overline{10})$ dimensional representations. Yet another example is if G is SO (10) and the $N = 2$ matter consists of four 16 dimensional representations.

Let us summarize the results of this section; an arbitrary $N = 2$ theory possessing spins less than or equal to one is finite above one loop and as a consequence the one-loop result for the β function is exact. Those theories for which the condition $\Sigma_i \, m_i T(R_i) = C_2(G)$ holds will be finite to all orders.

The results of this section have been confirmed by an explicit two-loop calculation of the infinities of an arbitrary $N = 2$ rigid supersymmetric theory (Howe and West, in preparation).

One can play similar games in other supersymmetric theories. One example is $N = 4$ conformal supergravity, the fields for which are contained in a superfield $W(x^\mu, \theta_{\alpha j})$ which satisfies the condition $\mathscr{D}_{\alpha i} W = 0$. The action for this theory is given by

$$\int d^4 x \, d^8 \theta \, W^2 + \text{HC} \tag{3.4}$$

Since this integral is not over the whole of superspace we may expect this theory to be finite above one loop. Finiteness at one loop has been found by explicit calculation (Fradkin and Tseytlin 1981).

It is interesting to note that neither the anomalies nor the non-renormalization arguments apply to supergravity. This is a consequence of the dimensional character of the gravitational coupling which destroys the dimensional analysis arguments vital to either proof. The $N = 8$ supergravity theory consists of the following $N = 4$ multiplets: one $N = 4$ supergravity multiplet, six $N = 4$ Yang–Mills multiplets and one $N = 4$ multiplet having spin $\frac{3}{2}$ and less. A superfield formulation of $N = 4$ supergravity coupled to six Yang–Mills multiplets is known (Howe *et al* 1982a). Using dimensional analysis to construct generic Feynman rules we find that this multiplet leads to divergences at three loops and above (Howe and West, unpublished discussion). This is somewhat ominous as this is exactly the order at which an $N = 8$ supergravity invariant counterterm (Kallosh 1981, Howe *et al* 1981, Howe *et al* in preparation) exists. Of course, it could be that the coupling of this multiplet to the remaining multiplet gives a miraculous cancellation, but this would not be in keeping with the pattern of $N = 4$ Yang–Mills.

4. A CLASS OF FINITE THEORIES GENERATED BY ADDING SOFT BREAKING TERMS TO $N = 4$ YANG–MILLS THEORY

As explained in the introduction there are considerable difficulties in attempting to construct a realistic model of particle physics based on the extended theories of rigid supersymmetry. One possible way to overcome these difficulties is to consider adding terms which explicitly break supersymmetry. The question then arises as to whether the addition of such terms destroys the finiteness of the theory. The answer to this question in the case of $N = 4$ Yang–Mills has been found in collaboration with A Parkes (Parkes and West 1982a,b). We have added all terms to $N = 4$ Yang–Mills theory with gauge, Lorentz and parity invariance and of dimension three or less. The necessary and sufficient conditions on these terms in order to maintain finiteness have been found and are the subject of this section.

The $N = 4$ Yang–Mills theory when decomposed into representations of $N = 1$ supersymmetry consists of three chiral multiplets with component fields $(A_j + \mathrm{i} B_j, \chi_{\alpha j}, F_j + \mathrm{i} G_j)$ and represented by a chiral superfield $\varphi_j(x, \theta); j = 1, 2, 3$ as well as one Yang–Mills multiplet with component fields $(A_\mu, \lambda_\alpha, D)$ and represented by a general real superfield $V(x^\mu, \theta_\alpha)$. The possible dimension 2 terms we can add are

$$A^2 - B^2 \qquad \text{and} \qquad A^2 + B^2 \qquad (4.1)$$

while those of dimension 3 are of the form

$$\bar\chi\chi \qquad \bar\lambda\lambda \qquad A(A^2 + B^2) \qquad A^3 - 3AB^2. \qquad (4.2)$$

As a first step to finding the necessary conditions for finiteness, we will find which infinities compatible with one of the $N = 1$ supersymmetries of the $N = 4$ Yang–Mills theory and the semi-simple nature of the gauge group could be induced by adding the above insertions. We note that any field corresponding to an Abelian subgroup will decouple. This is achieved by expressing the $N = 4$ Yang-Mills theory in terms of $N = 1$ superfields (Fayet 1979, Grisaru et al 1979, Caswell and Zanon 1982) and writing the explicit symmetry breaking terms by use of directional $N = 1$ superfields. This is the so-called spurion technique (Piguet et al 1980, Girardello and Grisaru 1982) and it has the advantage that it enables us to use $N = 1$ super-Feynman rules. As an example, let us consider adding a term of the form $A^2 - B^2$. We can rewrite this term in the following form

$$\text{Tr} \int d^4x \, d^2\theta \; N^{ij}\varphi_i\varphi_j + \text{HC} \qquad (4.3)$$

where $N^{ij} = \theta^2 \, \eta^{ij}$. Any contributions to the effective action must be integrals over the full superspace and so on dimensional grounds any divergent terms are necessarily of the form

$$\text{Tr} \int d^4\theta \, d^4x \; N^{ij} \, \varphi. \qquad (4.4)$$

This term, however, vanishes since $\text{Tr} \, \varphi = 0$, and consequently the insertion of $A^2 - B^2$ leads to no infinite η^{ij} dependent counterterms. In such a circumstance one can, order by order, use a η^{ij} independent renormalization scheme (Weinberg 1973) and so the theory will have the same infinities as $N = 4$ Yang–Mills. In other words, adding an $A^2 - B^2$ term maintains the finiteness of the theory to all orders. The application of this technique to the other insertions does, however, lead to possible infinities.

The next stage in the calculation is to evaluate which of these possible infinities occur at one loop. This takes into account the full residual symmetries of the $N = 4$ Yang–Mills theory and it involves a lengthy calculation using $N = 1$ super-Feynman rules. The reader is referred to Parkes and West (1982b) for more details. It was found, with the exception of fermion masses, whenever an infinity was possible it did indeed arise at one loop. The table overleaf gives the infinities which arise from the addition of a given insertion and it includes the possibility of infinities arising due to the addition of two insertions.

Clearly, the addition of none of these terms by themselves, except for the $A^2 - B^2$ term, leads to a finite theory. However it may be possible, in this very special theory, that for certain combinations of the terms the infinities could cancel. Examining the table shows that the fermion mass terms, $\bar\chi\chi$, lead, amongst others, to an $A(A^2 + B^2)$ infinity which can only be cancelled by adding an insertion of the form

Insertion \ Infinity produced	$A^2 - B^2$	$A^2 + B^2$	$\bar{\chi}\chi$	$\bar{\lambda}\lambda$	$A(A^2 + B^2)$	$A^3 - 3AB^2$
$A^2 - B^2$						
$A^2 + B^2$		✓				
$\bar{\chi}\chi$	✓	✓			✓	
$\bar{\lambda}\lambda$	✓	✓				✓
$A(A^2 + B^2)$	✓	✓			✓	
$A^3 - 3AB^2$	✓	✓				✓

$A(A^2 + B^2)$. In fact the cubic insertion one must add to obtain finiteness is uniquely fixed both in form and magnitude by the spinor mass term, $\bar{\chi}\chi$. Having cancelled this $A(A^2 + B^2)$ infinity the resulting $A^2 - B^2$ infinities cancel automatically. The $A^2 + B^2$ infinities cancel provided we add an $A^2 + B^2$ insertion and adjust its coefficient such that the supertrace of the masses squared is zero.

As a result we can add the following insertions and still maintain finiteness at one loop

$$\text{Tr}\{\mu_i^{\ s}(A^i A_s + B^i B_s) + m^{ij}\bar{\chi}_i\chi_j + 4g\, m^{ir}\, \varepsilon_{rjk}\, A_i([A^j, A^k] + [B^j, B^k])\} \quad (4.5)$$

where

$$\mu_k^{\ k} = -16 m^{ij} m_{ij}. \quad (4.6)$$

Examination of these insertions shows that they correspond to none other than the addition of $N = 1$ supersymmetric masses, that is they can also be written in the form

$$\text{Tr} \int \mathrm{d}^4 x\, \mathrm{d}^2\theta\, s^{ij}\, \varphi_i \varphi_j + \text{HC} \quad (4.7)$$

By the non-renormalization theorem (Wess and Zumino 1974, Iliopoulos and Zumino 1974, Ferrara *et al* 1974, Ferrara and Piguet 1975) this term is not renormalized and indeed there can be no infinite s^{ij} dependent contributions to the effective action. As such we can use s^{ij} independent renormalization constants and the addition of this term results in a theory that is finite not only at one loop, but to all orders (Parkes and West 1982a).

We can further add a mass term for the fourth spinor λ. Examining the table shows that the only possible way to cancel the resulting $A^3 - 3AB^2$ infinity is to introduce a term of this form as an insertion. The one-loop calculation of these infinities shows that the necessary cubic term one must add in order to maintain finiteness is again determined uniquely by the λ mass term. The $A^2 - B^2$ infinities then automatically cancel and the $A^2 + B^2$ infinities are absent provided the coefficient of the $A^2 + B^2$ insertion is readjusted so as to maintain the relation

$$\sum_j (-1)^{2j}\, m_j^2\, (2j + 1) = 0. \quad (4.8)$$

This infinity has only been proved to be absent at one loop, but because of the O(4) relation between this mass term and the mass terms for the other spinors, χ_{zi} we may be confident that it also holds to all orders.

Finally we note that there is another way to maintain finiteness. The addition of a term of the form $\mu_i^s (A^i A_s + B^i B_s)$ leads at one loop to the infinite counter term

$$- g^2 C_2 (G) \ln \Lambda \; \mu_s^s \; (A^i A_i + B^i B_i). \tag{4.9}$$

Clearly, finiteness is maintained if

$$\sum_s \mu_s^s = 0. \tag{4.10}$$

It is not clear whether this final result holds at higher loops. The way the infinities cancel guarantees that the final result is independent of the order in which the insertions are added to the action. Taking the masses to be diagonal we find that it is possible to add nine mass parameters of the ten possible independent mass parameters and still maintain finiteness, the constraint being that of equation (4.8).

It should be apparent from the above discussion that in the case of the $N = 2$ supersymmetric finite theories one can add $N = 1$ supersymmetric mass terms as well as $A^2 - B^2$ without destroying the finiteness of the theory. What other possibilities can occur is under study.

It will be interesting to explore the possible consequences of adding these terms. The following questions immediately spring to mind. Can one use these terms to break spontaneously the gauge symmetry and so give masses to the spin one particles? Further, what symmetry breaking patterns can emerge and can one split the masses of the mirror fermions from the observed fermions? On the more formal side can one give all the vectors a mass and so obtain an infrared and ultraviolet theory?

Similar results to those discussed have independently been found by Namazie, Salam and Strathdee using lightcone techniques (Namazie et al 1982).

5. DIMENSIONAL REDUCTION AND $N = 4$ YANG–MILLS

The work presented in this section has been performed in collaboration with D Olive (Olive and West 1982). In the introduction we noted that it was impossible, within the framework of perturbation theory, to break spontaneously $N = 4$ supersymmetry. Although non-perturbative breaking may occur, no such mechanism is known. Faced with this situation we may either try adding the explicit symmetry breaking terms which maintain finiteness, as discussed in the previous section, or we can consider the theory in a higher dimension and use a dimensional reduction scheme which breaks supersymmetry. An example of the latter procedure is the dimensional reduction scheme of Scherk and Schwarz (1979). Although dimensional reduction techniques use a possibly unphysical mechanism, namely the existence of extra dimensions, they have the advantage of being a procedure which is relatively free from ambiguity and in some cases they are known to preserve the original quantum properties of the theory. In this context it would be interesting to examine if the explicit breaking terms which maintain the finiteness of $N = 4$ Yang–Mills can be obtained from some dimensional reduction scheme.

Before discussing such a scheme let us establish the general strategy we wish to adopt. The reduction should break the gauge group, G and the supersymmetry at a scale which is very large, certainly very much larger than the weak interaction scale. The remaining breaking to $SU(3) \times SU(2) \times U(1)$ and beyond we regard as being generated by some other mechanism, presumably dynamical in origin. The success of the scheme, at least in terms of the limited aims of this discussion is judged by whether the observed fermions survive the breaking at the high energy and are massless.

We take the gauge group of the $N = 4$ Yang–Mills theory to be E_8. This is natural in this case as the fermions must lie in the adjoint representation and E_8 is the only group in which the adjoint representation is the lowest representation. Further E_8 is the highest member of a finite sequence of groups. The full sequence is given by E_8, E_7, E_6, $E_5 = SO(10)$, $E_4 = SU(5)$ and $E_3 = SU(2) \times SU(3)$. It is a remarkable fact that so many of the favored groups of particle physics should belong to one finite sequence. The other infinite sequences are $SU(N)$, $SO(N)$, $SO(2N)$, $SO(2N+1)$ and $Sp(2n)$. The adjoint representation of E_8 does indeed contain sufficient particles to account for three generations of observed fermions. We hope to break E_8 to another member of the sequence whilst keeping the observed fermions massless. Octonians may provide a very deep reason for the connection between $N = 4$ Yang–Mills and the group E_8. It is conceivable that $N = 4$ Yang–Mills may admit an octonionic formulation and this formulation may be connected to the finiteness of the theory.

Known dimensional reduction schemes which break supersymmetry are the one of Scherk and Schwarz (1979) and the scheme of Manton, Chapline and Forgacs (Forgacs and Manton 1980, Chapline and Manton 1981, Manton 1981). These two schemes are similar in that the field dependence on the extra dimensions is such that the Lagrangian becomes independent of the extra dimensions. This is achieved in the former case by using an internal rigid symmetry whereas in the latter case a gauge symmetry is exploited. One could also have an arbitrary dependence on the extra coordinates and perform a harmonic analysis (Salam and Strathdee 1981) to find the mass spectrum.

We will assume the extra dimensions to be a coset space S/R with coordinates θ^t and adopt the dimensional reduction scheme of Manton, Chapline and Forgacs (Forgacs and Manton 1980, Chapline and Manton 1981, Manton 1981). In this scheme the Lagrangian is made independent of the extra dimensions by demanding that the dependence of the fields on the extra dimensions can be compensated by a gauge transformation. This is implemented by the equation

$$\delta_{\xi_N} \mathscr{F} = D(W_N) \mathscr{F} \qquad (5.1)$$

where δ_{ξ_N} is the change in the field $\mathscr{F}$ induced by the movement $\theta^t \to \theta^t + \xi^t$ on S/R and $D(W_N)$ is a gauge transformation on $\mathscr{F}$ with parameter W_N.

In the case of a vector field $A_{\hat{\mu}}$ this equation becomes

$$-\xi_N{}^{\hat{\rho}}\partial_{\hat{\rho}} A_{\hat{\mu}} - \partial_{\hat{\mu}}\xi_N{}^{\hat{\rho}} A_{\hat{\rho}} = \partial_{\hat{\mu}}W_N - \mathrm{i}[W_N, A_{\hat{\mu}}]. \tag{5.2}$$

Equation (5.1) not only places restrictions on and determines the dependence of the fields on the extra dimensions, it also results in certain consistency conditions. This is a consequence of the fact that these are dim S equations whereas the fields depend on only dim S − dim R extra coordinates. The consistency conditions place restrictions on W_N as well as requiring that R be a subgroup of G. Having evaluated the restrictions on and field dependence of the fields we can evaluate the four-dimensional Lagrangian and examine its mass spectrum. It often happens that this four-dimensional Lagrangian will undergo spontaneous symmetry breaking; indeed if $S \subset G$ then spontaneous symmetry breaking is inevitable. More often than not the fermions do not survive the constraints of equation (5.1) and even if this is the case they then often acquire mass in the dimensional reduction or in the resulting spontaneous symmetry breaking in four dimensions.

Another way of discovering which fermions are massless is to examine the fermion equation of motion which has the form

$$\gamma^\mu \mathscr{D}_\mu \psi = -\gamma^r \mathscr{D}_r \psi \tag{5.3}$$

where $\mu = 0, \ldots, 3$ and $r = 5, \ldots, 10$ and the derivatives are gauge covariant as well as including any necessary spin connections required for the curvature in the extra dimensions. Iterating equation (5.3) we find that

$$\mathscr{D}_\mu \mathscr{D}^\mu \psi = -\mathscr{D}_r \mathscr{D}^r \psi + \tfrac{1}{4} R \psi - \Sigma^{rs}[F_{rs}, \psi] \tag{5.4}$$

where $R = R_{rs}{}^{rs}$ and F_{rs} are the curvature and Yang–Mills field strength respectively. The appearance of the last term in the above equation can affect the fermion spectrum; indeed without this term and if R is positive and constant there will be no massless fermions in four dimensions (Lichnerowicz 1963).

Using the dimensional reduction scheme outlined above to determine the field dependence on the extra dimensions and taking into account any scalar field expectation values induced by the spontaneous symmetry breaking in four dimensions we can evaluate equation (5.4) and obtain

$$\mathscr{D}_\mu \mathscr{D}^\mu \psi = (C_S - 3C_R)\psi + \tfrac{1}{16}(f_{rst})^2 \psi + \tfrac{1}{8} f_{trs}f_{tr's'}\gamma^{rsr's'}\psi \tag{5.5}$$

where C_S and C_R are the Casimirs of S and R and f_{rst} are the structure constants of S. The indices r, s and t refer to the generators outside R. For a symmetric space $f_{rst} = 0$ and we find that

$$\mathscr{D}_\mu \mathscr{D}^\mu \psi = (C_S - 3C_R)\psi. \tag{5.6}$$

Let us apply the general scheme given above to the theory in question, namely the ten-dimensional $N = 1$ supersymmetric E_8 Yang–Mills theory. We will choose the extra dimensions to be $\mathrm{SU}(2)/\mathrm{U}(1) \times (S^1)^4$. The $\mathrm{U}(1)$ must be identified in E_8 and we choose it to be the Q_3 factor in the $\mathrm{SU}(2)$ of the maximal subgroup $E_7 \times \mathrm{SU}(2)$ of E_8. The adjoint representation of E_8 breaks up under this $E_7 \times \mathrm{SU}(2)$ subgroup as follows

$$248 = (133, 1) \oplus (1, 3) \oplus (56, 2). \tag{5.7}$$

The constraints of equation (5.1) demand that of the vectors only the $(133, 1)$ and the $(1, 0)$ in the $(1, 3)$ survive leaving us with an $E_7 \times U(1)$ gauge theory. The ten-dimensional fermions are required to satisfy the constraint

$$\tfrac{i}{2}\, \Gamma^9 \, \psi = [Q_3, \psi] \tag{5.9}$$

where Γ^9 is the nine-dimensional helicity operator. From this equation we find that only half of the $(56, 2)$ survive.

Two of the surviving scalars then lead the resulting $E_7 \times U(1)$ gauge theory to be spontaneously broken to E_7. The massless modes of this theory are then only the E_7 vectors and a $2 \times 56_L$ plus a $2 \times 56_R$ of E_7 fermions as well as 4×133 scalars. The 56 dimensional representational of E_7 breaks into $E_5 = SO(10)$ representations as follows

$$56 = 16 + \overline{16} + 4 \times 1 + 2 \times 10. \tag{5.10}$$

This demonstrates that we are left with four generations of conventional fermions plus their mirror partners as well as some additional fermions. It is likely that these additional fermions will acquire a mass in any further breaking since they belong either to real representations or are singlets. A further breaking to $SO(10)$ was considered in Olive and West (1982).

This conclusion is also supported by considering the fermion equation of motion. Since $SU(2)/U(1)$ is a symmetric space it is sufficient to consider equation (5.6) which in this case becomes

$$\mathscr{D}_\mu \mathscr{D}^\mu \psi = [j(j+1) - 3j_3{}^2]\psi.$$

For the $(56, 2)$ we find that $j = \tfrac{1}{2}$ and $j_3 = \pm \tfrac{1}{2}$ and so the right-hand side vanishes and these fermions are massless.

Consequently we find that with initial group E_8 and coset space $SU(2)/U(1) \times (S^1)^4$ we can indeed achieve the strategy outlined in the beginning of this section, namely sufficient fermions have survived the breaking of both supersymmetry and E_8 gauge invariance to account for the number observed. The interactions between these fields are uniquely determined by the $N = 1$, $D = 10$ supersymmetric theory. The reader is referred to Olive and West (1982) for the details of this construction.

One interesting question is to what extent is the final E_7 theory realistic? In other words does it break to the correct $E_3 \times U(1) = SU(3) \times SU(2) \times U(1)$ theory? We have not investigated this, but perhaps the most important problem to be solved is giving the mirror fermions sufficient mass. Clearly, the couplings of the classical theory will change under radiative corrections as supersymmetry has been broken. However, the $SU_L(2) \times SU_R(2)$ internal symmetry of the four-dimensional theory will keep the observed fermions and their mirrors massless. A related question is whether the low energy content of the theory is $SU(3)$ asymptotically free. Neglecting the scalars which could acquire large masses and the additional fermions we find that the remaining fermion content $(4 \times 16_L + 4 \times \overline{16}_R)$ is indeed color asymptotically free.

Another question is how many other choices of gauge group G and coset space fulfil the strategy outlined? Equation (5.5) leads us to suspect that there are not too many possibilities.

Finally there is the question of how was the supersymmetry broken. This could have happened in two ways. One possibility is that the constraints of equation (5.1) violate supersymmetry in which case a more general field dependence on the extra coordinates would restore supersymmetry. The other possibility is that the coset space is not compatible with supersymmetry. One way to investigate this latter possibility is to embed the $D = 10$ $N = 1$ Yang–Mills theory in $D = 10$ supergravity and look for a Kaluza–Klein type solution (i.e. a solution of the $D = 10$ field equations) which has an $SU(2)/U(1) \times (S^1)^4$ solution. This unfortunately is not possible for $N = 1$ supergravity in $D = 10$ dimensions (Freedman *et al* 1983). In this context it is worth noting that dimensional reduction from ten-dimensional supergravity plus matter may have some advantages over dimensional reduction from eleven-dimensional supergravity. The reason is that the former possesses vector fields before dimensional reduction and these can be used to change the number of massless fermions as well as allowing the possibility of obtaining complex chiral representations in four dimensions without mirror fermions.

Another possibility is to couple $N = 4$ Yang–Mills to $N = 4$ supergravity in four dimensions in the most general way and induce the $N = 4$ supersymmetry to break spontaneously in a similar way to the methods used in $N = 1$ supergravity coupled to $N = 1$ super matter (Cremmer *et al* 1978, 1979, 1982, 1983).

Dimensional reduction of the ten-dimensional super-Yang–Mills theory was also independently considered in Chapline and Slansky (1982), although the strategy these authors adopted is different from that considered here.

The model presented in this section illustrates the fact that if supersymmetry is broken at some high energy either spontaneously or explicitly it is likely that the β function of the low energy effective theory will be non-zero. It would be interesting to explore how such a low energy theory can belong to a theory which at high energy has $\beta = 0$.

ACKNOWLEDGMENT

I would like to thank Bryce DeWitt for many happy hours of discussion which provided me with valuable insights into physics.

This article has stressed the importance of answering questions of principle (such as consistency) within the framework of quantum field theory. We are indebted to Bryce, who has advocated, for many years, that such questions must be answered.

REFERENCES

Achiman Y and Stech B 1978 *Phys. Lett.* **77B** 389
Brink L, Lindgren O and Nilsson B 1982 *Texas Preprint* UTTG-1-82
Brink L, Schwarz J and Scherk J 1977 *Nucl. Phys.* B **121** 77

Capper D and Ramon Medrano M 1976 *J. Phys. G: Nucl. Phys.* **2** 269
Caswell W E and Zanon D 1982 *Nucl. Phys.* **B 182** 125
Chapline C and Manton N S 1981 *Nucl. Phys.* **B 184** 391
Chapline C and Slansky R 1982 *Nucl. Phys.* **B 209** 461
Clark T, Piguet O and Sibold K 1977 *Ann. Phys.* **109** 418
—— 1978 *Nucl. Phys.* **B 143** 445
—— 1979 *Nucl. Phys.* **B 159** 1
Cremmer E, Ferrarara S, Girardello L and Van Proeyn A 1982 *Phys. Lett.* **116B** 231
—— 1983 *Nucl. Phys.* **B 212** 413
Cremmer E, Julia B, Scherk J, Ferrara S, Girardello L and van Nieuwenhuizen P 1978 *Phys. Lett.* **79B** 231
—— 1979 *Nucl. Phys.* **B 147** 1051
Deser S and Zumino B 1976 *Phys. Lett.* **62B** 335
Dirac P A M 1942 *Proc. R. Soc.* **180** 1
Fayet P 1976 *Nucl. Phys.* **B 113** 135
—— 1979 *Nucl. Phys.* **B 149** 137
Ferrara S, Iliopoulos J and Zumino B 1974 *Nucl. Phys.* **B 77** 41
Ferrara S and Piquet O 1975 *Nucl. Phys.* **B 93** 261
Ferrara S and Zumino B 1974 *Nucl. Phys.* **B 79** 413
—— unpublished
Forgacs P and Manton N S 1980 *Commun. Math. Phys.* **72** 15
Fradkin E S 1955 *Zh. Eksp. Teor. Fiz.* **28** 750
Fradkin E S and Tseytlin A A 1981 *Phys. Lett.* **110B** 117
Freedman D Z, Gibbons G W and West P C 1983 *Phys. Lett.* **124B** 491
Fritzsch H and Minkowski P 1975 *Ann. Phys.* **93** 193
Frohlich J 1982 *Nucl. Phys.* **B 20** 281
Georgi H and Glashow S L 1974 *Phys. Rev. Lett.* **32** 438
Girardello L and Grisaru M 1982 *Nucl. Phys.* **B 194** 55
Gliozzi F, Olive D and Scherk J 1977 *Nucl. Phys.* **B 122** 253
Grimm R, Sohnius M and Wess J 1978 *Nucl. Phys.* **B 133** 275
Grisaru M, Roček M and Siegel W 1979 *Nucl. Phys.* **B 159** 429
—— 1980 *Phys. Rev. Lett.* **45** 1063
Grisaru M and Siegel W 1982 *Nucl. Phys.* **B 201** 292
Gursey F, Ramond P and Sikivie P 1976 *Phys. Lett.* **60B** 177
Haag R, Lopsnzanki J and Sohnius M 1975 *Nucl. Phys.* **B 88** 61
Heisenberg W 1958 *Physics and Philosophy* (New York: Harper and Row)
't Hooft G 1982 *Phys. Lett.* **109B** 474
Howe P S, Nicholai H and van Proeyen A 1982a *CERN preprint* TH 3226
Howe P S, Stelle K S and Townsend P K 1981 *Nucl. Phys.* **B 191** 445
—— 1982b The relaxed hypermultiplet: and unconstrained $N = 2$ superfield theory *Nucl. Phys.* B to be published
—— in preparation
Howe P, Stelle K and West P 1983 A class of finite four-dimensional supersymmetric field theories *Santa Barbara preprint* NSF-ITP-83-09 *Phys. Lett.* B to be published
Howe P and West P *unpublished discussion*
—— in preparation
Iliopoulos J and Zumino B 1974 *Nucl. Phys.* **B 76** 1310
Kallosh R E 1981 *Phys. Lett.* **99B** 122
Kirzhnits D A and Linde A D 1978 *Phys. Lett.* **73B** 323

Koller J 1982 Unconstrained Prepotentials in Extended Superspace *Caltech preprint*-CALT-68-981

Landau L D and Pomeranchuk I 1955 *Dokl. Akad. Nauk.* **102** 489

Lee T D and Wick G C 1969 *Nucl. Phys.* B **9** 201

Lichnerowicx A 1963 *C. R. Acad. Sci. Paris* A-B **257** 7

Mandelstam S 1982 *Berkeley preprint* UCB-PTH-82115

Manton N S 1981 *Nucl. Phys.* B **193** 502

Mezincesu L 1979 *JINR Report* P2-12572

Namazie M A, Salam A and Strathdee J 1982 Finiteness of Broken $N = 4$ super-Yang–Mills theory *Trieste preprint*

Olive D and West P 1983 *Nucl. Phys.* B **217** 248

Parkes A and West P 1982a $N = 1$ supersymmetric mass terms in $N = 4$ supersymmetric Yang–Mills theory *King's College preprint, Phys. Lett.* B to be published

—— Finiteness and explicit supersymmetry breaking in the $N = 4$ supersymmetric Yang–Mills theory *King's College preprint*

Piguet O, Sibold K and Schweda M 1980 *Nucl. Phys.* B **174** 183

Salam A and Strathdee J 1974 *Phys. Lett.* **51B** 33

—— 1981 On Kaluza–Klein Theory *Trieste preprint* IC/81/211

Scherk J and Schwarz J 1979 *Phys. Lett.* **82B** 60

—— 1979 *Nucl. Phys.* B **153** 61

Sikivie P and Gursey F 1977 *Phys. Rev.* D **16** 816

Sohnius M 1980 *Nucl. Phys.* B **165** 483

Sohnius M, Stelle K and West P 1980a *Phys. Lett.* **92B** 123

—— 1980b *Nucl. Phys.* B **173** 127

Sohnius M and West P 1981 *Phys. Lett.* **100B** 245

Stelle K S 1982a Extended Supercurrents and the Ultraviolet finiteness of the $N = 4$ supersymmetric Yand–Mills theory in *Quantum Structure of Space and Time* ed M J Duff and C J Isham (Cambridge: Cambridge University Press)

Stelle K S 1982b in Proceedings of the Paris High Energy Conference, *Imperial College preprint*

Weinberg S 1973 *Phys. Rev.* D **8** 3497

Wess J and Zumino B 1974 *Phys. Lett.* **49B** 52

West P 1976 *Nucl. Phys.* B **106** 219

Witten E 1981 *Nucl. Phys.* B **188** 52

Cosmological Topological Supergravity

S DESER†

1. INTRODUCTION

Recently, the locally supersymmetric extension of topologically massive gravity (Deser *et al* 1982) in three space–time dimensions was constructed (Deser and Kay 1983). That model consisted of the usual supergravity sector plus a separately supersymmetric topological one, in which the original topological invariant of the gravitational theory acquired a fermionic companion. Each sector was separately invariant under the usual supergravity transformations, as might be expected from their different dimensionality and derivative orders. As for the bosonic case, only the sum of the sectors led to a dynamically nontrivial theory. We shall show here that, due to an initial conformal-type invariance of the free topological part which is miraculously maintained by the supersymmetric coupling, it is possible to extend the model further to include a supercosmological sector. The latter consists as usual of a cosmological term together with an explicit, non-topological, mass term for the fermion. This three-layered model presents an amusing interplay of invariances and of dynamics. In particular, the invariance restoration mentioned above involves initial separate conformal and γ transformations of gravity and fermi parts which fuse to a common invariance after coupling. On the other hand, while it is possible to do without either outer layer, the middle, supergravity, one is essential to bind the other two consistently.

Since the present work is dedicated to Bryce DeWitt on his 60th birthday, it is perhaps appropriate to mention that I have tried, through this simple model, at least to touch upon four of the many areas in which he has worked. As the title indicates, these include topology, gravity and supergravity gauge fields as well as their cosmological extensions. The first is left largely unexplored here, but whether one can define fermionic analogs of the classical notions of Pontryagin and Chern–Simons classes through the spin $\frac{3}{2}$ gauge field clearly merits further attention. The model itself, like its antecedents, provides a counter-example to many of our folklore notions in four dimensions, and should thereby help our understanding of the latter, as well as find application in other odd-dimensional (e.g. Kaluza–Klein) contexts.

† Supported in part by NSF grant PHY 8201094.

The paradoxical properties of three-dimensional models include the following (refer to the references given above for details):

(1) gauge fields acquire mass without losing invariance under (small) gauge transformations,

(2) the topological terms are required for the theory to have any dynamics at all,

(3) these terms are not automatically invariant under large gauge transformations, but require the mass/coupling constant ratio to be quantized; a sub-paradox here is that this quantization seems to be signature-dependent for gravity, a rare example of significant differences between Minkowski and Euclidean worlds,

(4) the theories, being massive, involve short-range interaction between their sources, yet preserve the long-range effects inherent in a flux integral definition of energy and spinorial charge,

(5) there is 'spin without spin,' to borrow from an eminent colleague of Bryce. The reduced actions depend on variables which do not have any of the spatial indices associated with spin, nor do they show any apparent trace of PT violation, yet they describe spin two, PT odd particles,

(6) these are genuine higher derivative models, but with none of the associated ghost models, tachyon or acausality behavior,

(7) finally, the present model also has the obvious merit of implying positive energy and stability of excitation both for itself and for its classical bosonic reduction, i.e. for cosmological topological gravity. It thus retains this essential feature of the conventional four-dimensional theories.

2. REVIEW AND CONVENTIONS

Our geometrical and spinorial conventions will be as follows. The covariant derivative D_μ of a spinor is ($\mu, v, a, b = 0, 1, 2$)

$$D_\mu \equiv \partial_\mu + \tfrac{1}{2}\omega_{\mu ab}\sigma^{ab} \equiv \partial_\mu + \tfrac{1}{2}\omega_{\mu a}\gamma^a. \tag{2.1}$$

The three-dimensional Dirac algebra is represented by real Pauli matrices obeying, with our signature $(+ + -)$,

$$\tfrac{1}{2}\{\gamma_a, \gamma_b\} = \eta_{ab} \qquad \tfrac{1}{2}[\gamma^a, \gamma^b] = \varepsilon^{abc}\gamma_c \tag{2.2}$$

and all spinors will be two-component Majorana fields. We will uniformly dualize all two-index antisymmetric tensors in terms of the Levi–Civita tensor (density) ε, with $\varepsilon^{012} = -\varepsilon_{012} = +1$, which means that $\varepsilon^{\mu\alpha\beta}\varepsilon_{\mu\lambda\sigma} = -\delta^{\alpha\beta}_{\lambda\sigma}$ and $\varepsilon^{\mu\alpha\beta}\varepsilon_{\mu\alpha\sigma} = -2\delta^\beta_\sigma$. Thus for any tensor,

$$\omega_{\mu a} \equiv \tfrac{1}{2}\varepsilon_a{}^{bc}\omega_{\mu bc} \qquad \omega_{\mu bc} = -\varepsilon_{bc}{}^a\omega_{\mu a}. \tag{2.3}$$

Because no ambiguity is possible in three dimensions, we need not use an explicit 'dual' label for either ω or the fermionic field strength below. The (dual of) the commutator of two derivatives defines the (dual of) the curvature:

$$\tfrac{1}{2}\varepsilon^{\alpha\mu v}[D_\mu, D_v] \equiv \tfrac{1}{2}R^{\alpha a}_{**}\gamma_a \tag{2.3a}$$

where the vector density $R^{\alpha a}_{**}$ is the double dual of the full curvature,

$$R^{\alpha a}_{**} \equiv \tfrac{1}{4}\varepsilon^{\alpha\mu\nu}\varepsilon^{abc}R_{\mu\nu bc} \qquad R_{\mu\nu bc} \sim +\partial_\mu\omega_{\nu bc} \qquad R_{\mu\nu bc} = \varepsilon_{\mu\nu\alpha}\varepsilon_{bca}R^{\alpha a}_{**}. \qquad (2.3b)$$

The Ricci tensor is defined as usual by $R_{\mu a} \equiv R_{\mu\nu ab}e^{\nu b}$, and its contraction, the curvature scalar density, is therefore given by

$$R = -2R^{\alpha a}_{**}e_{\alpha a}. \qquad (2.3c)$$

We recall in this connection that in three-space the full curvature is equivalent to the Ricci tensor or, as is manifest in (2.3b), to $R^{\alpha a}_{**}$, which is itself just the Einstein tensor density.

To complete the list of conventions, we define a cosmological constant Λ such that in a space of constant curvature,

$$R_{\mu\nu ab} = \Lambda(e_{\mu a}e_{\nu b} - e_{\mu b}e_{\nu a}) \qquad R^{\alpha a}_{**} = -\Lambda e\,e^{\alpha a}. \qquad (2.4)$$

Equivalently, a gravitational action of the form

$$I_E + I_\Lambda \equiv \int d^3x\,(-R + 2\lambda)\sqrt{-g} \qquad (2.5)$$

has as its field equations $G_{\mu\nu} + \Lambda g_{\mu\nu} = 0$, whose unique solution is a space of constant curvature, $R_{\mu\nu} = 2\Lambda g_{\mu\nu}$. We mention incidentally that even pure three-dimensional Einstein gravity, with or without cosmological constant, is a much richer theory than it seems (Deser *et al* 1983, Deser and Jackiw 1983).

3. THREE SUPERSYMMETRIC TIERS

Cosmological
The supercosmological term consists of a traditional cosmological constant and an associated fermion mass term:

$$I_C = I_\Lambda + I_m \equiv \int d^3x\,[2\Lambda e + \tfrac{1}{2}im\varepsilon^{\mu\alpha\nu}\bar\psi_\mu\gamma^a\psi_\nu e_{\alpha a}]. \qquad (3.1)$$

It is invariant under the purely algebraic transformation rules

$$\delta e_{\mu a} = i\bar\alpha\gamma_a\psi_\mu \qquad \bar\delta\psi_\mu = m\gamma_\mu\alpha \qquad (3.2)$$

if the relation

$$\lambda = -m^2 \qquad (3.3)$$

holds. All this (with the major exception that the mass term is PT odd!) is as in four dimensions and by (2.2) the mass term has the same $\bar\psi_\mu\sigma^{\mu\nu}\psi_\nu$ form; only the coefficient in (3.3) changes with dimension. Note that the dreibein's contribution to δI_m vanishes by the identity

$$\varepsilon^{\alpha\mu\nu}(\bar\psi_\mu\gamma^a\psi_\nu)\psi_\alpha \equiv 0 \qquad (3.4)$$

which follows immediately from a Fierz transformation.

Supergravity
The traditional supergravity action (Howe and Tucker 1978, Dereli and Deser 1978) is the direct dimensional reduction of the usual four-dimensional model, namely

$$I_{SG} = I_E + I_{3/2} = \int d^3x \, (-eR + i\varepsilon^{\mu\alpha\beta}\bar{\psi}_\mu D_\alpha \psi_\beta) \tag{3.5}$$

where D_α is the spinor covariant derivative of (2.1) and the connection includes the standard torsion. This action is invariant under the familiar

$$\delta\omega_{\mu a}(e, \psi) = -i\bar{\alpha}(\gamma_\mu f_a - \tfrac{1}{2} e_{\mu a}\gamma \cdot f). \tag{3.6a}$$

and, in the second-order formalism used here, this determines the connection's variation to be

$$\delta\omega_{\mu a}(e, \psi) = -i\bar{\alpha}(\gamma_\mu f_a - \tfrac{1}{2} e_{\mu a}\gamma \cdot f). \tag{3.6b}$$

We have introduced the notation

$$f^\mu \equiv \tfrac{1}{2}\varepsilon^{\mu\alpha\beta}(D_\alpha\psi_\beta - D_\beta\psi_\alpha) \tag{3.7}$$

for the dual of the field strength, which will be used exclusively hereafter.

Cosmological Supergravity
If one adds the actions (3.1) and (3.5) and the transformations (3.2) and (3.6), a new model arises. Here the total variations are

$$\delta e_{\mu a} = i\bar{\alpha}\gamma_a\psi_\mu \qquad \delta_T\psi_\mu = 2(D_\mu + \tfrac{1}{2}m\gamma_\mu)\alpha \equiv 2\bar{D}_\mu\alpha$$

$$\delta_T\omega_{\mu a}(e,\psi) = -i\bar{\alpha}(\gamma_\mu f_a - \tfrac{1}{2}e_{\mu a}\gamma \cdot f) + \tfrac{1}{2}im\bar{\alpha}(\varepsilon_{\mu a}{}^\nu\psi_\nu + \gamma_a\psi_\mu). \tag{3.8}$$

Note, incidentally that $[\bar{D}_\mu, \bar{D}_\nu]$ vanishes for constant curvature spaces with the correct value (3.3) of Λ. To check that (3.8) is an invariance of $(I_C + I_{SG})$ is relatively direct. The $\delta\psi$ variation $I_{3/2}$ cancels the $\delta\psi$ variation of I_m, with a little care needed to show that the pure torsion contribution, where D_ν acts on the dreibein in the mass term, also vanishes thanks to (3.4). Similarly, the $\delta\omega$ terms coming from $I_{3/2}$ do not contribute, also because of (3.4). Actually, it is guaranteed that the whole $\delta_T\omega$ contribution will vanish, since its coefficient is just the torsion relation which yields the correct $\omega(e, \psi)$ form from the first-order point of view.

The present theory by itself is just as empty dynamically as the one without cosmological constant, and for the same reason. However, as has been noted earlier (Dereli and Deser 1978) this is no longer the case if we drop the cosmological, but keep the explicit mass term. There results a massive single degree of freedom, as can be seen by Hamiltonian analysis of the system. Decomposing the spatial components of the free field into transverse and longitudinal parts,

$$\psi_i = \tfrac{1}{2}\varepsilon_i{}^j\,\hat{\partial}_j\eta + \hat{\partial}_i\chi \qquad \hat{\partial}_i \equiv \partial_i/\sqrt{-\nabla^2} \qquad i, j = 1, 2 \tag{3.9}$$

we note that $I_{3/2}$ can only depend on the two gauge invariant (under $\delta\psi_\mu = \partial_\mu\alpha$) combinations $(\eta, G \equiv \sqrt{-\nabla^2}\,\psi_0 - \dot{\chi})$, and it is known to have the simple form

$I_{3/2} = i \int d^3 x \bar{\eta} G$. The mass term is not invariant, but depends on all three variables:

$$\frac{im}{2} \int d^3 x \varepsilon^{\mu\alpha\beta} \bar{\Psi}_\mu \gamma_\alpha \psi_\beta = im \int d^3 x [\bar{\Psi}_0 (-\hat{V}) (\tfrac{1}{2}\eta + \gamma_0\chi) - \tfrac{1}{2}\bar{\eta}\gamma_0 \chi]. \qquad (3.10)$$

One finds that ψ_0 is a Lagrange multiplier whose variation constrains χ in terms of η in a time-local way, so that the theory reduces to that of a single degree of freedom, namely

$$I_{3/2} + I_m = \tfrac{1}{2}i \int d^3 x \bar{\eta} (\partial\!\!\!/ + m)\eta. \qquad (3.11)$$

This then is a non-topological way of realizing the same (free) massive spin $\frac{3}{2}$ action as in the (free) topological fermion case. The parity oddness of the mass term accounts for the single degree of freedom here.

Topological
We now define the third, topological, layer, refering to the original papers for details. Its action is the sum of the gravitational Chern–Simons invariant and its fermionic companion,

$$I_T = I_{cs} + I_{tf} = \mu^{-1} \int d^3 x \, (R^{\mu a}_{**} \omega_{\mu a} - \tfrac{1}{6} \varepsilon^{\mu\alpha\beta} \varepsilon^{abc} \omega_{\mu a} \omega_{\alpha b} \omega_{\beta c})$$

$$- \tfrac{1}{2}i\mu \int d^3 x \bar{f}^\mu \gamma^a \gamma^b f^\nu e_{\nu a} e_{\mu b} e^{-1}. \qquad (3.12)$$

It is invariant, as it stands, with respect to the normal transformations (3.6), with ω treated in second-order form. However, there is, or was, more. Prior to coupling, that is when ω was purely metric and the fermi action was in flat space, each part had a separate but related symmetry. Under conformal transformations, $\delta g_{\mu\nu} = \phi g_{\mu\nu}$, and for a torsion-free connection ($D_\mu e_{\nu a} = 0$),

$$\delta I_{cs} \equiv \int d^3 x C^{\mu\nu} \delta g_{\mu\nu} \rightarrow \int d^3 x C_\mu^{\ \mu} \phi = 0 \qquad (3.13)$$

as the Weyl tensor $C^{\mu\nu}$ is identically traceless. The corresponding characteristic of the free fermi action I_{tf} is an invariance under $\delta\psi_\mu = \gamma_\mu\beta(x)$. Both invariances cease to hold in presence of coupling, just because of torsion and curvature. For example, although one might think a harmless algebraic invariance such as $\delta\psi_\mu = \gamma_\mu\beta$ would survive coupling, it fails to do so for a whole series of reasons. In the free case, we have

$$\delta \int \bar{f}^\mu \gamma_\nu \gamma_\mu f^\nu = 2 \int \bar{f}^\mu \gamma_\nu \gamma_\mu \varepsilon^{\nu\lambda\sigma} \partial_\lambda \gamma_\sigma \beta = -4 \int \bar{f}^\mu \partial_\mu \beta \qquad (3.14)$$

where the last equality follows from γ algebra and the final term vanishes by the Bianchi identity upon integration by parts. However, in the presence of gravity, not only is $D_\mu f^\mu$ not zero, but also torsion contributes when the covariant derivative acts on the dreibeins implicit in the γs. So these two separate original invariances are definitely lost. However, a generalized γ invariance is miraculously retained. Because of torsion, associated with any ψ transformation is a change in ω as well; in particular, the $\overline{\delta}\omega$ associated with the $\overline{\delta}\psi$ of (3.2) is given in (3.8). Thus, one must also take into account the $\overline{\delta}\omega$ contributions both from the bosonic terms I_{cs} and from I_{tf}. It is this whole series of terms generated from the γ transformation of ψ which cancels in the end, and enables the final layer to be added, as we shall see next.

Cosmological Topological Supergravity

We will show here that the topological layer can be incorporated, i.e. that its variation with respect to the m dependent parts of (3.8),

$$\overline{\delta}(I_{cs}+I_{tf}) = (\overline{\delta}_\psi + \overline{\delta}_\omega)(I_{cs}+I_{tf})$$

$$= \int d^3x [(2R_{**}^{\alpha a} - \tfrac{1}{2}i\bar{f}^\mu \gamma_\nu \gamma_\mu \gamma^a \psi_\beta \varepsilon^{\nu\alpha\beta})\overline{\delta}\omega_{\alpha a} - i\bar{f}^\mu \gamma_\nu \gamma_\mu \varepsilon^{\nu\alpha\beta} D_\alpha \overline{\delta}\psi_\beta] \quad (3.15)$$

in fact does vanish. To be more explicit, there occurs a series of cancellations reminiscent of those in the supergravity sector itself, or in the topological one under the normal transformations (3.6). In particular, one has a cancellation of dangerous curvature terms into a pure torsion contribution which cancels the remaining variations. We find that

$$\overline{\delta}I_{cs} = im \int d^3x \, (R_{**}^{\mu a} \, \bar{\alpha}\gamma_a \psi_\mu + 2R_{***}^\nu \bar{\alpha}\,\psi_\nu)$$

$$\overline{\delta}_\psi I_{tf} = -im \int d^3x \, (R_{**}^{\mu a} \bar{\alpha}\,\gamma_a\psi_\mu - \bar{\alpha}\gamma^a \gamma_\mu \gamma_\nu f^\mu \varepsilon^{\nu\alpha\beta} \, D_\alpha e_{\beta a}) \quad (3.16)$$

$$\overline{\delta}_\omega I_{tf} = \frac{m}{4}\int d^3x \bar{f}^\lambda \gamma_\nu \gamma_\lambda \gamma_a \psi_\beta \varepsilon^{\nu\mu\beta} \, (\bar{\alpha}\psi_\sigma \varepsilon_\mu^{\ a\sigma} + 2\gamma^a \psi_\mu).$$

The apparent curvature term R_{***}^ν is a pure torsion effect, because by (2.3b) it is just the cyclic part of the curvature:

$$R_{***}^\nu \equiv \tfrac{1}{2}\varepsilon^{\nu\alpha\beta} R_{\alpha\beta}^{**} = \varepsilon^{\lambda\sigma\tau} R_{\lambda\sigma\tau}^{\ \ \nu}. \quad (3.17)$$

The latter is itself proportional to the curl of the torsion (cf Deser and Zumino 1976),

$$R_{***}^\nu = \tfrac{1}{4}i\,\bar{\psi}_\alpha \gamma^\nu f^\alpha. \quad (3.18)$$

Likewise, the torsion terms can be evaluated to be

$$\varepsilon^{\nu\alpha\beta} D_\alpha e_{\beta a} = \tfrac{1}{4}i\,\varepsilon^{\nu\alpha\beta}\,\bar{\psi}_\alpha \gamma_a \psi_\beta. \quad (3.19)$$

The rest of the work consists in Fierz transformations (here $\bar{\alpha}\beta\,\bar{\eta}\delta = -\tfrac{1}{2}(\bar{\alpha}\Gamma_A\delta)(\bar{\eta}\Gamma^A\beta)$, $\Gamma^A \equiv (1, \gamma^a)$) to a uniform $(\bar{\psi}\psi)(\bar{\alpha}f)$ pattern. When this is accomplished (it is easier than in four dimensions) there remains the unpromising combination

$$\bar{\alpha}\gamma_\lambda f^\mu [\bar{\psi}_\mu \gamma_\beta \psi_\alpha \varepsilon^{\alpha\lambda\beta} + \bar{\psi}^\lambda \gamma_\beta \psi_\sigma \varepsilon_\mu^{\ \beta\sigma} - \tfrac{1}{2}\bar{\psi}_\alpha \gamma_\mu \psi_\beta \varepsilon^{\lambda\alpha\beta}$$
$$- \tfrac{1}{2}\bar{\psi}_\sigma \gamma^\lambda \psi_\beta \varepsilon_\mu^{\ \sigma\beta} - \delta_\mu^\lambda (\varepsilon^{\alpha\beta\nu}\bar{\psi}_\alpha \gamma_\beta \psi_\nu)]. \quad (3.20)$$

To show it vanishes, one may either use brute force, i.e. check if any off-diagonal terms of the form $\bar{\psi}_1 \gamma^a \psi_2$ survive (there can be no diagonal ones since $\bar{\psi}_1 \gamma^a \psi_1 \equiv 0$), or more elegantly, use the fact that any totally antisymmetric three-tensor is proportional to ε. This is accomplished by rearranging (3.20) into the required form

$$\tfrac{1}{2}\bar{\alpha}\gamma_\lambda f_\mu \big[\varepsilon^{\alpha\lambda\beta}\,(\bar{\psi}^\mu\gamma_\beta\psi_\alpha + \bar{\psi}_\alpha\gamma^\mu\psi_\beta + \bar{\psi}_\beta\gamma_\alpha\psi^\mu) + (\lambda \leftrightarrow \mu)\big]$$

$$-\bar{\alpha}\gamma\cdot f\,(\varepsilon^{\alpha\beta\mu}\,\bar{\psi}_\alpha\gamma_\beta\psi_\mu) + \tfrac{1}{2}\bar{\alpha}\gamma_\lambda f^\mu\big[\varepsilon^{\alpha\lambda\beta}\,(\bar{\psi}_\mu\gamma_\beta\psi_\alpha - \bar{\psi}_\beta\gamma_\alpha\psi_\mu)$$

$$-\varepsilon_\mu{}^{\beta\alpha}\,(\bar{\psi}^\lambda\gamma_\alpha\psi_\beta - \bar{\psi}_\alpha\gamma_\beta\psi^\lambda)\big]. \tag{3.21}$$

The last terms vanish by antisymmetry, and the first ones just cancel the $\bar{\alpha}\,\gamma\cdot f$ term when the above ε property is used. This establishes the desired invariance of the three-tier model under the cosmological supersymmetry transformations.

4. DISCUSSION

One immediate consequence of this model is that it is energy-stable: one may simply rely on a recent discussion (Abbott and Deser 1982) of stability in four-dimensional cosmological gravity and supergravity, in terms of anti-DeSitter space as the vacuum state there, which generalizes the considerations of ($\Lambda = 0$) asymptotic flat-space physics. That is, for solutions approaching anti-DeSitter space asymptotically, the timelike Killing vector of this space enables one to define an energy. The latter is positive by the global algebra of the spinorial changes and total energy (essentially the usual $H = \mathrm{Tr}\,Q^2$ form). The only difference is that in our case, these generators are a bit more subtly defined as asymptotic flux integrals, both for the energy, (Deser *et al* 1982) and (as can similarly be shown) for the charges.

It is clear that the vertical sectors, i.e. the three-term purely bosonic or purely fermionic actions, are also separately consistent. For the former, the field equations $G_{\mu\nu} + \mu^{-1}C_{\mu\nu} + \Lambda g_{\mu\nu} = 0$ break up into a constant scalar curvature $R = 6\Lambda$ part together with a tracefree part independent of Λ. Conformal techniques might be useful here to decouple these two aspects further in a search for general solutions. The fermionic sector also poses an interesting question: there are now apparently two independent mass terms, one topological, the other normal. The excitation content can easily be investigated by Hamiltonian methods in flat space. We have not pursued this point, although we emphasize that in the coupled theory with cosmological constant, it is well known that the explicit mass term is there to retain gauge invariance.

Through the various layers, we have encountered an interesting hierarchy of transformations and of ex-invariances which are sufficiently reinstated by the coupling to make things work, in particular restoration of effective γ invariance in the topological sector. All this is reminiscent of supergravity itself, in which the original $\delta\psi_\mu = \partial_\mu\alpha$ invariance of the free fermion action is promoted to the $D_\mu\alpha$ invariance of the coupled theory when the gravitational variables also acquire a new transformation. It is also possible to ask whether any two out of the three sectors can be coupled consistently. We know that supergravity plus topological can, as can supergravity plus cosmological (although it is rather empty). What about topological–cosmological? It is not invariant under the pure $\bar{\delta}$ transformations, nor under the full ones. The derivative orders are too far apart, it seems. We suspect that theory to be trivial anyhow, since its bosonic sector, $I_{cs} + 2\Lambda \int d^3xe$ has field equations which equate the traceless part $C_{\mu\nu}$, to the pure trace, $g_{\mu\nu}$, leaving no dynamics in either part. The free fermionic sector is also trivial: Hamiltonian analysis shows absence of excitations there as well. It is only through the intermediary of I_{SG} that the two endsectors are permitted to communicate, i.e., to connect their radically different transformation properties. This fact may also be understood from the auxiliary field structure associated with each of the sectors (Cobb *et al* 1983).

Finally, the many standard questions connected with cosmological terms, including gauging in extended models (if they exist) and symmetry breaking mechanisms, remain to be studied in this lower-dimensional but intriguing world.

ACKNOWLEDGMENTS

I thank B Laurent and the Institute for Theoretical Physics of Stockholm University for hospitality while this work was completed.

REFERENCES

Abbott L F and Deser S 1982 *Nucl. Phys.* B **195** 76
Cobb W, Deser S and Gates J 1983 to be published
Dereli T and Deser S 1978 *J. Phys. A: Math. Gen.* **11** L27
Deser S and Jackiw R 1983 *Ann. Phys., NY* in the press
Deser S, Jackiw R and 't Hooft G 1983 *Ann. Phys., NY* in the press
Deser S, Jackiw R and Templeton S 1982 *Phys. Rev. Lett.* **48** 975
—— *Ann. Phys., NY* **140** 372
Deser S and Kay J H 1983 *Phys. Lett.* **120B** 97
Deser S and Zumino B 1976 *Phys. Lett.* **62B** 335
Howe P S and Tucker P W 1978 *J. Math. Phys.* **19** 869

Future Avenues in Supergravity

J G TAYLOR

1. INTRODUCTION

I am delighted and honored to be able to contribute to the volume dedicated to Bryce DeWitt on the occasion of his sixtieth birthday. His contribution to quantum gravity is so profound that it is difficult to know where to begin in referencing his specific papers on this or that topic. Yet without doubt his famous *Dynamical Theory of Groups and Fields* (DeWitt 1964) and his important *Physical Review* papers (DeWitt 1967 a,b,c) will long stand as very important texts in the subject. Many of the ideas contained there are still of great relevance and the associated problems are still being analyzed. In particular the crucial problem of how to handle the ultraviolet divergences is still unresolved.

There have recently been especially interesting possibilities in the area of supergravity (Freedman *et al* 1976, Deser and Zumino 1976). This involves the use of local supersymmetry (SUSY), so requires a companion spin $\frac{3}{2}$ fermi partner to the graviton (see van Nieuwenhuizen (1981) or Ferrara and Taylor (1982) for recent reviews of the subject). The attraction of supergravity to its practitioners is that it offers a hope for the amelioration of the ultraviolet divergences I mentioned above, due to cancellations between contributions between bosons and their SUSY fermi partners. This feature has been recognized for some time in supersymmetric Yang–Mills (SYM) theories as well as in supergravity (SGR) (see the review of Duff (1982) for a fuller discussion). The difficulty in discovering if maximally extended SYM or SGR are sensible enough to be even finite theories has been that the ultraviolet divergence cancellation mechanism inherent in SUSY theories can only be 'automated' by constructing the theory in a form involving fields defined on space–time extended by the inclusion of anti-commuting fermi variables. This so-called 'superspace' (not to be confused with the object defined and used in the canonical approach to quantum gravity) requires the use of complete irreducible representations (irreps) of the appropriate extended SUSY algebra S_N (N denoting the associated number of fermionic SUSY generators). However these are not identical to the physical fields used in the initial construction of N-extended supergravities or SYM theories. What was missing was the auxiliary fields of those various theories.

Auxiliary fields are required in order to balance the bosonic and fermionic degrees of freedom, $\partial_0(B)$ and $\partial_0(F)$, present in the theory. Whilst there is no mismatch when only physical fields satisfying their equations of motion (on-shell) are involved, this is not so off-shell. Thus for supersymmetric QED the physical fields are the photon and photino fields forming the multiplet of spins $\frac{1}{2}$ and 1 which we denote by $(\frac{1}{2}, 1)$. On-shell for this multiplet $\partial^0(B) = \partial^0(F) = 2$, but off-shell $\partial^0(B) = 3$, $\partial^0(F) = 4$. A further scalar field D is required off-shell to achieve a representation of global SUSY with a closed algebra. D may be eliminated trivially from the theory by virtue of its equations of motion, but only at the expense of losing the closure of the algebra.

Auxiliary fields were searched for actively since the construction of SYM and SGR theories, especially so that the associated superfield theories could be constructed. However so far they have only been discovered for $N = 1$ and 2 SYM theories and $N = 1$ and 2 SGR. There seems to be a barrier at $N = 3$ to this construction, blocking our ability to break through and construct superfield versions of $N = 4$ SYM and $N = 8$ SGR, the hoped for finite theories. There is a good reason for the existence of such an $N = 3$ barrier, which I will explain briefly in the next section. I will then turn to various ways in which the $N = 3$ barrier may be broached in subsequent sections, related to the use of 'relaxed' multiplets, to the use of central charges, and to the use of lightcone gauge methods. We will see that especially the first and the last of these approaches will lead us to $N \geqslant 3$; we already have a glimpse of the promised land on the other side of the barrier!

2. THE NO-GO THEOREMS

These theorems, proven originally in four dimensions (Rivelles and Taylor 1981, Taylor 1982a), but then extended to higher dimensions and simplified (Rivelles and Taylor 1983) are based on the use of the fundamental irrep of S_N to construct linearised N–SYM N–SGR. We use the notion of field redefinition rules (Rivelles and Taylor 1982), in which fields of different spin are combined either to become auxiliary or to produce the Einstein or Rarita–Schwinger Lagrangian. The simplest rule is obtained from the rewriting

$$Ap^2A - Bp^2B = (A - B)p^2(A + B) = CD \tag{2.1}$$

where $C = A - B$, $D = p^2(A + B)$; the fields C and D may be regarded as auxiliary fields, with equations of motion $C = D = 0$. Equation (2.1) may be re-expressed as

$$0_p - 0_p \approx 0 \tag{2.2}$$

where the left-hand side of equation (2.2) corresponds to the Lagrangians of the two scalars A and B, each being physical (of dimension L^{-1}) and the right-hand side corresponds to the auxiliary nature of the scalars C and D. A similar rule for fermions arises from

$$\bar{\psi}_1 \not{p} \psi_1 - \bar{\psi}_2 \not{p} \psi_2 = \bar{\lambda}_1 \lambda_2 \tag{2.3}$$

where $\lambda_1 = \psi_1 - \psi_2$, $\lambda_2 = \not{p}(\psi_1 + \psi_2)$; this may be expressed in a similar fashion to equation (2.2) as

$$\tfrac{1}{2} - \tfrac{1}{2} \approx 0. \tag{2.4}$$

Besides equation (2.4) we may take the annihilation rule which expresses the Rarita–Schwinger Lagrangian L_{RS} in terms of its independent spin $\frac{3}{2}$ and $\frac{1}{2}$ fields:

$$L_{RS} = \tfrac{3}{2} - \tfrac{1}{2} \tag{2.5}$$

The No-Go theorems are based on equating the total linearized Lagrangian for N–SYM or N–SGR in terms of contributions from irreps of N–SUSY. The former can be written as

$$\sum_{i=1}^{N} L_{RS}(\psi_{\mu i}) + \sum_{j=1}^{M} \psi_j \not{p} \psi_j + \sum_{l=0}^{\frac{1}{4}A} \bar{\chi}_l \eta_l \tag{2.6}$$

where $\frac{1}{4}A$ is the number of auxiliary spinors required, and M is the number of physical spinors with values $M = (0, 0, 1, 4, 11, 26, 56$ and $56)$ for the values of $N = (1, \ldots, 8)$ respectively. The re-expression of equation (2.6) in terms of contributions from irreps of N–SUSY can only be written as

$$\sum_m \bar{\tau} \not{p} \tau - \sum_n \eta \not{p} \eta \tag{2.7}$$

where m and n are the numbers of times the fundamental irrep is used. We equate the degrees of freedom of the positive and negative energy parts of equations (2.6) and (2.7). The fermi degrees of freedom of the fundamental irrep of N–SUSY, arising from $2N$ creation operators, say the positive or negative chiral projections of the N–SUSY generators, is $2^{2N-1}r_N$ ($r_N = \frac{1}{2}$ or 1 taking account of reality properties). We obtain

$$(m-n) = (M+N)/(2^{2N-1}r_N). \tag{2.8}$$

The right-hand side of equation (2.8) decreases exponentially in N, so that integer (non-zero) values of $(m-n)$ can only occur for small enough N. Inserting the values given above we find that only for $N = 1$ or 2 can $(m-n)$ be integer.

A similar result to the above can also be shown for N–SYM, again with the barrier at $N = 3$ being deduced. Extension of this analysis is also simple for higher dimensions, giving an even reduced barrier at $N = 2$. For $N = 1$ SGR in $d = 11$, $(m-n)$ is not an integer, so such a supergravity theory would require central charges beyond the extra seven dimensions usually regarded as available for such a construction. This is not so for $N = 1$ SGR in $d = 10$, which has recently been constructed, in a linearized superfield version (Howe *et al* 1982). We will see in the next section that this may be of use in the construction of $N = 8$ SGR. This is, however, the only case in dimension $d > 6$ of interest in giving effectively an N–SGR in $d = 4$; the $N = 1$ SGR theory in $d = 10$ reduces to $N = 4$ SGR plus six $N = 4$ SYM theories.

3. BYPASSING THE $N = 3$ BARRIER

The $N = 3$ barrier arose from equation (2.8) because the dimension of the representations of N–SUSY increase too rapidly with N. If we reduce the effective value of N, by being less ambitious and constructing N–SGR or N–SYM in terms of irreps of $\frac{1}{2}N$–SUSY, say, then we may be able to satisfy equation (2.8) satisfactorily. Indeed that is the case for all relevant values of N in $d = 4$; $N = 4$ SYM can be constructed in terms of $N = 2$ SUSY and $N = 8$ SGR in terms of $N = 4$ SUSY.

This spinor reduction can be achieved in one of at least three different ways. One is by means of central charges, which are additional generators in the SUSY algebra which commute with all of them. These extra generators have the dimension of mass, and so allow a Dirac-type of condition to be imposed on the N–SUSY generators $S_{\alpha i}$ themselves. This allows, say, the positively chiral projections of $S_{\alpha i}$ to be expressed in terms of the negatively chiral parts, so reducing the effective number of SUSY generators, and hence $N \to N/2$. This method has some possibilities for higher N (Taylor 1982b) though requires much further work to become really effective (see however, Rogers (1982)).

Another avenue is to explicitly only take $N/2$ SUSY as the basic supersymmetry. This requires the construction of 'relaxed' hypermultiplets for $N = 2$ and 4 to contain the 'matter' part to be added to the gauge part of $N = 2$ SYM or ($N = 4$ SGR plus six $N = 4$ SYM) respectively. The former of these programs has been completed (Howe *et al* 1982, Stelle 1982) and the latter is presently in the process of being so (Bufton and Taylor 1982). The former theory of $N = 4$ SYM has only $N = 2$ SUSY, but is finite to all orders. It is not clear that the latter will be so, but may allow further questions about $N = 8$ SGR to be analyzed, so is clearly of interest. A third avenue allowing for spin reduction is that using the lightcone gauge (Brink *et al* 1982, Mandelstam 1982, Taylor 1983a). It is that approach which we will now describe in more detail.

4. THE LIGHTCONE GAUGE APPROACH

This method uses the lightcone coordinates $x^{\pm} = (1/\sqrt{2})(x^0 \pm x^3)$ and transverse coordinates x^i ($i = 1, 2$). The similarly defined Dirac matrices γ^+, γ^- satisfy $\gamma^+ \gamma^- + \gamma^- \gamma^+ = 2$, so $P_+ = \frac{1}{2}\gamma^+\gamma^-$, $P_- = \frac{1}{2}\gamma^-\gamma^+$ are orthogonal projections. In a standard representation for the Dirac matrices these projectors extract either the 1st and 4th or 2nd and 3rd components of a spinor, and so clearly give spinor reduction. However the total internal symmetry associated with the extension indices i ($1 \leqslant i \leqslant N$) is unchanged, so that SU (N) symmetry is still preserved.

Lightcone dynamics is achieved by taking, for N–SYM, the vector gauge condition $A_- = 0$, and elimination of A_+ and ψ_- by the equations of motion (or equivalently by Gaussian integration). The resulting component action involves solely physical modes. These may then be put into superfield form using chiral superfields. In terms of the complex scalar $A = (1/\sqrt{2})(A_1 + iA_2)$ constructed from the physical transverse modes A_1, A_2 of the vector and the spinor ψ_+ we may define the chiral superfield ϕ as

$$\phi = A = \theta\psi \tag{4.1}$$

for $N = 1$, with correspondingly extra terms for $N = 2$ and 4. In the latter case we have, moreover, a reality condition of the form $D^4\phi = 4\partial_-^2\,\overline{\phi}$, where D is the supercovariant derivative related to the lightcone projection $S_{+\alpha}$ of the SUSY generators S_α.

The process of 'super-fieldization' can be achieved directly in terms of components, and gives the actions for $N = 1$ as (Taylor 1983b)

$$A_1 = \int d^4x\,d\theta\,d\overline{\theta}\,\mathrm{Tr}\,\{\overline{\phi}\partial - \phi\,\Box\,\partial_-\,\phi + ig\partial_-\,\overline{\phi}[\partial_-\,\phi,\overline{\partial}\phi]_- + \mathrm{HC}$$

$$\qquad\qquad - 8g^2\partial_-^{-1}\,[\partial_-\,\phi,\partial_-\,D\overline{\phi}]_-\,\partial_-^{-1}\,[\partial_-\,\overline{\phi},\partial_-\,D\overline{\phi}]_-\}.$$

(4.2)

Similar expressions arise for $N = 2$ and 4 SYM (Taylor 1983b, Brink *et al* 1982), each with only cubic and quartic interaction terms, though with only two or zero powers of derivatives in the interaction respectively. Superfield quantization can now be performed and the degree ∂^0 of ultraviolet divergence computed for a general supergraph. We find (Taylor 1983b)

$$\partial^0 = 4 - N \tag{4.3}$$

Reduction of this quantity can be achieved by extraction of powers of momenta onto external lines, an idea first suggested by Salam and Strathdee for $N = 4$ (Salam and Strathdee 1982), giving

$$\begin{aligned}\partial^0 &= 4 - N - E \quad (N = 1,2)\\ \partial^0 &< 0 \qquad\qquad\quad (N = 4)\end{aligned} \tag{4.4}$$

The finiteness of $N = 4$ SYM is thus proved by explicit calculation, as first suggested by Mandelstam (Mandelstam 1982).

The program has been applied to N–SGR (Taylor 1983c) where choice of the lightcone gauge allows all four of the constraint equations to be reduced to single variable differential equations in x^- and solved explicitly. The resulting component Lagrangian for N–SGR can then be 'super-fieldized' directly. This has been done explicitly and in closed form for $N \leqslant 6$ and to all orders in the gravitational coupling constant for $N = 8$ (except several terms at second order).

The resulting theory can then be analyzed as far as its ultraviolet divergence properties are concerned. We find (Taylor 1983b) the degree of divergence

$$\partial^0 = 4 - N + E + \sum_r (r-2)n_r \tag{4.5}$$

where there are n_r lines and E external lines. Further divergence reduction can now be performed to reduce this divergence degree but the last term in equation (4.5) cannot be removed and the theories still appear to have non-renormalizable ultraviolet divergences.

REFERENCES

Brink L, Lindgren O and Nilsson B E W 1982 '$N = 4$ super Yang–Mills Theory on the Lightcone' *Gothenburg preprint.*

Bufton G and Taylor J G 1982 to appear

Deser S and Zumino B 1976 *Phys. Lett.* **62B** 335–7

DeWitt B S 1964 'Dynamical Theory of Groups and Fields' in *Relativity, Groups and Topology* ed B S DeWitt and C DeWitt (New York: Gordon and Breach) pp 587–820

—— 1967a *Phys. Rev.* **160** 1113–48

—— 1967b *Phys. Rev.* **162** 1195–1239

—— 1967c *Phys. Rev.* **162** 1239–56

Duff M J 1982 *Supergravity '81* ed S Ferrara and J G Taylor (Cambridge: Cambridge University Press) pp 197–26

Ferrara S and Taylor J G ed 1982 *Supergravity '81* (Cambridge: Cambridge University Press)

Freedman D, van Nieuwenhuizen P and Ferrara S 1976 *Phys. Rev.* D **13** 3214–18

Howe P, Nicolai H and van Proeyen A 1982 *Phys. Lett.* **112B** 446

Howe P, Stelle K and Townsend P 1982 to appear

Mandelstam S 1982 'Lightcone Superspace and the Vanishing of the β function' to appear

van Nieuwenhuizen P 1981 *Phys. Rep.* **68** 192–398

Rivelles V O and Taylor J G 1981 *Phys. Lett.* **104B** 131–5

—— 1982 *J. Phys. A: Math: Gen.* **15** 2819

—— 1983 *Phys. Lett.* **121B** 37

Rogers A 1982 'A Full Superspace Action for Off-Shell N = 2 Supergravity' *Imperial College preprint* ICTP/82–83/3

Salam A and Strathdee J 1982 private communication

Stelle K S 1982 invited talk at the IHEPS Conference, Paris

Taylor J G 1982a *J. Phys. A: Math. Gen.* **15** 867

—— 1982b *Quantum Structure of Space and Time* ed M J Duff and C J Isham (Cambridge: Cambridge University Press) pp 283–98

—— 1983a *Phys. Lett.* **121B** 386

—— 1983b $N = 1$ and 2 Super-Yang–Mills in the Lightcone Gauge *King's College London preprint*

—— 1983c Extended Superspace Supergravities in the Lightcone Gauge *Proceedings Karpacz Winter School on Supergravity* (Singapore: World Scientific Publishing Co)

Dynamical Applications of the Gauge-Invariant Effective Action Formalism

STEPHEN L ADLER

1. INTRODUCTION

Functional integration techniques currently play a central role in the study of quantum field theory (see, e.g. Abers and Lee (1973) and Faddeev and Slavnov (1980)), and within the functional integration formalism, the effective action functionals

$$W[J] \qquad J = \text{an external source}$$

and

$$\Gamma[\bar{A}] \qquad \bar{A} = \text{a classical field}$$

play a prominent part. Originally, these functionals were introduced as tools for studying the perturbation expansion of field theory, where W serves as the generating functional for the connected n-point functions and Γ as the corresponding generating functional for their one-particle irreducible parts. The theory of these functionals in the case of scalar fields is straightforward, but it becomes more complicated in the case of gauge fields. In particular, the necessity for adding a gauge-fixing action in quantizing gauge field theories leads in general to an effective action $\Gamma[\bar{A}]$ which is not a gauge-invariant functional of $\bar{A}$. It is only recently, through an extension of the background field method of DeWitt (1967), that a simple construction has been given by 't Hooft (1975), Abbott (1981), Boulware (1981) and DeWitt (1981), which leads to a useful, manifestly gauge-invariant effective action Γ. A second, and related recent development (see Adler (1981, 1982a) and Adler and Piran (1982)), is the use of the effective action as a dynamical tool, for studying non-perturbative effects arising from radiative corrections in quantum field theory.

My aim in this article is to give a self-contained introduction to these two recent developments in the theory of the effective action functional, within the context of a simple example, consisting of an Abelian gauge field interacting with a light fermion species and with a set of infinitely massive test charges. In § 2 I give a detailed analysis of the problem of calculating the static potential of the test charges in this model, and show that V_{static} is given both by the standard Wilson (1974) loop formula and by a classical variational principle involving the gauge-invariant effective action functional Γ. This latter formulation can be viewed as a generalization of the usual methods of classical electrodynamics which includes the effects of quantum corrections. In § 3 I apply the effective action formulation to an analysis of the static potential in the Abelian gauge field model in the long distance and the extreme short distance limits. I conclude § 3 with a very brief discussion of the generalization of the effective action formalism to the case of a non-Abelian gauge field, and its application to the computation of the non-Abelian static potential in the long distance, or confining, limit.

2. THE STATIC POTENTIAL OF MASSIVE TEST CHARGES IN ABELIAN ELECTRODYNAMICS WITH A LIGHT FERMION SPECIES

Statement of the Model
The model which we will analyze is defined by the Lagrangian density

$$L = L_{\text{QED}} + L_J \tag{2.1}$$

with L_{QED} the usual Lagrangian density for Abelian spinor electrodynamics,

$$L_{\text{QED}} = -\tfrac{1}{4} F_{\mu\nu} F^{\mu\nu} + \bar{\psi}(i\slashed{D} - m)\psi \tag{2.2}$$

$$F_{\mu\nu} = \partial_\mu A_\nu - \partial_\nu A_\mu \qquad \slashed{D} = \gamma^\mu D_\mu \qquad D_\mu = \partial_\mu - ieA_\mu$$

and with L_J an additional coupling of the photon to a system of massive external test charges,

$$L_J = -A_0 J_0 \qquad J_0 = \sum_i Q_i \delta^3(x - x_i). \tag{2.3}$$

In Equation (2.2), A_μ and $F_{\mu\nu}$ are respectively the photon vector potential and field strength tensor, ψ is a spin $\tfrac{1}{2}$ field of charge $-e$ and mass m, and γ^μ are the Dirac matrices,† while in equation (2.3), Q_i and x_i are respectively the test charge magnitudes and coordinates. If the charge e were zero, the static potential of the test charges (defined in the usual way as the energy needed to assemble the test charge system from infinity) would be given by the Coulomb formula

$$V_{\text{static}}^{(e=o)} = \tfrac{1}{2} \sum_{i \neq j} \frac{Q_i Q_j}{4\pi |x_i - x_j|}. \tag{2.4}$$

When e is non-zero, the spin $\tfrac{1}{2}$ field acts as a polarizable medium which alters the static potential from the expression in equation (2.4), and our problem is to set up a formalism for calculating the modified static potential in the interacting theory.

Hamiltonian Formulation and Canonical Quantization
Corresponding to the Lagrangian density of equations (2.1)–(2.3), the canonical

---

† We use the $(+, -, -, -)$ metric and γ matrix conventions of Björken and Drell (1965); in particular, we have $A^0 = A_0$.

Hamiltonian is

$$H_c = \int d^3 x \left[\frac{\partial L}{\partial [\partial \psi_A / \partial x^0]} \frac{\partial \psi_A}{\partial x^0} + \frac{\partial L}{\partial [\partial A / \partial x^0]} \frac{\partial A}{\partial x^0} - L \right]$$

$$= \int d^3 x \left\{ \tfrac{1}{2} (\pi^2 + B^2) - \pi \cdot \nabla A_0 + A_0 J_0 + \bar{\psi} [\gamma \cdot (-i\nabla + eA) + m] \psi \right\} \qquad (2.5)$$

with

$$\pi^j = -E^j = \frac{\partial}{\partial x^j} A^0 + \frac{\partial}{\partial x^0} A^j \qquad (2.6)$$

the momentum conjugate to A^j. Since the Lagrangian density has no dependence on $\partial A^0 / \partial x^0$, the canonical momentum π^0 conjugate to A^0 vanishes identically. This means that A^0 is not a true dynamical variable, but rather plays the role of a Lagrange multiplier enforcing the charge conservation constraint

$$0 = \phi \equiv \nabla \cdot \pi + J_0. \qquad (2.7)$$

Following the standard procedure for constrained Lagrangian systems, we achieve a symmetrical reduction of the phase space by imposing the subsidiary condition

$$0 = \chi \equiv G \cdot A - f \qquad (2.8)$$

with G and f arbitrary functions. Differentiating equation (2.8) with respect to time and substituting equation (2.6) then gives

$$0 = G \cdot (\pi - \nabla A^0) + \frac{\partial G}{\partial x^0} \cdot A - \frac{\partial f}{\partial x^0}. \qquad (2.9)$$

Hence equations (2.6) and (2.8) determine A^0 in terms of the vector potentials and their conjugate momenta,

$$A^0 = (G \cdot \nabla)^{-1} (G \cdot \pi + \partial G / \partial x^0 \cdot A - \partial f / \partial x^0) \qquad (2.10)$$

while equations (2.8) and (2.9) imply that only two of the three vector potential components (call them $\hat{A}^{1,2}$) and two of the three conjugate momenta (call them $\hat{\pi}^{1,2}$) are independent dynamical variables.

Let us now recall some results from the general theory of constrained systems (see Faddeev and Slavnov (1980) pp 73–6). Let

$$S = \int dt \left[\sum_{i=1}^{n} p_i \dot{q}_i - h(p,q) - \sum_{\alpha=1}^{m} \lambda^\alpha \phi^\alpha (p,q) \right] \qquad m < n \qquad (2.11)$$

be a dynamical system in which the variables λ^α are Lagrange multipliers which enforce the constraints $\phi^\alpha = 0$. The constraints are assumed to form a closed linear algebra under the Poisson bracket operation with the Hamiltonian and with each other,

$$\{h, \phi^\alpha\} = c^{\alpha\beta} (p,q) \phi^\beta$$

$$\{\phi^\alpha, \phi^\beta\} = c^{\alpha\beta\gamma} (p,q) \phi^\gamma \qquad (2.12)$$

$$\{A, B\} \equiv \sum_{i=1}^{n} \left(\frac{\partial A}{\partial q_i} \frac{\partial B}{\partial p_i} - \frac{\partial A}{\partial p_i} \frac{\partial B}{\partial q_i} \right)$$

guaranteeing that the constraints are preserved in time under the dynamics generated by any member of the equivalence class of Hamiltonians of the form $h + \sum \lambda^\alpha \phi^\alpha$, for arbitrary λ^α. To achieve a symmetrical reduction of the phase space, we impose m subsidiary conditions

$$\chi^\alpha(p, q) = 0 \qquad \alpha = 1, \ldots, m \qquad (2.13)$$

which satisfy

$$\det|\{\phi^\alpha, \chi^\beta\}| \neq 0 \qquad (2.14a)$$

$$\{\chi^\alpha, \chi^\beta\} = 0 \qquad (2.14b)$$

leaving a $2n - 2m$ dimensional phase space containing the true dynamical variables $\hat{q}$, $\hat{p}$. Equation (2.14b) allows us to choose the original canonical variables q_i, p_i so that χ^α coincide with the first m variables q_i, giving a basis in which it is easy to prove the following two facts:

(i) The equations of motion for the original constrained system of equation (2.11),

$$\dot{p}_i + \frac{\partial h}{\partial q_i} + \sum_{\alpha=1}^{m} \lambda^\alpha \frac{\partial \phi^\alpha}{\partial q_i} = 0$$

$$\dot{q}_i - \frac{\partial h}{\partial p_i} - \sum_{\alpha=1}^{m} \lambda^\alpha \frac{\partial \phi^\alpha}{\partial p_i} = 0 \qquad (2.15)$$

$$\phi^\alpha = 0$$

imply that the variables $\hat{q}$, $\hat{p}$ obey an unconstrained Hamiltonian dynamics,

$$\dot{\hat{q}} = \frac{\partial \hat{h}}{\partial \hat{p}} \qquad \dot{\hat{p}} = -\frac{\partial \hat{h}}{\partial \hat{q}} \qquad (2.16)$$

with

$$\hat{h}(\hat{p}, \hat{q}) = h(p, q)\Big|_{\phi=0, \chi=0} \qquad (2.17)$$

(ii) When the constraints ϕ^α are varied, with the $\hat{q}$, the $\hat{p}$ and the subsidiary conditions χ^α all held fixed, the change in the Hamiltonian $\hat{h}$ is given by†

$$\delta\hat{h} = \sum_{\beta=1}^{m} \lambda^\beta \delta\phi^\beta. \qquad (2.18)$$

When specialized to the model of equations (2.5)–(2.10), the general results just quoted become the statements that (i) the dynamics of the canonical coordinates and momenta $\hat{q}^a, \hat{p}^a$ contained in the fields $\hat{A}^{1,2}, \hat{\pi}^{1,2}$ takes a standard Hamiltonian form, with the Hamiltonian $H \equiv \hat{h}$ given by

$$H(\hat{q}^a, \hat{p}^a) = H(\hat{A}^{1,2}, \hat{\pi}^{1,2}) = H_c\Big|_{\phi=0, \chi=0} \qquad (2.19)$$

† For a proof of (i), see Faddeev and Slavnov (1980) pp 73–6. In the notation of this reference, and using a summation convention for repeated indices, the proof of (ii) is as follows: By the chain rule we have $\delta\hat{h} = \partial h/\partial p^\alpha \, \delta p^\alpha$, where the p^α are the m dependent momenta which have been eliminated by inverting the constraints $\phi^\beta = 0$. Combining with the equation $\dot{q}^\alpha = \dot{\chi}^\alpha = 0 = \partial h/\partial p^\alpha + \lambda^\beta \partial \phi^\beta/\partial p^\alpha = 0$, we have $\delta\hat{h} = -\lambda^\beta \partial \phi^\beta/\partial p^\alpha \delta p^\alpha$. However, since the varied dependent momenta $p^\alpha + \delta p^\alpha$ are determined by solving the varied constraints $\phi^\beta + \delta\phi^\beta = 0$, just as the p^α were determined by solving the original constraints $\phi^\beta = 0$, we have

$$0 = (\phi^\beta + \delta\phi^\beta)\Big|_{p^\alpha + \delta p^\alpha} \quad \overset{-\phi^\beta}{\Big|_{p^\alpha}} = \delta\phi^\beta + \partial \phi^\beta/\partial p^\alpha \, \delta p^\alpha,$$

and so the formula for $\delta\hat{h}$ reduces to $\delta\hat{h} = \lambda^\beta \delta\phi^\beta$.

and (ii) when the external charge density J_0 is varied, with $\hat{A}^{1,2}$, $\hat{\pi}^{1,2}$ and χ held fixed, the corresponding change in H is

$$\delta H = \int d^3 x \, A^0 \delta J_0. \tag{2.20}$$

Because the $\hat{q}^a$, $\hat{p}^a$ and H form an unconstrained Hamiltonian system, we can now quantize as usual by postulating the equal-time commutation relations

$$[\hat{q}^a, \hat{q}^b] = [\hat{p}^a, \hat{p}^b] = 0 \qquad [\hat{q}^a, \hat{p}^b] = i\delta^{ab} \tag{2.21}$$

(together with the corresponding anticommutation relations for the fermion variables), giving a quantum-mechanical system in which all observables $O(\hat{A}^{1,2}, \hat{\pi}^{1,2}, \psi, \psi\dagger)$ evolve in time according to

$$i\frac{\partial O}{\partial x^0} = [O, H]. \tag{2.22}$$

The Static Potential of External Test Charges
There are two equivalent ways of defining the static potential $V_{\text{static}}[J_0]$ associated with the external charge distribution J_0. The first is simply

$$V_{\text{static}}[J_0] = E_0[J_0] = {}_{J_0}\langle 0|H|0\rangle_{J_0} \tag{2.23}$$

where $|0\rangle_{J_0}$ is the ground state, and $E_0[J_0]$ is the ground state energy, of the quantized theory in the presence of the source J_0. The second is the differential statement

$$\delta V_{\text{static}}[J_0] = \int d^3 x \, \bar{A}_0[J_0]\delta J_0 \tag{2.24a}$$

where

$$\bar{A}_0[J_0] = {}_{J_0}\langle 0|A_0|0\rangle_{J_0} \tag{2.24b}$$

is the ground state expectation of the scalar potential in the presence of the source J_0. To verify the equivalence of the two definitions, we apply first-order perturbation theory to equation (2.23), and then substitute equation (2.20) for δH, giving

$$\delta V_{\text{static}}[J_0] = \delta E_0[J_0] = {}_{J_0}\langle 0|\delta H|0\rangle_{J_0}$$
$$= {}_{J_0}\langle 0|\int d^3 x \, A_0 \delta J_0|0\rangle_{J_0} = \int d^3 x \, \bar{A}_0[J_0]\delta J_0. \tag{2.25}$$

Thus the static potential of the external test charge system, as defined by equation (2.23), has the conventional interpretation as the energy needed to assemble the test charge system from infinity.

In general, for a complex system it is a difficult task to get an accurate approximation for the ground state wavefunction, and so it is very useful to have basis-independent expressions for the quantities $V_{\text{static}}[J_0]$ and $\bar{A}_0[J_0]$. These can be obtained by introducing the partition function

$$Z = \text{Tr}[\exp(-iHT)] = \sum_n \exp[-iE_n T]$$
$$= \exp[-iE_0 T]\left[1 + \sum_{n \neq 0} \exp[i(E_0 - E_n)T]\right] \tag{2.26}$$

in terms of which we have

$$V_{\text{static}}[J_0] = E_0 = \frac{1}{-iT}\log Z + \frac{1}{iT}\log\left[1 + \sum_{n \neq 0} \exp\left[i(E_0 - E_n)T\right]\right]. \quad (2.27)$$

In the limit as $T \to \infty$, the second term in equation (2.27) vanishes by the Riemann–Lebesgue theorem, giving the desired formula

$$V_{\text{static}}[J_0] = \lim_{T \to \infty} \frac{1}{-iT}\log Z. \quad (2.28)$$

By repeating the reasoning of equation (2.25), we can similarly derive the formula

$$\bar{A}_0[J_0] = \lim_{T \to \infty} \frac{1}{-iT}\frac{\delta}{\delta J_0}\log Z. \quad (2.29)$$

These equations will serve as the basis for our further analysis of the problem of determining the static potential.

Functional Integral for Z, and the Wilson Loop Formula

Because canonical quantization of our model requires the imposition of a subsidiary (or 'gauge-fixing') condition, the formulas (2.23)–(2.29) have a complicated implicit dependence on the functions G and f introduced in equation (2.8). It will be useful to recast the equations in a form where the subsidiary condition is explicitly on view, so as to facilitate the study of gauge covariance properties. This can be done by using the following functional integral representation for the partition function (see Faddeev and Slavnov (1980) chapter 2, and Gross *et al* (1981) §II).

$$Z = N(T)^{-1} \oint d[A^\mu] d[\psi] d[\bar{\psi}] \Delta_{gf} \exp i(S + S_{gf})$$

$$S = \int_0^T dx^0 \int d^3x\, L \quad (2.30)$$

$$S_{gf} = \int_0^T dx^0 \int d^3x \tfrac{1}{2}(G \cdot A - f)^2$$

with L the total Lagrangian density of equation (2.1). The integral in equation (2.30) is a functional integral over all values of the classical potential A^μ, and of the classical Grassmann variables ψ, $\bar{\psi}$, satisfying the periodicity/antiperiodicity conditions (indicated by the circle on the integral sign)

$$A^\mu(x^0 = 0, x) = A^\mu(x^0 = T, x)$$
$$\psi(x^0 = 0, x) = -\psi(x^0 = T, x) \quad (2.31)$$
$$\bar{\psi}(x^0 = 0, x) = -\bar{\psi}(x^0 = T, x).$$

The action S is the classical action functional associated with L, while S_{gf} is a gauge-fixing action term corresponding to the subsidiary condition of equation (2.8). The gauge-invariant functional Δ_{gf} is the compensating determinant, which corrects the integral in equation (2.30) for the effect of the gauge-fixing term. It is given, in terms of the constraint ϕ of equation (2.7) and the subsidiary condition χ of equation (2.8), by the functional determinant

$$\Delta_{gf} = \det\left|\{\phi(x), \chi(y)\}\right| = \det\left|-G^j(y)\frac{\partial}{\partial x^j}\delta^3(x-y)\right| \tag{2.32}$$

and thus in our Abelian model (but not in general) is an A independent constant. Since the normalization constant $N(T)$ is independent of the source term J_0, it can only contribute an irrelevant constant to equation (2.28) for V_{static} and will be omitted in the subsequent discussion. It is important to emphasize that the derivation of equation (2.30) from the canonical Hamiltonian formalism given in Faddeev and Slavnov (1980) is unaffected by the presence of the source term J_0; hence equation (2.30) does not require J_0 to vanish, and in fact contains a complete description of the interaction of the photon and light fermion fields of our model with the external charge distribution.

When L in equation (2.30) is split into pieces L_{QED} and L_J as in equation (2.1), the action S splits into corresponding pieces

$$S_{QED} = \int_0^T dx^0 \int d^3x\, L_{QED}$$

$$S_J = -\int_0^T dx^0 \int d^3x\, A_0 J_0. \tag{2.33}$$

Let us consider now the special case in which the source J_0 contains only two equal and opposite point charges (corresponding to a massive particle–antiparticle pair),

$$J_0 = Q\delta^3(x-x_1) - Q\delta^3(x-x_2) \qquad |x_1 - x_2| = R \tag{2.34}$$

for which the action term S_J in equation (2.33) takes the form

$$S_J = Q \int_0^T dx^0 \left[A_0(x^0, x_2) - A_0(x^0, x_1)\right]. \tag{2.35}$$

Since the periodicity condition of equation (2.31) implies that

$$0 = \int_{x_1}^{x_2} dx \cdot \left[A(0, x) - A(T, x)\right] \tag{2.36}$$

equation (2.35) can be written as

$$S_J = Q \int_0^T dx^0 A_0(x^0, x_2) + Q \int_{x_2}^{x_1} dx \cdot A(T, x)$$

$$+ Q \int_T^0 dx^0 A_0(x^0, x_1) + Q \int_{x_1}^{x_2} dx \cdot A(0, x)$$

$$= Q \oint dx^\mu A_\mu(x) \tag{2.37}$$

with $\oint dx^\mu$ a line integral over the rectangular loop bounded by $x = x_1$, $x = x_2$,

$x^0 = 0$ and $x^0 = T$. Combining equations (2.28), (2.30), (2.33) and (2.37) gives the Wilson (1974) loop formula for the static potential of a massive particle–antiparticle pair,

$$V_{\text{static}}(R) = \lim_{T \to \infty} \frac{1}{-iT} \log \left\{ \oint d[A^\mu] d[\psi] d[\bar\psi] \right.$$

$$\left. \times \Delta_{gf} [\exp i(S_{\text{QED}} + S_{gf})] \left[\exp i Q \oint dx^\mu A_\mu(x) \right] \right\}. \quad (2.38)$$

There exists an extensive literature using the Wilson loop formula as the starting point for discussions of Abelian, and especially, non-Abelian statics.

Effective Action Formulation of the Statics Problem
Although equation (2.38) gives an exact expression for the static potential, it is very dissimilar in appearance from the formalism which one uses in the familiar case of classical electrostatics. Let us now construct an alternative method for computing the static potential in the model of equations (2.1)–(2.3), which closely resembles classical electrostatics in form, and which reduces directly to classical electrostatics when fermion vacuum polarization effects are neglected.

To proceed with this construction, let us introduce an effective action functional $W[J_\mu; \bar{A}]$ with a background field radiation gauge-fixing, defined by

$$\exp i W[J_\mu; \bar{A}] = \oint d[A^\mu] d[\psi] d[\bar\psi] \Delta_{gf} \exp i(S_{\text{QED}} + S_{gf}[\bar{A}] - \oint d^4x\, A^\mu J_\mu)$$

$$\oint d^4x \equiv \int_0^T dx^0 \int d^3x. \quad (2.39)$$

In equation (2.39), $\bar{A}^j$ is any background vector potential obeying the periodicity condition

$$\bar{A}^j(0, x) = \bar{A}^j(T, x) \quad (2.40)$$

while $S_{gf}[\bar{A}]$ corresponds to making the choices $G_j = \partial/\partial x^j, f = \nabla_j \bar{A}^j$ in equation (2.30), so that

$$S_{gf}[\bar{A}] = \oint d^4x \tfrac{1}{2} [\nabla \cdot (A - \bar{A})]^2. \quad (2.41)$$

Specializing to the case when $J = 0$ and the background potential $\bar{A}$ is time-independent, the right-hand side of equation (2.39) can be reduced to the right-hand side of the radiation-gauge version of equation (2.30), by making the time-independent gauge transformation

$$A_0 \to A_0 \qquad A \to A + \nabla \Lambda \qquad \psi \to \psi \exp(-ie\Lambda) \qquad \bar\psi \to \bar\psi \exp ie\Lambda$$

$$\Lambda = \frac{1}{\nabla^2}(\nabla \cdot \bar{A}). \quad (2.42)$$

Hence we can express the static potential in terms of the effective action functional W,

$$V_{\text{static}} = -\frac{1}{T} \lim_{T \to \infty} W[J_0, J = 0; \bar{A}]. \tag{2.43}$$

It is now straightforward (see Adler 1982a) to re-express the problem of calculating $W[J_0, J = 0; \bar{A}]$ in terms of a classical variational problem involving a gauge-invariant effective action functional $\Gamma[\bar{A}^\mu]$. Let us choose $\bar{A}^\mu$ to be the average of A^μ over the functional ensemble of equation (2.39),

$$\bar{A}^\mu = -\frac{\delta W}{\delta J_\mu} = \frac{\oint d[A^\nu] \dots \exp i\{S_{\text{QED}} + \dots\} A^\mu}{\oint d[A^\nu] \dots \exp i\{S_{\text{QED}} + \dots\}} \tag{2.44}$$

and introduce Γ by constructing the Legendre transform of W,

$$\Gamma = W[J_\mu; \bar{A}] + \oint d^4 x\, \bar{A}^\mu J_\mu. \tag{2.45}$$

Varying equation (2.45) and making use of equation (2.44), we obtain

$$\begin{aligned}
\delta\Gamma &= \oint d^4 x \left(\frac{\delta W}{\delta J_\mu} \delta J_\mu + \frac{\delta W}{\delta \bar{A}^j} \delta \bar{A}^j + \delta \bar{A}^\mu J_\mu + \bar{A}^\mu \delta J_\mu \right) \\
&= \oint d^4 x \left(\frac{\delta W}{\delta \bar{A}^j} \delta \bar{A}^j + J_\mu \delta \bar{A}^\mu \right).
\end{aligned} \tag{2.46}$$

Hence Γ is a functional of $\bar{A}^\mu$ alone, $\Gamma = \Gamma[\bar{A}^\mu]$, and satisfies the variational equation

$$\frac{\delta\Gamma}{\delta \bar{A}^j} = \frac{\delta W}{\delta \bar{A}^j} + J_j \qquad \frac{\delta\Gamma}{\delta \bar{A}^0} = J_0. \tag{2.47}$$

Specializing to the situation $J = 0$, $J_0 =$ time-independent, equation (2.47) implies that $\bar{A}^\mu$ is time-independent, and the gauge transformation argument of equations (2.41)–(2.43) then implies that $\delta W/\delta \bar{A}^j = 0$. Hence by solving

$$\frac{\delta\Gamma}{\delta \bar{A}^j} = 0 \qquad \frac{\delta\Gamma}{\delta \bar{A}^0} = J_0 \tag{2.48}$$

to determine $\bar{A}^j$, $\bar{A}^0$, and inverting the Legendre transform according to

$$W[J_0, J = 0; \bar{A}] = \Gamma[\bar{A}^\mu] - \oint d^4 x\, \bar{A}^0 J_0 \tag{2.49}$$

we can calculate the static potential V_{static}. The recipe of equations (2.43), (2.48) and (2.49) can be summarized in the compact formula

$$V_{\text{static}} = -\lim_{T \to \infty} \text{ext}_{A^\mu} \left\{ T^{-1} \Gamma[A^\mu] - \int d^3 x\, A^0 J_0 \right\} \tag{2.50}$$

with the notation $\text{ext}_{A^\mu}\{\ \}$ denoting the extremum of the curly bracket over all time-independent potentials A^μ.

An explicit functional integral formula for $\Gamma[\bar{A}^\mu]$ can be obtained by making the change of variable $A^\mu = \bar{A}^\mu + q^\mu$ in equation (2.39) and using equation (2.45), giving

$$\exp i\Gamma[\bar{A}^\mu] = \oint d[q^\mu]d[\psi]d[\bar{\psi}]\Delta_{gf}$$

$$\exp i\int d^4x\{L_{QED}[A^\mu = \bar{A}^\mu + q^\mu] + \tfrac{1}{2}(\nabla\cdot q)^2 - q^\mu J_\mu[\bar{A}^\lambda]\}. \qquad (2.51a)$$

The source current $J_\mu[\bar{A}^\lambda]$ is determined as a functional of $\bar{A}^\mu$ by solving equation (2.44), which in terms of the new variable q^μ takes the form

$$0 = \oint d[q^\mu]d[\psi]d[\bar{\psi}]\Delta_{gf}q_\sigma$$

$$\times \exp i\oint d^4x\{L_{QED}[A^\mu = \bar{A}^\mu + q^\mu] + \tfrac{1}{2}(\nabla\cdot q)^2 - q^\mu J_\mu[\bar{A}^\lambda]\}. \qquad (2.51b)$$

Let us now consider the effect on equation (2.51) of making a general, time-dependent change in the gauge of $\bar{A}^\mu$,

$$\bar{A}^\mu \to \bar{A}^\mu - \partial^\mu \Lambda. \qquad (2.52a)$$

Since the integration measure $d[\psi]\,d[\bar{\psi}]$ is invariant under the change of integration variable

$$\psi \to \psi\exp-ie\Lambda \qquad \bar{\psi} \to \bar{\psi}\exp ie\Lambda \qquad (2.52b)$$

the combined effect of the gauge change of equation (2.52a) and the change of variable of equation (2.52b) is to replace

$$L_{QED}[A^\mu = \bar{A}^\mu + q^\mu, \psi, \bar{\psi}] \qquad (2.53)$$

in equation (2.51) by

$$L_{QED}[A^\mu - \partial^\mu\Lambda = \bar{A}^\mu - \partial^\mu\Lambda + q^\mu, \psi\exp(-ie\Lambda), \bar{\psi}\exp(ie\Lambda)]. \qquad (2.54)$$

However, the gauge invariance of L_{QED} implies that equation (2.54) is equal to equation (2.53), and so we conclude that $\Gamma[\bar{A}^\mu]$ is a gauge-invariant functional of the potential $\bar{A}^\mu$. To summarize, the analysis of this section shows that in the Abelian model with light fermions of equations (2.1)–(2.3), the Wilson loop formula for the static potential is exactly equivalent to the classical effective action variational problem specified by equations (2.50) and (2.51).

The Reduction to Classical Electrostatics
The effective action $\Gamma[\bar{A}^\mu]$ includes all fermion vacuum polarization effects, and is thus a complicated object. Suppose, however, we are interested only in processes with a characteristic length scale l, with $l^{-1} \ll m$. Then virtual pair effects can be neglected, and we can approximate the effective action Γ by

$$\Gamma[\bar{A}^\mu] \approx \Gamma_{\text{no pair}}[\bar{A}^\mu]. \qquad (2.55)$$

The approximated effective action $\Gamma_{\text{no pair}}[\overline{A}^\mu]$ is given by formulas analogous to equation (2.51), in which the fermion functional integration and the fermion terms in the Lagrangian are neglected,

$$\left\{ \exp i \Gamma_{\text{no pair}}[\overline{A}^\mu] \atop 0 \right\} = \oint d[q^\mu] \Delta_{gf} \left\{ 1 \atop q_\sigma \right\}$$

$$\times \exp i \oint d^4x \left\{ L_{\text{Maxwell}}[A^\mu = \overline{A}^\mu + q^\mu] + \tfrac{1}{2} (\nabla \cdot q)^2 \right.$$

$$\left. - q^\mu J_\mu[\overline{A}^\lambda] \right\}.$$

$$L_{\text{Maxwell}}[A^\mu] = -\tfrac{1}{4} F_{\mu\nu} F^{\mu\nu}. \tag{2.56}$$

Since the Maxwell Lagrangian is a quadratic form in the potentials, by integrating by parts in the term linear in q^μ, and using the periodicity of $\overline{A}^\mu$ and q^μ, it is easy to derive the identity

$$S_{\text{Maxwell}}[\overline{A}^\mu + q^\mu] = \oint d^4x \, L_{\text{Maxwell}}[\overline{A}^\mu + q^\mu]$$

$$= S_{\text{Maxwell}}[\overline{A}^\mu] + S_{\text{Maxwell}}[q^\mu] + \oint d^4x \, q^\mu J_\mu[\overline{A}^\lambda] \tag{2.57}$$

where

$$J^0[\overline{A}^\lambda] = -\nabla^2 \overline{A}^0 - \partial/\partial x^0 \, (\partial/\partial x^j \, \overline{A}^j)$$

$$J^n[\overline{A}^\lambda] = -\nabla^2 \overline{A}^n + \partial/\partial x^n \, (\partial/\partial x^j \, \overline{A}^j) + \partial/\partial x^0 \, (\partial \overline{A}^0/\partial x^n + \partial \overline{A}^n/\partial x^0) \tag{2.58}$$

are the currents associated with the background potentials $\overline{A}^\mu$. When equation (2.57) is substituted into equation (2.56), the terms linear in q^μ cancel, and we then find (up to an irrelevant constant) that

$$\Gamma_{\text{no pair}}[\overline{A}^\mu] = S_{\text{Maxwell}}[\overline{A}^\mu] = \oint d^4x \, L_{\text{Maxwell}}[\overline{A}^\mu]. \tag{2.59}$$

Hence when fermion vacuum polarization effects are neglected, equation (2.50) reduces to

$$V_{\text{static}} = -\text{ext}_{A^\mu} \int d^3x \, \{ L_{\text{Maxwell}}[\overline{A}^\mu] - A^0 J_0 \}. \tag{2.60}$$

To see that equation (2.60) gives a variational reformulation of classical electrostatics, we note that since there is no source term for the vector potential, the integral on the right-hand side is stationary at

$$A = 0 \qquad A^0 = \sum_i \frac{Q_i}{4\pi |x - x_i|} \tag{2.61}$$

and when substituted back into equation (2.60), these potentials immediately yield equation (2.4).

3. DISCUSSION: THE LONG AND SHORT DISTANCE LIMITS OF THE STATIC POTENTIAL, AND GENERALIZATION TO THE NON-ABELIAN CASE

Let us now use the effective action formula of equation (2.50) to discuss the static potential of massive test charges in spinor quantum electrodynamics, in the long and short distance limits. In the limit of long distances the smallness of $\alpha = e^2/4\pi \approx 1/137$ permits us to use perturbation theory, and so the leading radiative correction to $\Gamma[A^\mu]$ comes from the second-order vacuum polarization graph,

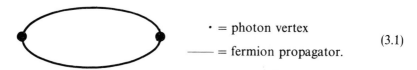

$$\cdot = \text{photon vertex}$$
$$\underline{\qquad} = \text{fermion propagator.}$$

(3.1)

A standard calculation then gives

$$\Gamma[A^\mu] = \int d^4x \{L_{\text{Maxwell}}[A^\mu] + L_{\text{vac. pol.}}[A^\mu] + O(\alpha^2)\} \qquad (3.2a)$$

with the effective Lagrangian density corresponding to the diagram of equation (3.1) given by the formula (see Schwinger (1951))

$$L_{\text{vac. pol.}}[A^\mu] = -\tfrac{1}{4}F_{\mu\nu}\left(1 - \frac{\alpha}{3\pi}\Box^2 \int_{4m^2}^{\infty} \frac{dt}{t}\frac{\rho(t)}{t+\Box^2}\right)F^{\mu\nu}$$

$$\Box^2 = (\partial/\partial x^0)^2 - \nabla^2 \qquad \rho(t) = (1 + 2m^2/t)(1 - 4m^2/t)^{1/2}. \qquad (3.2b)$$

Specializing to time-independent kinematics and substituting equation (3.2) into equation (2.50), we get

$$V_{\text{static}} = -\text{ext}_{A^0} \int d^3x \left[\tfrac{1}{2}\nabla A^0 \cdot \left(1 + (\alpha/3\pi)\nabla^2 \int_{4m^2}^{\infty} \frac{dt}{t}\frac{\rho(t)}{t-\nabla^2}\right)\nabla A^0 - A^0 J_0\right]$$

$$= -\int d^3x \tfrac{1}{2}J_0(\nabla^2)^{-1}\left(1 + (\alpha/3\pi)\nabla^2 \int_{4m^2}^{\infty} \frac{dt}{t}\frac{\rho(t)}{t-\nabla^2}\right)^{-1} J_0$$

$$= \tfrac{1}{2}\sum_{i\neq j} \frac{Q_i Q_j}{4\pi}\left(1/r_{ij} + (\alpha/3\pi)\int_{4m^2}^{\infty} \frac{dt}{t}\rho(t)\frac{\exp -t^{1/2}r_{ij}}{r_{ij}}\right) \qquad (3.3)$$

$$+ \text{self-energies} + O(\alpha^2)$$

$$r_{ij} = |\mathbf{x}_i - \mathbf{x}_j|$$

giving the standard Uehling vacuum polarization correction to the static potential. We note that although the effective action of equation (3.2) is spatially non-local, it is still only quadratic in the field strength and hence leads to linear effective field equations, explaining why equation (3.3) takes the form of a superposition of Yukawa potentials.

In the short distance limit, $e^2 F_{\mu\nu} F^{\mu\nu}/m^4$ becomes large, and so we can no longer use the smallness of α to justify the neglect of the effective action contributions

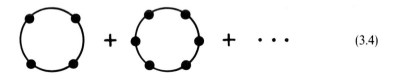

$$\tag{3.4}$$

relative to that of equation (3.1). Consequently, in the limit of short distances the effective field equations become highly nonlinear. In the approximation in which non-localities of the effective action are neglected, the nonlinear contributions to the effective action are determined by renormalization group arguments (see Migdal (1973) Matinyan and Savvidy (1978), and Pagels and Tomboulis (1978)) to be

$$\Gamma[A^\mu] \approx \int d^4x \, \frac{-\tfrac{1}{4} e^2 F_{\mu\nu} F^{\mu\nu}}{e^2_{\text{running}}[(-\tfrac{1}{4} e^2 F_{\mu\nu} F^{\mu\nu})^{1/2}]} \tag{3.5a}$$

with $e^2_{\text{running}}(x)$ the running coupling constant

$$e^2_{\text{running}}(x) = \frac{e^2}{1 - (\alpha/3\pi) \log(x/m^2) + \dots}. \tag{3.5b}$$

Again specializing to time-independent kinematics and substituting equation (3.5) into equation (2.50), we get

$$V_{\text{static}} = -\text{ext}_{A^0} \int d^3x \left\{ \tfrac{1}{2} (\nabla A^0)^2 \left[1 - \frac{\alpha}{3\pi} \log(e|\nabla A^0|/m^2) + \dots \right] - A^0 J_0 \right\} \tag{3.6}$$

and the variation over A^0 implicit in equation (3.6) now gives a nonlinear differential equation for A^0,

$$\nabla \cdot D = J_0 \qquad D = E\varepsilon(E) \qquad E = -\nabla A^0$$

$$\varepsilon(E) = 1 - \frac{\alpha}{3\pi} \log(eE/m^2) + \dots \, . \tag{3.7}$$

Let us briefly study the solutions of equations (3.6) and (3.7) in the special case in which J_0 contains a single isolated charge Q at $x = 0$ and a compensating spherical shell of charge $-Q$ at infinity,

$$J_0 = Q\delta(x) + \text{spherical shell} - Q \text{ at infinity}. \tag{3.8}$$

In this case, discussed by Greenman and Rohrlich (1973) and by Migdal (1973), equation (3.7) is satisfied by the spherically symmetric Ansatz

$$D = \frac{Q\hat{r}}{4\pi r^2} \qquad E = \frac{Q(r)\hat{r}}{4\pi r^2} \qquad r = |x| \qquad (3.9a)$$

with $Q(r)$ obtained by solving the transcendental equation

$$Q = Q(r)\varepsilon(eQ(r)/(4\pi r^2)). \qquad (3.9b)$$

As discussed in detail by Migdal (1973), the physical interpretation of $Q(r)$ is that it is the charge lying within a sphere of radius r, which is larger than Q because vacuum polarization effects in an Abelian gauge theory are charge-screening. Moreover, Migdal (1973) shows that when

$$\frac{eQ(r)}{4\pi} \gg 1 \qquad (3.10)$$

the approximation of neglecting non-local behavior in the effective action becomes self-consistent. To see why this should be so, we note that according to equation (3.7), the one-loop local but nonlinear correction to ε behaves as

$$-\frac{\alpha}{3\pi} \log X_1 \qquad X_1 = \frac{eQ(r)}{4\pi r^2 m^2} \qquad (3.11a)$$

while from equations (3.2)–(3.3), the one-loop non-local correction behaves as

$$-\frac{\alpha}{3\pi} \log X_2 \qquad (3.11b)$$

$$X_2 = \frac{(\nabla \log A^0)^2}{m^2} \sim \frac{1}{r^2 m^2}$$

and so equation (3.10) is just the condition for $X_2 \gg X_1$. Hence, *although the Abelian vacuum behaves as a linear but spatially non-local medium at large distances, its character changes at very short distances, where instead it behaves as a medium which is effectively local but nonlinear.*

Let us conclude with a brief sketch of how the results obtained above are modified in the non-Abelian case. In a non-Abelian gauge theory, the Wilson loop formula of equation (2.38) has a natural generalization in which the exponential on the far right in equation (2.38) is replaced by

$$\text{Tr}\left[P \exp i \oint dx^\mu A_\mu(x) \right] \qquad (3.12)$$

with P a path ordering symbol, and Tr a trace over the quark color matrix structure implicit in the non-Abelian A_μ. When this color matrix structure is treated in quasi-Abelian approximation, so that the path ordering can be neglected, the non-Abelian Wilson loop formula (which is exact) can be transformed into an approximate effective action variational principle for V_{static}, closely resembling equation (3.50) in structure. Because non-Abelian gauge theories are asymptotically free, the qualitative behavior of the effective action $\Gamma[A^\mu]$ at short and at large distances is just the reverse (see Adler (1983)) of what we found in the Abelian case: *at short distances, the non-Abelian vacuum behaves as a linear but spatially non-local medium, while at very large distances, the non-Abelian vacuum behaves as a medium which is effectively local but nonlinear.* Since quark confinement is a large distance effect, the nonlinearity of the non-Abelian vacuum can be expected to play an important role. Indeed, work over the last few years has shown that by using the gauge-invariant effective action formalism to analyze the confinement problem, and by taking into account the nonlinearity of the non-Abelian vacuum, one obtains a simple and physically compelling picture both of the origin of a linear static potential at large distances (see Adler (1981)), and of total flux confinement (see Adler and Piran (1982)).

ACKNOWLEDGMENTS

I wish to thank N H Kuiper for the hospitality of the Institut des Hautes Études Scientifiques, and J Iliopoulos for that of the École Normale Supérieure, during the summer of 1982.

This work was supported by the United States Department of Energy under Grant No DE-AC02-76ERO2220.

REFERENCES

Abbott L F 1981 *Nucl. Phys.* B **185** 189
Abers E S and Lee B W 1973 *Phys. Rep.* **9** 1
Adler S L 1981 *Phys. Rev.* D **23** 2905
—— 1982 *Phys. Lett.* **110B** 302
—— 1983 *Nucl. Phys.* B **217** 381
Adler S L and Piran T 1982 *Phys. Lett.* **113B** 405
Björken J D and Drell S D 1965 *Relativistic Quantum Fields* (New York: McGraw Hill)
Boulware D G 1981 *Phys. Rev.* D **23** 389
DeWitt B S 1967 *Phys. Rev.* **162** 1195
—— 1981, *Quantum Gravity II* ed C J Isham, R Penrose and D Sciama (Oxford: Oxford University Press)
Faddeev L D and Slavnov A A 1980 *Gauge Fields* (Reading, Mass.: Benjamin/Cummings)
Greenman M and Rohrlich F 1973 *Phys. Rev.* D **8** 1103
Gross D J, Pisarski R D and Yaffe L G 1981 *Rev. Mod. Phys.* **53** 43
't Hooft G 1975 in *Acta Universitatis Wratislavensis No. 368, XIIth Winter School of Theoretical Physics in Karpacz Feb –March 1975; Functional and Probabilistic Methods in Quantum Field Theory* Vol 1
Migdal A B 1973 *Nucl. Phys.* B **52** 483
Matinyan S G and Savvidy G K 1978 *Nucl. Phys.* B **134** 539
Pagels H and Tomboulis E 1978 *Nucl. Phys.* B **143** 485
Schwinger J 1951 *Phys. Rev.* **82** 664
Wilson K G 1974 *Phys. Rev.* D **10** 2445

Liouville Field Theory: a Two-Dimensional Model for Gravity?

R JACKIW

1. INTRODUCTION

Unphysical two-dimensional field theories have served us well by providing insight and understanding of quantum field theoretical possibilities, which can then be developed for physical, four-dimensional models. Of course the complexity of physically realistic theories precludes obtaining complete solutions; consequently information gleaned from the simpler, lower-dimensional setting can be enormously helpful. For example, dynamical gauge symmetry breaking was first understood in two-dimensional massless electrodynamics—the Schwinger model (Schwinger 1962, Farhi and Jackiw 1982); large-N behavior of non-Abelian gauge theories was illuminated by the solution of two-dimensional quantum chromodynamics ('t Hooft 1974); recent successes of field-theoretic semiclassical methods were first seen in the two-dimensional soliton phenomenon (Jackiw 1977, Rajaraman 1982).

Four-dimensional quantum gravity—Bryce DeWitt's foremost preoccupation—remains a thoroughly 'un-understood' theory; indeed some even doubt that it is well defined. Consequently, it is natural to seek instruction in lower dimensions, but one is disappointed to learn nothing: in three dimensions, the proportionality of the Riemann tensor to the Ricci tensor renders the source-free theory trivial[†] (this is analogous to two-dimensional vector gauge theories, which also are trivial without sources); in two dimensions, the Einstein tensor vanishes, and the theory cannot even be formulated (the Hilbert–Einstein action is a topological invariant).

General relativity is a theory in which the field's dynamics is invariant against arbitrary redefinitions of space–time coordinates. Conventional and non-trivial gravitational dynamics for such a field do not exist in dimensions lower than four. Here I shall discuss a two-dimensional field theory which admits a very large, though not arbitrarily large, class of coordinate transformations as symmetry operations.

[†] Discussions of three-dimensional gravity are to be found in Staruszkiewicz (1963), Deser *et al* (1982a,b, 1983), Deser and Jackiw (1983).

The model is the Liouville theory, governed by the (Minkowski-space) Lagrange density,

$$\mathscr{L} = \tfrac{1}{2}\partial_\mu \Phi \partial^\mu \Phi - \frac{m^2}{\beta^2} e^{\beta\Phi} \tag{1.1}$$

which gives the following equation of motion.

$$\Box\Phi + \frac{m^2}{\beta} e^{\beta\Phi} = 0. \tag{1.2}$$

In our units, the velocity of light is 1, and the action $\int d^2x \mathscr{L}$ is dimensionless. Then the field Φ also is dimensionless, as is β which may be taken positive, without loss of generality. The quantity m^2, with dimension of $[\text{mass}]^2$, is also taken positive so the energy density

$$\mathscr{E} = \tfrac{1}{2}\dot\Phi^2 + \tfrac{1}{2}\Phi'^2 + \frac{m^2}{\beta^2} e^{\beta\Phi} \tag{1.3}$$

is manifestly positive.

The Liouville equation was first posited and solved by Liouville, in the nineteenth century, during a study of vortices. Also its geometrical significance was soon appreciated: if a two-dimensional metric tensor is written in the form $\eta_{\mu\nu} = e^{\beta\Phi}g_{\mu\nu}$, where $g_{\mu\nu}$ is the flat Minkowski or Euclidean metric, then the curvature scalar is $R = \beta e^{-\beta\Phi}g^{\mu\nu}\partial_\mu\partial_\nu\Phi$; consequently when Φ satisfies (1.2), the two-dimensional curvature is constant (Liouville 1853, Bianchi 1879, Bateman 1944).

In modern times Liouville's equation has arisen in the study of solitons and instantons (Witten 1977, Chia 1977, Dolan 1977), and more recently in reformulations of the dual string model (Omnès 1977, Barbashov *et al* 1979, Polyakov 1981a). In this essay, dedicated to Bryce DeWitt, I shall summarize the classical and quantized model with the suggestion that one may here find some ideas relevant to gravity theory.

Evidently, the Liouville theory is a completely integrable field theory—Liouville integrated it! We may ask therefore how the complex of properties associated with other integrable field theories—namely infinitely many conservation laws, inverse scattering method, soliton phenomena, action-angle variables—are realized here. As we shall see, a version of these is found in the Liouville model, but in ways different from previous examples†.

The field equation (1.2) possesses a Bäcklund transformation, connecting its solutions to those of the two-dimensional wave equation.

$$\partial_+(\Phi - \phi_0) = \frac{m}{\beta}\alpha \exp \tfrac{1}{2}\beta(\Phi + \phi_0) \tag{1.4a}$$

$$\partial_-(\Phi + \phi_0) = -\frac{m}{\beta}\frac{1}{\alpha}\exp\tfrac{1}{2}\beta(\Phi - \phi_0). \tag{1.4b}$$

Φ solves (1.2) if $\Box\phi_0 = 0$, with α being an arbitrary parameter. The general solution

† This essay is mostly based on D'Hoker and Jackiw (1982)

of the free wave equation involves two arbitrary functions;

$$\Box \phi_0 = 0 \Rightarrow \phi_0(x) = \phi_0^+(x^+) + \phi_0^-(x^-)$$

$$x^\pm = \frac{1}{\sqrt{2}}(x^0 \pm x^1) \tag{1.5}$$

the general solution of the Liouville equation is obtained by integrating the first-order differential equations (1.4) with ϕ_0 as in (1.5). The formula, already given by Liouville (1853), also uses two arbitrary functions.

$$\Phi(x) = \frac{1}{\beta} \ln \frac{F'(x^+)G'(x^-)}{(1 + (m^2/4)F(x^+)G(x^-))^2}. \tag{1.6}$$

This expression is invariant against the replacement of F and G by their fractional transforms.

$$\frac{m}{2}F \to \frac{\frac{1}{2}\gamma mF - \delta}{\frac{1}{2}\varepsilon mF + \eta} \qquad \frac{m}{2}G \to \frac{\frac{1}{2}\eta mG - \varepsilon}{\frac{1}{2}\delta mG + \gamma} \qquad \gamma\eta + \delta\varepsilon \neq 0. \tag{1.7}$$

2. CONFORMAL SYMMETRY

Since both the wave equation and the Liouville equation are completely integrable, one expects that there exists an infinite number of constants of motion, and indeed such a set can be given by using conformal symmetry.

Under a conformal transformation the two-dimensional space–time coordinates x^μ change infinitesimally into

$$x^\mu \to x^\mu + \delta_f x^\mu$$

$$\delta_f x^\mu = -f^\mu(x) \tag{2.1}$$

where f^μ is a two-dimensional conformal Killing vector, i.e. it satisfies

$$\partial_\mu f_\nu + \partial_\nu f_\mu - g_{\mu\nu}\partial_\alpha f^\alpha = 0. \tag{2.2}$$

Any function f^μ solving this equation has $+$ component depending only on x^+, and $-$ component on x^-.

$$f^+ = f^+(x^+)$$

$$f^- = f^-(x^-). \tag{2.3}$$

The infinitesimal transformations obey a composition law,

$$[\delta_f, \delta_g] = \delta_h \tag{2.4}$$

with f, g and h conformal Killing vectors, the last given by the Lie bracket of the former two.

$$h^\mu = f^\alpha \partial_\alpha g^\mu - g^\alpha \partial_\alpha f^\mu. \tag{2.5}$$

To appreciate the conformal properties of the Liouville theory, it is useful to contrast them with those of the non-interacting theory. We therefore review these well known results (Ferrara *et al* 1972, Fubini *et al* 1973).

A. Free Theory
The action for the free Lagrangian

$$\mathcal{L} = \tfrac{1}{2}\partial_\mu \phi \partial^\mu \phi \Rightarrow \Box \phi = 0 \tag{2.6}$$

is invariant against conformal transformations, provided the infinitesimal field change is the Lie derivative of the field.

$$\delta_f \phi = f^\alpha \partial_\alpha \phi. \tag{2.7}$$

The conserved currents, expressed in terms of a symmetric and traceless energy–momentum tensor,

$$\theta_{\mu\nu} = \partial_\mu \phi \partial_\nu \phi - \tfrac{1}{2} g_{\mu\nu} \partial_\alpha \phi \partial^\alpha \phi \tag{2.8}$$

take the familiar Bessel-Hagen form (Bessel-Hagen 1921).

$$J_f{}^\mu = \theta^{\mu\nu} f_\nu. \tag{2.9}$$

Since $\theta_{\mu\nu}$ is traceless and conserved when ϕ solves the wave equation,

$$\phi(x) = \phi^+(x^+) + \phi^-(x^-) \tag{2.10}$$

it follows that θ_{--} depends only on x^-, and θ_{++} only on x^+. Consequently, two sets of constants of motion may be defined

$$Q_f^- = \int dx^- f^-(x^-) \theta_{--}(x^-) \tag{2.11a}$$

$$Q_f^+ = \int dx^+ f^+(x^+) \theta_{++}(x^+)$$

$$Q_f = Q_f^+ + Q_f^-. \tag{2.11b}$$

When a canonical formalism for the theory is constructed in lightcone coordinates, which are natural for the problem at hand, it is necessary to postulate both equal $-x^+$ and equal $-x^-$ Poisson brackets, owing to the conformal invariance of the theory (see also § 3 below).

$$\{\phi(x), \phi(y)\}\big|_{x^+ = y^+} = \tfrac{1}{4}\varepsilon(x^- - y^-) \tag{2.12a}$$

$$\{\phi(x), \phi(y)\}\big|_{x^- = y^-} = \tfrac{1}{4}\varepsilon(x^+ - y^+) \tag{2.12b}$$

Equivalently, since ϕ is a wave field, we have

$$\{\phi^-(x^-), \phi^-(y^-)\} = \tfrac{1}{4}\varepsilon(x^- - y^-) \tag{2.13a}$$

$$\{\phi^+(x^+), \phi^+(y^+)\} = \tfrac{1}{4}\varepsilon(x^+ - y^+) \tag{2.13b}$$

$$\{\phi^+(x), \phi^-(y)\} = 0. \tag{2.13c}$$

The Poisson bracket of two quantities A and B, which are functionals of $\phi^\pm$, is defined by

$$\{A, B\} = \tfrac{1}{4}\int dz\,dz'\,\varepsilon(z - z')\left[\frac{\delta A}{\delta\phi^+(z)}\frac{\delta B}{\delta\phi^+(z')} + \frac{\delta A}{\delta\phi^-(z)}\frac{\delta B}{\delta\phi^-(z')}\right]. \quad (2.14)$$

All the usual properties hold: the bracket is linear and anti-symmetric in (A, B) and satisfies the Jacobi identity.

One readily shows, with the help of the brackets (2.13), that bracketing the field with a charge generates the transformation (2.7).

$$\{Q_f, \phi\} = f^\alpha \partial_\alpha \phi. \quad (2.15)$$

Moreover, the Poisson brackets of the charges (2.11) reproduce the composition law (2.4), provided surface integrals vanish (we shall always omit them).

$$\{Q_f^\pm, Q_g^\pm\} = -Q_h^\pm \quad (2.16a)$$

$$\{Q_f^+, Q_g^-\} = 0. \quad (2.16b)$$

The massless, free theory also admits another (trivial) symmetry: the field may be shifted by an arbitrary wave field.

$$\delta\phi = \Omega$$

$$\square\Omega = 0 \Rightarrow, \Omega(x) = \Omega^+(x^+) + \Omega^-(x^-). \quad (2.17)$$

The conserved current is

$$j_\Omega^\mu = \partial^\mu\phi\,\Omega - \phi\,\partial^\mu\Omega. \quad (2.18)$$

Here again two independent charges may be defined.

$$q_\Omega^- = 2\int dx^-\,\Omega^-\partial_-\phi^- \quad (2.19a)$$

$$q_\Omega^+ = 2\int dx^+\,\Omega^+\partial_+\phi^+$$

$$q_\Omega = q_\Omega^+ + q_\Omega^-. \quad (2.19b)$$

One may combine the field translation symmetry with the conformal symmetry. An especially interesting choice for Ω, in view of our later discussion of the Liouville theory, is proportional to the divergence of the conformal Killing vector, which satisfies the wave equation. Thus, we consider the charges

$$\tilde{Q}_f = Q_f + \frac{1}{\gamma}q_{\partial f}. \quad (2.20)$$

They generate an inhomogeneous symmetry transformation on ϕ, with γ an arbitray constant.

$$\{\tilde{Q}_f, \phi\} = f^\alpha \partial_\alpha \phi + \frac{1}{\gamma}\partial_\alpha f^\alpha. \quad (2.21)$$

Although the Poisson bracket algebra no longer closes,

$$\{\tilde{Q}_f, \tilde{Q}_g\} = -\tilde{Q}_h + \frac{1}{\gamma^2}\Delta(f, g)$$

$$\Delta(f, g) = \int dx^-\,[f^-\partial_-^3 g^- - g^-\partial_-^3 f^-] + \int dx^+\,[f^+\partial_+^3 g^+ - g^+\partial_+^3 f^+] \quad (2.22)$$

we may nevertheless adopt the transformation law (2.21) as the realization of conformal transformations on the field ϕ, because the additional termΔ in (2.22) is independent of dynamical variables—it is merely a center for the infinite dimensional Lie algebra of the two-dimensional conformal group†.

Finally, let us observe that a current for the combined transformation, takes the Bessel-Hagen expression,

$$\tilde{J}^{\mu}_f = \Theta^{\mu\nu} f_\nu \tag{2.23}$$

with $\Theta^{\mu\nu}$ an improved, traceless energy–momentum tensor, which differs from the canonical one (2.8) by a superpotential (Callan *et al* 1970)

$$\Theta_{\mu\nu} = \theta_{\mu\nu} + \frac{2}{\gamma}(g_{\mu\nu}\Box - \partial_\mu\partial_\nu)\phi$$

$$= \partial_\mu\phi\partial_\nu\phi - \tfrac{1}{2}g_{\mu\nu}\partial_\alpha\phi\partial^\alpha\phi - \frac{2}{\gamma}\partial_\mu\partial_\nu\phi. \tag{2.24}$$

($\tilde{J}^{\mu}_f$ differs from $J^{\mu}_f + \frac{1}{\gamma}j^{\mu}_{of}$ by a superpotential.) It follows that

$$\tilde{Q}^{\pm}_f = \int dx^{\pm} f^{\pm}(x^{\pm}) \left[\partial_{\pm}\phi^{\pm}\partial_{\pm}\phi^{\pm} - \frac{2}{\gamma}\partial^2_{\pm}\phi^{\pm} \right]. \tag{2.25}$$

B. Liouville Theory

The Liouville theory is also conformally invariant, as is seen from the fact that the explicit solution (1.6) is form invariant against the transformation

$$x^+ \to y^+(x^+)$$

$$x^- \to y^-(x^-)$$

$$\Phi(x) \to \Phi(y) - \beta^{-1}\ln(\partial_+ y^+)(\partial_- y^-). \tag{2.26}$$

The Lagrangian (1.1) changes by a total derivative under the following infinitesimal transformation, which we define to be the conformal transformation of the Liouville field Φ.

$$\delta_f\Phi = f^\alpha\partial_\alpha\Phi + \beta^{-1}\partial_\alpha f^\alpha. \tag{2.27}$$

Thus, we see that while the free theory is separately invariant against the usual conformal transformation, as well as against the field translation (γ arbitrary), in the Liouville model the two must be combined ($\gamma = \beta$) to achieve the symmetry transformation.

Owing to conformal invariance we expect the conserved currents to be given in terms of a traceless energy–momentum tensor. This is indeed possible, but the

---

† It is interesting to recall that in the quantized theory there is a further, positive contribution to the center of the Poisson bracket algebra owing to anomalous commutators between components of the energy–momentum tensor; the quantal term is of the same form as Δ. Thus, even the conventional charges Q_f, when quantized, do not reproduce the classical algebra (2.4); see Ferrara *et al* (1972), Fubini *et al* (1973).

canonical Noether tensor must be improved (Callan *et al* 1970)†.

$$\Theta_{\mu\nu} = \partial_\mu \Phi \partial_\nu \Phi - g_{\mu\nu} \left(\frac{1}{2} \partial_\alpha \Phi \partial^\alpha \Phi - \frac{m^2}{\beta^2} e^{\beta\Phi} \right) + \frac{2}{\beta} (g_{\mu\nu} \Box - \partial_\mu \partial_\nu) \Phi \qquad (2.28)$$

$$\Theta^\mu_\mu = \frac{2m^2}{\beta^2} e^{\beta\Phi} + \frac{2}{\beta} \Box \Phi = 0. \qquad (2.29)$$

The current is

$$J^\mu_f = \Theta^{\mu\nu} f_\nu. \qquad (2.30)$$

Again, two sets of charges may be defined, since Θ_{--} depends only on x^- and Θ_{++} only on x^+.

$$Q^\pm_f = \int dx^\pm f^\pm (x^\pm) \Theta_{\pm\pm} (x^\pm)$$

$$= \int dx^\pm f^\pm \left(\partial_\pm \Phi \partial_\pm \Phi - \frac{2}{\beta} \partial^2_\pm \Phi \right). \qquad (2.31)$$

Just as in the free-field case, canonical Poisson brackets must be postulated at equal $-x^+$ and equal $-x^-$.

$$\{\Phi(x), \Phi(y)\}\big|_{x^+ = y^+} = \tfrac{1}{4} \varepsilon (x^- - y^-) \qquad (2.32a)$$

$$\{\Phi(x), \Phi(y)\}\big|_{x^- = y^-} = \tfrac{1}{4} \varepsilon (x^+ - y^+). \qquad (2.32b)$$

With (2.32), one verifies that Q_f generates the transformation law (2.27),

$$Q_f = Q^+_f + Q^-_f$$

$$\{Q_f, \Phi\} = f^\alpha \partial_\alpha \Phi + \frac{1}{\beta} \partial_\alpha f^\alpha \qquad (2.33)$$

as well as the fact that the algebra of equation (2.22) is satisfied.

$$\{Q_f, Q_g\} = -Q_h + \frac{1}{\beta^2} \Delta(f, g). \qquad (2.34)$$

The (improved) Liouville energy–momentum tensor may be written in terms of a free field. To this end we define

$$\phi^+ (x^+) \equiv 1/\beta \ln F' (x^+)$$
$$\phi^- (x^-) \equiv 1/\beta \ln G' (x^-)$$
$$\phi(x) \equiv \phi^+ (x^+) + \phi^- (x^-) \qquad (2.35)$$

and substitute into (2.28) the solution (1.6), expressed in terms of (2.35). This results in a formula for $\Theta_{\mu\nu}$, which coincides with the (improved) free-field energy–momentum tensor in equation (2.24), with $\gamma = \beta_0$.

† The improved tensor for this problem, as well as the Poincaré and dilatation charges, is constructed in Dhordzhadze *et al* (1978, 1979) and Pogrebkov (1979, 1980).

Thus we see that Liouville theory is invariant against arbitrary coordinate transformations which do not mix x^+ with x^-, and this symmetry is responsible for the solvability of the model (D'Hoker and Jackiw 1982).

3.   CANONICAL EQUIVALENCE TO A FREE THEORY

It is our experience with integrable two-dimensional theories that one can explicitly solve the Hamiltonian–Jacobi equation which canonically maps the nonlinear theory onto a linear one. This is also true for the Liouville model. Indeed three different procedures may be employed: the first makes use of the explicit solution; the second, of the Backlünd transformation; the third, of the inverse scattering method.

A. Explicit Solution as a Canonical Transformation
With the help of the definitions (2.35), the explicit solution (1.6) may be written in terms of the free field ϕ.

$$\Phi = \phi - \frac{2}{\beta} \ln\left[1 + m^2/4 \left(\frac{1}{\partial_+} \exp \beta\phi^+ \right)\left(\frac{1}{\partial_-} \exp \beta\phi^- \right) \right]$$

$$= \phi - \frac{2}{\beta} \ln\left(1 + (m^2/2\square) \exp \beta\phi \right). \tag{3.1}$$

Here $1/\partial_+$, $1/\partial_-$ and $1/\square = 1/2\partial_+\partial_-$ are Green's functions which will be specified presently. One may view equation (3.1) as a transformation from the old set of variables Φ to a new set ϕ, with ϕ satisfying a linear, i.e. a free equation. We have already seen that this transformation makes the (improved) Φ field's energy–momentum tensor take on the free (improved) form. Consequently, the Liouville Hamiltonian is mapped into the free Hamiltonian. It remains to verify that the transformation (3.1) is canonical. To this end, we postulate canonical, lightcone brackets for, $\phi^\pm$,

$$\{\phi^+(x^+), \phi^+(y^+)\} = \tfrac{1}{4}\varepsilon(x^+ - y^+) \tag{3.2a}$$

$$\{\phi^-(x^-), \phi^-(y^-)\} = \tfrac{1}{4}\varepsilon(x^- - y^-) \tag{3.2b}$$

$$\{\phi^+(x^+), \phi^-(y^-)\} = 0 \tag{3.2c}$$

and take the Green's functions to be of the Yang–Feldman form, so that we have the choice between

$$\frac{1}{\partial}f = \int_{-\infty}^{x} dy f(y) \quad \text{and} \quad \frac{1}{\partial}f = -\int_{x}^{\infty} dy f(y). \tag{3.3}$$

In both cases the transformation is canonical. It may therefore be written as

$$\Phi(x) = \phi(x) - \frac{2}{\beta} \ln\left(1 + \frac{m^2}{2} \int d^2 y\, G(x - y) \exp \beta\phi(y) \right) \tag{3.4}$$

$$\square\phi = 0$$

with $G(x - y) = \frac{1}{2}g(x^+ - y^+)g(x^- - y^-)$ and either $g(z) = \theta(z)$ or $g(z) = -\theta(-z)$. For both Green's function, one may verify that the lightcone algebra of equation (2.32), as well as the equal-time algebra

$$\{\Phi(x), \Phi(y)\}\big|_{x^0 = y^0} = \{\dot\Phi(x), \dot\Phi(y)\}\big|_{x^0 = y^0} = 0 \tag{3.5a}$$

$$\{\dot\Phi(x), \Phi(y)\}\big|_{x^0 = y^0} = \delta(x^1 - y^1) \tag{3.5b}$$

is satisfied when Φ is given by equation (3.4) and ϕ obeys equation (3.2) (D'Hoker and Jackiw 1982).

Equation (3.4) puts into evidence the very transparent significance of our transformation. Observe that the field equation (1.2) has a formal Yang–Feldman solution (Yang and Feldman 1950)

$$\Phi(x) = \phi(x) - \frac{m^2}{\beta}\int d^2 y\, G(x - y)\exp\beta\Phi(y). \tag{3.6}$$

Iterating this in powers of m^2 produces,

$$\Phi(x) = \phi(x) - \frac{m^2}{\beta}\int d^2 y\, G(x - y)\exp\beta\phi(y) + \frac{m^4}{\beta^2}\int d^2 y\, G(x - y)\exp\beta\phi(y)$$

$$\times \int d^2 z\, G(y - z)\exp\beta\phi(z) + \ldots \tag{3.7}$$

which is the same power series as the one obtained by expanding the logarithm in equation (3.4).

B. Bäcklund Transformation as a Canonical Transformation

It is well known that a Bäcklund transformation can be viewed as a canonical transformation. The way this works here is best seen in an equal-time formulation. We present (1.4) as

$$\Pi = \phi_0' + (\sqrt{2}m/\beta)\exp\tfrac{1}{2}\beta\Phi \sinh\left(\tfrac{1}{2}\beta\phi_0 + \ln\alpha\right)$$

$$\Pi_0 = \Phi' - (\sqrt{2}m/\beta)\exp\tfrac{1}{2}\beta\Phi \cosh\left(\tfrac{1}{2}\beta\phi_0 + \ln\alpha\right) \tag{3.8}$$

where $\Pi = \dot\Phi$, $\Pi_0 = \dot\phi_0$. It is recognized that there exists a generating functional for this transformation,

$$F(\Phi, \phi_0) = \int dx\left[\phi_0'\Phi + \frac{2\sqrt{2}m}{\beta^2}(\exp\tfrac{1}{2}\beta\Phi)\sinh\left(\tfrac{1}{2}\beta\phi_0 + \ln\alpha\right)\right] \tag{3.9}$$

$$\Pi = \frac{\delta F}{\delta\Phi} \qquad \Pi_0 = -\frac{\delta F}{\delta\phi_0} \tag{3.10}$$

which proves that (3.8) and equivalently (1.4) is a canonical transformation (Braaten et al 1982).

It should be noted that the free field ϕ occurring in the explicit solution (3.4) does not coincide with ϕ_0, the free field in the Bäcklund transformation. Rather there is a non-local relation between the two (see D'Hoker and Jackiw (1982))

C. Inverse Scattering Method

The inverse scattering procedure is uncommonly intricate in the Liouville theory owing to the conformal invariance of the model. We work again in lightcone coordinates and observe that the following matrix Lax pair, involving the Pauli matrices σ_i, is consistent if and only if the Liouville equation holds (Andreev 1976).

$$L = -i\sigma_2 \partial_- - \frac{\beta}{2}\partial_-\Phi\sigma_1$$

$$B = (\sigma_1 + i\sigma_2)\frac{m^2}{8\lambda}\exp\beta\Phi \qquad (3.11a)$$

$$L\psi = \lambda\psi$$

$$\partial_+\psi = B\psi. \qquad (3.11b)$$

However, the implementation of the inverse scattering procedure is not straightforward. Normally one specifies the initial data, by giving values to the 'potential' $\partial_-\Phi$ in the eigenvalue equation $L\psi = \lambda\psi$, and then one determines the evolution of the data from $\partial_+\psi = B\psi$. The complication in the present attempt to execute these steps is best seen from a second-order formulation of equation (3.11), which is appropriate to the case $\lambda \neq 0$. The spinor

$$\psi = \begin{pmatrix} u \\ v \end{pmatrix} \qquad \begin{aligned} u &= -\left(\partial_- + \frac{\beta}{2}\partial_-\Phi\right)w \\ v &= \lambda w \end{aligned} \qquad (3.12)$$

solves $L\psi = \lambda\psi$, provided w satisfies a Schrödinger-like second-order equation.

$$(-\partial_-^2 + V)w = \lambda^2 w$$

$$V = \frac{\beta^2}{4}(\partial_-\Phi)^2 - \frac{\beta}{2}\partial_-^2\Phi. \qquad (3.13)$$

The evolution of the scattering data for equation (3.13) is however trivial: the matrix $\frac{1}{2}(\sigma_1 + i\sigma_2)$ is nilpotent, hence

$$(\sigma_1 + i\sigma_2)\partial_+\psi = 0 \Rightarrow \partial_+ w = 0 \qquad (3.14)$$

i.e. the scattering data contained in w does not evolve. This result may be further understood by recognizing that according to the Liouville equation $\partial_+ V = 0$; hence the combination of fields contained in V does not evolve. Only the $\lambda = 0$ mode can escape the above result, since the evolution equation is not defined at $\lambda = 0$. We conclude that the Lax pair (3.11) is inadequate for constructing solutions to the Liouville equation.

The reason for this pathology derives from the conformal invariance of the equation, and from the consequent decoupling of the x^+ and x^- motions. In order to specify completely a unique solution to equation (1.2), it is not sufficient to give data $\Phi(x_0^+, x^-)$ at fixed $x^+ = x_0^+$; also it is necessary to give $\Phi(x^+, x_0^-)$ at fixed $x^- = x_0^-$. Only in this way are the two arbitrary functions in the general solution (1.6) determined. (Similar behavior of the free wave equation is readily recognized.) Specifying V in (3.13) is equivalent to fixing $\Phi(x_0^+, x^-)$, which however must be supplemented by $\Phi(x^+, x_0^-)$. This is also the reason why equal x^+ Poisson brackets must be supplemented by those at equal x^-.

We proceed by choosing $x_0^- = \pm \infty$, and fixing Φ for all x^+ at that asymptotic value. Observe first that in order (3.13) define a 'Schrödinger' equation with scattering, one must demand that V converge towards a finite asymptote as $x^- \to \pm \infty$. Moreover, the values at $x^- = \pm \infty$ must be equal so that the inverse scattering method may be straightforwardly applied. Hence we shall assume that $\beta^2(\partial_- \Phi)^2 \xrightarrow[x^- \to \pm \infty]{} 4V_\infty > 0$. This condition still leaves open the question whether $\partial_- \Phi$ tends towards the same limit as $x^- \to \pm \infty$ or whether the limits are opposite in sign. The differential equation requires the opposite limits: the derivative term $\partial_+ \partial_- \Phi$ vanishes asymptotically, so also must $\exp \beta \Phi$. Consequently the boundary conditions at $x^+ = \pm \infty$ must be chosen as

$$\lim_{x^- \to +\infty} [\beta\Phi(x^+, x^-) + 2\sqrt{V_\infty}\, x^- - \phi_{+\infty}(x^+)] = 0 \qquad (3.15a)$$

$$\lim_{x^- \to -\infty} [\beta\Phi(x^+, x^-) - 2\sqrt{V_\infty}\, x^- - \phi_{-\infty}(x^+)] = 0. \qquad (3.15b)$$

Consideration of the Liouville equation shows that V_∞ is independent of x^+ but $\phi_{\pm\infty}(x^+)$ may vary with x^+. Of course both functions $\phi_{\pm\infty}(x^+)$ cannot be chosen arbitrarily; the differential equation determines one from the other.

We may view $\phi_{+\infty}(x^+)$ (or $\phi_{-\infty}(x^+)$) as additional data specifying Φ at fixed x_0^-, namely at positive (or negative) infinity. This supplements the initial data which specify Φ at fixed x_0^+. The boundary conditions (3.15) have the property that the potential energy density $m^2/\beta^2 \exp \beta\Phi$ vanishes in the limit $|x^-| \to \infty$. Furthermore they also insure that a normalizable zero mode always exists for the equation $L\psi = \lambda\psi$.

$$\psi \alpha \begin{pmatrix} 1 \\ 0 \end{pmatrix} \exp \beta\Phi/2 \qquad (3.16)$$

Next we must replace the Lax pair (3.11). More specifically, we need another evolution operator $\tilde{B}$, instead of B, which will determine the evolution of all the scattering data and which is not singular at $\lambda = 0$. The following non-local choice is made (Andreev 1976).

$$\tilde{B}\psi(x^-; \lambda) = \frac{m^2}{8} [\exp \tfrac{1}{2}\beta\Phi(x^+, x^-)] \int_{-\infty}^{x^-} dy^- [\exp \tfrac{1}{2}\beta\Phi(x^+, y^-)] (1 + \sigma_3)\psi(y^-; \lambda)$$

$$(3.17a)$$

(We do not exhibit explicitly the x^+ dependence of ψ.) The compatability condition on the system

$$L\psi = \lambda\psi \qquad (3.17b)$$

$$\partial_+ \psi = \tilde{B}\psi \qquad (3.17c)$$

is still the Liouville equation, provided that $\psi(-\infty;\lambda)\exp[\frac{\beta}{2}\Phi(x^+,-\infty)] = 0$. Since the boundary conditions imply that $\Phi(x^+, \pm\infty) = -\infty$, this condition is satisfied.

Although the system (3.17) is not equivalent to (3.11), the latter is a consequence of the former when $\lambda \neq 0$. To see this, suppose $L\psi = \lambda\psi$ ($\lambda \neq 0$) and $\partial_+\psi = B\psi$. Then

$$\partial_+\psi(x^-;\lambda) = \frac{m^2}{8\lambda} \int_{-\infty}^{x^-} dy^- \left(\exp\frac{\beta}{2}[\Phi(x^+,y^-)+\Phi(x^+,x^-)]\right)(1+\sigma_3)$$

$$\times \left[-i\acute{\sigma}_2\partial_{y^-}\frac{\beta}{2}\partial_-\Phi(x^+,y^-)\right]\psi(y^-;\lambda) \tag{3.18}$$

simplifies with the help of boundary condition (3.15) to give $\partial_+\psi = B\psi$. So the Lax pairs (3.11) and (3.17) are identical when $\lambda \neq 0$, but only the non-local one is defined with $\lambda = 0$.

Having found an appropriate Lax pair, albeit a non-local one, the inverse scattering procedure may be completed. The details have been given elsewhere, and will not be repeated here, owing to their lengthy character. The result produces, in the conventional way, a solution of the Liouville equation equivalent to (1.6); an infinite set of constants of motion which encompass (2.31); and a canonical mapping of the interacting theory onto a non-interacting one. An important result is that the canonical variables are unconstrained, as in a free-field theory (D'Hoker and Jackiw 1982).

4. SOLITONS

Complete integrability of a field theory is usually associated with the existence of soliton solutions, like in the sine–Gordon or non-linear Schrödinger examples. However, the conventional soliton phenomenon does not occur in the Liouville equation[†].

If we look for static solutions to (1.2), we are led to the equation

$$\phi_s'' = (m^2/\beta) \exp \beta\phi_s. \tag{4.1}$$

The only homogeneous 'vacuum' solution is $\phi_s = -\infty$, which corresponds to zero energy and action. Position dependent solutions take the form

$$\phi_s(x^1) = -\frac{1}{\beta} \ln \frac{m^2}{2\varepsilon} \sinh^2\sqrt{\varepsilon}\,(x^1 - x_0^1) \tag{4.2a}$$

where x_0^1 is the constant of integration associated with translation invariance, and ε is a second integration constant, which may be positive, negative or zero. For negative ε, there are periodically spaced singularities; otherwise only $x^1 = x_0^1$ is singular. For $\varepsilon = 0$, (4.2a) becomes

$$\phi_s(x^1) = -\beta^{-1} \ln (m^2/2) (x^1 - x_0^1)^2. \tag{4.2b}$$

[†] Liouville 'solitons' are disucssed by Andreev (1976) and Barbashov *et al* (1979).

The energy density (1.3) is singular at $x^1 = x_0^1$ (we take $\varepsilon \geqslant 0$)

$$\mathcal{E} = \frac{4\varepsilon}{\beta^2 \sin^2 \sqrt{\varepsilon} \, (x^1 - x_0^1)} + \frac{2\varepsilon}{\beta^2} \tag{4.3a}$$

and there are no finite energy soliton solutions, but one sees that solutions for $\varepsilon > 0$ have greater energy than those at $\varepsilon = 0$. However, the improve energy densiy, given by the time–time component of the tensor in equation (2.28), is finite and vanishes at $\varepsilon = 0$

$$\Theta_{00} = (2/\beta^2)\varepsilon. \tag{4.3b}$$

Moreover, the solutions are stable. If we write $\phi(x) = \phi_s(x^1) + \exp - i\omega t \, \delta\phi(x^1)$ then $\delta\phi$ satisfies with $x_0^1 = 0$

$$\delta\phi''(z) + \frac{2}{\sinh^2 z} \delta\phi(z) = \frac{\omega^2}{\varepsilon} \delta\phi(z) \qquad z = \sqrt{\varepsilon} \, x^1 \qquad \varepsilon > 0 \tag{4.4a}$$

$$-\delta\phi''(z) + \frac{2}{z^2} \delta\phi(z) = \omega^2 \delta\phi(z) \qquad z = x^1 \qquad \varepsilon = 0. \tag{4.4b}$$

These Schrödinger-like equations have positive, continuous eigenvalues, hence ω is real. For example in the $\varepsilon = 0$ case, the eigenmodes are

$$\delta\phi_\omega(z) = \frac{1}{\sqrt{\pi}} \left(\frac{\sin\omega z}{\omega z} - \cos\omega z \right) \qquad \omega > 0. \tag{4.5}$$

These functions are orthonormal, but complete only on the half-interval.

It is interesting that there are no normalizable zero-eigenvalue modes. Hence the propagator of the small oscillations may be constructed.

$$G(x,y;t) = \tfrac{1}{2} \int_0^\infty \frac{d\omega}{\omega} \exp - i\omega|t| \, [\delta\phi_\omega(x)\delta\phi_\omega(y)]. \tag{4.6a}$$

For $\varepsilon = 0$, (4.6a) becomes, with the help of (4.5),

$$G(x,y;t) = -\frac{1}{4\pi} + \frac{1}{8\pi} \left[\left(\frac{t^2 - z_-^2}{2xy} - 1 \right) \ln (t^2 - z_-^2 - i0) \right.$$
$$\left. - \left(\frac{t^2 - z_+^2}{2xy} + 1 \right) \ln (t^2 - z_+^2 - i0) \right]$$
$$z_\pm = x \pm y \tag{4.6b}$$

which may also be presented in terms of a Legendre function of the second kind

$$G(x,y;t) = \frac{1}{4\pi} Q\left(\frac{t^2 - x - y^2}{2xy} - i0 \right). \tag{4.6c}$$

However, owing to the half-line completeness of the functions (4.5), G is a Green's function on the half-line.

$$(\square_{t,x} + 2/x^2)G(x,y;t) = -i\delta(t)\left[\tfrac{1}{2}\delta(x - y) + \tfrac{1}{2}\delta(x + y) \right]. \tag{4.6d}$$

5. QUANTUM THEORY

There are no apparent obstacles to a canonical equal-time or lightcone quantization of the Louville theory, where Poisson brackets are replaced by commutators. Of course, a solution requires developing calculational techniques, approximate ones if need be.

A semiclassical analysis is available in view of the canonical transformations, discussed in § 3, which map the Liouville theory onto a free theory. We conclude that the highly excited states, for which the semiclassical method is accurate, form a continuous spectrum, as in a free-field theory. This is consistent with conformal invariance.

Perturbation theory in β for vacuum matrix elements is not possible owing to the term linear in $(1/\beta)\Phi$, which appears when $(m^2/\beta^2) \exp \beta\Phi$ is expanded in powers of β. This tadpole cannot be removed by a field shift, since shifting Φ only redefines m^2.

Perturbation theory in m is ultraviolet finite after mass renormalization (for $\beta^2 < 4\pi$) (Albeverio and Hoëgh-Krohn 1974, Albeverio *et al* 199), but it is severely infrared divergent. These divergences are to be expected, since m^2 is dimensionful. Moreover, a power series in m^2 cannot be uniformly defined, since shifting $\beta\Phi$ by $\ln m^2$ removes m^2 from the theory.

We interpret these failures of perturbative computations for vacuum amplitudes as evidence for the failure of the assumption which underlying perturbation theory that there exists a normalizable, translationally invariant ground (vacuum) state. This is substantiated by the quantum version of equation (1.2): if there were a translationally invariant and normalizable state $|0\rangle$, $\langle 0|\square\Phi|0\rangle$ would vanish, as well as $\langle 0|(m^2/\beta)\exp \beta\Phi|0\rangle$; but $(m^2/\beta)\exp \beta\Phi$ is a positive definite operator (even when normal ordered), and this leads to a contradiction (D'Hoker and Jackiw 1982).

The same conclusion emerges from a loop expansion for the effective potential. This expansion does not require selecting a ground state initially; no infrared divergences appear and a well defined expression is obtained. A two-loop calculation for the effective potential gives†

$$V(\phi) = \frac{m^2}{\beta^2} \exp \beta\phi \left\{ 1 + \frac{\hbar\beta^2}{8\pi} \left[\left(\ln \frac{\Lambda^2}{m^2} - \beta\phi \right) + C_1 \right] \right.$$
$$\left. + \frac{1}{2} \left(\frac{\hbar\beta^2}{8\pi} \right)^2 \left[\left(\ln \frac{\Lambda^2}{m^2} - \beta\phi \right)^2 + C_2 \right] \right\} + 0(\hbar^3) \qquad (5.1a)$$
$$C_1 = 1, \; C_2 = 3.05208 \ldots$$

Here Λ is an ultraviolet cutoff. To the same degree of accuracy the effective potential may be written as

$$V(\phi) = (m_r^2/\tilde{\beta}^2) \exp \tilde{\beta}\phi$$
$$\tilde{\beta}^{-1} = \beta^{-1} + (\hbar\beta/8\pi) \qquad (5.1b)$$

where m_r^2 is a renormalized mass.

---

† Loop expansions for effective potentials were first considered by DeWitt (1972). The method of computation used here is that in Jackiw (1974).

In fact one may show that (5.1*b*) is exact (Goldstone, unpublished). The argument
is as follows. Let us choose for the quantum Hamiltonian the expression

$$H = \int dx : \tfrac{1}{2}\Pi^2 + \tfrac{1}{2}(\Phi')^2 + \frac{\mu^2}{\beta^2} \exp \beta\Phi : \big|_\mu \qquad (5.2)$$

where the normal ordering is performed with respect to a free field of mass μ. The
effective potential is the minimal expected value of H in a normalized state,
constrained so that the expectation value of Φ is ϕ: $V(\phi)L = \langle H \rangle$, $\phi = \langle \Phi \rangle$. Here
L is the (infinite) length of the spatial interval. On dimensional grounds it follows
that $V(0) = \mu^2 C$, where C is a constant. Moreover, $V(\phi)$ may be obtained from $V(0)$
by shifting Φ by ϕ. But such a shift merely redefines μ^2 in (5.2). Re-normal ordering,
so that the new ordering mass coincides with the new coefficient of $\exp \beta\Phi$, produces
(5.1*b*) (Goldstone, unpublished).

The vacuum is determined by the minimum of $V(\phi)$, but we see tha there is no such
minimum (save at $\phi = -\infty$), which again shows that no translationally invariant,
normalizable ground state exists.

Is there a translationally invariant but non-normalizable ground state? We
suspect not; since the classical configuration of lowest energy requires a negatively
infinite value for the dynamical variable, it would take the system an infinite time to
reach such a configuration. Correspondingly the zero-energy quantum state should
not belong to the spectrum.

Such peculiar behavior does not violate any quantum-mechanical principles, as
the following quantum-mechanical example shows. Consider a one-dimensional
problem with potential $m^2/\beta^2 \exp \beta q$. The Schrödinger equation possesses a
continuous energy spectrum with eigenfunctions

$$\psi_E(q) = (4/\pi\beta \sinh\pi v)^{1/2} K_{iv}(\sqrt{8m/\hbar\beta^2} \exp \beta q/2)$$

where $v = \sqrt{8E/\hbar\beta}$ and K is the modified Bessel function (Bethe and Jackiw 1968).
While the states with $E \neq 0$ are continuum normalized, there is no state for $E = 0$,
since $\psi_{E=0}$ vanishes.

There is another intriguing possibility: the ground state is not invariant against
spatial translations. We saw in §4 that classical, translationally non-invariant
solutions possess infinite energy. Perhaps one can 'renormalize' this infinity, and
build a quantum field theory on a degenerate family of 'ground' states $|x_0^1\rangle$ for
which

$$\langle x_0^1 | \Phi(x) | x_0^1 \rangle = -\frac{1}{\beta} \ln \frac{m^2}{2}(x^1 - x_0^1)^2 + \text{quantum corrections.} \qquad (5.3)$$

Here, it is encouraging that no zero-eigenvalue (Goldstone) mode is present
in the small oscillations and the propagator may be constructed (compare (4.5) and
(4.6)). Consequently, infrared divergences, which would restore translational
symmetry, may not be obstacles. However, the absence of a complete set of
oscillations leaves us uncertain about how to proceed with the full quantum theory.

The fact that the quantum effective potential differs from the classical one merely
by the replacement of β by $\tilde{\beta}$, suggests that the conformal symmetry survives

quantization, and that the conformal charges (2.31) remain conserved in the quantum theory, provided β is replaced by $\tilde{\beta}$. That this is indeed so has been established by a normal ordering analysis for the theory on a finite interval. Moreover, the quantum commutation relations of the conformal charges follow (2.34), with $\tilde{\beta}$ replacing β and with a further contribution of $(1/48\pi)\Delta$ to the center of the algebra (Curtright and Thorn 1982).

Of course, if the translationally non-invariant quantum field theory mentioned above can be completely developed, the full conformal invariance would be spontaneously broken. Nevertheless, there would remain the invariance of the background field (5.3): a three-parameter $O(2,1)$ Lie group consisting of time translations, dilatations and special conformal transformations in the time direction.

6. CONCLUSION

The large coordinate (conformal) invariance produces much unexpected behavior in the Liouville classical and quantal field theory. While the classical theory is solvable, the usual properties associated with complete integrability are present only in modified form. Quantization apparently preserves conformal invariance, but no coordinate symmetry, which reduces the full conformal group to an $O(2,1)$ subgroup. (One can show that an $O(2,1)$ group also is the maximal symmetry of the most general solution (1.6), not only of the static solution (4.2b). This is related to the fact that the isometry group for two-dimensional surfaces of constant curvature is $SO(2,1)$.)

One cannot help but wonder whether further elucidation of this geometrically significant model in two dimensions would illuminate four-dimensional gravity. Specifically I suggest that two-dimensional gravitational dynamics not be described by the Einstein equation

$$R_{\mu\nu} - \tfrac{1}{2}\eta_{\mu\nu} R + \Lambda\eta_{\mu\nu} = 0. \tag{6.1}$$

Since the two-dimensional Ricci tensor, $R_{\mu\nu}$, is identically equal to $\tfrac{1}{2}\eta_{\mu\nu} R$, equation (6.1) has only the unacceptable solution $\eta_{\mu\nu} = 0$ when the cosmological constant Λ is non-zero, and for vanishing Λ, $\eta_{\mu\nu}$ is undetermined. Rather two-dimensional gravity should be governed by Liouville dynamics, which may also be presented in purely geometrical form

$$R + \Lambda = 0. \tag{6.2}$$

When coordinates are chosen so that the metric is conformally flat $\eta_{\mu\nu} = \exp(\beta\Phi)g_{\mu\nu}$ (this may always be achieved by a coordinate transformation) equations (6.2) and (1.2) coincide, with $\Lambda = \mu^2$. Moreover the proposed field equation (6.2) may be obtained from an action, which although not entirely geometric, is generally covariant

$$I = \int d^2x \, N \, (-\det \eta_{\mu\nu})^{1/2} \, (R + \Lambda). \tag{6.3}$$

Variation of the scalar field N (a Lagrange multiplier) produces (6.2), while variation of $\eta_{\mu\nu}$ gives an equation for N, but puts no further constraints on $\eta_{\mu\nu}$. The action (6.3) has the further appealing feature that it arises by dimensional reduction from the three-dimensional Einstein–Hilbert action, provided all dependence on the third component is suppressed, η_{02} and η_{12} are set to zero and N^2 is identified with $-\eta_{22}$.

Finally, it is worth noting that there exists a super-symmetric generalization of the Liouville theory (Polyakov 1981b), and that much of what we have presented here holds in the super symmetric model as well (D'Hoker 1983).

Since this article was written the following further developments have taken place.

(1) The translationally non-invariant approach to the quantum Liouville model mentioned at the end of §5, has been found. The theory is quantized on an SO (2, 1) invariant manner and the invariance group of the vacuum is the SO (2, 1) group not the Poincaré group. Thus the ground state is de Sitter-like (D'Hoker and Jackiw 1983 a, b).

(2) The proposal that Liouville dynamics replace Einstein dynamics for two-dimensional gravity has been independently made by C Teitelboim in his contribution to this volume and by Banks and Susskind (unpublished).

ACKNOWLEDGMENT

This research is supported in part by the US Department of Energy under contract DE-AC02-76ERO3069. The manuscript was prepared during a visit to the Institute for Theoretical Physics, Santa Barbara CA. I am grateful for their hospitality.

REFERENCES

Albeverio, S, Gallavotti G and Hoëgh-Krohn R 1979 *Phys. Lett.* **83B** 177
Albeverio S and Hoëgh-Krohn 1974 *J. Funct. Anal.* **16** 39
Andreev V 1976 *Teor. Mat. Fiz.* **29** 213 (1976 *Theor. Math. Phys.* **29** 1027)
Banks T and Susskind L unpublished
Barbashov B, Nesterenko V and Chervyakov A 1979 *Teor. Mat. Fiz.* **40** 15 (1979 *Theor. Math. Phys.* **40** 572)
Bateman H 1944 *Partial Differential Equations of Mathematical Physics* (New York: Dover)
Bessel-Hagen E 1921 *Math. Ann.* **84** 258
Bethe H and Jackiw R 1968 *Intermediate Quantum Mechanics* (Reading, MA: Benjamin/Cummings)
Bianchi L 1879 *Ann. Sci. Norm. Super. Pisa, Ser.* 1 **2** 26
Braaten E, Curtright T and Thorn C 1982 *Phys. Lett.* **118B** 115
Callan C, Coleman S and Jackiw R 1970 *Ann. Phys., NY* **59** 42
Chia Kwei Peng 1977 *Sci. Sin.* **20** 345
Curtright T and Thorn C 1982 *Phys. Rev. Lett.* **48** 1309, 1768

Deser S and Jackiw R 1983 *Ann. Phys., NY* in the press

Deser S, Jackiw R and 't Hooft G 1983 *Ann. Phys., NY* in the press.

—— 1982b *Ann. Phys., NY* **140** 372

DeWitt B S 1972 in *Magic Without Magic: John Archibald Wheeler, a Collection of Essays in Honor of His 60th Birthday* ed J Klauder (San Francisco: Freeman)

D'Hoker E 1983 *Phys. Rev. D* in the press

D'Hoker E and Jackiw R 1982 *Phys. Rev. D* **26** 3517

—— 1983a *Phys. Rev. Lett.* **50** 1719

—— 1983b *Phys. Rev. D* in the press

Dhordzhadze G, Pogrebkov A and Polivanov M 1978 *Dokl. Acad. Nauk.* **243** 318 (1978 *Sov. Phys.–Dokl.* **23** 828)

—— 1979 *Teor. Mat. Fiz.* **40** 221 (1979 *Theor. Math. Phys.* **40** 706)

Dolan L 1977 *Phys. Rev. D* **15** 2337

Farhi E and Jackiw R 1982 *Dynamical Gauge Symmetry Breaking* (Singapore: World Scientific)

Ferrara S, Gatto R and Grillo A 1972 *Nuovo Cimento* A **12** 959

Fubini S, Hanson A and Jackiw R 1973 *Phys. Rev. D* **7** 1732

Goldstone J unpublished

't Hooft G 1974 *Nucl. Phys.* B **75** 461

Jackiw R 1974 *Phys. Rev. D* **9** 1686

—— 1977 *Rev. Mod. Phys.* **49** 681

Liouville J 1853 *J. Math. Pure Appl.* **18** 71

Omnès R 1977 *Nucl. Phys.* B **149** 269

Pogrebkov A 1979 *Dokl. Acad. Nauk.* **244** 873 (1979 *Sov. Phys.–Dokl.* **24** 105)

—— 1980 *Teor. Mat. Fiz.* **45** 161 (1980 *Theor. Math. Phys.* **45** 951)

Polyakov A 1981a *Phys. Lett.* **103B** 207

—— 1981b *Phys. Lett.* **103B** 211

Rajaraman R 1982 *Solitons and Instantons* (Amsterdam: North-Holland)

Schwinger J 1962 *Phys. Rev.* **128** 2425

Staruszkiewicz 1963 *Acta. Phys. Pol.* **24** 735

Witten E 1977 *Phys. Rev. Lett.* **38** 121

Yang C N and Feldman D 1950 *Phys. Rev.* **79** 972

Towards a Quantum Theory without 'Quantization'

DAVID DEUTSCH

1. INTRODUCTION: QUANTUM THEORY VERSUS QUANTIZATION THEORY

Bryce DeWitt is a philosophical realist. He believes that the world exists objectively and that the task of physics is to obtain as true as possible a description of it. He has always contended that we should take seriously the assertions of our theories, to 'push them to their limits' until they either fail or yield new insights into the nature of reality. He has opposed *ad hoc* attempts to reformulate newer, better but uncomfortable theories in terms of the formalism, kinematics and ontology of familiar but obsolete theories. Thus for him, general relativity is a theory of the dynamics of space–time geometry, not just another field theory on a Minkowskian flat space–time background. And quantum theory is an objective theory of parallel interfering universes, not of a sequence of subjective classical experiences.

Quantum theory has been extraordinarily slow in freeing itself from the apron strings of its classical ancestors. In championing and developing Everett's interpretation (DeWitt and Graham 1973) DeWitt has been instrumental in the exorcism of classical concepts from the *interpretation* of quantum theory. But in the more important matter of formalism we still know of no other way of constructing quantum theories than 'quantization', a set of semi-explicit *ad hoc* rules for making a silk purse (a quantum theory) out of a sow's ear (the associated classical theory). And even DeWitt, who originated most of the ideas presented in this article, now appears to acquiesce in this.

I believe that quantization will have to go before further progress is made at the foundations of physics.

Perhaps the reason is best illustrated by analogy: suppose that in an elementary chemistry textbook, in the chapter on combustion, no mention were made of oxygen. Instead, the chapter begins with a detailed exposition of the theory of phlogiston. It then explains that this theory is now known to be false, but that a better theory may be constructed from it by means of 'chemicalization rules': 'phlogiston must be thought of formally as occupying a negative volume', and so forth. These rules are numerous, ramshackle and without independent motivation but (in experienced hands) they do correctly predict the results of experiments. Usually. If chemistry were really in the state indicated by such a textbook, it would bode ill for the future of the subject. Progress would be halted until chemists stopped thinking in terms of phlogiston and someone invented a theory of oxygen.

To base the theory of quantum fields $\hat{\varphi}_i$ on that of classical fields φ_i is like basing chemistry on phlogiston or general relativity on Minkowski space–time: it can be done, up to a point, but it is a mistake; not only because the procedure is ill defined and the resulting theory of doubtful consistency, but because the world isn't really like that. *No classical fields φ_i exist in nature.* Like phlogiston, they were participants in obsolete physical theories. Only when the quantum formalism contains no reference to classical theory can we hope to understand what it says. And only then can we hope to improve upon it.

It might be objected that the Correspondence Principle gives the classical theory of φ_i a special place in the quantum theory of $\hat{\varphi}_i$. But this is just a confusion created by quantization theory. As we shall see, the Correspondence Principle can be formulated without reference to classical theory.

2. QUANTUM THEORY WITHOUT CLASSICAL THEORY

I shall now try to set up the formalism of quantum theory without referring to classical concepts noting, but not always solving, the problems as they arise.

An observable $\hat{\varphi}$ is something whose numerical value can in principle be measured, or could be measured if the requisite measuring apparatus were present at the appropriate time(s) and place(s). The set $\{\varphi\}$ of possible outcomes of a measurement of $\hat{\varphi}$ is called the spectrum of $\hat{\varphi}$. If the spectrum of $\hat{\varphi}$ is independent of time, $\hat{\varphi}$ is said to have 'no explicit time dependence'. In quantum theory, observables $\hat{\varphi}$ are represented by Hermitian operators and the values $\{\varphi\}$ are their eigenvalues.

The observables in nature are quantum fields; that is, parametrized sets of observables $\hat{\varphi}_i$. The index i represents a set of parameters such as space–time coordinates, tensor indices, internal symmetry indices and enumeration indices. At present it is customary always to include a time coordinate t among the parameters. The fact that something with as fundamental a physical significance as the time appears only as a parameter is somewhat unsatisfactory and may be eliminated in future formulations (Page and Wootters 1982) of quantum theory.

The state of the world is represented by a unit vector, the state vector $|\psi\rangle$, in the Hilbert space spanned by the eigenvectors of any maximal commuting set of observables. This is not the place to describe the mechanism by which the formal structure of vectors and Hermitian operators is asserted by quantum theory to correspond to reality. The reader is referred to the work of Everett (DeWitt and Graham 1973; Deutsch 1980). Suffice it to say that the *interpretation* of quantum theory has been shown to require no classical element. When the same has been done for the *formalism*, quantum theory will have come of age.

The dynamics of quantum fields are generated by a principle of stationary action. The action is an *operator*-valued functional

$$\hat{S}[\hat{\varphi}_i] \tag{2.1}$$

of the field observables. Under an infinitesimal variation

$$\hat{\varphi}_i \rightarrow \hat{\varphi}_i + \delta\hat{\varphi}_i \tag{2.2}$$

$$\hat{S} \rightarrow \hat{S} + \sum_i \frac{\delta\hat{S}}{\delta\hat{\varphi}_i}\,\delta\hat{\varphi}_i. \tag{2.3}$$

An ambiguity of notation arises in (2.3) (which can serve as a formal definition of the functional derivative): it is not clear how the three pairs of implied operator indices are to be connected and summed over. It is necessary to make these indices explicit, writing (2.3) as

$$S_{\hat{\alpha}} \rightarrow S_{\hat{\alpha}} + \frac{\delta S_{\hat{\alpha}}}{\delta\varphi_{\hat{\beta}i}}\,\delta\varphi_{\hat{\beta}i} \tag{2.4}$$

where each quantum index $\hat{\alpha}$ stands for a Hermitian pair of Hilbert space labels. I have also adopted the Einstein summation convention both for the quantum indices and for the generalized coordinates i.

The quantum principle of stationary action is not as straightforward as the classical one. It is not in general possible for the full variation $\delta S_{\hat{\alpha}}/\delta\varphi_{\hat{\beta}i}$ to vanish. The indices show that the resulting system of dynamical equations would be over-determined. The correct principle has the form

$$\frac{\delta S_{\hat{\alpha}}[\hat{\varphi}]}{\delta\varphi_{\hat{\beta}i}}\,X_{\hat{\beta}i}{}^j[\hat{\varphi}] = 0 \tag{2.5}$$

for some functional $\hat{X}_i{}^j[\hat{\varphi}]$, which is equivalent to the requirement that the action be stationary, not under general variations in $\hat{\varphi}_i$, but only under variations of the form

$$\delta\varphi_{\hat{\alpha}i} = X_{\hat{\alpha}i}{}^j\delta\varphi_j \tag{2.6}$$

where the $\delta\varphi_i$ are suitable infinitesimal c-number test functions. It is not known in general how to choose the functional $\hat{X}_i{}^j[\hat{\varphi}]$. Schwinger (1953), Peierls (1952), and DeWitt (1967) all make the natural choice (for Boson fields)

$$\hat{X}_i{}^j = \delta_i{}^j\hat{1} \tag{2.7}$$

which would mean that the action is to be stationary under pure c-number variations in the fundamental field. For certain well studied systems this reproduces the same quantum theory as 'canonical quantization'. But it cannot be the correct choice in general because the variations (2.6) are not in general compatible with the algebra of the operators $\hat{\varphi}_i$. Fermion fields are an obvious example since anticommutators are not invariant under c-number variations. A simple Boson example is the case where i runs from 1 to 3 and $\hat{\varphi}_i$ represents the ith angular momentum component $\hat{L}_i(t)$ of a spin-1 system,

$$[\hat{L}_i(t), \hat{L}_j(t)] = i\varepsilon_{ij}{}^k\hat{L}_k(t). \tag{2.8}$$

Variations $\delta L_i \hat{1}$ are incompatible with (2.8). One compatible choice would be

$$\delta \hat{L}_i = i\varepsilon_i{}^{jk} \delta L_j(t) \hat{L}_k(t) \tag{2.9}$$

corresponding to

$$X_{\hat{\alpha}i}{}^j[\hat{L}] = i\varepsilon_i{}^{jk} L_{\hat{\alpha}k}. \tag{2.10}$$

Since different variations $\hat{X}_i{}^j \delta\varphi_j$ generate different stationary action principles and different dynamics, it follows that quantum theory is not covariant under coordinate transformations in configuration space, at least not in any sense known at present. This is in marked contrast to classical theory where the variational principle

$$\frac{\delta S[\varphi]}{\delta \varphi_i} = 0 \tag{2.11}$$

implies

$$\frac{\delta S}{\delta \chi_i} = \frac{\delta S}{\delta \varphi_j} \frac{\delta \varphi_j}{\delta \chi_i} = 0. \tag{2.12}$$

In the quantum case

$$\frac{\delta S_{\hat{\alpha}}}{\delta \chi_{\hat{\beta}i}} X_{\hat{\beta}i}{}^j = \frac{\delta S_{\hat{\alpha}}}{\delta \phi_{\hat{\beta}r}} \frac{\delta \phi_{\hat{\beta}i}}{\delta \chi_{\hat{\gamma}j}} X_{\hat{\gamma}j}{}^k \tag{2.13}$$

which cannot in general be required to vanish whenever $(\delta \hat{S}/\delta \phi_{\hat{\beta}i}) X_{\hat{\beta}i}{}^j$ does.

Coordinate invariance in the base space (i.e. parameter space) can of course still be maintained in quantum theory, but gauge invariance (at least for non-Abelian gauge groups) cannot. What is to become of these cherished invariances of classical field theory—of what property of the quantum theory they are the limiting cases—is an open question. Perhaps the quantum action principle is invariant only under some special class of transformations. Or more interestingly, perhaps the $\hat{X}_i{}^j$ itself suffers changes under coordinate transformations in configuration space, such as to preserve coordinate invariance. This raises the possibility of a more general action principle

$$\frac{\delta S_{\hat{\alpha}}}{\delta \phi_{\hat{\beta}i}} X_{\hat{\beta}i}{}^j + \Gamma^{\hat{\beta}j}{}_{\hat{\alpha}} S_{\hat{\beta}} = 0. \tag{2.14}$$

Could this be regarded as a 'covariant functional derivative' with 'connection coefficients' $\hat{\Gamma}^i$ (DeWitt, private communication)?

3. SOME ELABORATION OF THE PURE QUANTUM THEORY

The dynamical equations (2.5) cannot be solved unless the algebra of the operators $\hat{\varphi}_i$ is given. In quantization theory the algebra is determined by setting the commutator of any two observables equal to i times their classical Poisson bracket (as generalized by Peierls and DeWitt). But, as I shall now show, this will not work in the true quantum theory.

Following Peierls and DeWitt I first construct the theory of small disturbances corresponding to the stationary action principle (2.5). If $\hat{\varphi}_i$ satisfies (2.5) and $\delta\hat{\varphi}_i$ is the infinitesimal disturbance in $\hat{\varphi}_i$ caused by a variation $\delta\hat{S}[\hat{\varphi}]$ in the form of the action then

$$F_{\hat{\alpha}}^{j\hat{\beta}i}\delta\varphi_{\hat{\beta}i} = -\frac{\delta\delta S_{\hat{\alpha}}}{\delta\varphi_{\hat{\beta}i}}X_{\hat{\beta}i}^{j} \tag{3.1}$$

where

$$F_{\hat{\alpha}}^{j\hat{\beta}i}[\hat{\varphi}] = \frac{\delta^2 S_{\hat{\alpha}}[\hat{\varphi}]}{\delta\varphi_{\hat{\beta}i}\delta\varphi_{\hat{\gamma}k}}X_{\hat{\gamma}k}^{j} + \frac{\delta S_{\hat{\alpha}}[\hat{\varphi}]}{\delta\varphi_{\hat{\gamma}k}}\frac{\delta X_{\hat{\gamma}k}^{j}}{\delta\varphi_{\hat{\beta}i}}. \tag{3.2}$$

(3.1) is the quantum equation of small disturbances. $F_{\hat{\alpha}}^{j\hat{\beta}i}$ is a bi-operator which we may abbreviate as $\hat{\hat{F}}$. (I have ignored the possibility that $\hat{X}_i^j$ might depend on the form of the action.) Like the dynamical equation itself, equation (3.1) is soluble only given a knowledge of the operator algebra. But in any case a Green's function theory can be based on the equation

$$\hat{\hat{F}}\,\hat{\hat{G}} = -\hat{\hat{1}} \tag{3.3}$$

i.e.

$$F_{\hat{\alpha}}^{j\hat{\gamma}k}G_{\hat{\gamma}k}^{\hat{\beta}}{}_{l} = -\delta_{\hat{\alpha}}^{\hat{\beta}}\delta_{l}^{j}. \tag{3.4}$$

Although the Cauchy problem for operator differential equations is in general more complicated than for scalars (because, for example, operators cannot in general be required to vanish at infinity), this should not affect the Green's function theory because there is presumably no reason why a *variation* $\delta\hat{\varphi}_i$ should not vanish in, say, the remote past. Therefore in particular there should exist a unique retarded Green's function $\hat{\hat{G}}^-$ which satisfies equation (3.3). As in the classical theory, the right and left inverses of $\hat{\hat{F}}$ are equal,

$$\hat{\hat{G}}^-\,\hat{\hat{F}} = -\hat{\hat{1}} \tag{3.5}$$

and much of the classical Green's function theory can be carried over to the quantum case just by putting extra hats and indices in appropriate places. For example, defining

$$D^-_{\hat{A}}\hat{B} = \lim_{\varepsilon \to 0}\delta^-_{\varepsilon\hat{A}}\hat{B} \tag{3.6}$$

where $\delta^-_{\delta\hat{S}}\hat{B}$ is the retarded disturbance in $\hat{B}$ produced by a variation $\delta\hat{S}$ in the action, we have

$$D^-_{\hat{A}}B_{\hat{\alpha}} = \frac{\delta B_{\hat{\alpha}}}{\delta\varphi_{\hat{\beta}j}}G_{\hat{\beta}j}^{-\hat{\gamma}}{}_{k}\frac{\delta A_{\hat{\gamma}}}{\delta\varphi_{\hat{\delta}l}}X_{\hat{\delta}l}^{k}. \tag{3.7}$$

There is, however, one important exception: the operator $\hat{\hat{F}}$ in the quantum case is not in general self-adjoint, i.e.

$$F_{\hat{\alpha}}^{j\hat{\gamma}k} \neq F^{\hat{\gamma}k}{}_{\hat{\alpha}}{}^{j} \tag{3.8}$$

(where raising and lowering of quantum indices denotes Hermitian conjugation). This has the consequence that the Peierls–Poisson–DeWitt bracket

$$(\hat{A}, \hat{B}) = D_{\bar{A}}\hat{B} - D_{\bar{B}}\hat{A} \qquad (3.9)$$

based on equation (3.7) does not obey the Jacobi identities and cannot therefore be consistently identified with $-i$ times the commutator as convention (and DeWitt) would have it. Peierls was aware of this problem and suggested that extra terms might be added to his bracket to restore the consistency of his quantization scheme. No one has yet found a way of doing this.

The Peierls–Poisson–DeWitt bracket method of specifying the operator algebra is closely related to the Schwinger variational principle which states that

$$\delta \langle \text{out} | \text{in} \rangle = i \langle \text{out} | \delta \hat{S} | \text{in} \rangle \qquad (3.10)$$

under a variation $\delta\hat{S}$ in the form of the action, where $|\text{out}\rangle$ and $|\text{in}\rangle$ are states corresponding to fixed eigenvalues of observables with no explicit time dependence and confined to the future and past respectively of $\delta\hat{S}$ (it is assumed that $\delta\hat{S}[\hat{\varphi}]$ is constructed from field quantities confined to some parameter space region of effectively finite duration). Using retarded boundary conditions we have

$$\delta | \text{in} \rangle = 0 \qquad (3.11)$$

so equation (3.10) implies

$$\delta | \text{out} \rangle = -i\delta\hat{S} | \text{out} \rangle \qquad (3.12)$$

and hence if $\hat{A}[\hat{\varphi}]$ is an 'out' observable,

$$\delta\hat{A} = -i[\delta\hat{S}, \hat{A}]. \qquad (3.13)$$

But $\delta\hat{A}$ can also be represented dynamically *via* the theory of small disturbances

$$\delta\hat{A} = D_{\bar{\delta S}}\hat{A}. \qquad (3.14)$$

Since the commutator is anti-symmetric in its arguments and $D_{\bar{A}}\delta S$ vanishes, it follows that

$$[\delta\hat{S}, \hat{A}] = i(D_{\bar{\delta S}}\hat{A} - D_{\bar{A}}\delta\hat{S}) \qquad (3.15)$$

which, since $\delta\hat{S}$ and $\hat{A}$ are effectively arbitrary, is the quantum analog of the Peierls expression. Unfortunately, advanced boundary conditions give a different answer

$$[\delta\hat{S}, A] = i(D_{A}^{+}\delta\hat{S} - D_{\delta S}^{+}\hat{A}). \qquad (3.16)$$

This signals an inconsistency in the dynamics implied by equation (3.10). The c-number analogs of equation (3.15) and equation (3.16) are indeed identical, but equation (3.8) causes the reciprocity relation between $D_{A}^{+}\hat{B}$ and $D_{\bar{B}}\hat{A}$ to fail in the quantum case. The argument that equations (3.15) and (3.16) differ only by factor-ordering ambiguities seems particularly hollow in this instance, but it does imply that equation (3.15) is true to the lowest order in $\hbar$.

It therefore appears that the Schwinger principle is, as it stands, inconsistent in the full quantum theory. Again it is not known how to modify it.

4. PERTURBATION THEORY AND THE CORRESPONDENCE PRINCIPLE

Fortunately, many of the physical predictions of quantum theory can be obtained without a complete knowledge of the operator algebra. This is the reason why quantization theory, in spite of its cavalier treatment of the 'factor-ordering problem' can have a measure of empirical success. And it is the reason why a c-number theory can exist as a limiting case. The fact that factor-ordering 'ambiguities' are of order $\hbar$ in perturbative expansions of physical quantities is an expression of the Correspondence Principle. To see how this convenient property arises, let us develop the archetypal perturbation method of quantum field theory, the *background field method*.

The objective of background field schemes is always to describe as much as possible of the system in terms of c-number fields. In particular, Schwinger (Schwinger 1953, DeWitt 1965, 1982) found that much can be learned about a quantum theory by investigating the effect of adding a linear source term

$$\hat{S} \to \hat{S} + J^i \hat{\varphi}_i \tag{4.1}$$

to the action functional. The J^i are c-number 'external sources'. Schwinger's starting point was his variational principle equation (3.10) which I shall assume is true to a sufficiently high order in $\hbar$ to make the following meaningful.

Equation (3.10) implies directly

$$\langle \text{out}| T(\hat{A}[\hat{\varphi}])|\text{in}\rangle = \vec{A}\left[\frac{\delta}{\delta iJ}\right]\langle \text{out}|\text{in}\rangle \tag{4.2}$$

where T is the time ordering symbol and $\vec{A}$ is the same functional of the $\delta/\delta iJ_i$ as $\hat{A}$ is of the $\hat{\varphi}_i$.

DeWitt introduced in addition to the external source J^i a second c-number field φ_i, a so-called 'classical solution' of the dynamical equations (2.5). $\varphi_i\hat{1}$ will serve as a zeroth approximation to $\hat{\varphi}_i$. Let us consider only the case where $\hat{X}_i{}^j = \hat{1}\delta_i{}^j$. Then the classical solution is defined to satisfy

$$\frac{\delta\hat{S}[\varphi_i\hat{1}]}{\delta\varphi_{\hat{\beta}j}} 1_{\hat{\beta}} = -J_i\hat{1}. \tag{4.3}$$

(For non-trivial $\hat{X}_i{}^j$ one would have to define the 'classical solution' as a q-number such as $\varphi_i Y_{\hat{\alpha}j}^i$, which would raise interesting interpretational problems. This has not been explored to my knowledge.) φ_i is not really a solution of equation (4.3) because $\varphi_i\hat{1}$ will not satisfy any non-trivial commutation relations which, as I have said, must tacitly accompany equation (4.3).

Provided that $\hat{S}$ acquires its operator character solely *via* the $\hat{\varphi}_i$, φ_i will also be the solution of an associated c-number ('classical') variational problem, for the action functional $S[\varphi_i]$ where

$$S_{\hat{\alpha}}[\varphi_i\hat{1}] = S[\varphi_i]1_{\hat{\alpha}}. \tag{4.4}$$

Writing

$$\hat{\varphi}_i = \varphi_i \hat{1} + \hat{\phi}_i \tag{4.5}$$

we now regard $\hat{\phi}_i$ as formally 'small' in a perturbative scheme for solving equation (2.5). The intuitive justification for this is that $\varphi_i \hat{1}$ differs from the true solution only because all the commutators $[\hat{\varphi}_i, \hat{\varphi}_j]$ have been set to zero in solving the dynamical equations (3.3)—and these commutators ought to be proportional to positive powers of $\hbar$. Nevertheless one can at most hope to represent $\hat{\varphi}_i$ by such a perturbation expansion over some finite part of its (usually) infinite spectrum. For the term $\varphi_i \hat{1}$ will not dominate an unbounded operator, however many powers of $\hbar$ the latter contains. However in regimes where $\hbar$ is 'small', we may continue

$$-J^i 1_{\hat{a}} = \frac{\delta S_{\hat{a}}[\varphi \hat{1}]}{\delta \varphi_{\hat{\beta}i}} 1_{\hat{\beta}} + \frac{\delta^2 S_{\hat{a}}[\varphi \hat{1}]}{\delta \varphi_{\hat{\gamma}j} \delta \varphi_{\hat{\beta}i}} 1_{\hat{\beta}} \hat{\phi}_{\hat{\gamma}j} + \frac{1}{2} \frac{\delta^3 S_{\hat{a}}[\varphi \hat{1}]}{\delta \varphi_{\hat{\delta}k} \delta \varphi_{\hat{\gamma}j} \delta \varphi_{\hat{\beta}i}} 1_{\hat{\beta}} \hat{\phi}_{\hat{\gamma}j} \hat{\phi}_{\hat{\delta}k} + \cdots . \tag{4.6}$$

Contrary to first appearances, the operator content of the first three functional derivatives in this expansion, given again that $\hat{S}$ depends on no operator independent of the $\hat{\varphi}_i$, is trivial and *independent of the algebra of the* $\hat{\varphi}_i$. This is important for quantization theory since the expansion is useless after the first term containing a factor-ordering ambiguity. In terms of the classical action $S[\varphi]$,

$$-J^i \hat{1} = \frac{\delta \hat{S}[\hat{\varphi}]}{\delta \varphi_{\hat{\beta}i}} 1_{\hat{\beta}} = \frac{\delta^2 S[\varphi]}{\delta \varphi_i \delta \varphi_j} \hat{\phi}_j + \frac{1}{2} \frac{\delta^3 S[\varphi]}{\delta \varphi_i \delta \varphi_j \delta \varphi_k} \hat{\phi}_j \hat{\phi}_k + \cdots . \tag{4.7}$$

We are now in a position to derive the centerpiece of quantization theory, the Feynman functional integral formula (DeWitt 1965, 1982), not as an exact theorem but as a remarkable approximation to the true theory. The last term in equation (4.7) can be rewritten as

$$\frac{1}{2} \frac{\delta^3 S[\varphi]}{\delta \varphi_i \delta \varphi_j \delta \varphi_k} T(\hat{\phi}_k \hat{\phi}_j) + \frac{1}{2} T\left(\frac{\delta^3 S[\varphi]}{\delta \varphi_i \delta \varphi_j \delta \varphi_k} \theta(j, k) [\hat{\varphi}_k, \hat{\varphi}_j] \right) + \cdots . \tag{4.8}$$

Hence

$$-J^i \hat{1} = T\left(\frac{\delta S[\hat{\varphi}]}{\delta \varphi_i} + \frac{1}{2} \frac{\delta^3 S[\hat{\varphi}]}{\delta \varphi_i \delta \varphi_j \delta \varphi_k} \theta(j, k) [\hat{\varphi}_k, \hat{\varphi}_j] \right) + \cdots . \tag{4.9}$$

But since we are supposing that the Peierls expression for the commutator is correct to leading order in $\hbar$,

$$-J^i \hat{1} = T\left(\frac{\delta S[\hat{\varphi}]}{\delta \varphi_i} + \frac{i}{2} \frac{\delta^3 S[\hat{\varphi}]}{\delta \varphi_i \delta \varphi_j \delta \varphi_k} G^{+kj}[\hat{\varphi}] \right) + \cdots \tag{4.10}$$

where G^{+ij} is the *c-number* advanced Green's function. The various functionals of φ_i in (4.9) and (4.10) are promoted to operators by evaluating them with the $\hat{\varphi}_i$s in any order! Finally we have

$$-J^i = T\left(\frac{\delta}{\delta \varphi_i} \left(S[\varphi] + \frac{i}{2} \ln \det G^+[\varphi] \right) \Big|_{\varphi \to \hat{\varphi}} \right). \tag{4.11}$$

Still following DeWitt we now express $\langle \text{out}|\text{in}\rangle_{J_i}$ as a functional Fourier transform

$$\langle \text{out}|\text{in}\rangle = \int F[\varphi]\exp\left(iJ^i\varphi_i\right)D[\varphi]. \tag{4.12}$$

Functional integration by parts and equations (4.2) and (4.11) give successively

$$\int \left(\frac{\delta}{\delta\varphi_i} F[\varphi]\right)\exp\left(iJ^j\varphi_j\right)D[\varphi] = -J^i\langle \text{out}|\text{in}\rangle \tag{4.13}$$

$$= \left\langle \text{out}\left| T\left(\frac{\delta}{\delta\varphi_i}\left(S[\varphi] + \frac{i}{2}\ln\det G^+[\varphi]\right)\right)\right|_{\varphi\to\hat{\varphi}}\right|\text{in}\right\rangle \tag{4.14}$$

$$= \left[\frac{\delta}{\delta\varphi_i}\left(S[\varphi] + \frac{i}{2}\ln\det G^+[\varphi]\right)\right]_{\varphi_j\to\frac{\delta}{\delta iJ^j}}\langle \text{out}|\text{in}\rangle \tag{4.15}$$

$$= \int \frac{\delta}{\delta\varphi_i}\left(S[\varphi] + \frac{i}{2}\ln\det G^+[\varphi]\right)F[\varphi]\exp\left(iJ^j\varphi_j\right)D[\varphi]. \tag{4.16}$$

It follows that

$$\frac{\delta}{\delta\varphi_i} F[\varphi] = i\frac{\delta}{\delta\varphi_i}\left(S[\varphi] + \frac{i}{2}\ln\det G^+[\varphi]\right)F[\varphi] \tag{4.17}$$

$$F[\varphi] \propto \exp\left(iS[\varphi]\right)\left(\det G^+[\varphi]\right)^{-1/2} \tag{4.18}$$

and

$$\langle \text{out}|\text{in}\rangle \propto \int \exp i\left(S[\varphi] + J^j\varphi_j\right)\left(\det G^+[\varphi]\right)^{-1/2}D[\varphi]. \tag{4.19}$$

CONCLUSION

We have seen how classical theory arises as a zeroth approximation and quantization theory as a first approximation to real quantum theory. How fortunate it is that we can say so much about the approximations to a theory the true nature of which is still so mysterious.

ACKNOWLEDGMENT

This paper is supported in part by NSF grant PHY 7826592

REFERENCES

Deutsch D 1980 *Quantum Theory as a Universal Physical Theory* (University of Texas Report) unpublished
DeWitt B S 1965 *Dynamical Theory of Groups and Fields* (New York: Gordon and Breach)
—— 1967 *Phys. Rev.* **162** 1195
—— 1982 Lecture notes
DeWitt B S and Graham N 1973 *The Many-Worlds Interpretation of Quantum Mechanics* (Princeton: Princeton University Press)
Page D M and Wootters W 1982 *Evolution Without Evolution* (University of Texas Report)
Peierls R E 1952 *Proc. R. Soc.* A **214** 143
Schwinger J 1953 *Phys. Rev.* **91** 713

On Quantum Gravity and the Many-Worlds Interpretation of Quantum Mechanics

LEE SMOLIN

The motivation for the general program outlined is, of course, a desire ultimately to attack the problem of the role played by gravitation in the quantum domain. No apology will be made for this motivation, although needless to say, recent experiments have had nothing to do with it. In the author's opinion it is sufficient that the problem is there, like the alpinist's mountain. Beyond that, however, the historical development of physics teaches a suggestive lesson in this connection, namely, that the existence of any fundamental theoretical structure which is far from having been pushed to its logical mathematical conclusions is a situation which may have great potentialities.
Bryce DeWitt (1957)

1. INTRODUCTION

As the history of the development of classical mechanics shows, the period after the introduction of a new kind of physical theory is marked by a kind of schizophrenia in which energetic development of the theory takes place side by side with impassioned and serious debate over the foundational and epistemological problems raised by the introduction of the new theory. In the case of classical mechanics these problems concerned, among other things, the meaning of absolute space and the use of action at a distance. Perhaps not surprisingly, these questions were not, at the time, resolved, although discussion of them died down after a time while a working consensus was establised which allowed physicists and mathematicians to proceed with the development of mechanics, leaving consideration of the foundational difficulties to the philosophers.

Of course the philosophers could not resolve what were, after all, questions of physics, but they clarified and sharpened the questions and kept discussion of them alive until new physical theories were developed, such as the theory of electro-magnetism and the general theory of relativity, which to some extent resolved them.

Similarly, the introduction of quantum mechanics was attended by much discussion of foundational and epistemological questions. As in the case of Newtonian mechanics, a working consensus, called the Copenhagen or 'orthodox' interpretation, was quickly reached which has allowed physicists to proceed with the development of quantum mechanics without concerning themselves with the foundational difficulties. Thus, with few exceptions, it is true, and has been since the 1930s, that physicists tend either to concern themselves with the problems which are well posed within the accepted framework of quantum mechanics, ignoring the foundational problems, or they become specialists in the problems in the foundations of quantum mechanics and concentrate on the problems there, usually in exactly the same terms that they were first posed in the twenties, without regard to more recent developments.

One of these rare exceptions is Bryce DeWitt who, in addition to his many important contributions to the development of quantum field theory and its application to gravitation, has played an important role in the debate over the foundational problems in quantum mechanics. The particular role that he has played has been as a vigorous advocate of an interpretation of quantum mechanics which has come to be called the many-worlds interpretation. This interpretation was first proposed by Hugh Everett III in 1957, but it did not attract a great deal of attention until it became the subject of several articles published by Bryce between 1967 and 1973. It is perhaps a measure of the effect of Bryce's advocacy that since that time this interpretation has become rather popular among physicists, and in particular among those who, apart from their interest in the many-worlds interpretation, have not been primarily concerned with the foundational problems in quantum mechanics.

However, its popularity is not the only reason for the presentation of a critical study of the many-worlds interpretation at this time. Anyone who has looked over Bryce's work as a whole cannot fail to be impressed by the consistency of Bryce's attitude towards the problem of quantum gravity and how it might be resolved, and by his resulting concentration, from his early papers up to the present, on a single program directed towards the solution of the problem of quantum gravity along the lines envisioned by him. That his advocacy of the many-worlds interpretation is a part of his overall approach to the problem of quantum gravity becomes clear if we consider the main goal which lies behind his approach to the problem of quantum gravity.

This main goal, as I understand it from his writings (1957, 1962, 1967, 1979) and conversations, is to construct, given the mathematical framework of quantum mechanics as it stands and given, for the dynamics of the gravitational field, the unmodified Einstein–Hilbert action, a completely consistent quantum theory of gravity. Moreover, Bryce proposes that this theory will turn out to be completely finite, involving no infinite renormalizations or additional counterterms and also perhaps avoiding the problem of space–time singularities in classical general relativity.

This might be thought of as a rather ambitious program, but it is also the most parsimonious, as it requires us not to invent new theories, but rather to work only with the materials as given to understand whether a consistent theory can be constructed from them and, if so, what that theory says. Further, even if one did not expect it to work there is a methodological argument, sometimes stressed to me by Bryce, that this is still the best program to pursue, because, in the absence of experiment, one is more likely to achieve progress if one works first to understand the solution to the problem as given, before dissipating energy pursuing the many different possible alterations of the problem. Indeed, it is perhaps a tribute to Bryce's insight into the problem of quantum gravity that if a complete solution to the problem has not yet been found following the lines laid out in his work, none of us who have pursued the many deviations from it can claim to have done better.

So we must ask, why are the problems in the foundations of quantum mechanics relevant for this program of constructing a quantum theory of gravity? The reason is that when gravity comes into the picture some of the old solutions to the foundational problems become invalid and new foundational problems are created. Of course the attitude that one takes towards the reawakening of the foundational problems for quantum-gravitational phenomena depends very much on one's over-all attitude towards both the problem of quantum gravity and the foundational problems themselves. For example, if one believes that the foundational problems are indications that the quantum theory as it stands gives an incomplete description of phenomena, then any indications that the theory may need to be modified to incorporate a particular class of phenomena might be considered a hopeful sign. On the other hand if one is interested in bringing gravitation into the framework of quantum mechanics as given, as Bryce seems most definitely to be, then one must find an interpretation of quantum mechanics that resolves both the new and the old foundational problems, even in situations where gravitational effects may be present and even dominant. This, as we shall see, is what the many-worlds interpretation, to the extent that it is satisfactory in all other respects, has a possibility of doing.

In order to set the stage for the consideration of the many-worlds interpretation I begin in the next section with a short discussion of the foundational problems which are raised by attempts to understand phenomena in which both quantum and gravitational effects should be important. I then proceed in the following section to a discussion of the many-worlds interpretation.

2. FOUNDATIONAL PROBLEMS IN QUANTUM GRAVITY

It is possible, if one restricts oneself to asking questions only about the comput-ability of S-matrix elements for the scattering of a small number of quanta in flat, empty, space–time, to do a great deal of work on quantum gravity without running into any problems of principle or difficulties of interpretation. However, there are many other kinds of questions that one can ask in both general relativity and quantum theory, and as soon as one starts to ask *physical* questions about situations in which quantum and gravitational effects are both strong one runs into a number of difficulties and puzzles that touch on the fundamentals of the formalism and interpretation of quantum theory. These include questions about the evolution of cosmologies in which there may be no asymptotically flat regions, questions about the interpretation of measurements made in regions in which the gravitational field is strong and there are no prefered coordinate systems, questions about the evolution and final outcome of evaporating black holes and questions about the actual measurability of the quantum state of the gravitational field even in weak field situations.

Some of these foundational difficulties which are relevant to the considerations of the many-world interpretation are described in what follows.

1) The question of where to stand in quantum cosmology. In some of the interpretations of quantum mechanics, including all of the versions of the so-called orthodox or Copenhagen interpretation, the observer must be placed outside the quantum system under consideration. Further, in the accounts of the relationship between theory and experiment given by these interpretations the process of measurement is not describable by the usual dynamical laws that govern the evolution of quantum systems but is subject to special laws which apply only at the interface between the quantum system and the observer or, in Bohr's account, the quantum and the classical realm. From this point of view it would seem to be impossible to treat problems in cosmology within the context of quantum physics, especially if the universe under consideration is closed, because there is then nowhere the observer may be and still be outside the system. This problem might seem to be only of academic interest were it not for the existence of certain basic problems in cosmology, such as the flatness problem, the horizon problem and the problem of explaining the very high degree of symmetry in the observable universe, the explanation of which, many cosmologists currently believe, may involve recourse to quantum effects.

Indeed, this problem is mentioned in the papers of both Everett (1957) and Wheeler (1957) as one of the original motivations for the many-worlds interpretation.

2) The problem of the measurability of the quantized gravitational field. This problem has a long history: it was raised originally by Rosenfeld (1957) who incorrectly came to the conclusion that an extension of the analysis of Bohr and Rosenfeld (1933) to the gravitational field would show that quantum effects of the gravitational field are not measurable. However this conclusion was corrected by Bryce DeWitt who, having made what is to my knowledge the first translation of the article of Bohr and Rosenfeld into English, performed the analysis for the gravitational field in great detail (DeWitt 1962). His analysis showed that quantum effects in the gravitational field are measurable as long as the test bodies employed are large compared to the Planck dimensions.

More recently, however, a further problem has emerged. An analysis by the author (Smolin 1982a, 1983b) indicates that, unlike the case of the electromagnetic field, it is impossible to construct any device, or combination of devices, by the use of which the quantum state of the linearized gravitational field could be reliably measured. That is, while, as shown by DeWitt, quantum-gravitational effects are in general measurable, there seems to be no way of giving empirical content to the statement, 'the quantum state of the linearized gravitational field is $|\psi\rangle$'. Instead, the most precise specification that can be given for the state of the gravitational field as a result of measurements is in terms of a mixed state, in which probabilities are assigned to different possible pure states.

3) The problem of the loss of information during black hole evaporation. This problem is discussed in a number of different places (Hawking 1976, 1982, Smolin 1982b) and will only be recalled here. The problem is that, according to the arguments of Hawking, it seems likely that information concerning the quantum states of all of the long ranged fields is irreversibly lost during the process of black hole evaporation so that if, before the formation of the black hole, the fields are in a given pure state the best prediction that can be made concerning their state after the black hole has formed and evaporated is in terms of a mixed state which specifies a distribution of probabilities for various states that the field might actually be in.

It should be mentioned that it is possible to imagine scenarios in which the black hole evaporates but the information is not lost. These scenarios depend on certain assumptions about the behavior of the quantized gravitational field in extremely strong fields and in particular depend on the assumption that quantum effects *always* prevent the formation of singularities in space–time.

4) The problem of the lack of a covariant definition of particle number in quantum field theory. Again, this is a problem that has been widely discussed in the literature and will only be mentioned here. For more information the reader may consult Fulling (1973), Unruh (1976), Sanchez (1981), DeWitt (1975) or Smolin (1982b). The problem is that the Fock basis for the Hilbert space of a quantum field, according to which a given state of the field is decomposed into a superposition of states each containing definite numbers of quanta, is dependent upon the choice of a prefered coordinate system. The choice of a Fock basis is invariant under Lorentz transformations but not under any more general set of coordinate transformations. Thus, in the presence of a gravitational field which does not admit the introduction of a prefered global coordinate system there is an ambiguity in how a particular quantum state is to be decomposed in terms of states with definite numbers of quanta or, conversely, in how an observation indicating the presence of a particular collection of quanta is to be represented in terms of a quantum state.

3. THE MANY-WORLDS INTERPRETATION OF QUANTUM MECHANICS

We will now turn from the problems of gravitation to a discussion of the problems concerning the interpretation of quantum mechanics. I will introduce the many-worlds interpretation through a discussion of the problem which was most important to its original motivation: the measurement problem of quantum mechanics. In order that we shall be clear in the following about what is being assumed and what is being argued I begin with a list of those principles of quantum theory which are *not* brought into question in the following discussion.

1) The state of a quantum system, S, is given its most precise description by associating to it a one dimensional subspace $|\psi\rangle$ of a Hilbert space, $\mathcal{H}_S$, $|\psi\rangle$ is called the quantum state of the system, S.

2) The superposition principle. By the linearity of $\mathcal{H}_S$, if $|\psi_1\rangle$ and $|\psi_2\rangle$ are possible states for the system, S, then so is any linear combination, $\alpha|\psi_1\rangle + \beta|\psi_2\rangle$.

3) Observables $\bar{A}$ of the system S are represented by self-adjoint linear operators A on $\mathcal{H}_S$.

4) The possible values which may be obtained by the measurement of the observable $\bar{A}$ are given by the eigenvalues, a_i of the operator A. Further, if and only if a system is in an eigenstate $|a_i\rangle$ of an observable $\bar{A}$, satisfying $A|a_i\rangle = a_i|a_i\rangle$, we say that the observable $\bar{A}$ takes the definite value a_i in the state $|a_i\rangle$.

5) The dynamics of an *isolated* quantum system is given by the Hamiltonian, H, a linear Hermitian operator on $\mathcal{H}_S$, which generates time evolution of the quantum state, $|\psi\rangle$ according to the Schrödinger equation,

$$i\hbar \frac{\partial|\psi\rangle}{\partial t} = H|\psi\rangle. \tag{3.1}$$

The *problem* of measurement arises as follows. Measurement in quantum mechanics is usually thought of as a process which involves an intervention of some macroscopic measuring device into the evolution of the quantum system S. Thus, during the time that the measurement is taking place the quantum system is not an isolated system and its evolution cannot be described by the Schrödinger equation. Now we know from 4) that the result of the measurement of an observable $\bar{A}$ will be one of the eigenvalues of the operator A and thus, in principle, all that we need to know about the measuring device and its interaction with the system is that it is constructed in such a way that it allows us to determine, for each interaction of the measuring device with the system, a unique a_i which can be considered the result of the measurement.

However, suppose that we wish to inquire into the state of the system S after the measurement has occurred. Given the macroscopic nature of the detector and the complexity of its interaction with the quantum system S we might expect that this is, in general, a difficult question to answer. It is then remarkable that in a large number of cases it is observed experimentally that a simple postulate is sufficient to describe the effect of the interaction with the measuring apparatus on the system.

P) After the measurement of the observable $\bar{A}$ the system is found to be in the eigenstate $|a_i\rangle$ of the operator A associated with the eigenvalue a_i which is found to be the result of the measurement.

P, which is called the projection postulate, is confirmed by the observed *repeatability* of measurements on quantum systems. That is, for observables $\bar{A}$ such that $[H, A] = 0$ a series of repeated measurements of $\bar{A}$ are found all to give the same a_i as the result. The only exception is if one interposes into the series the measurement of another observable $\bar{B}$ such that $[A, B] \neq 0$. Indeed, it is because of this observed property of repeatability that the projection postulate plays an important role in establishing the correspondence between quantum theory and experiment in most practical cases.

Of course, given that the measuring device is just a very complicated arrangement of elementary quantum systems, one may think that the projection postulate should be no more than a convenient and approximate device which saves us the trouble of computing, from the laws of physics, the actual effect of the interaction with the measuring device on the quantum system S. In order to realize this idea we must consider that the measuring device, M, is to be included with the quantum system S in an isolated quantum system whose evolution we shall study by means of the Schrödinger equation. To describe the quantum state of such a quantum system, $M + S$, consisting of two subsystems, M and S, we need the following additional principle.

6) The Hilbert space $\mathscr{H}_{M+S}$ associated with the state of a system consisting of two subsystems, M and S, is the direct product of the Hilbert space for the subsystems,

$$\mathscr{H}_{M+S} = \mathscr{H}_M \otimes \mathscr{H}_S. \tag{3.2}$$

As the measuring device is macroscopic we cannot hope to describe its precise quantum state. However, it will be enough if the Hilbert space, $\mathscr{H}_M$, can be decomposed into orthogonal subspaces

$$\mathscr{H}_M = \mathscr{H}_1 \oplus \mathscr{H}_2 \oplus \ldots \tag{3.3}$$

such that, if the system S is initially in an eigenstate of the observable, $\bar{A}$, say $|a_i\rangle$, then the result of the measurement interaction is that the combined system is in one of the states,

$$|\psi_{M+S}\rangle = |M(a_i)\rangle \otimes |a_i\rangle \tag{3.4}$$

where

$$|M(a_i)\rangle \in \mathscr{H}_i. \tag{3.5}$$

Thus, if we know that the system S has been prepared to be initially in some eigenstate of A we can tell the result of the measurement of the observable $\bar{A}$ by which of the subspaces of (3.4) the state of the apparatus falls in after the measurement has been completed.

Now, the measurement problem arises in the following way. By the superposition principle we also need to consider the case in which the system S is initially in a state which is a superposition of the eigenstates $|a_i\rangle$ of the observable being measured,

$$|\psi_S\rangle = \sum_i c_i |a_i\rangle. \tag{3.6}$$

Then, by the linearity of the Schrödinger equation, the state of the combined system of S plus the measuring device after the measurement has taken place will be

$$|\psi_{M+S}\rangle = \sum_i c_i |M(a_i)\rangle \otimes |a_i\rangle. \tag{3.7}$$

This result is problematic for a number of reasons.

1) The result of the interaction of the measuring apparatus and system is not a *measurement* of the observable $\bar{A}$ in the sense that the result (3.7) does not select one eigenvalue a_i which could be called the result of the measurement. This only occurs in the case that the system is known beforehand to be in some eigenstate of the observable $\bar{A}$.

2) This result is apparently inconsistent with the projection postulate, because the result of the measurement on the state of the system S is not to project it into one of the eigenstates of the observable $\bar{A}$.

Instead, the system of S plus the apparatus is in a state which may be described as a superposition of states, each one of which is of the form (3.4) which describes a state in which S is in an eigenstate of A and the state of the apparatus is in the corresponding subspace.

These problems are far more than mere conceptual difficulties because without the ability to secure one particular a_i as the result of each given interaction of the quantum system S with the measuring apparatus and, in most circumstances, without the projection postulate, it is impossible to discuss the correspondence between quantum theory and experiment. Further, as the projection postulate describes what is actually observed the theory cannot be correct if it gives results which are in contradiction with it.

In addition, the result (3.7) is disturbing because it goes against our common sense view of the correspondence between the formalism of physics and our perception of reality. This common sense view may be expressed as follows.

CS) Regardless of whatever difficulties may attend the determination of the states of microscopic systems we directly perceive, at least to some degree of approximation, the values of some observables associated with macroscopic objects, for example the approximate position of macroscopic objects and the readings on macroscopic pointers, dials and video displays. These observables are always perceived to have definite values. Furthermore, whatever difficulties attend the description of the states of microscopic systems we should expect that a correct theory should give us a description of the states of many simple macroscopic systems that is in agreement with our direct perception of them.

From these apparent difficulties and contradictions most accounts of the connection between experiment and the formalism of quantum mechanics draw the following conclusions,

i) From 1) and 2) it follows that the process of measurement, by which is meant any process by which the experimenter obtains definite values for the measurement of observables of quantum systems in order to compare them with the results of calculations, cannot be described entirely within the framework of quantum dynamics itself. The process of measurement must be seen as an interruption of the evolution of an isolated quantum system and cannot be described by the incorporation of the measurement system into a larger quantum system.

ii) Further, given i), the projection postulate, which has empirical content and is verified experimentally, cannot be derived from the basic principles of quantum mechanics, 1) to 6) above, but must remain an independent postulate of the theory.

This remains true no matter how detailed an understanding of the physics of the measurement apparatus we may have.

In addition, from the apparent contradiction of the description of the measurement process by quantum dynamics with the common sense view one is tempted to conclude something like,

iii) The states of macroscopic systems are not described by quantum mechanics. In particular the superposition principle, while necessary for the description of microscopic systems seems to contradict the evidence of our senses when applied to macroscopic systems.

A large number of different responses have been offered to this situation, however, it is not within the scope of this article to go into them. (The interested reader is refered to the comprehensive discussion in the book of Jammer 1974.) Instead we turn directly to a discussion of the many-worlds interpretation, which begins with the following assertons.

M1) As macroscopic systems are just large collections of microscopic quantum systems quantum mechanics should apply to microscopic systems as it does to each of their subsystems and constituents.

M2) There can be no *a priori* physical distinction between interactions that are used for the purposes of measurement and any other interactions between physical systems, hence the analysis above leading to equation (3.7) must be correct.

These two assertions seem to contradict 1) the projection postulate, 2) the assertion that the result of a measurement is some single definite value, 3) the assertion that the observables of macroscopic systems always have definite values and 4) the assertion that our direct perception of macroscopic observables, for example, the words in front of me as I write this, should be simply correlated with the states of the relevant macroscopic systems. The beauty of the many-worlds interpretation, whether one agrees with it or not, is that it is possible to deny all four of these apparently irrefutable assertions and still give an account of quantum mechanics that allows us to compare the results of quantum theory with experiments.

The literature on the many-worlds interpretation is not large. It consists of the original papers by Everett (1957, 1973) together with a comment by Wheeler (1957). There are then a number of papers by Bryce DeWitt (1967, 1970, 1971) and the paper of Graham (1973), who was a student of DeWitt. In addition some of the same ideas were advanced independently by Cooper and Van Vechten (1969), who, however, do not agree with the most radical conclusion of the proponents of the many-worlds interpretation. These papers are all found in a book, edited by DeWitt and Graham (1973). In addition the central theorem which makes possible the formulation of the many-worlds interpretation was independently found by Hartle (1968), each of whom, however, advocate an interpretation quite different from the many-worlds interpretation.

To my knowledge, the literature critical of the many-worlds interpretation is even slimmer: it consists of the replies to Bryce's *Physics Today* article, (Ballentine *et al* (1971)) and an article by Ballentine (1973). Bell (1981) argues for an understanding of the many-worlds interpretation as a modification of the de Broglie–Bohm pilot wave theory.

In spite of the small number of papers concerning the many-worlds interpretation, reading this literature can be somewhat confusing for the reader who first approaches the subject. This is for two reasons: first, because, as in the case with some of the other accounts of the relationship between experiment and the formalism of quantum mechanics, the principal authors, Everett, Wheeler, Graham and DeWitt make somewhat different assertions. (Although in this they are much better than the authors of the Copenhagen or 'orthodox' interpretation, who on examination are found to be proposing very different and contradictory interpretations.) Second, one finds in each of the papers statements of very different kinds together in the discussions. There are straightforward deductions of new results within the formalism of the quantum mechanics. There are also statements of a very broad and philosophical character. Finally, there are assertions of relationships of statements of the second kind to those of the first kind. In order to present the situation in as clear a light as possible I will proceed by discussing first the two straightforward consequences of the basic formalism that are found in the papers on the many-worlds interpretation but which have a validity independent of any particular interpretation of the formlism.

The first of the new results concerning the formalism of quantum theory is that the repeatability of quantum measurements can be deduced, without use of the projection postulate, from a new and perhaps less drastic postulate concerning the relationship between the formalism and experiments. This postulate might be called the postulate of *contingent correlations.*

C) If S is a quantum system consisting of two subsystems S_1 and S_2 such that

$$\mathcal{H}_S = \mathcal{H}_{S_1} \otimes \mathcal{H}_{S_2} \tag{3.8}$$

then given any basis of states $|a_i\rangle \in \mathcal{H}_{S_1}$ for the system S_1 any state $|\psi_{1+2}\rangle \in \mathcal{H}_S$ for the combined system S may be decomposed as

$$|\psi_S\rangle = \sum_i |q_i\rangle \otimes |S_2(q_i)\rangle. \tag{3.9}$$

Here, $|S_2(a_i)\rangle$ is called the state of S_2 relative to the state $|a_i\rangle$ of S_1. The postulate of contingent correlations is then that the state of the form of equation (3.9) is to be interpreted as the conjunction of the *contingent* statements that *if S_1 is found by experiment to be in the state $|a_i\rangle$, then* the system S_2 will be found in the state $|S_2(a_i)\rangle$.

Of course, what this postulate does not do is to give us an account of how one associates with the result of an experiment the statement that a system is in a given state. But as such statements are instructions for how one makes a correspondence between a mathematical formalism and actual experiments this notion must remain undefined within the formal structure of any given theory. (This is, instead, a task for the interpretation of the theory, as will be discussed in the next section.) What this postulate does do is enable us to derive the repeatability of quantum measurements without use of the projection postulate. For it follows that if we introduce a second measuring device M' that also performs a measurement of the observable $\bar{A}$ on the system S we will have, by linearity, if the initial state of S was an eigenstate of A, that the state of the system $S + M + M'$ after the measurement will be

$$|\psi_{M'+M+s}\rangle = |M'(a_i)\rangle \otimes |M(a_i)\rangle \otimes |a_i\rangle \qquad (3.10)$$

where $|M'(a_i)\rangle$ is defined analogously to equation (3.5). It then follows, again by linearity, that if the initial state of S was a superposition of eigenstates of A as in equation (3.6) that the final state of the system $S + M + M'$ will be in the state

$$|\psi_{M'+M+s}\rangle = \sum_i c_i |M'(a_i)\rangle \otimes |M(a_i)\rangle \otimes |a_i\rangle. \qquad (3.11)$$

By the postulate C we can deduce from this the statement that *if* the system M is found to be in the state $|M(a_i)\rangle$, which implies that S is in the state $|a_i\rangle$, *then* the second system M' will be found in the state $|M'(a_i)\rangle$, which corresponds to S being in the same state a_i. But this is just a statement of the repeatability of quantum measurements.

Thus, given the postulate C we can derive the property of repeatability of quantum measurements without the need for the projection postulate P. Thus we can replace the projection postulate P with the postulate C which, while it makes reference to observation, does not in any way distinguish between measurement interactions and any other interaction between quantum systems. Thus, given C, the observed repeatability of quantum measurements can be consistent with the two assertions, M1 and M2 above.

Of course, if we are able to make the statement that the measuring device M is definitely in one of the states $|M(a_i)\rangle$ then we can conclude that all future measurements made on the system $M + S$, or just on the system S, will turn out as if the projection postulate were true. In this sense the projection postulate can, if one wishes, be recovered from the postulate C together with an assertion that the measuring instrument is in a definite state. The postulate C is, however, weaker than the projection postulate in that it does not contradict the use of quantum dynamics to describe the combined system $M + S$. (Indeed it requires it.) At the same time C is completely sufficient for the comparison of quantum theory with experiment.

Many readers will no doubt have noticed that one important principle was omitted from the list of basic principles of quantum mechanics given above. This is the statement of the probability interpretation.

PI) Given an ensemble of N systems identical to S which are all prepared in the same given state $|\psi\rangle$, then if one makes on each of the systems a measurement of $\bar{A}$ the frequency with which a given result, a_i, appears in the list of results is given by,

$$|\langle\psi|a_i\rangle|^2. \qquad (3.12)$$

It is by means of this postulate that the notion of *probability* is introduced into the quantum theory. Notice that the particular notion of probability which is introduced in this way and, indeed, the only notion of probability which is relevant for the connection between the formalism of quantum mechanics and experiment, is the notion of probability as relative frequency.

The second technical result involved with the many-worlds interpretation is the discovery, first by Finkelstein (1963), and later by Graham (1973) and Hartle (1968), that the postulate PI is not independent of the previously given basic principles of quantum mechanics 1) to 6), but in fact can be derived as a consequence of them. In order to do this Graham and Hartle explicitly introduce an operator which corresponds to the measurement of relative frequency for the outcome of a given measurement on a quantum system. To do this let us assume that we have N copies of a system S, denoted by $S^N = S + S + S + \ldots$, which we can all prepare in the same state $| \psi \rangle$. The state of S^N is then given by

$$| \psi^N \rangle = | \psi \rangle \otimes | \psi \rangle \otimes \ldots \otimes | \psi \rangle. \tag{3.13}$$

$$\underbrace{\hspace{4cm}}_{N \text{ times}}$$

Let us suppose that we make a measurement of the observable $\bar{A}$ on each of the N copies of S. Now we want to construct an operator F^k which will measure the relative frequency with which the particular value a_k is found in the collection of results of these N experiments on systems prepared in the identical states $| \psi \rangle$. In order to do this we cannot, of course, use the statement of the probability interpretation PI, which we are trying to derive, but we can only use the principle 2) that the possible values for the result of a measurement of F^k must be one of its eigenvalues. F^k will be defined to act on states of the form $| \Psi \rangle = | \psi_1 \rangle \otimes | \psi_2 \rangle \otimes \ldots \otimes | \psi_N \rangle$. Such states should certainly be eigenstates of F^k if each of the states $| \psi_\alpha \rangle$ is one of the eigenstates of A, for $\alpha = 1, \ldots, N$ with the eigenvalue being the number of times the particular eigenstate $| a_k \rangle$ occurs in $| \Psi \rangle$. By linearity, the operator F^k is then determined to be

$$F^k = \sum_{i_1 \ldots i_N} | a_{i_1} \rangle \otimes \ldots \otimes | a_{i_N} \rangle \left[\sum_{\alpha=1}^{N} \frac{\delta_{k i_\alpha}}{N} \right] \langle a_{i_N} | \otimes \ldots \otimes \langle a_{i_1} |. \tag{3.14}$$

We can now state the theorem of Finkelstein. For the state $| \psi^N \rangle$ defined above one can show that

$$\lim_{N \to \infty} F^k | \psi^N \rangle = | \langle a_k | \psi \rangle |^2 | \psi^N \rangle \tag{3.15}$$

where $\langle a_k | \psi \rangle$ is the inner product in $\mathcal{H}$. Thus, in the limit in which an experiment is performed on an infinite number of systems prepared in identical states any state $| \psi^N \rangle$ is an eigenstate of the relative frequency operator F^k with eigenvalue $| \langle a_k | \psi \rangle |^2$. Thus, the statement of the probability interpretation PI has been derived from the other principles of quantum mechanics. The reader is encouraged to refer to the papers of Finkelstein (1963), Graham (1973) and Hartle (1968) for the derivation of equation (3.15).

The two results we have just discussed, the derivation of the repeatability of quantum measurements from the postulate of contingent correlations C and the derivation of the probability interpretation PI from the other axioms in quantum theory are straightforward results in quantum theory. They do not, in themselves, constitute or indicate any particular interpretation of quantum mechanics or account of the relationship between the results of experiments and the formalism of quantum theory. They do make possible a number of different interpretations, as we shall now discuss.

The simplest interpretation of quantum theory which is made possible by these two results I shall call the minimal relative state interpretation.

MRS) The quantum state of a composite system is to be interpreted as corresponding, not to the actual state of the system, but only to the list of contingent statements generated from the state according to C above. In particular, quantum states of individual systems have no empirical content. Correspondence with experiment is to be achieved by explicitly incorporating the measurement apparatus into the quantum state. In this way the predictions of quantum mechanics can, using C and the Schrödinger equation, be expressed as contingent statements refering to the states of macroscopic preparations and measuring apparatuses at different times.

Here the definition of a macroscopic system is a system sufficiently large and well defined that the values of some of its dynamical variables can be ascertained by us directly by looking at them. This definition is admittedly informal, but as some kind of informal definition is needed at some stage in the process to connect the formal mathematics with real experiments that people do, and as this definition is consistent with our experience that the dynamical variables of some large objects have definite values that we can ascertain by looking at them, nothing more complicated than this is needed.

As far as statements about probabilities for various outcomes to occur, we shall require that such statements only be made about measurements made on ensembles of identical systems, in which case the notion of probability will be explicitly introduced in terms of measurements of relative frequency, as constructed above. Probabilistic statements cannot be made concerning the outcome of single measurements on individual systems.

The important point about this minimal relative state interpretation is that it is sufficient to allow us to compare computations done in quantum theory with the results of any possible experiments which might be made. However, because only contingent statements are made, and not any assertions concerning definite states of affairs, and because statements about probability can only be introduced for systems which are ensembles of identical systems, some of the various difficulties with the interpretation of quantum theory are avoided. In particular, because the connection with measurement is made through the postulate of contingent correlation, C, rather than through the projection postulate, P, the measurement problem is avoided.

If one takes the view that physics is only about the correlation of observations made on macroscopic measuring apparatus than this minimal relative state interpretation is perhaps the most satisfactory of the available possibilities. In particular, it is perhaps superior to Bohr's in that it avoids the necessity of drawing a line which divides the quantum realm from the classical realm. Quantum dynamics, as given by the Schrödinger equation may be applied to macroscopic as well as to microscopic variables, and to the interaction between them. This interpretation requires instead the more informal, but obviously true, assertion that there are some observables in the world which take definite values that are directly apparent to our senses. This suffices because, according to this interpretation, quantum theory is not about statements that in particular instances such observables are seen to take one particular value or another, but is only about the contingent statements which related different possible outcomes of these observations to each other.

Another interpretation in which these technical developments play a role, and in particular the reduction of the probability interpretation to non-probabilistic axioms, is that given in the paper of Hartle (1968), in which these results are used to argue for the conclusion that the quantum state is not an objective correlate of the actual physical state of microscopic systems.

I mention the existence of these other interpretations to stress that these technical developments, which do make the many-worlds interpretation possible, do not particularly lead to or call for the many-worlds interpretation. What leads to the many-worlds interpretation are the assertions that we began with, M1 and M2, together with these technical developments, together with the following assertion.

RQS) The actual, objective, state of affairs of a given individual quantum system is exactly represented by its quantum state. In this assertion is contained the assumption that there is an objective state of affairs that can be said to constitute the reality of an individual quantum system at each given time and that the possible such states of affairs for a given system S can be put in direct one-to-one correspondence with the one dimensional subspaces of a given Hilbert space, $\mathscr{H}_S$. It is beyond the scope of this article to make a deep inquiry as to how these assertions are meant, suffice it to say that they are meant in exactly the same way it was meant in the late nineteenth and early twentieth centuries that the molecules in a monatomic gas are at a given time in some actual objective state, which is exactly given by a point in the phase space of the gas.

This shall be called the postulate of the reality of the quantum state, or RQS. Before proceeding it is important to make the point that this postulate cannot in any way be derived from the formalism of quantum mechanics, or from any *a priori* or methodological analysis concerning how quantum mechanics is used. Its independence from both the formal structure of quantum mechanics and the practice of the use of quantum mechanics by real physicists is clear from the fact that it is contradicted by many of the interpretations of quantum mechanics which have been offered, and which are themselves completely consistent with both the formal structure and the use of quantum mechanics.

However, if we presume RQS, together with M1 and M2 we may draw the following conclusions. Recall that after the interaction of a microscopic system S with the measuring system M the state of the total system $M + S$ is of the form of equation (3.7). According to RQS the physical reality of the system $M + S$ is exactly described by this quantum state (3.7). Recall that by postulating C we found that (3.7) could be interpreted as a collection of contingent statements of the form, '*If* the apparatus is in the state $|M(a_i)\rangle$, *then* the system is in the state $|a_i\rangle$'. By RQS we must interpret equation (3.7) as being, instead, a description of the actual reality of the system $M + S$. In particular there is no reason why any particular element of the superposition in equation (3.7) could be construed as having a privileged role in the working out of this description. However, we know, from our own experience, the kind that we use in verifying the contingent statements that C asserts we can make, that we observe one and only one of the contingent statements allowed by C as the true state of affairs. There is thus an apparent contradiction between RQS and our observation of the true state of affairs for M. If we hold to RQS then we must draw the conclusion that our experience does not indicate the actual state of affairs for the system $M + S$. Instead, whatever interpretation we may give to the particular elment of the superposition (3.9) that is associated with the single outcome picked out by our observation of M, the same interpretation must be given to all of the elements of the superposition (3.9).

This is then the consequence of holding RQS together with M1 and M2. Whatever description we may give to a particular element of the superposition, based on our observations, must be given to all of the elements of the superposition. In particular, if we wish to regard our having observed a particular a_i to be the outcome of the experiment as an actual event in the world then it follows that we must also regard our having observed each of the other possibilities as also being actual events in the world. In this way we see that the many-worlds interpretation is the view which follows from the simultaneous assertion of RQS, M1 and M2.

At this point one is tempted to conclude that it should be possible to prove that RQS and M1 and M2 cannot all be simultaneously true. However, any attempt to carry this out is eventually based on an assertion that there is only one of a particular individual, or at least of the experience or observations of that individual in the world at a particular time. This is, from most points of view, a reasonable assertion. However, the proponents of the many-worlds interpretation choose to maintain these postulates and instead deny that there can be in the world at any one time only one of each individual, or his or her experience. Instead they choose to assert that to the extent that I, or my experience or observation, of a particular outcome may be considered real or actual, so must I consider all the other possible I's observing all of the other possible outcomes. Briefly, to the extent there is a contradiction between the postulates RQS and M1 and M2 on one hand and the evidence of my senses concerning the state of reality of the world around me the proponents of the many-worlds interpretation choose to assert the former and deny the latter.

This is not to say, as the proponents of the many-worlds interpretation are quick to point out, that there is a direct contradiction between my experience, *per se*, and the assertion of RQS, M1 and M2. By an extension of the kinds of arguments given above to derive the repeatability of quantum measurements from the postulate of contingent correlations, C, one can show that each of the individuals into which I have become have no communication with any of the others, and indeed, no way of detecting the existence of the others. What is contradicted is any assertion that what I see around me has any direct correlation with the actual state of affairs of the world, which was expressed by the common-sense postulate CS above.

At this point it is tempting to take the discussion onto levels ontological, metaphysical, methodological, or perhaps just plain metaphorical, to consider the question of whether one should believe these conclusions of RQS or not. However, I will confine myself, for the time being, to the assertion that, given the existence of the minimal relative state interpretation, MRS, which agrees with all of the conclusions of the many-worlds interpretation besides those that follow from affirming RQS (in particular in that it allows us to assert M1 and M2 and thus resolve the measurement problem) the many-worlds interpretation is certainly not forced on us by anything other than the desire to take the reality of the quantum state, together with the superposition principle, seriously in the macroscopic domain. Thus, if one is going to take the obviously radical step of denying CS and accepting the conclusions necessitated by holding simultaneously RQS, M1 and M2 one should have compelling reasons. In order to see just what the reasons are that one might have for taking this step I will, in §5, examine the relative advantages of the many-worlds interpretation and the minimal relative state interpretation, particularly with respect to the problems involving gravitation which were mentioned in §2. However, before we come to this we must examine, in more detail, how the correspondence between theory and observation is made in each of the interpretations we have discussed here.

4. THE ROLE OF OBSERVATION IN THE MANY-WORLDS AND IN THE MINIMAL RELATIVE STATE INTERPRETATIONS

An interpretation of quantum mechanics, besides resolving, or at least making us comfortable with, the various well known difficulties in the foundations of quantum mechanics, must give a complete and satisfactory account of how the correspondence is made between calculations which are carried out within the formalism of quantum mechanics and measurements which are performed by real physicists. This is the task of the present section.

We shall consider first the account of measurement given by the minimal relative state interpretation. Recall that according to the minimal relative state interpretation a quantum state for a composite system, of the form

$$|\psi_{1+2}\rangle = \sum_i c_i |a_i\rangle \otimes |a_i'\rangle \qquad (4.1)$$

is to be interpreted only as standing for a list of contingent statements of the form 'If observable A of system 1 has the definite value a_i then an observable A' of system 2 has the definite value a_i'.' In order to complete the description of the correspondence between quantum theory and experiment the minimal relative state formalism must assume that there exist some observables for which it is possible to make assertions of the form 'The observable O has the definite value x.' Included in this class are assumed to be observables giving us the position of dials and pointers and the images on CRT screens of measuring devices that are read by experimental physicists. Given the existence of some observables about which it can be asserted that they take definite values in at least some situations, the minimal relative state interpretation tells us how the results of quantum-mechanical calculations can be compared with experiments.

Notice that the minimal relative state interpretation does not give us any indication of why such observables exist, or what distinguishes them from other observables. However, as the interpretation asserts that the role of quantum theory is only to supply us with contingent statements concerning correlations between observables then it need not be considered a drawback of the interpretation if it cannot answer these kinds of questions. At the same time there is, of course, no obstacle to trying to construct an explanation, within the quantum theory itself, of why certain kinds of observables, for example observables of certain kinds of macroscopic systems in thermal equilibrium, might have special properties. But such an account of what distinguishes the observables of macroscopic measuring devices from other kinds of observables is not necessary for the completeness of the minimal relative state interpretation. Furthermore, as under the minimal relative state interpretation quantum mechanics makes only contingent assertions, there can be no harm in appending to it statements asserting that a certain observable is taking a certain value, as long as such assertions do not logically contradict any of the contingent assertions following from the quantum theory.

The situation is, however, not so simple in the case of the many-worlds interpretation. Recall that the many-worlds interpretation is essentially equivalent to the minimal relative state interpretation together with the assertion of the reality of the quantum state, RQS above. According to the RQS the quantum state represents not contingent propositions, but all true assertions concerning the actual physical state of the physical system in question. In particular, if the process in question is a measurement process then both the measuring instrument and the experimenter must be included in the quantum state description, if the physics is to explain the correlation between the state of the microscopic system and what the experimenter sees looking at the measuring instrument. Thus, in the many-worlds interpretation it is not possible to append to the quantum state description any statements asserting that particular observables are taking particular definite values. As we noted above the many-worlds interpretation achieves this by asserting that *all* of the statements of the form 'The observable O of the measuring apparatus takes the value x', or even 'The experimenter perceives that the measured value is x' are true (so long as they are possibilities in the sense that they are the consequence of some contingent statement that would be read from the quantum state by the minimal relative state interpretation).

How then does one actually implement the correspondence between experiment and the results of calculations in the many-worlds interpretation? While the minimal relative state interpretation tells us that the state for a composite system, of the form (3.11), can be read as a series of contingent statements, each of the form '*If* the observable A of the system S has the value a *then* the observable M of the measuring instrument has the value $M(a)$', the many-worlds interpretation asserts that each of the elements of the superposition is to read as a statement of the form 'The observable A of S has the value a and the observable M has the value $M(a)$.' Then the many-worlds interpretation asserts that the conjunction of all of these statements is true.

(One usually introduces at this point in the discussion the expression that the different statements are all true, but each is true of a different 'branch' of the wavefunction. While this is a convenient way of speaking it is not necessary. The statement that the wavefunction can be considered as having different branches is equivalent to the statement that while the wavefunction can be interpreted as implying the conjunction of statements that each assert that a given observable is taking different particular values, the rule for the construction of these statements will never force us to a statement that we, or a second measuring device acting as our surrogate, observe that a particular observable is taking on different values simultaneously. The disadvantage of using the language of branches is that it suggests that some kind of dynamical mechanism is taking place, in addition to the evolution of the wavefunction, in which, as time goes on and initially isolated systems come together and interact, the universe is 'splitting' into more and more 'branches'.)

In the case that an experiment consists of a number of devices for preparation and measurement of the quantum system then the quantum state in question will be a composite state describing a number of different subsystems and each element of the superposition will be interpreted as a complicated statement of the form 'The observable A of S has the value a and the observables M, M', M'' etc of the various measuring instruments have the values $M(a)$, $M'(a)$, $M''(a)$ etc.' The many-worlds interpretation then asserts that the conjunction of all of these statements, one for each element in the decomposition of the wavefunction, will be true.

Now the result of an experiment is a statement of the form 'The values of the various observables M, M', M'' etc are x, x', x'' etc.' Thus it is clear that given the result of a measurement, together with the conjunction of statements that, given the quantum state, the many-worlds interpretation asserts are true, we can ask if the result of the experiment agrees with, or contradicts, any of the elements of the conjunction of statements which correspond to the given quantum state. In this way calculations in quantum mechanics can be compared with experiment.

All of this is fine so long as attention is restricted to the different possible values that a commuting set of observables could have. However, how are we to deal with the fact that the state function can be decomposed with respect to the eigenstates of different sets of operators which do not commute with each other? For example we can say, according to the many-worlds interpretation, 'The spin of the silver atom is up and the spin of the silver atom is down.' Both statements are deemed to be true, but are asserted to correspond to 'different branches of the universe'. Similarly, we can also say 'The spin of the silver atom is left and the spin of the silver atom is right.' However, we cannot say, 'The spin of the silver atom is up and it is down and the spin of the silver atom is left and it is right.' Thus if, according to the many-worlds interpretation, the quantum state is to be considered as corresponding to a conjunction of true assertions, each corresponding to the values that a particular observable takes in different 'branches', such a description must be made with respect to a given set of commuting observables, and is not consistent with the description which is generated by an incompatible set of commuting observables.

Thus, if it is meaningful to make assertions of the form 'The observable A of S has the value a while the observable M of the measuring system has the value $M(a)$ *and* the observable A of S has the value b while the observable M of the measuring apparatus has the value $M(b)$' such statements must be made with respect to a given decomposition of the state function with respect to a given commuting set of operators and are not consistent with other statements derived from other incompatible sets of operators. Thus if the many-worlds interpretation is to assert that statements not only of the form 'the quantum state of the system is $|\psi\rangle$' are true, but also that statements about certain observables taking certain values are true (granted that statements that assert that the observables also take different values are also true) then it must be assumed, at least implicitly, that a particular choice of basis has somehow been picked out.

Of course, one possible response is that quantum mechanics makes no such assertion concerning observables taking particular values, but only makes assertions concerning which quantum state a particular system is in. The difficulty with this is that experiments never result in statements of the form 'The quantum state of the system is such and such'. Real experiments only result in statements of the form that certain observables are taking certain definite values, and if the many-worlds interpretation does not allow us to deduce from the quantum state statements which are also of this form it does not allow us to compare the results of calculations with experiments.

Thus, to compare quantum mechanics with experiments, the many-worlds interpretation must allow us to assert, given the quantum state of a system, statements which are the conjunction of statements in which certain observables are asserted to take certain definite values. And in order to do this a particular privileged basis for the Hilbert space must be chosen.

We must then ask how this basis is to be chosen in particular cases. In order to be able to compare the theory with experiment it is clear that the operators with respect to which the prefered basis is chosen must include the observables of the measuring instruments which are being used in a particular experiment. Further, if we wish to compare the theory with our everyday observations of the world it seems that these observables must include many of the macroscopic observables for the objects around us.

It is certainly possible to choose such a basis in any given situation. However, in doing so we must ask two questions. First, does the theory itself give us any indication that this basis is to be prefered to any other, or is the choice of this basis, which is necessary to compare the theory with experiment, an additional, *ad hoc* assumption which is being added to the theory from outside of it? Second, if the latter is the case, are we in this way restoring to macroscopic measuring instruments and their observables the privileged role the elimination of which had been one of the goals of setting up the many-worlds interpretation?

These questions are not addressed in the literature on the many-worlds interpretation, (although a related point was raised in the paper of Ballentine (1973)) and I am not going to attempt to give a definitive answer to them here. However, the situation, as it stands, is that the many-worlds interpretation cannot give a satisfactory account of how calculations in quantum theory may be compared with experiments unless there are appended to it additional *ad hoc* hypotheses which restore the privileged role which observers and measuring instruments play in most of the other interpretations of quantum theory.

5. ADVANTAGES OF THE MANY-WORLDS INTERPRETATION

As we saw above, the many-worlds interpretation differs from the strictly operational and much more modest minimal relative state interpretation by the assertion of the postulate of the reality of the quantum state. Correspondingly, we shall first list here the advantages of the many-worlds interpretation which are shared with the minimal relative state interpretation (and might perhaps be shared by other interpretations that assert that quantum dynamics is applicable in the macroscopic domain but deny the reality of the quantum state). After this, we shall list the advantages of the many-worlds interpretation that depend on the assertion of the reality of the quantum state, together with the assertion of the applicability of quantum theory in the macroscopic domain, and thus are advantages only of the many-worlds interpretation.

Advantages shared by the many-worlds interpretation and the minimal relative state interpretation

1) The projection postulate is not needed for constructing the correspondence between the results of calculations in quantum theory and experiment.

2) The notion of probability need not be introduced in the statement of the basic principles of quantum mechanics, but emerges naturally when one considers explicitly the quantum-mechanical description of systems containing many identical copies of a given system.

3) Quantum kinematics and dynamics, including the quantum state description and the Schrödinger equation, are applicable equally to microscopic and macroscopic systems, as well as to the interactions between them.

4) The concept of measurement does not play a fundamental role in the formulation of the dynamics of the theory. In particular, the interaction of microscopic systems with macroscopic measuring apparatuses can, and indeed must, be treated by the Schrödinger equation. Collapse of the wavefunction need never be invoked.

At the same time, the concept of measurement does come in to each of the interpretations in an apparently irreducible way at the level of the correspondence between the formalism and observations. The minimal relative state interpretation assumes that there exist certain special observables about which we can assert from observation that they take definite values. In order to connect the formalism to experiments the many-worlds interpretation requires that a prefered basis be chosen, with regard to which true assertions can be made. However, it may be that this is preferable to bringing in the notion of measurement at the level of kinematics or dynamics.

5) Since both the many-worlds interpretation and the minimal relative state interpretation allow the observer to be treated as a part of the quantum system they allow problems in cosmology, in which the whole universe is included in the physical system, to be treated by quantum mechanics.

Advantages of the many-worlds interpretation not shared by the minimal relative state interpretation

1) By affirming the principle of the reality of the quantum state, RQS, unambiguously for macroscopic as well as microscopic physics the many-worlds interpretation restores to quantum physics the classical notion that the mathematical elements of a physical theory should be in one-to-one correspondence with the elements of physical reality. (For a nice formulation of this notion see Einstein, Podolsky and Rosen (1935).)

2) Similarly, as the Schrödinger equation is a deterministic equation for the quantum state, by affirming RQS, as opposed to, for example, affirming the reality of particles in quantum physics, the many-worlds interpretation restores to quantum physics the classical notion that physics should give a deterministic description of the evolution of physical reality.

3) We saw above that both the many-worlds interpretation and the minimal relative state interpretation allow the whole universe, including the observer, to be treated as a quantum system. However, the minimal relative state interpretation does require that somewhere in the universe there be macroscopic objects and observers who can ascertain the values of at least some of the observables of the macroscopic systems. Without the presence of observers and macroscopic systems the minimal relative state interpretation has no way to give meaning to the strictly contingent statements that according to it are the only content of the quantum state. Thus the minimal relative state interpretation cannot be used to study the cosmology of the very early universe, or any other situation where the physical conditions prevent the existence of observers. Of course one could attempt to get around this problem by postulating the existence of some hypothetical kind of material out of which observers could be constructed even in these extreme regimes. But the many-worlds interpretation, by simply asserting the RQS, does not have this problem and thus allows us to meaningfully describe, the formalism of quantum mechanics, situations in which it is hard to imagine the presence of observers.

4) The many-worlds interpretation allows the resolution of the problem of the ambiguity of the definition of the particle content of the quantum state, mentioned in our list of foundational problems involving quantum mechanics and gravity. As long as there is no ambiguity in the definition of particle number one can choose to believe either in the reality of the quanta themselves (as for example localized particle-like entities that obey non-classical, non-local, dynamical laws), which are what are actually observed in experiments, or in the reality of the quantum state itself, believing that the quanta are only artifacts of the measurement procedure. However, once one considers the description of quantum states in regions of space–time where there is no prefered coordinate systems to pick out a natural definition of particle number, one has to make a choice. One can take a more operational view and continue to believe in the reality of the quanta, in which case one has the problem of the lack of a covariant distinction between quantum and thermal effects. Or, on the other hand, one could choose to believe in the reality of the quantum state, denying the fundamental importance of any particular represen-

tation in terms of operators for definite numbers of quanta. By asserting the RQS the many-worlds interpretation makes it possible to take the latter course.

5) Similarly, by asserting the RQS the many-worlds interpretation resolves the difficulty of the impossibility of determining by experiment the precise quantum state of the gravitational field. For if one believes in the RQS then one can assert that the impossibility of determining experimentally which quantum state the gravitational field is in does not prevent one from asserting that there is a definite quantum state that the field is in. The advantages and disadvantages of being able to do this are perhaps not unlike the advantages and disadvantages of those interpretations of quantum mechanics that assert the reality of the electron orbit even after Heisenberg's Uncertainty Principle.

Thus we see that, while the many-worlds interpretation has a number of advantages, all of those having to do with the traditional problems in the foundations of quantum mechanics, the measurement problem, the role of the projection postulate, the role of probability in quantum mechanics, etc, are shared with the minimal relative state interpretation. The advantages which are enjoyed uniquely by the many-worlds interpetation are those which depend on the ontological postulate RQS. The advantages of asserting RQS are, first, that it allows us to maintain a belief in the determinism and one-to-one connection with physical reality of physical theory and, second, that it allows us to avoid certain puzzles involving gravitation and quantum theory which, under a different interpretation of quantum mechanics, would continue to trouble us.

6. CONCLUSION

In order to put this, or any discussion of foundational problems in physical theory, in the proper light it is important to recall the larger task that we, as physicists, are engaged in and how discussions of foundational problems fit into it. The particular task that we are concerned with here is the construction of a theory which incorporates both quantum and gravitational phenomena. Now it is possible that this can be achieved by a straightforward development within the present formal structure of quantum mechanics and it is possible that it cannot be achieved without a radical revision in our basic ideas which would lead to an overthrow of the quantum theory. At present we must say that we simply do not know how it will actually turn out, if we did we would have already solved the problem.

At the same time it is necessary to have a point of view in order to choose a direction in which to proceed, especially when faced with such a difficult problem. In this kind of situation an interpretation of a theory under development is not a passive thing, but rather plays an important role in the evaluation of what are important research problems and productive directions to pursue. In the particular case of the problem of quantum gravity an interpretation which asserts that the quantum theory is essentially correct as formulated, and without any unresolved foundational difficulties, tends to reinforce efforts to construct a quantum theory for gravity based on the established principles of quantum mechanics. This, I believe, is the importance of the many-worlds interpretation. If someone succeeds in constructing a satisfactory conventional quantum theory of gravity the many-worlds interpretation, through the postulate of the reality of the quantum state, will allow the theory to avoid what, by other interpretations of quantum theory, would be serious problems.

If, on the other hand, the truth is that the understanding of gravitation on a microscopic level will lead to radical modifications in our basic ideas about quantum theory, then any interpretation that asserts that quantum mechanics is complete and satisfactory as it stands is likely to be counterproductive and misleading. It may even cover up and obscure difficulties that are important clues to how quantum theory may need to be modified in order to incorporate gravitation. For example, as the many-worlds interpretation dissolves the puzzles resulting from the ambiguity in the definition of particle number in gravitational fields and the unmeasurability of the quantum state of the gravitational field it dissolves also the new insights into the connection between quantum and gravitational phenomena that come from reflecting on these puzzles.

If one is thus concerned with looking for modifications or extensions of the quantum theory which will resolve the puzzles involving gravitational theory then one needs a more provisional interpretation of quantum mechanics, one which does not commit us to ontological statements like the postulate of the reality of the quantum state. For this reason I believe that what I have called here the minimal relative state interpretation might be helpful. In particular, by asserting that the quantum state makes only contingent assertions concerning correlations between systems which are, or have been, in interaction, it suggests that the theory might be completed by the introduction of hidden variables which also express relations or correlations between systems, rather than being more precise descriptions of elementary systems in isolation. A formulation of a hidden variable theory along these lines, that also sheds some light on the puzzles in gravitational theory, has also been made by the author (Smolin 1983a).

As we do not know which direction will lead to the correct theory, we cannot make a decision as to which interpretation of quantum mechanics is correct. What we can do is to hope that different individuals will take the different points of view as the basis of their research so that investigation of the implications of each can be vigorously pursued, leaving it to the future to decide which of them will have been the more fruitful. (See also the work of Deutsch (1981) and Kochen (1983).)

ACKNOWLEDGMENTS

I would like to thank Rafael Sorkin for pointing out an error in a draft of this paper and Ted Jacobsen for bringing to my attention the paper of Finkelstein (1963). I would also like to thank Steven Christensen for encouraging me to take the many-worlds interpretation as the subject of this contribution. This research was supported in part by the Department of Energy under Contract No. DE-AC02-76ER02220.

REFERENCES

Ballentine L E 1973 *Found. Phys.* **3** 229

Ballentine L E, DeWitt B S, Grever J, Koga T, Pearle P, Sachs M and Walker E H 1971 *Phys. Today* **24** 36

Bell J S 1981 in *Quantum Gravity II* ed C J Isham, R Penrose and D W Sciama (Oxford: Oxford University Press)

Bohr N and Rosenfeld L 1933 *Det. K. Danske. Vidensk. Selsk. Mat.-Fys. Medr.* **12** 8

Cooper L N and Van Vechten D 1969 *Am. J. Phys.* **37** 1212

Deutsch D 1981 *Quantum Theory as a Universal Physical Theory* University of Texas preprint

DeWitt B S 1957 *Rev. Mod. Phys.* **29** 377

—— 1962 in *Gravitation: an introduction to current research* ed L Witten (New York: John Wiley)

—— 1967a *Phys. Rev.* **160** 1113

—— 1967b *Phys. Rev.* **162** 1195

—— 1967c *Phys. Rev.* **162** 1239

—— 1970 *Phys. Today* **23** 30

—— 1971 in *Proceedings of the International School of Physics 'Enrico Fermi', Course IL: Foundations of quantum mechanics* ed B D'Espagnet (New York: Academic)

—— 1975 *Phys. Rep.* **19** 297

—— 1979 in *General relativity, An Einstein Centenary survey* ed S W Hawking and W Israel (Cambridge: Cambridge University Press)

DeWitt B S and Graham N 1973 ed *The many-worlds interpretation of quantum mechanics* (Princeton: Princeton University Press)

Einstein A, Podolsky B and Rosen N 1935 *Phys. Rev.* **47** 777

Everett H III 1957 *Rev. Mod. Phys.* **29** 454

—— 1973 in *The many-worlds interpretation of quantum mechanics* ed B S DeWitt and N Graham (Princeton: Princeton University Press)

Finkelstein D 1963 *Trans. NY Acad. Sci.* **25** 621

Fulling S A 1973 *Phys. Rev.* D **7** 2850

Graham N 1973 in *The many-worlds interpretation of quantum mechanics* ed B S DeWitt and N Graham (Princeton: Princeton University Press)

Hartle J B 1966 *Am. J. Phys.* **36** 704

Hawking S W 1976 *Phys. Rev.* D **13** 2460

—— 1982 *Commun. Math. Phys.* **87** 153

Jammer M 1974 *The philosophy of quantum mechanics: The interpretation of quantum mechanics in historical perspective* (New York: John Wiley)

Kochen S 1983 *The Interpretation of Quantum Mechanics* Princeton University preprint

Rosenfeld L 1957 In *Conference on the role of gravitation in physics* ed C Morette-DeWitt

Sanchez N 1981 *Phys. Rev.* D **24** 2100

Smolin L 1982a *On the intrinsic entropy of the gravitational field* IAS preprint

—— 1982b *On the nature of quantum fluctuations and their relation to gravitation and the principle of inertia* IAS preprint

—— 1983a *Gen. Relativ. Grav.* to appear

—— 1983b *Derivation of Quantum Mechanics from a Deterministic non-local Hidden Variable Theory* IAS preprint

Unruh W G 1976 *Phys. Rev.* D **14** 870

Wheeler J A 1957 *Rev. Mod. Phys.* **29** 463

The Positivity of the Jacobi Operator on Configuration Space and Phase Space

C R DOERING and C DeWITT-MORETTE

FOREWORD

The small disturbance equation (alias the Jacobi equation) is a tool often used by Bryce DeWitt. We point out an unexpected difference between the configuration space version of the Jacobi operator (with given boundary conditions) and its phase space version: the positivity of the former does not imply the positivity of the latter. Consequently, a Gaussian measure whose covariance is a Green's function of a Jacobi operator defined on configuration space does not necessarily have a counterpart on the space of paths which take their values in phase space. Of course the positivity of the covariance is not an issue if one works with Feynman path integrals (as opposed to Wiener integrals) with complex Gaussians.

1. INTRODUCTION

Gaussian measures play an important role in functional integration. A Gaussian measure is characterized by its covariance and in Wiener integration, as opposed to Feynman integration, the covariance must be a *positive* symmetric quadratic form. If one is interested in a path integral written formally as

$$\int \exp(-S[q]) \mathscr{D}q \qquad (1.1)$$

with

$$S[q] = \int L(q, \dot{q}) \, dt \qquad (1.2)$$

or as

$$\int \exp(-S[q, p]) \mathscr{D}p \mathscr{D}q \qquad (1.3)$$

with

$$S[q, p] = \int p\,dq - H(q, p)\,dt \tag{1.4}$$

the covariance of the relevant Gaussian is a Green's function of the Jacobi operator with boundary conditions determined by the domain of integration. We recall that the Jacobi operator (alias the small disturbance operator, alias the operator of geodetic deviation) is obtained from the second variation of the action. For instance, on the space of paths with values in the configuration space, $q: T \to \mathbb{R}^n$, $T = [t_a, t_b]$, the expansion of the action around the classical path Q is

$$S[Q + q] = S[Q] + \text{(boundary terms)} + \tfrac{1}{2}S''[Q]qq + \ldots \tag{1.5}$$

and the Jacobi operator $\mathscr{I}[Q(t)] \equiv \mathscr{I}_t$ is the differential operator (of second-order for Lagrangians that depend quadratically on velocities) defined by

$$S''[Q]qq = \int_{t_a}^{t_b} q(t)(\mathscr{I}_t q(t))\,dt \equiv \langle q, \mathscr{I}q \rangle. \tag{1.6}$$

On the space of paths (q, p) with values in phase space, $\mathbb{R}^{2n}$, the expansion of the action around the classical path (Q, P) is

$$S[Q + q, P + p] = S[Q, P] + \text{(boundary terms)} + \tfrac{1}{2}(q, p)S''[Q, P]\begin{pmatrix} q \\ p \end{pmatrix} + \ldots \tag{1.7}$$

and the phase space Jacobi operator is defined by the equation

$$(q, p)S''[Q, P]\begin{pmatrix} q \\ p \end{pmatrix} = \int_{t_a}^{t_b} (q(t), p(t))\mathscr{I}_t^{(\text{p.s.})}\begin{pmatrix} q(t) \\ p(t) \end{pmatrix} dt \tag{1.8}$$

where $\mathscr{I}^{(\text{p.s.})}$ is a $2n \times 2n$ matrix built up with first-order differential operators.

Although one might expect the configuration space version and the phase space version of the path integrals for simple systems, say systems without constraints, to be totally equivalent, we shall find it not to be so. Even for a free particle in $\mathbb{R}^n$ there is an important difference: given equivalent boundary conditions, the configuration space Jacobi operator may have only positive eigenvalues whereas the corresponding phase space Jacobi operator has both positive and negative eigenvalues. It follows that the Green's function of the configuration space Jacobi operator can be used as a covariance, whereas the Green's function of the corresponding phase space Jacobi operator cannot.†

In the following, we consider the Lagrangian‡

$$L = \tfrac{1}{2}\dot{q}^2 - V(q) \tag{1.9}$$

---

† Path integration using Green's functions of the phase space Jacobi operator has been developed in DeWitt-Morette *et al* (1979) and in DeWitt-Morette *et al* (1982). The path integrals in these papers are defined in terms of prodistributions and the non-positivity of the Green's function is not an issue.

‡ See for instance DeWitt-Morette *et al* (1979) for the configuration and phase space Jacobi operators for arbitrary systems.

and the associated Jacobi operator in configuration space (§ 2) and in phase space (§ 3). The implications of the results are discussed in § 4.

2.   THE JACOBI OPERATOR IN CONFIGURATION SPACE

For a Lagrangian of the form (1.9), the Jacobi operator is

$$\mathcal{J}_t = -\frac{d^2}{dt^2} - V''(Q(t)). \tag{2.1}$$

If the classical path $Q(t)$ satisfies boundary conditions of the form

$$\cos \theta_a Q(t_a) + \tau \sin \theta_a \dot{Q}(t_a) = \alpha_a$$
$$\cos \theta_b Q(t_b) + \tau \sin \theta_b \dot{Q}(t_b) = \alpha_b \tag{2.2}$$

the path integral is defined over the space† Ω of paths q which satisfy the homogeneous boundary conditions

$$\cos \theta_a q(t_a) + \tau \sin \theta_a \dot{q}(t_a) = 0$$
$$\cos \theta_a q(t_b) + \tau \sin \theta_b q(t_b) = 0. \tag{2.3}$$

The covariance $G(t, s)$ of a Gaussian measure μ defined on this space must satisfy (for $t_a \leqslant s \leqslant t_b$)

$$\cos \theta_a G(t_a, s) + \tau \sin \theta_a \dot{G}(t_a, s) = 0$$
$$\cos \theta_b G(t_b, s) + \tau \sin \theta_b \dot{G}(t_b, s) = 0 \tag{2.4}$$

since

$$G(t, s) = \int_\Omega q(t)q(s)d\mu(q). \tag{2.5}$$

The Fourier transform of μ is

$$\mathcal{F}\mu(q') \equiv \int_\Omega \exp\left(i\langle q', q \rangle\right)d\mu(q)$$
$$= \exp\left(-\frac{1}{2}\int_{t_a}^{t_b}\int_{t_a}^{t_b} q'(t)G(t, s)q'(s)dsdt\right) \tag{2.6}$$

where $q' \in \Omega'$, the dual of Ω. Note that $\Omega \subset \Omega'$ and the inner product is formally

$$\langle q', q \rangle = \int_{t_a}^{t_b} q'(t)q(t)dt. \tag{2.7}$$

---

† The boundary conditions on q have to be vanishing boundary conditions for Ω to be a vector space.

The restrictions on the possible choices of $G(t, s)$ can be stated in terms of the Gaussian measures on $\mathbb{R}^n$ induced by linear maps from Ω into $\mathbb{R}^n$. In general, a linear map

$$P:\Omega \to \mathbb{R}^n \qquad (2.8)$$

may be written

$$P^i(q) = \langle q'^i, q \rangle \qquad (2.9)$$

for any set $\{q'^i\}$ of n members of Ω'. The induced measure on $\mathbb{R}^n$ is the Gaussian of covariance

$$A^{ij} = \langle q'^i, Gq'^j \rangle = \int_{t_a}^{t_b} \int_{t_a}^{t_b} q'^i(t)G(t, s)q'^j(s)\,ds\,dt. \qquad (2.10)$$

As is well known, a covariance matrix on $\mathbb{R}^n$ must be positive-definite

$$x_i A^{ij} x_j > 0 \qquad \forall\, x \in \mathbb{R}^n \qquad x \neq 0 \qquad (2.11)$$

and symmetric

$$A^{ij} = A^{ji}. \qquad (2.12)$$

The covariance operator $G(t, s)$ must then be positive

$$\langle q', Gq' \rangle > 0 \qquad \forall\, q' \in \Omega' \qquad q' \neq 0 \qquad (2.13)$$

and symmetric

$$G(t, s) = G(s, t). \qquad (2.14)$$

The operator $G(t, s)$ will be positive if and only if all of its eigenvalues are positive. If the covariance is a Green's function of the Jacobi operator:

$$\mathscr{I}_t G(t, s) = \delta(t, s) \qquad (2.15)$$

then $G(t, s)$ is positive if and only if all the eigenvalues of $\mathscr{I}_t$ are positive.

The one-dimensional Jacobi operator (2.1) along with the boundary conditions (2.3) defines a Sturm–Liouville problem which has been studied extensively.† As long as $V''(Q(t))$ is a continuous function of t in the interval $[t_a, t_b]$, the eigenfunctions of such operators form a complete orthogonal set and the eigenvalues λ_n may be ordered

$$0 \leqslant \lambda_1 < \lambda_2 < \lambda_3 < \ldots. \qquad (2.16)$$

If the operator (and boundary conditions) admit the vanishing eigenvalue, then there is no Green's function. The conclusion is thus: if the Green's function (2.14), (2.15) exists for the Jacobi operator (2.1), whose domain is the space of paths satisfying (2.3), then it is an acceptable covariance for a Gaussian measure.

For the free particle $(V \equiv 0)$, the Green's function exists if the boundary conditions are *not* of two types:

$$1) \; \dot{q}(t_a) = 0 \qquad \dot{q}(t_b) = 0 \qquad (2.17)$$

---

† See for instance Ritger and Rose (1968).

or

$$2) \quad q(t_a) + (T + \tau)\dot{q}(t_a) = 0$$
$$q(t_b) + \tau\dot{q}(t_b) = 0 \tag{2.18}$$

where $T = t_b - t_a$ and τ is arbitrary. To see this, note that the general solution to the eigenvalue zero problem is

$$q_0(t) = c_1 t + c_2. \tag{2.19}$$

Applying the boundary conditions (2.3),

$$0 = c_1(t_a \cos\theta_a + \tau \sin\theta_a) + c_2 \cos\theta_a$$
$$0 = c_1(t_b \cos\theta_b + \tau \sin\theta_b) + c_2 \cos\theta_b$$

so that q_0 does not vanish identically if and only if the determinant of the coefficients vanishes. This leads to the condition

$$\tan\theta_a = \tan\theta_b + T/\tau \tag{2.20}$$

which implies (2.17) or (2.18).

3. THE JACOBI OPERATOR IN PHASE SPACE

For a Lagrangian of the form (1.9), the Hamiltonian is

$$H(p, q) = \tfrac{1}{2}p^2 + V(q) \tag{3.1}$$

and the second variation of the action is

$$\tfrac{1}{2}(q, p)S''[Q, P]\begin{pmatrix} q \\ p \end{pmatrix} = \frac{1}{2}\int_{t_a}^{t_b} (q(t), p(t)) \begin{pmatrix} -V''(Q(t)) & -\dfrac{d}{dt} \\ \dfrac{d}{dt} & -1 \end{pmatrix}\begin{pmatrix} q(t) \\ p(t) \end{pmatrix} dt. \tag{3.2}$$

If the classical solution $(Q(t), P(t))$ satisfies the boundary conditions

$$\cos\theta_a Q(t_a) + \tau \sin\theta_a P(t_a) = \alpha_a$$
$$\cos\theta_b Q(t_b) + \tau \sin\theta_b P(t_b) = \alpha_b \tag{3.3}$$

then the path integral is defined on the space Ω of path (q, p) which satisfies

$$\cos\theta_a q(t_a) + \tau \sin\theta_a p(t_a) = 0$$
$$\cos\theta_b q(t_b) + \tau \sin\theta_b p(t_b) = 0. \tag{3.4}$$

If the variables of integration (q, p) are considered jointly Gaussian random processes with the covariance defined by

$$\begin{pmatrix} -V''(Q(t)) & -\dfrac{d}{dt} \\ \dfrac{d}{dt} & -1 \end{pmatrix}\begin{pmatrix} G_1(t, s) & G_2(t, s) \\ G_3(t, s) & G_4(t, s) \end{pmatrix} = \begin{pmatrix} \delta(t, s) & 0 \\ 0 & \delta(t, s) \end{pmatrix} \tag{3.5}$$

with

$$\cos \theta_a G_1(t_a, s) + \tau \sin \theta_a G_3(t_a, s) = 0$$
$$\cos \theta_b G_1(t_b, s) + \tau \sin \theta_b G_3(t_b, s) = 0$$
$$\cos \theta_a G_2(t_a, s) + \tau \sin \theta_a G_4(t_a, s) = 0$$
$$\cos \theta_b G_2(t_b, s) + \tau \sin \theta_b G_4(t_b, s) = 0 \tag{3.6}$$

and

$$G_1(t, s) = G_1(s, t)$$
$$G_2(t, s) = G_3(s, t)$$
$$G_4(t, s) = G_4(s, t) \tag{3.7}$$

then, as is the case in configuration space, the eigenvalues of the Jacobi operator must all be positive. In the free particle case it is easy to check for the existence of the vanishing eigenvalue: the conditions on the boundary conditions are analogous to (2.17) and (2.18). The free particle Jacobi operator in phase space admits a vanishing eigenvalue if the boundary conditions are of the form

$$1) \; p(t_a) = 0 \qquad p(t_b) = 0 \tag{3.8}$$

or

$$2) \; q(t_a) + (\tau + T)p(t_a) = 0$$
$$q(t_b) + \tau p(t_b) = 0. \tag{3.9}$$

The question is not closed, however. The Jacobi operator on phase space does not define a Sturm–Liouville problem and we find that in general it admits both positive and negative eigenvalues. Consider, for example, the position to momentum transition for the free particle. The boundary conditions are

$$q(t_a) = 0$$
$$p(t_b) = 0. \tag{3.10}$$

The eigenvalues of the eigenvalue problem

$$\begin{pmatrix} 0 & -\dfrac{d}{dt} \\ \dfrac{d}{dt} & -1 \end{pmatrix} \begin{pmatrix} q(t) \\ p(t) \end{pmatrix} = \lambda \begin{pmatrix} q(t) \\ p(t) \end{pmatrix} \tag{3.11}$$

are

$$\lambda_n^{\pm} = -\tfrac{1}{2} \pm \sqrt{\tfrac{1}{4} + \omega_n^2} \tag{3.12}$$

where

$$\omega_n = (n + \tfrac{1}{2})\pi / T \qquad n = 0, 1, 2, \ldots. \tag{3.13}$$

There are both positive as well as negative eigenvalues. The Green's function exists (for $t_a = 0$, $t_b = T$):

$$\begin{pmatrix} G_1(t, s) & G_2(t, s) \\ G_3(t, s) & G_4(t, s) \end{pmatrix} = \begin{pmatrix} \inf(t, s) & \theta(t, s) \\ \theta(s, t) & 0 \end{pmatrix} \tag{3.14}$$

but this is clearly not an acceptable covariance. Indeed,

$$\int_\Omega p(t)p(s)\mathrm{d}\mu(q, p) = G_4(t, s) = 0 \tag{3.15}$$

implies that

$$p(t) = 0 \tag{3.16}$$

(when p and q are real valued random processes), but

$$\int_\Omega q(t)p(s)\mathrm{d}\mu(q, p) = G_2(t, s) \neq 0 \tag{3.17}$$

which is a contradiction.

4. DISCUSSION

Hamilton's equations and the Jacobi equation in phase space are the systems of first-order differential equations equivalent to the second-order differential equations given by the Euler–Lagrange equation and the Jacobi equation in configuration space. The same is not true of the eigenvalue equations of the Jacobi operators in configuration space and phase space.† The system of first-order equations equivalent to

$$\left(-\frac{\mathrm{d}^2}{\mathrm{d}t^2} - V''(Q(t))\right)q(t) = \lambda q(t) \tag{4.1}$$

is

$$\dot{q}(t) = p(t)$$
$$\dot{p}(t) = -[\lambda + V''(Q(t))]q(t) \tag{4.2}$$

which is not equivalent to

$$\begin{pmatrix} -\{V''(Q(t)) + \lambda\} & -\dfrac{\mathrm{d}}{\mathrm{d}t} \\ \dfrac{\mathrm{d}}{\mathrm{d}t} & -(1+\lambda) \end{pmatrix}\begin{pmatrix} q(t) \\ p(t) \end{pmatrix} = 0. \tag{4.3}$$

Thus it is not surprising that the spectrum is different in configuration space and phase space. By formulating the problem in phase space, it has changed already at the classical level. In configuration space $q(t)$ determines $\dot{q}(t)$ whereas in phase space, q and p are related only through the equations of motion. When we consider arbitrary paths in phase space, we include paths where p and $\dot{q}$ may be in opposite directions. This is the case for eigenvectors $(q(t), p(t))$ of the Jacobi operator in phase space with eigenvalues less than -1.

† The authors are grateful for remarks made by J Dollard on this point.

In conclusion, this simple example shows the caution one has to exercise when comparing Lagrangian and Hamiltonian path integrals in Euclidean regimes: the Wiener measure is a measure on the space of paths taking their values in a flat configuration space. It has no counterpart on the space of paths taking their values in the corresponding phase space.

ACKNOWLEDGMENT

This work was supported in part by NSF Grant PHY-81-07381.

REFERENCES

DeWitt-Morette C, Maheshwari A and Nelson B 1979 *Phys. Rep.* **50** 255–372.
DeWitt-Morette C and Zhang T R 1982 'A Feynman–Kac Formula in Phase Space with Application to Coherent State Transitions' *Preprint, University of Texas*
Ritger P and Rose N 1968 *Differential Equations with Applications* (New York: McGraw-Hill)

On Complete Group Covariance without Torsion

LEOPOLD HALPERN

1. INTRODUCTION

Riemann's discovery of the complementarity of physical and geometrical laws in the description of nature stimulated investigations on modified geometries to achieve simplifications of physics. *The Principles of Mechanics* by Heinrich Hertz (1896) contains an attempt to geometrize all physical forces. The discovery of special relativity which unites transformations of space and time in one semi-simple invariance group was however a prerequisite for the breakthrough along these lines in Einstein's general theory of relativity. Einstein's theory of gravitation does not really prefer one particular invariance group—the principle of equivalence, one of its heuristic foundations—can be formulated in accord with a whole family of invariance groups besides the Poincaré group. For example the De Sitter groups SO (4, 1) and SO (3, 2) (Halpern 1977b). The physics inserted into the geometrical spaces of gravitation—the right-hand side in Einstein's equations, which is an 'alien to the theory', is however always formulated in accord with the Poincaré group as invariance group. Dirac had given a formulation of the equations of matter fields covariant with respect to the De Sitter groups and the conformal groups (Dirac 1935, 1936). Lubkin was probably the first who attempted to modify general relativity to De Sitter covariance (Lubkin 1971). His ideas on the modified symmety appear to be deeper going than those of many more recent authors who consider it only as an artifice to return to Poincare invariance in the limit where the radius of the universe grows infinite.

Geometries are frequently related to certain groups of transformations; it appears thus that also invariance groups can be fitted into the systems of mutually complementary geometries and physics. One can imagine the fundamental laws being formulated originally under completely different physical conditions than those encountered by Galileo and thus subject to different invariance groups which determine their differential equations. Considering sufficiently vast domains where these conditions don't hold any more rigorously, the presence of forces and fields has to be postulated to take the responsibility for these violations. They will in general differ considerably from those introduced by us under different initial conditions. Einstein has succeeded to generalize the local laws of Galileo and Newton for the gravitational phenomena to all the univese. If we find however any ambiguity in the choice of the invariance group under local conditions we should not hesitate to explore the physical implications when the symmetry of the altered invariance group begins to be broken—irrespective of how screwed up they may appear at first, we could obtain valuable suggestions from it†. In a recent lecture (Halpern 1982d) this procedure has been compared to a fairy tale which has a pre-established outlook dealing with life. Usually events happen in it that appear obviously in disaccord with experience—yet under closer contemplation such events are often recognized to do justice to reality in a deeper sense than our limited experience would let us expect.

In such a spirit we have started to explore a modification of a general relativity which is in accord with complete De Sitter covariance. This means, the right-hand side of Einstein's equations should be given De Sitter covariance to the degree that also all phenomena irreconcilable with Poincaré group physics in any limit, be admissible.

To achieve this, we start from the group manifold and construct space–time as a factor space with respect to a subgroup. A continuous invariance group in our view has to have a metric to be meaningful or it has to have a preferred coordinate system which in a way corresponds to a metric. Semi-simple groups have a natural metric and the De Sitter group is even simple. To consider the effects of complete group covariance, we project geometrical constructions like metric and orbits of subgroups on the base manifold and try to identify it with physical quantities there. We start with the orbits of one dimensional subgroups, some of which are identifiable with particle trajectories. Some others can however not be identified—we don't know of free spiral trajectories in space–time. Here the part of the fairy tale starts which seems in disaccord with reality. Namely we can not discard such orbits while accepting the others; in our case of a simple group everything is consistently interrelated. We don't want to reveal the conjectured solution to this puzzle already here. The reader is invited to study this mathematically rigorous fairy tale—a theory dictated by mathematics, not however by mathematicians. Unfortunately, the derivation of the results require a mathematical apparatus that not every physicist uses. One can however learn the results in the middle of §4 and §5 and then see how it was obtained.

The Kaluza–Klein theory of §4, starting from the natural metric on the group manifold seems recently to have found applications to the benefit of supersymmetry. The approach differed however originally from the methods of other approaches using the group manifold.

Previous papers on the subject by the author contained sometimes inaccuracies and errors in the mathematical formalism, which should not prejudice the physical

---

† How difficult the task can be to bring physical laws based on an unsuitable invariance group in accord with observation can be seen from the attempts of physicists around Lenard and Stark to formulate a non-Einsteinian German physics based on the Galileo group.

ideas. The interest in the outcome was too intense to await the mastery of the game of patience that many exact presentations on the mathematics of gauge theories constitute.

We treat here a special case of the theory—that of vanishing torsion. The content of the theory is considerably altered if torsion does not vanish. Introduction of torsion may remove contradictory behavior of the spin model of § 3 in strong gravitational fields.

2.   THE GROUP MANIFOLD AND ITS GEOMETRY

The theory in its final four-dimensional form can be expressed in the formalism of classical general relativity. A purpose of this article is to show its close relationship to the structure of Lie groups. The fundamentals of the differential geometry of Lie groups are thus a prerequisite. These are expressed most concisely with the help of fibre bundles. The space available here limits us to a brief reminder in introducing basic facts. Although theorems from the theory of fibre bundles are quoted, the reader hardly needs them for the examples considered because their bundles are trivial. A short lucid self-consistent introduction to the subject is presented in Nomizu (1956). It is treated in connection with gauge theory in Choquet-Bruhat *et al* (1982). Many details in the classical formulation are presented in Eisenhart (1933). The author started his venture in group theory and its relation to gauge theory from B DeWitt's Les Houches lecture (DeWitt 1963) in which many of the fundamentals can be found.

We consider an r-dimensional Lie group G which acts transitively on a k-dimensional manifold B as a group of transformations. Then in general G has a closed $(r-k)$-dimensional Lie subgroup H which leaves one element of B invariant. B is then homeomorphic to the coset space G/H and a natural map π from G onto B can be defined. (The group manifold of G is the bundle space of a principal fibre bundle $P(G, \pi, H, B)$ with base B and group and typical fibre H (Steenrod 1974).)

We keep in particular the case of one of the De Sitter groups $SO(3, 2)$ or $SO(4, 1)$ for G and the Lorentz group $SO(3, 1)$ for H as an example in mind. B can then be identified with the manifold of the corresponding De Sitter universe.

The multiplication of group elements $c = ab$ in a local chart reads:

$$c^t = \phi^t(a, b) = \phi^t(a \ldots a^r, b \ldots b^r). \tag{2.1}$$

Differentiation yields

$$dc^t = V(a, b)_u{}^t \, da^u + W(a, b)_u{}^t \, db^u. \tag{2.1a}$$

Associativity: $\phi(a \cdot h, b) = \phi(a, h \cdot b)$ implies that the r one forms

$$A^T(x) = W_u{}^t(x^{-1}, x) \, dx^u \tag{2.2}$$

are invariant with respect to left translations $G \times G \to G$: $L_a x = a \cdot x$

$$L_a{}^* A^T = A^T \tag{2.2a}$$

and the r one forms

$$\bar{A}^T(x) = V_u{}^t(x, x^{-1}) \, dx^u \tag{2.2b}$$

are right invariant.

Any left invariant two form can be expanded in terms of the exterior products of pairs of left invariant forms with coefficients which are left invariant null forms and thus constants. The exterior derivative d commutes with $L_a{}^*$ so that the relation

$$dA^T + \tfrac{1}{2} c_U{}^T{}_V A^U A^V = 0 \tag{2.3}$$

is fulfilled for constant $c_U{}^T{}_V = -c_V{}^T{}_U$ (structure constants) because all the members are left invariant. In components, with commas denoting derivatives:

$$A^T{}_{k,i} - A^T{}_{i,k} + \tfrac{1}{2} c_U{}^T{}_V (A_i{}^U A_k{}^V - A_k{}^U A_i{}^V) = 0. \tag{2.3'}$$

The right invariant forms fulfill then the corresponding equation

$$d\bar{A}^T - \tfrac{1}{2} c_U{}^T{}_V \bar{A}^U \bar{A}^V = 0 \tag{2.3a}$$

with the same structure constants. Equations (2.3) and (2.3a) are called Maurer–Cartan equations. The r left invariant vector fields A_s dual to the forms A^T: $A^T(A_s) = \delta_s{}^T$ form bases to the tangent space of the manifold. They fulfill

$$[A_U, A_V] = c_U{}^S{}_V A_S \tag{2.3''}$$

correspondingly the right invariant vectors $\bar{A}_s$ dual to the $\bar{A}^T$ fulfill:

$$[\bar{A}_U, \bar{A}_V] = -c^S{}_{UV}\bar{A}_S \tag{2.3a''}$$

$$[A_U, \bar{A}_V] = 0. \tag{2.3b}$$

A Lie group is simple if it has no proper invariant subgroup and semi-simple if it has no Abelian invariant subgroup. The components of the Cartan–Killing metric γ on the manifold of the group G is given in an unholonomic frame by:

$$\gamma_{ST} = c_S{}^U{}_V c_T{}^V{}_U \tag{2.4}$$

or in local coordinates:

$$\gamma_{uv} = A_u{}^S \gamma_{ST} A_v{}^T. \tag{2.4a}$$

The Ricci tensor of this metric on the group manifold fulfills the relation:

$$R_{uv} = \tfrac{1}{4} \gamma_{uv}. \tag{2.5}$$

The author has re-interpreted this relation into

$$R_{uv} - \tfrac{1}{2} \gamma_{uv} R + \tfrac{1}{8} (r - 2) \gamma_{uv} = 0 \tag{2.6}$$

that means the metric fulfills Einstein's equations in r-dimensions with a cosmological member (Halpern 1980, 1981, 1982 a, b, c). This relation led to a method to generalize the metric from that of the global group manifold to a situation where right-hand members of equation (2.6) exist. Generalized Kaluza–Klein theories (Kaluza 1921, Klein 1926, DeWitt 1963) make use of such equations. The author showed that the group manifold fulfills the conditions of a special, peculiar Kaluza–Klein theory.

Locally the group manifold is homeomorphic to $B \times H$. The mapping on the points of the group manifold H depends, however, on the particular local trivialization of P. One can always choose a basis of r left invariant tangent vectors A_S on P such that the last $(r - k)$ vectors belong to the subgroup H.

We shall henceforth denote all indices running from $1 \ldots k$ by letters in the alphabet before L and indices running from $k + 1 \ldots r$ by letters from L till inclusive q. Indices running from $1 \ldots r$ (dimension of group manifold) are denoted by letters after q in the alphabet. We shall also adjust the summation convention to this rule: $A_e b^e$ or $A_E b^E$ implies thus summation from $1 \ldots k$ (dimension of B) only and $A_m b^m$ implies summation from $k + 1 \ldots r$. We shall use this convention usually without further warning.

For a local trivialization of P a coordinate system can be introduced such that the $r - k$ basis vectors A_M depend only on the $r - k$ coordinates x^n and all components A^i_M vanish. B is the space of left cosets of H. Right translations give rise to adjoint transformations of the left invariant forms (and coadjoint transformations of left A_S on P such that the last $(r - k)$ vectors belong to the subgroup H.

$$R_a{}^* A^S(x) = \bar{A}_u{}^S(a^{-1}) A_v{}^u(a^{-1}) A^V(x) = ad(a^{-1}) A^S(x) \qquad (2.2c)$$

and similarly

$$L_a{}^* \bar{A}^S(x) = A_u{}^S(a) \bar{A}_v{}^u(a) \bar{A}^V(x) = ad(a) \bar{A}^S(x). \qquad (2.2d)$$

According to equation (2.3b) the left invariant tangent vectors A_S are the generators of right translations and the right invariant vectors $\bar{A}_S$ the generators of left translations. Equation (2.3'') gives the infinitesimal coadjoint transformations of left invariant vectors, etc.

Adjoint transformations leave the metric γ (equation (4a)) invariant:

$$[A_S, \gamma] = 0 \qquad (2.7)$$

Applied to $S > k$ this implies that a projection of the contravariant metric tensor of γ from P on B by the differential of π is independent of the x^m and thus uniquely defined. If the subgroup H is a semi-simple Lie subgroup then a non-singular metric g on $B = G/H$ can be defined as the projection of the metric γ. The basis vectors A_S can be chosen everywhere orthogonal with respect to the metric γ. As a result of equation (2.3) one finds that the orbit of every one-dimensional subgroup of G on P is a geodesic.

The metric γ defines a horizontal vector space on P, perpendicular to the vertical vector space spanned by the A_M. The horizontal vectors in an orthogonal basis are thus the A_E. The connection form ω is given by

$$\omega = A^M(x) \hat{A}_M. \qquad (2.8)$$

$\hat{A}_M$ are the elements of the Lie algebra pertaining to the left invariant vector fields A_M. One has $\omega(A_E) = 0$. The required transformation properties of ω follow from equation (2.2c), provided all $c_M{}^P{}_E$ vanish. The curvature two form F of the connection is given by Cartan's structure equations:

$$F = d\omega + [\omega, \omega] \qquad (2.9)$$

and its coordinate components:

$$F_{ik} = (A^M{}_{k,i} - A^M{}_{i,k} + A_i{}^P c_P{}^M{}_Q A_k{}^Q)\, \hat{A}_M \tag{2.9a}$$

$c^M{}_{PQ}$ are here the structure constants of the subgroup H only. The right-hand member therefore differs from the left-hand member of equation (2.3') for $T = M$. Because the latter vanishes, the curvature F for the group manifold does in general not vanish. One has:

$$F_{ik} = -c_E{}^M{}_F\, A_i{}^E A_k{}^F A_M. \tag{2.9b}$$

$A^M c_M{}^E{}_F$ can be related to a linear connection on the frame bundle with $\theta = A^E A_E{}^i$ the soldering form.

The equation (2.3) for $T = E$ says then that the connection defined on the group manifold is torsion free. An orthogonal or pseudo-orthogonal H implies also the vanishing of the covariant derivative of the metric so that the connection is then Riemannian.

3.   THE SHADOW PLAY OF PHYSICS ON SPACE–TIME

Our aim is to explore the physical implications of an invariance group G in all its aspects. We may try to describe nature with its help—but only if undue mutilations of the mathematics can be avoided. If the invariance group dominates the physics, we expect that the geometrical properties of the group manifold relate to physical properties manifested in space and time. We have already seen at the end of the last section that this is true for the metric on the base manifold B which is a projection from P. It also is true for free particle orbits. The projection of the orbits of one-dimensional subgroups of G with horizontal tangent vectors, are geodesics on B. (We have seen that these orbits themselves are geodesics on P.) What about the $(r - k)$-parameter family of one-dimensional orbits that have tangent vectors which have vertical components? Do we have to discard them? This may be justifiable for the Poincaré group. Semi-simple groups do however have a closer knit structure that cannot easily be submitted to such separations.

The equations of the orbit of a one-dimensional subgroup with tangent vector:

$$\dot{x}^t = A^t{}_S c^S \qquad (c^U = \text{const}) \tag{3.1}$$

is a geodesic on P

$$\ddot{x}^t + \{ {}^t_{uv} \}\, \dot{x}^u \dot{x}^v = 0 \tag{3.1a}$$

with the integral

$$\gamma_{UV}\, c^U c^V = \text{constant} = \pm \tfrac{1}{0}. \tag{3.1b}$$

The projection on B involves the curvature if the tangent vector is not horizontal:

$$\ddot{x}^i + \{ {}^i_{jk} \}\, \dot{x}^j \dot{x}^k = F^{Mi}{}_k\, \dot{x}^k c^N \gamma_{NM} \tag{3.2}$$

where the same parameter as in (3.1a) can be used because of (3.1). The metric of equation (3.2) is g whereas that of equation (3.1a) is γ. The coefficients F^M of the curvature form which occur in (3.2) depend on the point of the fibre:

$$\frac{\partial}{\partial x^n} F^M = c_P{}^M{}_Q A_n{}^Q F^P \tag{2.9b}$$

their values must be chosen on the points of the geodesic for the c^M to remain constant along the geodesic. The fibre bundle $P(G, \pi H, H)$ is a reduction of the frame bundle (the principal fibre bundle of the general linear group in k-dimensions). This is in general true for orthogonal or pseudo-orthogonal groups H. A connection on the frame bundle (linear connection) is therefore determined by the connection on P. We can express the linear connection with the help of the structure constants which are generators of adjoint transformations of H. For a given local trivialization of P determined by a certain local cross-section, we obtain a connection form $\bar{\omega}$ in this local trivialization. We can write:

$$\bar{\omega} = A_i{}^M (x^j) \hat{A}_M \tag{3.3}$$

and for the corresponding linear connection:

$$\bar{\omega}_L{}^E{}_F = A_i{}^M (x^j) c_M{}^E{}_F. \tag{3.3a}$$

In the same way one can relate the two forms F on the base manifold (field strengths)

$$\bar{F} = d\bar{\omega} + [\bar{\omega}, \bar{\omega}]. \tag{3.4}$$

We can now choose a local cross-section on which lay the points of an interval of the geodesic. We shall have $F = F(\bar{x})$ for these points. Let us assume here that the torsion vanishes and that H is pseudo-orthorgonal. We can then conclude that the connection must be Riemannian and the curvature $\bar{F}_L$ be given in terms of the Riemann tensor:

$$\bar{F}_{Lik}{}^j{}_h = (A^M{}_{k,i} - A^M{}_{i,k} + c_I{}^M{}_J A^I{}_i A^J{}_k) c_M{}^E{}_F A^j{}_E A^F{}_h. \tag{3.4a}$$

One finds from the vanishing torsion:

$$\gamma_{BH} c_K{}^B{}_M A_i{}^M A_l{}^i = \gamma_{HKI} \tag{2.3b}$$

where γ_{HKI} is the Ricci coefficient of rotation and it follows that

$$\bar{F}_{Lik}{}^j{}_h = R^j{}_{hki}. \tag{3.4b}$$

The right-hand member of equation (3.2) can now be expressed as

$$R_{jhk}{}^i \dot{x}^k s^{jh} \tag{3.5}$$

where s^{jh} are the components of an anti-symmetric tensor which for the case of an orthogonal or pseudo-orthogonal group can be written as:

$$s_{jh} = -c_{JHM} c^M A_j{}^J A_h{}^H \tag{3.5a}$$

Equation (3.2) can thus be expressed on the base manifold. The $A^j{}_j$ are now regarded as the components of tetrades on B. They can be arbitrarily chosen at an initial point but we must not forget that they are considered as the projections of the k horizontal base vectors of P. To obtain the value of s_{jh} at other points of the trajectory we have to transport the initial tetrad along the trajectory in accord with equation (3.2) (not parallel in the Riemannian sense). c^M do not change along the trajectory and the tetrads undergo only rotations without changing their absolute values. The term (3.5) is well known from theories of classical spin motion (Papapetrou 1951). If one considers c^M as the spin, one has the peculiar situation that the particle acts as if it possessed spin, yet by its motion produces that spin so that terms described in Papapetrou's equations are missing. It behaves thus more like elementary particle spin which is seen as produced by internal motion. Equation (3.2) shows that the trajectory resembles that of a charged particle in an electromagnetic field. The particle acts also as a source of such a field. The equations of motion have been examined in previous papers (Halpern 1979).

The cosmological member which occurs in Einstein's equations for the manifold of a semi-simple Lie group is of order unity. It can be related to the radius of the De Sitter universe on the base. We adopt thus units where also this radius is of order unity and so is the gravitational constant. The magnitudes related to the quantum of action are then correspondingly small. One can choose $G = \frac{1}{8}\pi$ and $c = 1$, both dimensionless. h has then the dimension of a length squared. We do not consider here a quantization of the theory. We observe however that the radius of the spiraling orbits is comparable to the de Broglie wavelength corresponding to the frequency of the inverse period.

4. THE GENERAL THEORY

The generalization to a Kaluza–Klein theory in r dimensions demands the following modifications of the formalism. The bundle space P is no more the group manifold but it retains the character of a principal fibre bundle with group H. The manifold has now a metric γ which is a solution of Einstein's equations in r dimensions which now may have source terms and the same cosmological member. The solutions must be restricted to allow the projection of the metric on the base manifold. This is achieved by keeping $(r - k)$ o& the r equations (2.7): they read now:

$$[A_M, \gamma] = 0. \qquad (2.7')$$

The A_M are still a base of the left invariant vectors of H which are given on each fibre and form thus the vertical vector fields of P. We can still choose horizontal vectors perpendicular to the vertical vectors. We denote a base of these by B_E. The forms dual to the r vector fields are denoted by A^M and B^E. We must of course request that $[A_M, B_E]$ is horizontal. We may even retain:

$$[A_M, B_E] = C^H{}_{ME} B_H \qquad (2.3''')$$

without restricting the generalized metric too much. We abandon thus only the previous commutation relations between horizontal vectors. The belief is spread at present that a general relativistic theory must admit solutions of various topologies.

However this is in no way proved. One may argue in the present case that a theory based as far as possible on De Sitter symmetry should admit only solutions of the topology of the De Sitter universe. The Lagrangian in r dimensions is the Einstein Lagrangian with a cosmological member:

$$\mathcal{L} = \sqrt{\gamma}\,[R(\gamma) + \lambda]. \tag{4.1}$$

We introduce besides this a point particle Lagrange function which gives the geodesic trajectories on P.

$$L = \gamma_{uv}\,\dot{x}^u\,\dot{x}^v. \tag{4.2}$$

The theory can be brought into a four-dimensional form $(k = 4)$ because of equations (2.7') and (2.3'''). One finds:

$$\mathcal{L}^{(k)} = \sqrt{g}\,[R(g) + \lambda' + \tfrac{1}{4}\,F^M_{\ ik}\,F^{Nik}\gamma_{MN}] \tag{4.1a}$$

where R is there the curvature invariant in k dimensions.

$$L^{(k)} = g_{ik}\,\dot{x}^i\,\dot{x}^k + 2\,A_i^{\ M}\,\dot{x}^i\,c_M. \tag{4.2a}$$

One can show that the c^M for a geodesic in r dimensions, defined by $c^M A_M^{\ n} = \dot{x}^n$ are still constants—of course the corresponding c^E are not, but $\gamma_{UV}\,c^U\,c^V = \text{constant}$ and thus also $\gamma_{EF}\,c^E\,c^F$ is constant. The projection of the geodesic equation on B is also with the more general metric still of the form of equation (3.2).

The Lagrangians (4.1a, 4.2a) can be varied with respect to the metric g, the potentials $A_i^{\ M}\,(x^k)$ and the $\dot{x}^k$. We notice that one can find solutions for vanishing sources for which $F^M = 0$ so that test particles would not describe the additional mode of motion. This is against the spirit of this work. We would like to retain the conditions found on the group manifold as far as we can in the generalization. What we can do is to retain the vanishing of torsion—in components:

$$B^E_{\ k,\,i} - B^E_{\ i,\,k} + c_M^{\ E}{}_H(A_i^{\ M}\,B_k^{\ H} - A_k^{\ M}\,B_i^{\ H}) = 0. \tag{4.3}$$

This leads again to the analog of equation (3.5a), now expressed with tetrads B_E^i.

The right-hand member of the geodesic equations assumes again the form of equation (3.5). The last term of the Lagrangian of equation (4.1a) is now modified to a term quadratic in the Riemann tensor:

$$\mathcal{L}^{(k)} = \sqrt{g}\,[R(g) + \lambda' + \tfrac{1}{4}\,R_{hijk}\,R^{hijk}]. \tag{4.1b}$$

Many solutions, especially of the sourceless equations for a Lagrangian with the non-linear term of (4.1b) are the same as for the linear Lagrangian only. One may even formulate the theory as a gauge theory with the linear part alone. Equation (4.2a) becomes

$$L = g_{ik}\,\dot{x}^i\,\dot{x}^k + 2\,c^M c_M^{\ EF}\,\gamma_{EFI}\,B_i^{\ I}\,\dot{x}^i \tag{4.2b}$$

with the Ricci rotation coefficient γ_{EFI}. Similarly, as with the Dirac wave equation in curved space, the Ricci rotation coefficients occur in this Lagrangian and one has to vary with respect to the tetrads instead of the metric (Halpern 1977a). If the model describes elementary particle spin that can not be identified with orbital angular momentum then it ought to be half integral. Only consistent quantization could show which discrete values the c^M may assume.

5. BRIEF SUMMARY AND DISCUSSION OF RESULTS

Guided by the beautiful structure of semi-simple Lie groups, a theory of (multi-dimensional) Kaluza–Klein type has been constructed in which the natural metric of the group manifold itself is a sourceless solution of the field equations with cosmological member. Point particles in this theory can perform other motions than geodesics, especially spiral orbits resembling that of a charge in a magnetic field. Motivated by the idea to explore the consequences of group covariance to the very last instead of adjusting it by neglections to our experience, the 'physics' dictated by the group is explored. The De Sitter groups are uniquely suited as candidates because they give rise to the Lorentz group as a gauge group and the De Sitter universes as space–time solutions of the sourceless equations. The Lorentz group is a subgroup of the general linear group so that torsion is defined and absent on the particular sourceless solution. A restriction of the general solutions of the Kaluza–Klein theory to the absence of torsion allows to reduce the theory to a purely metric theory on space–time. The Lagrangian equivalent to the Einstein Lagrangian (with cosmological member) in the higher dimensions is found to be the Einstein Lagrangian plus a term quadratic in the curvature tensor on space–time. Many solutions, especially vacuum solutions are thus the same as in general relativity. The Lagrangian of a point particle is found to have an additional term resembling that of a spin tensor interacting with the gravitational field. This results in the above mentioned anomalous motion for velocities with 'vertical' components in the higher dimensions. The equations of motion show the interaction of the 'spin tensor' with the metric, but not the term relating temporal variations of the spin tensor to the velocity of the particle. This is however related to the fact that the anomalous motion of the particle is itself the cause of the spin tensor and the latter consequently has no contraction with the (observable) averaged velocity of the particle. The extension of the anomalous motion reduces the order of a De Broglie wave if the spin frequency is that of the particle energy. The conclusion (since long considered by the author) is that the non-geodesic orbits of one-dimensional subgroups of the De Sitter group on space–time yield a model of classical motion of spinning (elementary) particles. A combination of 'inner' and conventional motion. Spin is the only quantum number with properties analogous to those of mechanics and the De Sitter groups which occur necessarily—even uniquely in our approach, treat momentum and angular momentum on an equal footing—which has led us to expect such a result. Remarkably the 'spin property' occurs here in a theory where we have excluded torsion—it may even disappear in certain natural generalizations of the theory where torsion is present.†

---

† We may not be able to do without torsion to describe the behavior of spin in strong gravitational fields.

The fact that the present metric approach leads to a theory where also torison could naturally occur shows that the post-Newtonian parametrization should not be considered as the only guideline for alternative theories of gravitation. The theory opens a new outlook to the problem of quantization which however cannot be discussed in the present context. Can one obtain all these results also from the Poincaré group, which of course does also admit the Lorentz group as gauge group? Only with great sacrifices in consistency and beauty of the approach. As repeatedly stressed the De Sitter groups are, from the point of view adopted here, a unique way to go. One is able to continue to the conformal group which then includes the electromagnetic properties.

In spite of the tempting analogy, severe caution is necessary when trying to identify the anomalous motion with a classical model of spin in nature. The curvature of the universe is so much smaller than that near the earth without causing observable modifications of spin properties. The model has to be completed before any application can be envisaged or rejected.

REFERENCES

Choquet-Bruhat Y, DeWitt-Morette C and Dillard-Bleick M 1982 *Analysis, Manifolds and Physics* Revised edn (Amsterdam: North-Holland.)

DeWitt B 1963 in *Relativity Groups and Topology* Les Houches Summer Institute, ed B DeWitt and C DeWitt (New York: Gordon Breach)

Dirac P A M 1935 *Ann. Math.* **36** 657

—— 1936 *Ann. Math.* **37** 429

Eisenhart L P 1933 *Continuous Groups of Transformations* (Princeton: Princeton University Press)

Halpern L 1977a Lecture at Bonn Symposium on Differential Geometry and Physics July 1975. Appeared in 'Springer Lecture Notes in Mathematics' **570** 355 ed K Bleuler and A Reetz *Florida State University Preprint* 1975

—— 1977b *J. Gen. Relativ. Grav.* **8** 623

—— 1979 *Int. J. Theor. Phys.* **18** 845

—— 1980 *Preprint Clausthal Summer Institute* July 1980

—— 1981 Lecture at the Vth Nathiagali Summer Institute June 1980 ed R Riazudin, *Int. J. Theor. Phys.* **20** 297

—— 1982a *Lecture at Conference on Differential Geometric Methods in Physics ICTP June 1981 Proceedings*

—— 1982b Lecture at Meeting on Group Theory and Physics Canterbury Sept 1981 appeared in *Physica* **114A** 146

—— 1982c Lecture of Symposium in Honour of 80th Birthday of P A M Dirac, New Orleans May 1981, appeared in *Int. J. Theor. Phys.* **21** 791

—— 1982d Lecture Symposium on Gauge Theories and Gravitation, Nara Japan Aug 1982, to appear in *Springer Tracts on Physics* ed N Nakanishi

Hertz H 1896 *Collected works Vol. III* 'Prinzipien der Mechanik (New York: Macmillan)

Kaluza T 1921 *Sitz. Preuss. Ahad. Wiss. Berlin* p 966

Klein O 1926 *Z. Phys.* **37** 895

Lubkin E 1971 *Relativity and Gravitation Symposium Haifa 1969* ed G B Kuper and A Peres (New York: Gordon and Breach)

Nomizu K 1956 *Lie Groups and Differential Geometry* (Publ of the Mathematical Society of Japan) 2

Papapetrou A 1951 *Proc. R. Soc.* A. **209** 248

Steenrod N 1974 *The Topology of Fibre Bundles* (Princeton: Princeton University Press)

Bryce DeWitt's Publications

1. 'Point Transformations in Quantum Mechanics' *Phys. Rev.* **85** 653–61 (1952).
2. (With Cecile M DeWitt) 'The Quantum Theory of Interacting Gravitational and Spinor Fields' *Phys. Rev.* **87** 116–22 (1952).
3. 'State Vector Normalization in Formal Scattering Theory' *Phys. Rev* **100** 905–11 (1955).
4. 'The Operator Formalism in Quantum Perturbation Theory' *University of California Radiation Laboratory Publication No. 2884* (1955) 280 pp
5. 'Transition from Discrete to Continuous Spectra' *Phys. Rev.* **103** 1565–71 (1956).
6. 'Dynamical Theory in Curved Spaces. I. A Review of the Classical and Quantum Action Principles,' *Rev. Mod. Phys.* **29** 377–97 (1957).
7. 'Principal Directions of Current Research Activity in the Theory of Gravitation' *J. Astronaut.* **4** 23–8 (1957).
8. 'Principal Directions in Current Research on Gravitation' in *Advances in Astronautical Sciences*, Vol. 1 (New York: Plenum) (1957).
9. 'The Scientific Uses of Large Space Ships' *General Atomic Report GAMD 965* (1959) 40 pp.
10. (With Robert W. Brehme) 'Radiation Damping in a Gravitational Field' *Ann. Phys., NY* **9** 220–59 (1960).
11. 'Freinage du à la Radiation d'une Particule dans un Champ de Gravitation' in *Les Théories Relativistes de la Gravitation, Colloque International organisé a Royaumont du 21 au 27 juin 1959* (Paris: Centre National de la Recherche Scientifique) (1962) pp 335–43.
12. 'Invariant Commutators for the Quantized Gravitational Field' *Phys. Rev. Lett.* **4** 317–20 (1960).
13. 'Quantization of Fields with Infinite-Dimensional Invariance Groups' *J. Math. Phys.* **2** 151–62 (1961).
14. 'Invariant Commutators for the Quantized Gravitational Field' in *Recent Developments in General Relativity* (London: Pergamon) (1962) pp 175–89.
15. 'Quantum Theory without Electromagnetic Potentials' *Phys. Rev.* **125** 2189–219 (1962).
16. (With Ryoyu Utiyama) 'Renormalization of a Classical Gravitational Field

Interacting with Quantized Matter Fields' *J. Math. Phys.* **3** 608–18 (1962).

17. 'Quantization of Fields with Infinite-Dimensional Invariance Groups. II. Anticommuting Fields' *J. Math. Phys.* **3** 625–36 (1962).

18. 'Definition of Commutators via the Uncertainty Principle' *J. Math. Phys.* **3** 619–24 (1962).

19. 'Quantization of Fields with Infinite-Dimensional Invariance Groups. III. Generalized Schwinger–Feynman Theory' *J. Math. Phys.* **3** 1073–93 (1962).

20. 'The Quantization of Geometry' in *Gravitation: An Introduction to Current Research* ed L Witten (New York: Wiley) (1963) ch 8, pp 266–381.

21. (With Cécile M DeWitt, ed) *Relativity, Groups and Topology, 1963 Les Houches Lectures* (New York: Gordon and Breach) (1964) 929 pp.

22. 'The Quantization of Geometry' in *Proceedings on the Theory of Gravitation, Conference in Warszawa and Jablonna 25–31 July 1962* (Warszawa: PWN-Editions Scientifiques de Pologne) (1964) pp 131–47.

23. 'Gravity' in *Advances in Space Science and Technology*, Vol. VI, ed Frederick I Ordway, III (New York: Academic) (1964) pp 1–37.

24. (With Cécile M DeWitt) 'Falling Charges' *Physics* **1** 3–20 (1964).

25. 'Theory of Radiative Corrections for Non-Abelian Gauge Fields' *Phys. Rev. Lett.* **12** 742–6 (1964).

26. 'Gravity: A Universal Regulator?' *Phys. Rev. Lett.* **13** 114–8 (1964).

27. *Dynamical Theory of Groups and Fields* (New York: Gordon and Breach) (1965) 248 pp.

28. 'Superconductors and Gravitational Drag' *Phys. Rev. Lett.* **16** 1092–3 (1966).

29. 'Quantum Theory of Gravity. I. The Canonical Theory' *Phys. Rev.* **160** 1113–48 (1967).

30. 'Quantum Theory of Gravity. II. The Manifestly Covariant Theory' *Phys. Rev.* **162** 1195–239 (1967).

31. 'Quantum Theory of Gravity. III. Application of the Covariant Theory' *Phys. Rev.* **162** 1239–56 (1967).

32. 'Eversion of the 2-Sphere' in *Battelle Recontres: 1967 Lectures in Mathematics and Physics*, ed C M DeWitt and J A Wheeler (New York: Benjamin) (1968) pp 546–57.

33. 'The Everett–Wheeler Interpretation of Quantum Mechanics' in *Battelle Recontres: 1967 Lectures in Mathematics and Physics* ed C M DeWitt and J A Wheeler (New York: Benjamin) (1968) pp 318–32.

34. 'Spacetime as a Sheaf of Geodesics in Superspace' in *Relativity: Proceedings of the Relativity Conference in the Midwest* ed M Carmeli, S I Fickler and L Witten (New York: Plenum Press) (1970) pp 359–74.

35. 'Quantum Mechanics and Reality' *Phys. Today* **23** 30–5 (1970).

36. (With L E Ballentine, P Pearle, E H Walker, M Sachs, T Koga and J Gerver) 'Quantum Mechanics Debate' *Phys. Today* **24** 36–44 (1971).

37. 'Quantum Theories of Gravity' *Gen. Relativ. Grav.* **1** 181–9 (1970).

38. 'The Many Universes Interpretation of Quantum Mechanics' in *Proceedings of the International School of Physics 'Enrico Fermi' Course IL: Foundations of Quantum Mechanics.* (New York: Academic) (1971).

39. (With R Neill Graham) 'Resource Letter 1QM-1 on the Interpretation of Quantum Mechanics' *Am. J. Phys.* **39** 724–33 (1971).

40. 'Covariant Quantum Geometrodynamics' in *Magic Without Magic: John Archibald Wheeler* ed J R Klauder (San Francisco: Freeman) (1972) pp 409–40.

41. Ed, *Lectures on Electrodynamics,* J R Oppenheimer (New York: Gordon and Breach) (1970) x + 164 pp.

42. (With Cécile M DeWitt, ed) *Black Holes, 1972 Les Houches Lectures* (New York: Gordon and Breach) (1973) xii + 574 + 161 pp.

43. (With Neill Graham, ed) *The Many-Worlds Interpretation of Quantum Mechanics* (Princeton: Princeton University Press) (1973) 250 pp.

44. (With F Estabrook, H Wahlquist, S Christensen, L Smarr, and E Tsiang) 'Maximally Slicing a Black Hole' *Phys. Rev.* D 7 2814–7 (1973).

45. (With R A Matzner and A H Mikesell) 'A Relativity Eclipse Experiment Refurbished' *Sky and Telesc.* **47** 301–6 (1974).

46. 'The Texas Mauritanian Eclipse Expedition' in *Gravitation and Relativity: Proceedings of the 7th International Conference on General Relativity and Gravitation, Tel Aviv University, June 1974* (Jerusalem: Keter Publishing House) (1975) pp 81–6.

47. 'Quantum Field Theory in Curved Space' in *Particles and Fields–1974: AIP Conference Proceedings No. 23, Particles and Fields Subseries No. 10* (New York: American Institute of Physics) (1975) pp 660–88.

48. 'Quantum Field Theory in Curved Spacetime' *Phys. Rep.* **19** 295–357 (1975).

49. (With other members of the Texas Mauritanian Eclipse Team) 'Gravitational Deflection of Light: Solar Eclipse of 30 June 1973 I. Description of Procedures and Final Results.' *Astron. J.* **81** 452–4 (1976).

50. (With L Smarr, A Cadez and K Eppley) 'Collision of Two Black Holes: Theoretical Framework' *Phys. Rev.* D **14** 2443 (1976).

51. 'Gravitational Deflection of Light. Solar Eclipse of 30 June 1973' in *Albert Einstein's Theory of General Relativity* ed G Tauber (New York: Crown Publishers) (1979) pp 125–6.

52. 'Quantum Gravity: The New Synthesis' in *General Relativity, An Einstein Centenary Survey* ed S W Hawking and W Israel (Cambridge: Cambridge University Press) (1979), pp 680–745.

53. (With C F Hart and C J Isham) 'Topology and Quantum Field Theory' in *Themes in Contemporary Physics: Essays in Honour of Julian Schwinger's 60th Birthday* (Amsterdam: North-Holland) (1979). Also *Physica* **96A** pp 197–211 (1979).

54. 'The Formal Structure of Quantum Gravity' in *Recent Developments in Gravitation, Cargèse 1978,* ed M Lévy and S Deser (New York: Plenum) (1979) pp 275–322.

55. 'Quantum Gravity' in *On the Path of Albert Einstein* ed B Kursunoglu, A Perlmutter and L F Scott (New York: Plenum) (1979) pp 127–43.

56. 'A Gauge Invariant Effective Action' in *Quantum Gravity II* ed C J Isham, R Penrose and D W Sciama (Oxford: Oxford University Press) (1981) pp 449–87.

57. 'Approximate Effective Action for Quantum Gravity' *Phys. Rev. Lett.* **47** 1647–50 (1981).
58. (With P van Nieuwenhuizen) 'Explicit Construction of the Exceptional Superalgebras F(4) and G(3)' *J. Math. Phys.* **23** 1953–63 (1982).

Index